GENOMICS AND PROTEOMICS

Principles, Technologies, and Applications

GENOMICS AND PROTEOMICS

Principles, Technologies, and Applications

Edited by

Devarajan Thangadurai, PhD, and Jeyabalan Sangeetha, PhD

AAP APPLE ACADEMIC PRESS

Apple Academic Press Inc.	Apple Academic Press Inc.
3333 Mistwell Crescent	9 Spinnaker Way
Oakville, ON L6L 0A2	Waretown, NJ 08758
Canada	USA

© 2015 by Apple Academic Press, Inc.
First issued in paperback 2021
Exclusive worldwide distribution by CRC Press, a member of Taylor & Francis Group
No claim to original U.S. Government works

ISBN-13: 978-1-77463-537-7 (pbk)
ISBN-13: 978-1-77188-114-2 (hbk)

Library of Congress Control Number: 2015936001

Library and Archives Canada Cataloguing in Publication

Genomics and proteomics : principles, technologies, and applications / edited by Devarajan Thangadurai, PhD, and Jeyabalan Sangeetha, PhD.

Includes bibliographical references and index.
ISBN 978-1-77188-114-2 (bound)
1. Genomics. 2. Proteomics. I. Thangadurai, D., author, editor II. Sangeetha, Jeyabalan, author, editor

| QH447.G45 2015 | 572.8'6 | C2015-902137-5 |

Apple Academic Press also publishes its books in a variety of electronic formats. Some content that appears in print may not be available in electronic format. For information about Apple Academic Press products, visit our website at **www.appleacademicpress.com** and the CRC Press website at **www.crc-press.com**

ABOUT THE EDITORS

Devarajan Thangadurai, PhD

Devarajan Thangadurai, PhD, is Senior Assistant Professor at Karnatak University in South India; President of the Society for Applied Biotechnology; and General Secretary for the Association for the Advancement of Biodiversity Science. In addition, Dr. Thangadurai is Editor-in-Chief of several journals, including *Biotechnology, Bioinformatics and Bioengineering; Acta Biologica Indica; Biodiversity Research International*; and the *Asian Journal of Microbiology*. He received his PhD in Botany from Sri Krishnadevaraya University in South India. During 2002–2004, he worked as CSIR Senior Research Fellow with funding from the Ministry of Science and Technology, Government of India. He served as Postdoctoral Fellow at the University of Madeira, Portugal; University of Delhi, India; and ICAR National Research Centre for Banana, India. He is the recipient of the Best Young Scientist Award with a Gold Medal from Acharya Nagarjuna University and the VGST-SMYSR Young Scientist Award of the Government of Karnataka, Republic of India. He has edited/authored fifteen books including *Genetic Resources and Biotechnology* (3 vols.), *Genes, Genomes and Genomics* (2 vols.) and *Mycorrhizal Biotechnology* with publishers of national and international reputation.

Jeyabalan Sangeetha, PhD

Jeyabalan Sangeetha, PhD, is the UGC Kothari Postdoctoral Fellow at Karnatak University in South India. She earned her a BSc in Microbiology and PhD in Environmental Sciences from Bharathidasan University, Tiruchirappalli, Tamil Nadu, India. She holds also an MSc in Environmental Sciences from Bharathiar University, Coimbatore, Tamil Nadu, India. She is the recipient of the Tamil Nadu Government Scholarship and the Rajiv Gandhi National Fellowship of University Grants Commission, Government of India, for her doctoral studies. She has published approximately 20 manuscripts detailing the effect of pollutants on the environment, and has organized conferences, seminars, workshops, and lectures. Her main research interests are in the areas of environmental microbiology and environmental biotechnology, with particular emphasis on solid waste management, environmental impact assessment, and microbial degradation of hydrocarbons. Her scientific and community leadership have included serving as an editor of the journal *Biodiversity Research International* and Secretary for the Society for Applied Biotechnology.

CONTENTS

List of Contributors .... *ix*

List of Abbreviations .... *xiii*

Preface .... *xvii*

1. **Different Gene Products with a Similar Role in Neuronal Defense Against Oxidative Stress – Heme Oxygenase System** 1

   Zhi-Gang Jiang and Hossein A. Ghanbari (USA)

2. **Embracing Complexity: Searching for Gene-Gene and Gene Environment Interactions in Genetic Epidemiology** 19

   Alison A. Motsinger and David M. Reif (USA)

3. **Elucidation of Proto-Oncogene C-Abl Function with the Use of Mouse Models and the Disease Model of Chronic Myeloid Leukemia** 59

   Jenny F. L. Chau and Baojie Li (Singapore)

4. **A Theory on the Molecular Nature of Post-Zygotic Reproductive Isolation** 95

   Francisco Prosdocimi (Brazil)

5. **Next Generation Sequencing and Microbiome Evaluation: Molecular Microbiology and Its Impact on Human Health** 121

   A. Focà, M. C. Liberto, and N. Marascio (Italy)

6. **Proteomics and Prostate Cancer** 143

   Jae-Kyung Myung and Marianne D. Sadar (Canada)

7. **RNA Interference Therapeutics – Past, Present and Future** 175

   Jayapal Manikandan, Mahmood Rasool, Muhammad Imran Naseer, Kothandaraman Narasimhan, Laila Abdullah Damiati, Kalemegam Gauthaman, Sami Bahlas, and Peter Natesan Pushparaj (Australia, Kingdom of Saudi Arabia)

8. **Molecular Mechanisms of Hepatitis C Virus Entry – Impact of Host Cell Factors for Initiation of Viral Infection** 189

   Mirjam B. Zeisel and Thomas F. Baumert (France)

9. **Molecular Phylogenetics: A Tool for Elucidation of Evolutionary Processes from Biological Data** 203

   Martina Talianova (Czech Republic)

10. **New Insights into the Role of Protein Prenylation and Processing in Plant Development, Phytohormone Signaling, and Secondary Metabolism** 251

Dring N. Crowell and Devarajan Thangadurai (India, USA)

11. **Mangomics: Information Systems Supporting Advanced Mango Breeding**...281

David J. Innes, Natalie L. Dillon, Heather E. Smyth, Mirko Karan, Timothy A. Holton, Ian S. E. Bally, and Ralf G. Dietzgen (Australia)

12. **Environmental Genomics: The Impact of Transgenic Crops on Soil Quality, Microbial Diversity and Plant-Associated Communities**..... 309

Jeyabalan Sangeetha, Devarajan Thangadurai, Muniswamy David, Roopa T. Somanath, Abhishek C. Mundaragi, and Digambarappa P. Biradar (India)

13. **Recent Advances in Biotechnology and Genomic Approaches for Abiotic Stress Tolerance in Crop Plants** 333

Mirza Hasanuzzaman, Rajib Roychowdhury, Joydip Karmakar, Narottam Dey, Kamrun Nahar, and Masayuki Fujita (Bangladesh, India, Japan)

14. **Molecular and Genomic Approaches for Microbial Remediation of Petroleum Contaminated Soils** 367

Jeyabalan Sangeetha, Devarajan Thangadurai, Muniswamy David, and Jadhav Shrinivas (India)

Index...405

LIST OF CONTRIBUTORS

Sami Bahlas
Rheumatology Department, College of Medicine, King Abdulaziz University, Jeddah - 21589, Kingdom of Saudi Arabia

Ian S. E. Bally
Horticulture and Forestry Sciences, Queensland Department of Agriculture, Fisheries and Forestry, Mareeba, Australia

Thomas F. Baumert
Université Louis Pasteur, Strasbourg, France

Digambarappa P. Biradar
University of Agricultural Sciences, Dharwad, Karnataka, 580005, India

Jenny F. L. Chau
The Institute of Molecular and Cell Biology, Proteos, 61 Biopolis Drive, Singapore138673, Republic of Singapore

Dring N. Crowell
Department of Biology, Indiana University - Purdue University Indianapolis, 723 West Michigan Street, Indianapolis, IN 46202, USA

Laila Abdullah Damiati
King Fahd Medical Research Center, Faculty of Applied Medical Sciences (P.O. Box: 80216), King Abdulaziz University, Jeddah - 21589, Kingdom of Saudi Arabia

Muniswamy David
Department of Zoology, Karnatak University, Dharwad, Karnataka, 580003, India

Narottam Dey
Department of Biotechnology, Visva-Bharati, Santiniketan-731235, West Bengal, India

Ralf G. Dietzgen
Queensland Alliance for Agriculture and Food Innovation, The University of Queensland, St Lucia, Australia

Natalie L. Dillon
Horticulture and Forestry Sciences, Queensland Department of Agriculture, Fisheries and Forestry, Mareeba, Australia

A. Focà
Microbiology, Department of Health Sciences, School of Medicine, University "Magna Graecia", Catanzaro, Italy

Masayuki Fujita
Laboratory of Plant Stress Responses, Department of Applied Biological Science, Kagawa University, 2393 Ikenobe, Miki-cho,Kita-gun, Kagawa 761-0795, Japan

Kalemegam Gauthaman
Center of Excellence in Genomic Medicine Research, Faculty of Applied Medical Sciences (P.O. Box: 80216), King Abdulaziz University, Jeddah - 21589, Kingdom of Saudi Arabia

Hossein A. Ghanbari
Panacea Pharmaceuticals, Inc., 207 Perry Parkway, Suite 2, Gaithersburg, MD 20877, USA

Mirza Hasanuzzaman
Department of Agronomy, Faculty of Agriculture, Sher-e-Bangla Agricultural University, Dhaka, 1207, Bangladesh

Timothy A. Holton
Queensland Alliance for Agriculture and Food Innovation, The University of Queensland, St Lucia, Australia

David J. Innes
Crop and Food Sciences, Queensland Department of Agriculture, Fisheries and Forestry, St Lucia, Australia

Zhi-Gang Jiang
Panacea Pharmaceuticals, Inc., 207 Perry Parkway, Suite 2, Gaithersburg, MD 20877, USA

Mirko Karan
Horticulture and Forestry Sciences, Queensland Department of Agriculture, Fisheries and Forestry, Mareeba, Australia

Joydip Karmakar
Department of Biotechnology, Visva-Bharati, Santiniketan, 731235, West Bengal, India

Baojie Li
The Institute of Molecular and Cell Biology, Proteos, 61 Biopolis Drive, Singapore138673, Republic of Singapore

M.C. Liberto
Microbiology, Department of Health Sciences, School of Medicine, University "Magna Graecia", Catanzaro, Italy

Jayapal Manikandan
Faculty of Science, The University of Western Australia (M011), 35 Stirling Highway, Australia

N. Marascio
Microbiology, Department of Health Sciences, School of Medicine, University "Magna Graecia", Catanzaro, Italy

Alison A. Motsinger
Bioinformatics Research Center, Department of Statistics, North Carolina State University, Raleigh, NC, USA

Abhishek C. Mundaragi
Department of Botany, Karnatak University, Dharwad, Karnataka, 580003, India

Jae-Kyung Myung
British Columbia Cancer Agency, Genome Sciences Centre, 675 West 10th Avenue, Vancouver, British Columbia, V5Z 1L3, Canada

Kamrun Nahar
Laboratory of Plant Stress Responses, Department of Applied Biological Science, Kagawa University, 2393 Ikenobe, Miki-cho,Kita-gun, Kagawa 761-0795, Japan

Kothandaraman Narasimhan
Center of Excellence in Genomic Medicine Research, Faculty of Applied Medical Sciences (P.O. Box: 80216), King Abdulaziz University, Jeddah - 21589, Kingdom of Saudi Arabia

Muhammad Imran Naseer
Center of Excellence in Genomic Medicine Research, Faculty of Applied Medical Sciences (P.O. Box: 80216), King Abdulaziz University, Jeddah - 21589, Kingdom of Saudi Arabia

Francisco Prosdocimi
Laboratório de Biodados, Depto. Bioquímica e Imunologia, ICB-UFMG, Av. Antônio Carlos 6627 CP 486, 31.270-010 BH-MG, Brazil

Peter Natesan Pushparaj
Center of Excellence in Genomic Medicine Research, Faculty of Applied Medical Sciences (P.O. Box: 80216), King Abdulaziz University, Jeddah - 21589, Kingdom of Saudi Arabia

Mahmood Rasool
Center of Excellence in Genomic Medicine Research, Faculty of Applied Medical Sciences (P.O. Box: 80216), King Abdulaziz University, Jeddah - 21589, Kingdom of Saudi Arabia

David M. Reif
National Center for Computational Toxicology, U.S. Environmental Protection Agency, Research Triangle Park, NC, USA

Rajib Roychowdhury
Department of Biotechnology, Visva-Bharati, Santiniketan, 731235, West Bengal, India

Marianne D. Sadar
British Columbia Cancer Agency, Genome Sciences Centre, 675 West 10th Avenue, Vancouver, British Columbia, V5Z 1L3, Canada

Jeyabalan Sangeetha
Department of Zoology, Karnatak University, Dharwad, Karnataka, 580003, India

Jadhav Shrinivas
Department of Zoology, Karnatak University, Dharwad, Karnataka, 580003, India

Heather E. Smyth
Queensland Alliance for Agriculture and Food Innovation, The University of Queensland, St Lucia, Australia

Roopa T. Somanath
Department of Botany, Karnatak University, Dharwad, Karnataka, 580003, India

Martina Talianova
Laboratory of Plant Developmental Genetics, Institute of Biophysics, Czech Academy of Sciences, Kralovopolska 135, 61265 Brno, Czech Republic

Devarajan Thangadurai
Department of Botany, Karnatak University, Dharwad, Karnataka, 580003, India

Mirjam B. Zeisel
Inserm, U748, Strasbourg, France

LIST OF ABBREVIATIONS

AIC	Akaike's Information Criterion
ALF	Acute Liver Failure
ALS	Amyotrophic Lateral Sclerosis
ANOVA	Analysis of Variance
ARE	Antioxidant Response Element
BCR	Breakpoint Cluster Region
BIC	Bayesian Information Criterion
CART	Classification and Regression Trees
CID	Collision-Induced Dissociation
CML	Chronic Myelogenous Leukemia
CNS	Central Nerve System
CPM	Combinatorial Partitioning Method
CRFs	Circulating Recombinant Forms
CSS	Chi-Square Subset
DBD	DNA-Binding Domain
DICE	Detection of Informative Combined Effect
DM	Dobzhansky-Muller
EBV	Epstein-Barr Virus
EGF	Epidermal Growth Factor
ERK	Extracellular Signal-Regulated Kinase
FBAT	Family Based Association Test
FDA	Food and Drug Administration
FDR	False Discovery Rate
FGS	First-Generation Sequencing
FITF	Focused Interaction Testing Framework
FRET	Fluorescence Resonance Energy Transfer
GA	Genetic Algorithms
GLMs	Generalized Linear Models
GPNN	Genetic Programming Neural Networks
GST	Glutathione S-Transferase
GTR	Time Reversible Models
HIV	Human Immunodeficiency Virus
HMP	Human Microbiome Project
HO	Heme Oxygenase
HTS	High-Throughput Sequencing

HWE	Hardy-Weinberg Equilibrium
IMAC	Immobilized Metal Affinity Chromatography
IR	Ionizing Irradiation
IREs	Iron Regulatory Elements
ITF	Interaction Testing Framework
JNK	c-Jun N-Terminal Kinase
KSHV	Kaposi's Sarcoma-Associated Herpesvirus
LBD	Ligand-Binding Domain
LCM	Laser Capture Microdissection
LCS	Longest Common Sequence
LDL	Low Density Lipoproteins
LEL	Large Extvcelullar Loop
MALDI	Matrix-Assisted Laser Desorption/Ionization
MANCOVA	Multivariate Analysis of Covariance
MAPK	Mitogen-Activated Protein Kinase
MCI	Mild Cognitive Impairment
MDR	Multifactor Dimensionality Reduction
MKK	MAP Kinase Kinase
MP	Maximum Parsimony
MRE	Metal Responsive Element
MRSA	Methicillin-Resistant *Staphylococcus aureus*
MudPIT	Multidimensional Protein Identification Technology
NES	Nuclear Export Signal
NFT	Neurofibrillary Tangles
NGS	Next-Generation Sequencing
NLS	Nuclear Localization Signals
NN	Neural Networks
NTA	Nickel-Nitrilotriacetic Acid
NTD	N-Terminal Domain
ODNs	Oligodeoxyribonucleic Acids
PBN	Phenyl-N-Tert-Butyl Nitrone
PCA	Principal Components Analysis
PCR	Polymerase Chain Reaction
PDGF	Platelet-Derived Growth Factor
PDM	Parameter Decreasing Method
PDT	Pedigree Disequilibrium Test
PRP	Patterning and Recursive Partitioning
PSA	Prostate-Specific Antigen
PTP	Protein Tyrosine Phosphatases
RF	Random Forests

RI	Reproductive Isolation
RISC	RNA-Induced Silencing Complex
ROS	Reactive Oxygen Species
RPM	Restricted Partition Method
SCI	Spinal Cord Injury
SEL	Small Extracellular Loop
SELDI	Surface-Enhanced Laser Desorption/Ionization
SERS	Surface-Enhanced Raman Spectroscopy
SGS	Second-Generation Systems
SILAC	Stable Isotope Labeling by Amino Acids in Cell Culture
SNPs	Single Nucleotide Polymorphisms
SOD	Superoxide Dismutase
SURPI	Sequence-Based Ultra-Rapid Pathogen Identification
TAD	Transcription Activation Domain
TAP	Tandem Affinity Purification
TDT	Transmission Disequilibrium Test
TGS	Third-Generation Sequencing
TOF	Time of Flight
UDPS	Ultra-Deep Pyrosequencing
UTR	Untranslated Region

PREFACE

Genomics is the comprehensive study of genetic information of a cell or organism including gene sequences in living organisms, the function of specific genes, the interactions among different genes, and the control and regulation of gene expression through activation and suppression of target genes. The human genome has been the biggest project undertaken to date but there are many research projects around the world trying to map the gene sequences of other organisms. Many diseases due to single gene defects have already been identified. New data obtained by human genome sequencing will help scientists better understand multifactorial diseases such as asthma, diabetes, heart disease and cancer. Genomic mapping enables us to develop new preventative and therapeutic approaches to the treatment of disease and understand the mechanics of cell biology. Genomics is generating a lot of excitement not only in the scientific research institutes and pharmaceutical companies but also in the financial and insurance worlds. Proteomics is the large-scale study of proteins, particularly their structures and functions. It describes the qualitative and quantitative comparison of proteomes under different conditions to further unravel biological processes.

Genomics and proteomics are promising scientific fields of current society. In this book, different gene products with a similar role in neuronal defense against oxidative stress were discussed by Zhi-Gang Jiang and Hossein A. Ghanbari in Chapter 1. Gene-gene and gene-environment interactions in genetic epidemiology were discussed by Alison A. Motsinger and David M. Reif in Chapter 2. Jenny F. L. Chau and Baojie Li describe the elucidation of proto-oncogene c-Abl function with the use of mouse models and the disease model of chronic myeloid leukemia in Chapter 3. Francisco Prosdocimi overviewed theory on the molecular nature of postzygotic reproductive isolation in Chapter 4. A. Focà, M. C. Liberto and N. Marascio described next generation sequencing, microbiome evaluation, molecular microbiology and its impact on human health in Chapter 5. Proteomics and prostate cancer was demonstrated by Jae-Kyung Myung and Marianne D. Sadar in Chapter 6. Jayapal Manikandan and others have described RNA interference therapeutics in Chapter 7. Molecular mechanisms of hepatitis C virus entry were discussed by Mirjam B. Zeisel and Thomas F. Baumert in Chapter 8. Molecular phylogenetics for elucidation of evolutionary processes from biological data was discussed by Martina Talianova in Chapter 9. The role of protein prenylation and processing in plant development, phytohormone signalling and secondary metabolism were discussed by Dring N. Crowell and Devarajan Thangadurai in Chapter 10. In Chapter 11, Ralf G. Dietzgen and his collaborators expound on the advantage of information systems like mangomics to support advanced mango breeding. In Chapter 12, Jeyabalan Sangeetha and her team overviewed the role of environmental genomics with special reference to understanding the impact of transgenic crops on soil quality, microbial diversity, and plant-associated communities. Mirza Hasanuzzaman and his colleagues have discussed biotechnological and genomic approaches for abiotic stress tolerance in crop plants in Chapter 13. Finally, molecular and genomic approaches of microbial remediation of petroleum contaminated soils have been discussed in Chapter 14.

The publication of this book was not solely the effort of only two of us. We owe a great debt of gratitude to all those who have so kindly contributed and deserves special credit for putting their valuable time and effort. In addition, we gratefully acknowledge Sandy Jones Sickels, Vice President, and Ashish Kumar, Publisher and President, at Apple Academic Press, Inc., for their keen interest and effort in bringing out this publication. Collectively, we hope the work will prove to be most useful in the understanding of past, present, and future scientific and technological innovations in these genomic and proteomic era.

— **Devarajan Thangadurai, PhD, and Jeyabalan Sangeetha, PhD**

CHAPTER 1

DIFFERENT GENE PRODUCTS WITH A SIMILAR ROLE IN NEURONAL DEFENSE AGAINST OXIDATIVE STRESS – HEME OXYGENASE SYSTEM

ZHI-GANG JIANG and HOSSEIN A. GHANBARI

CONTENTS

1.1 Introduction 2

1.2 Oxidative Stress and Neurodegenerative Diseases 3

1.3 Members of the HO Antioxidative System 5

1.4 Distribution of Members of the HO Antioxidative System in the CNS 6

1.5 Function of the HO Antioxidative System 6

1.6 Expression and Activity Regulations of the HO Antioxidative System 7

1.7 HOs in Neurodegenerative Diseases 9

1.8 The HO Antioxidative System as a Drug Target 10

1.9 Conclusion and Future Perspectives..11

Keywords 11

References 12

1.1   INTRODUCTION

During lifespan, an organism receives oxidative stress from time to time. Whether cells can survive oxidative insult depends on the balance between the strength of the oxidative stress and the protective efficacy of intracellular antioxidative systems. Prevention of neuronal cell death from oxidative attack appears more pivotal than that for any other proliferative cell types since neurons are a terminally differentiated cell type and rarely regenerate. In addition, the brain is significantly vulnerable to reactive oxygen species (ROS)-induced damage due to its high rate of oxygen consumption and relative weakness of antioxidant systems (Coyle and Puttfarcken, 1993). There exists several different antioxidative systems in neurons, for example, superoxide dismutase (SOD) that converts superoxide anion radical ($O_2^{\cdot-}$) to oxygen and hydrogen peroxide (H_2O_2), glutathione reductase/peroxidase (GR/GP) that destroys hydrogen peroxide to form water, and heme oxygenase (HO) that catalyzes heme to generate the antioxidative product biliverdin (Halliwell, 2006). The microsomal HO system is the most effective mechanism for degradation of heme and generation of biliverdin, which is further reduced to bilirubin (Barañano and Snyder, 2001). Cellular depletion of bilirubin by RNA interference markedly augments tissue levels of ROS and causes apoptotic cell death (Barañano et al., 2002). Heme as a substrate of HO is present in neurons as well. It is at the core of numerous hemoproteins including cytochrome P450 isoforms CYP3A11 and CYP3A13 (Hagemeyer et al., 2003), cytochrome c (Kuchar et al., 2004), nitric oxide synthases (Abu-Soud et al., 1995), and neuroglobin (Burmester and Hankeln, 2004). Neuroglobin is a recently discovered minomeric globin with high affinity for oxygen and preferential localization to vertebrate brain. Stress can release free-heme from hemoproteins (Srisook et al., 2005). The existence of intracellular heme is essential for antioxidant activity of HO (Foresti et al., 2001). To date, three catalytically active isozymes HO-1, HO-2 and HO-3 have been identified. HO-1 and HO-2 proteins are different gene products and have little in common in primary structure, regulation, and tissue distribution (Cruse and Maines, 1988). The predicted amino acid structure of HO-3 differs from both HO-1 and HO-2 although HO-3 shares a very high level of homology with HO-2 (approximately 90%). In addition, purified HO-3 protein does not cross react with polyclonal antibodies to either rat HO-1 or HO-2 (Maines, 1997). The function of HO-3 is still unclear. Although HO-1 and HO-2 are products of different genes, they show a similar role in catalyzing heme to produce intracellular antioxidants with an essential function for neurons to defend against oxidative stress.

1.2 OXIDATIVE STRESS AND NEURODEGENERATIVE DISEASES

Intracellular ROS accumulation mediates pathogenesis of both acute (e.g. stroke) and chronic [exampled by Alzheimer's disease (AD)] neurodegenerative diseases. Ischemic stroke is a fatal cerebrovascular disease. A stroke occurs when a blood vessel that carries oxygen and nutrients to the brain is blocked, such as by a thrombolic or embolic clot. Cerebral ischemia followed by reperfusion activates numerous intracellular toxic pathways that lead to cell death. N-methyl-D-aspartate receptor (NMDAr) activation by excitatory amino acid glutamate plays a primary role in this process. The activation of NMDAr causes extracellular free calcium Ca^{2+} influx into the cell (Zipfel et al., 2000) and results in intracellular ROS accumulation (Chan, 2001). Excessive $Ca2^+$ accumulated in mitochondria interrupts the electron transport chain and collapses the mitochondrial membrane potential (Zhang et al., 1990; Rego et al., 2000). Thus free electrons are accumulated in the mitochondria, react with oxygen that is available following reperfusion, and result in the production of superoxide. The superoxide is further processed to produce the hydroxyl radical by a Fenton reaction (Won et al., 2002). An increase in intracellular nitric oxide following the activation of NMDAr also occurs. The nitric oxide rapidly reacts with superoxide to form peroxynitrite (Dawson et al., 1991; Kiedrowski et al., 1992). Both intracellular free calcium ($[Ca^{2+}]i$) and ROS can induce the mitochondrial permeability transition (MPT; Bernardi et al., 1992; Kowaltowski et al., 1996; Halestrap et al., 2002; Gunter et al., 2004). As a consequence, there is mitochondrial failure culminating in either apoptotic or necrotic cell death. The major role of ROS in medicating ischemic neurotoxicity is also evidenced by therapies using antioxidants. The spin-trap alpha-phenyl-N-tert-butyl nitrone (PBN) reduces infarct size and prevents a secondary mitochondrial dysfunction due to reperfusion, probably scavenging free radicals at the blood-endothelial cell interface (Kuroda and Siesjö, 1997). Other antioxidants such as α-lipoic acid and vitamin E have also been shown to reduce infarct volume in cerebral ischemia (Packer et al., 1997; van der Worp et al., 1998).

AD is usually a late-onset disease. Its early manifestations are memory deficits and cognitive impairment, including temporal and geographic disorientation, impaired judgment and problem solving and reduced language capability (Faber-Langendoen et al., 1988). In later stages of AD, behavioral and personality changes appear (Rubin et al., 1987; Swearer et al., 1988). Later symptoms include motor dysfunction and dementia (Morris et al., 1989; Romanelli et al., 1990; Forstl and Kurz, 1999). The average life expectancy of a patient is 8-10 years following diagnosis. The disease is pathologically hallmarked by brain atrophy, following gradual cell loss in the central nerve system (CNS). The pathologic findings characteristic of AD are formation of extracellular senile plaques

(Dickson, 1997; Selkoe, 2001) containing amyloid β_{1-42} and β_{1-40} (Aβ) peptides, as well as bundles known as neurofibrillary tangles (NFT, Sisodia *et al.*, 1990; Dowjat *et al.*, 2001; Stoothoff and Johnson, 2005), consisting of paired helical filaments of the hyperphosphorylated microtubule-associated protein tau. These features are observed with increasing prevalence as the disease progresses. Another feature of AD, common to other neurodegenerative diseases, is oxidative stress and the neurological damage associated with it (Markesbery and Carney, 1999; Smith *et al.*, 2000). Oxidative stress has been shown to play an important role in sporadic AD, which accounts for the vast majority of cases. ROS occurrence is the earliest event in the progress of AD (Perry *et al.*, 2000; Nunomura *et al.*, 2001). An oxidized nucleoside derived from RNA, 8-hydroxyguanosine (8OHG), and an oxidized amino acid, nitrotyrosine significantly increase in vulnerable neurons of AD patients. An investigation into the relationship between neuronal 8OHG, nitrotyrosine, histological, and clinical variables including Aβ-containing senile plaques and NFT, duration of dementia and apolipoprotein E (ApoE) genotype revealed that oxidative damage is an early-stage event in the process of neurodegeneration in AD. ROS-induced oxidative damage initiates the development of cognitive disturbances and pathological features observed in AD. Increased levels of oxidative damage and Mild Cognitive Impairment (MCI), which is believed to be one of the earliest stages of AD, are coexisted (Ding *et al.*, 2007). A decline in protein synthesis capabilities occurs in the same brain regions which exhibit increased levels of oxidative damage in AD subjects, while protein synthesis may be one of the earliest cellular processes disrupted by oxidative damage in AD. In contrast to sporadic AD, familial AD has been linked to amyloid beta protein precursor (AβPP) and presenilin gene mutations. Lines of evidence have demonstrated that Aβ formation promotes ROS production. Aβ activates the prooxidative enzyme NADPH-dependent oxidase, leading to the production of O_2^- (Behl *et al.*, 1994). H_2O_2 is directly generated during the process of Aβ aggregation (Bush *et al.*, 2003). Aβ can convert molecular oxygen into H_2O_2 by reducing divalent metal ions (Fe^{2+}, Cu^{2+}) (Lynch *et al.*, 2000; Behl and Moosmann, 2002). Aβ induces lipid peroxidation and subsequent production of cytotoxic aldehyde 4-hydroxynonenal (4-HNE) (Mark *et al.*, 1995), which impairs membrane Ca^{2+} pumps and enhances influx through voltage-dependent and ligand-gated calcium channels (Mattson and Chan, 2003). Thus Aβ formation may also worsen Ca^{2+} homeostasis.

When the level of oxidative stress in a cell exceeds the normal capacity of its protective mechanisms, oxidative stress can result in cell damages, including lipid, DNA/RNA and protein oxidations (Sayre *et al.*, 1997; Gabbita *et al.*, 1998; Nunomura *et al.*, 1999; Smith *et al.*, 2007). To remove excessive ROS and avoid ROS-induced damage, as mentioned as above, neurons use intrinsic antioxidative mechanisms, including enzymatic systems of SOD, GR/GP, and HO. The HO system is an interesting one in which different isoforms of HO come from different genes and cooperates to play an important role in neuronal defense.

1.3   MEMBERS OF THE HO ANTIOXIDATIVE SYSTEM

The HO system consists of three forms identified to date: the oxidative stress-inducible protein HO-1 or heat shock protein 32 (HSP32), and the constitutive isozymes HO-2 and HO-3. HO-1 is ubiquitous, and its mRNA and activity can be increased several-fold by heme, other metalloporphyrins, transition metals, and stimuli that induce cellular stress. Thus, HO-1 is recognized as a major heat shock/stress-response protein. HO-1 has an apparent molecular weight of 30 kDa, loses 30 % of its activity when heated at 60°C for 10 min, and precipitates in ammonium sulfate at 0–35 % saturation. In contrast, HO-2 has a molecular weight of 36 kDa, loses 80 % of its activity when heated as above, and precipitates in ammonium sulfate at 35–60 % saturation (Braggins *et al.*, 1986; Maines *et al.*, 1986; Trakshel *et al.*, 1986). Evidence that HO-1 and HO-2 are products of different genes is also demonstrated by their primary amino acid compositions. Three cysteine/cystine residues that appeared in HO-2 do not show in HO-1 (Cruse and Maines, 1988). Although human HO-1 and HO-2 share 43 % amino acid sequence identity (Yoshida *et al.*, 1988; Mc Coubrey *et al.*, 1992; Ishikawa *et al.*, 1995), HO-1 is composed of 288 amino acids whereas HO-2 consists of 316 amino acids (Yoshida *et al.*, 1988). HO-1 lacks a signal peptide, but contains a hydrophobic segment of 22 amino acid residues at the carboxyl terminus. In contrast, HO-2 contains two copies of the heme-binding site, a dipeptide of cysteine and praline (CP motif) (Shibahara, 2003). In addition, human HO-1 and HO-2 genes are separately located at chromosome 22q12 and 16p13.3 (Kutty *et al.*, 1994).

HO-3 is the product of a single transcript of approximately 2.4 kb and can encode a protein of approximately 33 kDa. The HO-3 transcript is found in the spleen, liver, thymus, prostate, heart, kidney, brain, and testis and is the product of a single-copy gene. The predicted amino acid structure of HO-3 differs from both HO-1 (HSP32) and HO-2 but is closely related to HO-2 (approximately 90%). *Escherichia coli* expressed and purified HO-3 protein does not cross react with polyclonal antibodies to either rat HO-1 or HO-2, is a poor heme catalyst, and displays hemoprotein spectral characteristics. The predicted protein has two heme regulatory motifs that may be involved in heme binding (Mc Coubrey *et al.*, 1997). However, a separate study suggests that HO-3 is a processed pseudo-gene product derived from HO-2 transcripts (Hayashi *et al.*, 2004).

1.4   DISTRIBUTION OF MEMBERS OF THE HO ANTIOXIDATIVE SYSTEM IN THE CNS

The two isozymes HO-1 and HO-2 display significant differences in tissue distribution. HO-1 under normal conditions is present in the whole brain at the limit of immunodetection and is localized in select neuronal populations. However, HO-1 1.8 kb transcript and protein (approximately 32 kDa) are increased in response to stressful stimuli, primarily in nonneuronal cell populations (Ewing and Maines, 1997). In contrast, HO-2 immunoreactive cells are only neurons in the CNS (Yamanaka *et al.*, 1996). In a separate study, HO-1 and HO-2 transcripts were found in both neurons and astrocytes while HO-3 transcript was uniquely found in astrocytes of hippocampus, cerebellum, and cortex (Scapagnini *et al.*, 2002).

1.5   FUNCTION OF THE HO ANTIOXIDATIVE SYSTEM

Both HO-1 and HO-2 catalyze oxidation of heme to biologically active molecules: iron, a gene regulator; biliverdin, an antioxidant; and carbon monoxide, a heme ligand and a promising and potentially significant messenger molecule (Figure 1.1; Maines, 1997). Products of the HO reaction have important effects: carbon monoxide is a potent vasodilator, which is thought to play a key role in the modulation of vascular tone, especially in the liver under physiological conditions, and in many organs under 'stressful' conditions associated with HO-1 induction. The 'free' iron increases oxidative stress and regulates the expression of many mRNAs (e.g., DCT-1, ferritin, and transferrin receptor) by affecting the conformation of iron regulatory protein (IRP)-1 and its binding to iron regulatory elements (IREs) in the 5'- or 3'-UTRs of the mRNAs (Kietzmann *et al.*, 2003). Biliverdin and its product bilirubin, formed in most mammals, are potent antioxidants. Bilirubin is neuroprotective at nanomolar concentrations (Doré *et al.*, 1999a) and protects cells from a 10,000-fold excess of H_2O_2 (Barañano *et al.*, 2002). HO-2 seems to be more significant in antioxidation than HO-1. Although the levels of oxidatively modified proteins of HO-1 and HO-2 cells in response to t-BuOOH toxicity are identical, the level of oxidatively modified proteins in HO-2 cells is less than that of HO-1 cells in response to H_2O_2 toxicity. Subcellular distribution shows that HO-2 and NADPH-cytochrome P_{450} reductase that are essential for HO activity in degradation of heme are colocalized in the microsome, whereas HO-1 is partially present in the microsome. HO-2 transfected cells are more resistant than HO-1 transfected neurons to H_2O_2 (Kim *et al.*, 2005). Under normal conditions, HO-2, but not HO-1, can be clearly detected in the rat brain (Trakshel *et al.*, 1988).

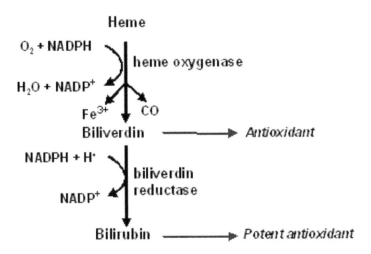

FIGURE 1.1 The metabolic pathway of heme.

1.6 EXPRESSION AND ACTIVITY REGULATIONS OF THE HO ANTIOXIDATIVE SYSTEM

There are several potential regulatory elements in the 5'-untranslated region (UTR) of HO-1, including activator protein 1 (AP-1), metal responsive element (MRE), oncogene c-myc/max heterodimer binding site (Myc/Max), antioxidant response element (ARE), and GC box binding (Sp1) sites. HO-1 responds to a myriad of other stress conditions, such as cytokines, hormones, volatile anesthetics, thiol-reactive substances, heavy metals, endotoxin, and UV radiation (Ryter *et al.*, 2006, 2007). HO-1 gene expression is also induced by various oxidative stress stimuli including sodium arsenite. Using deletion-reporter gene constructs, sites that mediate the arsenite-dependent induction of HO-1 were mapped, and components of the extracellular signal-regulated kinase (ERK) and p38 (a homolog of the yeast HOG1 kinase), but not c-Jun N-terminal kinase (JNK), mitogen-activated protein kinase (MAPK) pathways were reported to be involved in arsenite-dependent upregulation (Elbirt and Bonkovsky, 1999). The role of arsenite for HO-1 gene regulation was also investigated in primary rat hepatocytes. In this cell type, the JNK inhibitor SP600125 decreased sodium arsenite-mediated induction of HO-1 mRNA expression. Over expressions of JNK, MAP kinase kinase (MKK) 3, and p38gamma up-regulated HO-1 expression, whereas p38alpha, beta, and delta decreased the production of HO-1. CRE/AP-1 element (-668/-654) was shown as a binding site for c-Jun, a target of the JNK pathway, while E-box (-47/-42) manifested as a binding site for MKK3, p38 isoform and c-Max. Thus, the HO-1 CRE/AP-1 element mediates HO-1 gene induction via activation of JNK/c-Jun whereas p38 isoforms act through a different mechanism via the E-box (Kietzmann *et al.*, 2003). Oxidative stress greatly affects HO-1 expression. The expression profile of HO-1 seems to depend on cell types and stress strength. Hydrogen peroxide at a concentration of 0.14-0.7μM for 30 or 60 min increases content of immunoreactive HO-1 in cultured rat forebrain astrocytes by sevenfold within 3 h after exposure. In contrast, the same concentration of H_2O_2 fails to induce HO-1 expression in neurons (Dwyer *et al.*, 1995). When the concentration of H_2O_2 reaches 25μM, HO-1 expression shows a time-dependent increase in neurons (Chun *et al.*, 2001). Furthermore, HO-1 is down-regulated under certain circumstances, such as thermal-stressed human erythroblastic cell line YN-1-0-A, interferon γ-treated human gioblastoma cell line T989G, and hypoxia-insulted human endothelial cells and astrocytes (Okinaga *et al.*, 1996; Takahashi *et al.*, 1999; Nakayama *et al.*, 2000).

In contrast, HO-2 is present chiefly in the brain and testes and is generally believed to be an uninducible or constitutive protein (Kietzmann *et al.*, 2003). However lines of evidences have demonstrated that the activity and expression of this enzyme is changeable. In the past, the only known regulator of HO-2 is adrenal glucocorticoids (Weber *et al.*, 1994; Maines *et al.*, 1996). Corticosterone treatment (40 mg/kg, 20 days) increases the 1.3- and 1.9- kb HO-2 mRNA (Weber *et al.*, 1994). Then, studies revealed that HO-2 achieves enhanced catalytic activity following phosphorylation by protein kinase C or treatment with phorbol esters, and increased production of bilirubin under oxidative stress conditions (Doré *et al.*, 1999a). Through a yeast two-hybrid screen, calmodulin was identified as a potential regulator of HO-2 activity as well. Calmodulin binds with nanomolar affinity to HO-2 in a calcium-dependent manner, resulting in a threefold increase in catalytic activity.

Mutations within the binding site block calmodulin binding and calcium-dependent stimulation of enzyme activity *in vitro* and in intake cells. The calcium mobilizing agents ionomycin and glutamate stimulate endogenous HO-2 activity in primary cortical culture (Boehning *et al.*, 2004). In experimental spinal cord injury (SCI), HO-2 mRNA levels were elevated proximal (above) to the site of injury and more prominently at 16 h post SCI while HO-1 mRNA levels were enhanced distal (below) to the site of injury at both time points (Panahian and Maines, 2001). The protein profiles for HO-1 and HO-2 showed similar patterns as mRNA distributions. By comparing HO-2 mRNA levels in cognitively unimpaired and impaired adult and aged rats, both young and aged cognitively impaired rats showed increased expressions in hippocampi compared with aged cognitively unimpaired rats, while no difference was found in cortices between all animal groups (Law *et al.*, 2000). Under certain circumstances, HO-2 is also down-regulated. Real time PCR revealed low levels of

HO-1 and HO-2 mRNA present in placenta and deciduas of early gestation CBA/J mice exposed to stress or interleukin 12 (Zenclussen *et al.*, 2002). HO-2 expression was also suppressed in human pathologic pregnancies (Zenclussen *et al.*, 2003). HO-2 protein level was determined in rat penile tissue of different ages and was found to decrease during aging (Hu and Han, 2006). A 48-h hypoxic insult reduced expression levels of HO-2 mRNA and protein in human cell lines by shortening the half-life of HO-2 mRNA from 12 to 6 h (Zhang *et al.*, 2006). A chronic restraint stress decreased HO-2 protein levels in hippocampal neurons (Chen *et al.*, 2005). Our recent results demonstrated that HO-2 protein, but not mRNA, was dose-dependently reduced following an increase in H_2O_2 concentration when neurons were maintained under a culture condition with a low level of antioxidant (unpublished data). A coordination of HO-1 and HO-2 gene expressions has recently been reported where HO-2 may down-regulate the expression of HO-1 (Ding *et al.*, 2006). Down-regulation of HO-2 expression with siRNA technique results in induction of HO-1 expression at both mRNA and protein levels via activating the HO-1 gene promoter and prolongs the half-life of HO-1 mRNA, whereas knockdown of HO-1 expression has no significant effect on HO-2 expression. This study also demonstrates that HO-2 is a potent heme metabolic enzyme since HO-2 knockdown causes heme accumulation when exposed to exogenous hemin although HO-1 expression is up-regulated. HO-3 has very low activity; its physiological function probably involves heme binding only (Mc Coubrey *et al.*, 1997; Kietzmann *et al.*, 2003).

1.7   HOs IN NEURODEGENERATIVE DISEASES

Both HO-1 and HO-2 expression decrease during aging. Age-related reductions in HO, especially in HO-2, proteins are found in select brain regions including the hippocampus and the substantia nigra, which are involved in the high order cognitive processes of learning and memory (Ewing and Maines, 2006). Aged animals also demonstrate decreased stress response to hypoxic/hyperthermia stress in the HO, especially in HO-2, expressions.

HOs appear to be an important intrinsic neuroprotective factors in acute neurodegeneration. HO-2 plays an important role in protecting ischemic neurodegeneration. HO-2 deletion (HO-2$^{-/-}$) leads to increased neurotoxicity in brain culture and increased neural damage following transient cerebral ischemia in intact mice. In contrast, stroke damage is not significantly altered in HO1(-/-) mice (Doré *et al.*, 1999a, b, 2000). Cerebral ischemia can also result in ROS accumulation and apoptosis in cerebral vascular endothelial cells. In an *in vitro* system using serum withdrawal to mimic *in vivo* ischemia, quiescent HO-2$^{-/-}$ cells showed greater basal apoptosis than wild-type cells and reduced cell resistance to serum withdrawal (Parfenova *et al.*, 2006). Thus HO-2 appears to be an essential endogenous antioxidative factor in the both neurons and cerebral vascular endothelial cells. Recently, a study demonstrated that HO-2 is also a crucial neuroprotective enzyme against collagenase-induced intracerebral hemorrhage (Wang *et al.*, 2006). The roles of HO-1 and HO-2 in protecting neurodegeneration induced by traumatic brain injury were also reported (Chang *et al.*, 2003). HO-2 activity from injured HO-2 knockout mice was significantly less than that of HO-2 wild types, despite the induction of HO-1 expression after the traumatic brain injury.

In AD, Aβ is generated from AβPP. AβPP can bind to either HO-1 or HO-2 and further suppress HO bioactivity (Doré, 2002). AβPP with mutations linked to familial AD provides substantially greater inhibition of HO activity than wild-type AβPP. Cortical neurons from transgenic mice expressing Swedish mutant AβPP greatly reduced bilirubin levels, establishing that mutant AβPP inhibits HO activity *in vivo*. Furthermore, oxidative neurotoxicity is markedly greater in cerebral cortical cultures from AβPP Swedish mutant transgenic mice than wild-type culture (Takahashi *et al.*, 2000).

1.8   THE HO ANTIOXIDATIVE SYSTEM AS A DRUG TARGET

Many antioxidants have been used in experimental and clinical treatment of neurodegeneration. Those include vitamins, coenzyme Q10, melatonin, ebselen, spin-trap scavenging agents, N-acetylcysteine, glutathione, metal ion chelators, uric acid, creatine lazaroids, nicaraven, etc. (Gilgun-Sherki *et al.*, 2002). Since HO-1 expression is inducible, many efforts in suppressing oxidative stress-induced damage have recently focused on inducing HO-1 expression. HO-1 expression can be induced with dietary antioxidants α-lipoic acid (in broccoli, spinach, tomatoes), cafestol and kahweol (in coffee), carnosol (in rosemary), curcumin (in turmeric), resveratrol (in grape), selenium (in cereals and fish), and sulphoraphane (in broccoli and sprouts) (Ogborne *et al.*, 2004). Pharmacologic inducers of HO-1 were recently reviewed (Li *et al.*, 2007). These inducers include simvastatin, lovastatin, NO donors, organic nitrates, NO, aspirin, 15-Epi-lipoxin-A_4 analog, AZD 3582, probucol, adrenomedullin, atrial natriuretic peptide, D-4F, curcumin, rosolic acid, caffeic acid phenethyl ester, sulforaphane, carnosol, piceatannol paclitaxel, rapamycin, ethyl ferulate, 1,2,3,4,6-Penta-o-galloyl-beta-D-glucose, insulin and isoproterenol.

Normal intracellular levels of HO-2 are important to cells for defense against oxidative stress, and therefore preservation of basic levels of HO-2 is essential for protecting against ROS-induced neurodegeneration. Our recent research demonstrated that the neuroprotective compound PAN-811 (Jiang *et al.*, 2006) completely blocks oxidative stress-induced neurotoxicity while preserving HO-2 protein to a normal level present in noninsulted control neurons (unpublished data).

1.9   CONCLUSION AND FUTURE PERSPECTIVES

HO-1 and HO-2 are derived from different genes but show a similar role in catalyzing heme to produce the antioxidants biliverdin and bilirubin, and in protecting oxidative stress-induced neurotoxicity. HO-1 and HO-2 mainly distribute in astrocytes and neurons under normal condition respectively. HO-2 plays an important role in maintenance of heme and ROS homeostasis whereas HO-1 reacts to environmental stress changes. A cooperation of them is shown by effects of HO-2 on the gene expression of HO-1. The important role of HO-1 and HO-2 in eliminating excessive intracellular ROS is evidenced by potent activity of their catalyzing product bilirubin. Therefore, HOs are important intracellular targets for neuroprotective drug development. One strategy is induction of HO−1 expression and other will be preserving HO-2 protein level. Due to the multifaceted mechanisms of neurodegenerative disease, blocking toxic pathways of oxidative stress alone may not be sufficient for treatment. Thus co-administration of an antioxidative drug with other neuroprotectant(s) with different targets (e.g. NMDA receptor antagonist) would achieve superior efficacy in treatment of neurodegenerative disease.

KEYWORDS

- **Alzheimer's disease**
- **Bilirubin**
- **Calmodulin**
- **Fenton reaction**
- **Gene mutations**
- **Glutathione reductase**
- **Heme oxygenase system**
- **HO antioxidative system**
- **Ischemia**
- **Isozymes**
- **Mild cognitive impairment**
- **Neurodegenerative diseases**
- **Neuronal cell death**
- **Neuroprotective factor**
- **Oxidative stress**
- **Reactive oxygen species**

- **Real time PCR**
- **RNA interference**
- **siRNA technique**
- **Superoxide dismutase**

REFERENCES

Abu-Soud, H. M.; Wang, J.; Rousseau, D. L.; Fukuto, J. M.; Ignarro, L. J.; Stuehr, D. J. Neuronal nitric oxide synthase self-inactivates by forming a ferrous-nitrosyl complex during aerobic catalysis. *J Biol Chem* 1995, 270, 22997-23006.

Barañano, D. E.; Snyder, S. H. Neural roles for heme oxygenase: Contrasts to nitric oxide synthase. *Proc Natl Acad Sci USA* 2001, 98, 10996-11002.

Baranano, D. E.; Rao, M.; Ferris, C. D.; Snyder, S. H. Biliverdin reductase: A major physiologic cytoprotectant. *Proc Natl Acad Sci USA* 2002, 99, 16093-16098.

Behl, C.; Davies, J. B.; Lesley, R.; Schubert, D. Hydrogen peroxide mediates amyloid protein toxicity. *Cell* 1994, 77, 817-827.

Behl, C.; Moosmann, B. Oxidative nerve cell death in Alzheimer's disease and stroke: Antioxidants as neuroprotective compounds. *Biol Chem* 2002, 383, 521-536.

Boehning, D.; Sedaghat, L.; Sedlak, T. W.; Snyder, S. H. Heme oxygenase-2 is activated by calcium-calmodulin. *J Biol Chem* 2004, 279, 30927-30930.

Braggins, P. E.; Trakshel, G. M.; Kutty, R. K.; Maines, M. D. Characterization of two heme oxygenase isoforms in rat spleen: Comparison with the hematin-induced and constitutive isoforms of the liver. *Biochem Biophys Res Commun* 1986, 141, 528-533.

Bernardi, P.; Vassanell, S.; Veronese, P.; Colonna, R.; Szabo, I.; Zoratti, M. Modulation of the mitochondrial permeability transition pore. Effect of protons and divalent cations. *J Biol Chem* 1992, 267, 2934-2939.

Bush, A. I.; Masters, C. L.; Tanzi, R. E. Copper, beta-amyloid, and Alzheimer's disease: Tapping a sensitive connection. *Proc Natl Acad Sci USA* 2003, 100, 11193-11194.

Burmester, T.; Hankeln, T. Neuroglobin: A respiratory protein of the nervous system. *News Physiol Sci* 2004, 19, 110-113.

Chan, P. H. Reactive oxygen radicals in signaling and damage in the ischemic brain. *J Bereb Blood Flow Metab* 2001, 21, 2-14.

Chang, E. F.; Wong, R. J.; Vreman, H. J.; Igarashi, T.; Galo, E.; Sharp, F. R.; Stevenson, D. K.; Noble-Haeusslein, L. J. Heme oxygenase-2 protects against lipid peroxidation-mediated cell loss and impaired motor recovery after traumatic brain injury. *J Neurosci* 2003, 23, 3689-3696.

Chen, Z.; Xu, H.; Haimano, S.; Li, X.; Li, X. M. Quetiapine and venlafaxine synergically regulate heme oxygenase-2 protein expression in the hippocampus of stressed rats. *Neurosci Lett* 2005, 389, 173-177.

Chun, H. S.; Gibson, G. E.; De Giorgio, L. A.; Zhang, H.; Kidd, V. J.; Son, J. H. Dopaminergic cell death induced by MPP(+), oxidant and specific neurotoxicants shares the common molecular mechanism. *J Neurochem* 2001, 76, 1010-1021.

Coyle, J. T.; Puttfarcken, P. Oxidative stress, glutamate, and neurodegenerative disorders. *Science* 1993, 262, 689-695.

Cruse, I.; Maines, M. D. Evidence suggesting that the two forms of heme oxygenase are products of different genes. *J Biol Chem* 1988, 263, 3348-3353.

Dawson, V. L.; Dawson, T. M.; London, E. D.; Bredt, D. S.; Snyder, S. H. Nitric oxide mediates glutamate neurotoxicity in primary cortical culture. *Proc Natl Acad Sci USA* 1991, 88, 6368-6371.

Dickson, D. W. The pathogenesis of senile plaques. *J Neuropathol Exp Neurol* 1997, 56, 321-339.

Ding, Q.; Dimayuga, E.; Keller, J. N. Oxidative damage, protein synthesis, and protein degradation in Alzheimer's disease. *Curr Alzheimer Res* 2007, 4, 73-79.

Ding, Y.; Zhang, Y. Z.; Furuyama, K.; Ogawa, K.; Igarashi, K.; Shibahara, S. Down-regulation of heme oxygenase-2 is associated with the increased expression of heme oxygenase-1 in human cell lines. *FEBS J* 2006, 273, 5333-5346.

Doré, S.; Takahashi, M.; Ferris, C. D.; Zakhary, R.; Hester, L. D.; Guastella, D.; Snyder, S. H. Bilirubin, formed by activation of heme oxygenase-2, protects neurons against oxidative stress injury. *Proc Natl Acad Sci USA* 1999a, 96, 2445-2450.

Doré, S.; Sampei, K.; Goto, S.; Alkayed, N. J.; Guastella, D.; Blackshaw, S.; Gallagher, M.; Traystman, R. J.; Hurn, P. D.; Koehler, R. C.; Snyder, S. H. Heme oxygenase-2 is neuroprotective in cerebral ischemia. *Mol Med* 1999b, 5, 656-663.

Doré, S.; Goto, S.; Sampei, K.; Blackshaw, S.; Hester, L. D.; Ingi, T.; Sawa, A.; Traystman, R. J.; Koehler, R. C.; Snyder, S. H. Heme oxygenase-2 acts to prevent neuronal death in brain cultures and following transient cerebral ischemia. *Neuroscience* 2000, 99, 587-592.

Doré, S. Decreased activity of the antioxidant heme oxygenase enzyme: Implications in ischemia and in Alzheimer's disease. *Free Radic Biol Med* 2002, 32, 1276-1282.

Dowjat, W. K.; Wisniewski, H.; Wisniewski, T. Alzheimer's disease presenilin-1 expression modulates the assembly of neurofilaments. *Neuroscience* 2001, 103, 1-8.

Dwyer, B. E.; Nishimura, R. N.; Lu, S. Y. Differential expression of heme oxygenase-1 in cultured cortical neurons and astrocytes determined by the aid of a new heme oxygenase antibody. Response to oxidative stress. *Brain Res Mol Brain Res* 1995, 30, 37-47.

Elbirt, K. K.; Bonkovsky, H. L. Heme oxygenase: Recent advances in understanding its regulation and role. *Proc Assoc Am Physicians* 1999, 111, 438-447.

Ewing, J. F.; Maines, M. D. Histochemical localization of heme oxygenase-2 protein and mRNA expression in rat brain. *Brain Res Protoc* 1997, 1, 165-174.

Ewing, J. F.; Maines, M. D. Regulation and expression of heme oxygenase enzymes in aged-rat brain: Age related depression in HO-1 and HO-2 expression and altered stress-response. *J Neural Transm* 2006, 113, 439-454.

Faber-Langendoen, K.; Morris, J. C.; Knesevich, J. W.; La Barge, E.; Miller, J. P.; Berg, L. Aphasia in senile dementia of the Alzheimer type. *Ann Neurol* 1988, 23, 365-370.

Foresti, R.; Goatly, H.; Green, C. J.; Motterlini, R. Role of heme oxygenase-1 in hypoxia-reoxygenation: Requirement of substrate heme to promote cardioprotection. *Am J Physiol Heart Circ Physiol* 2001, 281, H1976-1984.

Forstl, H.; Kurz, A. Clinical features of Alzheimer's disease. *Eur Arch Psychiatry Clin Neurosci* 1999, 249, 288-290.

Gabbita, S. P.; Lovell, M. A.; Markesbery, W. R. Increased nuclear DNA oxidation in the brain in Alzheimer's disease. *J Neurochem* 1998, 71, 2034-2040.

Gilgun-Sherki, Y.; Rosenbaum, Z.; Melamed, E.; Offen, D. Antioxidant therapy in acute central nervous system injury: Current state. *Pharmacol Rev* 2002, 54, 271-284.

Gunter, T. E.; Yule, D. I.; Gunter, K. K.; Eliseev, R. A.; Salter, J. D. Calcium and mitochondria. *FEBS Lett* 2004, 567, 96-102.

Halestrap, A. P.; Mc Stay, G. P.; Clarke, S. J. The permeability transition pore complex: Another view. *Biochimie* 2002, 84, 153-166.

Halliwell, B. Oxidative stress and neurodegeneration: Where are we now? *J Neurochem* 2006, 97, 1634-1658.

Hagemeyer, C. E.; Rosenbrock, H.; Ditter, M.; Knoth, R.; Volk, B. Predominantly neuronal expression of cytochrome P450 isoforms CYP3A11 and CYP3A13 in mouse brain. *Neuroscience* 2003, 117, 521-529.

Hayashi, S.; Omata, Y.; Sakamoto, H.; Higashimoto, Y.; Hara, T.; Sagara, Y.; Noguchi, M. Characterization of rat heme oxygenase-3 gene. Implication of processed pseudogenes derived from heme oxygenase-2 gene. *Gene* 2004, 336, 241-250.

Hu, H. L.; Han, R. F. The concentration of HO-2 and CO in rat penile tissue of different ages. *Zhonghua Nan Ke Xue* 2006, 12, 424-427.

Ishikawa, K.; Takeuchi, N.; Takahashi, S.; Matera, K. M.; Sato, M.; Shibahara, S.; Rousseau, D. L.; Ikeda-Saito, M.; Yoshida, T. Heme oxygenase-2. Properties of the heme complex of the purified tryptic fragment of recombinant human heme oxygenase-2. *J Biol Chem* 1995, 270, 6345-6350.

Jiang, Z. G.; Lu, X. C.; Nelson, V.; Yang, X.; Pan, W.; Chen, R. W.; Lebowitz, M. S.; Almassian, B.; Tortella, F. C.; Brady, R. O.; Ghanbari, H. A. A multifunctional cytoprotective agent that reduces neurodegeneration after ischemia. *Proc Natl Acad Sci USA* 2006, 103, 1581-1586.

Kiedrowski, L.; Costa, E.; Wroblewski, J. T. Glutamate receptor agonists stimulate nitric oxide synthase in primary cultures of cerebellar granule cells. *J Neurochem* 1992, 58, 335-341.

Kietzmann, T.; Samoylenko, A.; Immenschuh, S. Transcriptional regulation of heme oxygenase-1 gene expression by MAP kinases of the JNK and p38 pathways in primary cultures of rat hepatocytes. *J Biol Chem* 2003, 278, 17927-17936.

Kim, Y. S.; Zhuang, H.; Koehler, R. C.; Doré, S. Distinct protective mechanisms of HO-1 and HO-2 against hydroperoxide-induced cytotoxicity. *Free Radic Biol Med* 2005, 38, 85-92.

Kowaltowski, A. J.; Castilho, R. F.; Vercesi, A. E. Opening of the mitochondrial permeability transition pre by uncoupling or inorganic phosphate in the presence of Ca^{2+} is dependent on mitochondrial-generated reactive oxygen species. *FEBS Lett* 1996, 378, 150-152.

Kuchar, J.; Hausinger, R. P. Biosynthesis of metal sites. *Chem Rev* 2004, 104, 509-525.

Kuroda, S.; Siesjö, B. K. Reperfusion damage following focal ischemia: Pathophysiology and therapeutic windows. *Clin Neurosci* 1997, 4, 199-212.

Kutty, R. K.; Kutty, G.; Rodriguez, I. R.; Chader, G. J.; Wiggert, B. Chromosomal localization of the human heme oxygenase genes: Heme oxygenase-1(HMOX1) maps to chromosome 22q12 and heme oxygenase-2 (HMOX2) maps to chromosome 16p13.3. *Genomics* 1994, 20, 513-516.

Law, A.; Doré, S.; Blackshaw, S.; Gauthier, S.; Quirion, R. Alteration of expression levels of neuronal nitric oxide synthase and haem oxygenase-2 messenger RNA in the hippocampi and cortices of young adult and aged cognitively unimpaired and impaired Long-Evans rats. *Neuroscience* 2000, 100, 769-775.

Li, C.; Hossieny, P.; Wu, B. J.; Qawasmeh, A.; Beck, K.; Stocker, R. Pharmacologic induction of heme oxygenase-1. *Antioxid Redox Signal* 2007, 9, 1-13.

Lynch, T.; Cherny, R. A.; Bush, A. I. Oxidative processes in Alzheimer's disease: The role of abeta-metal interactions. *Exp Gerontol* 2000, 35, 45-451.

Maines, M. D. The heme oxygenase system: A regulator of second messenger gases. *Annu Rev Pharmacol Toxicol* 1997, 37, 517-554.

Maines, M. D.; Trakshel, G. M.; Kutty, R. K. Characterization of two constitutive forms of rat liver microsomal heme oxygenase. Only one molecular species of the enzyme is inducible. *J Biol Chem* 1986, 261, 411-419.

Maines, M. D.; Eke, B. C.; Zhao, X. Corticosterone promotes increased heme oxygenase-2 protein and transcript expression in the newborn rat brain. *Brain Res* 1996, 722, 83-94.

Mark, R. J.; Hensley, K.; Butterfield, D. A.; Mattson, M. P. Amyloid beta-peptide impairs ion-motive ATPase activities: Evidence for a role in loss of neuronal Ca2+ homeostasis and cell death. *J Neurosci* 1995, 15, 6239-6249.

Markesbery, W. R.; Carney, J. M. Oxidative alterations in Alzheimer's disease. *Brain Pathol* 1999, 9, 133-146.

Mattson, M. P.; Chan, S. L. Neuronal and glial calcium signaling in Alzheimer's disease. *Cell Calcium* 2003, 34, 385-397.

Mc Coubrey, W. K.; Jr, Ewing, J. F.; Maines, M. D. Human heme oxygenase-2: Characterization and expression of a full-length cDNA and evidence suggesting that the two HO-2 transcripts may differ by choice of polyadenylation signal. *Arch Biochem Biophys* 1992, 295, 13-20.

Mc Coubrey, W. K.; Huang, T. J.; Maines, M. D. Isolation and characterization of a cDNA from the rat brain that encodes hemoprotein heme oxygenase-3. *Eur J Biochem* 1997, 247, 725-732.

Morris, J. C.; Drazner, M.; Fulling, K.; Grant, E. A.; Goldring, J. Clinical and pathological aspects of parkinsonism in Alzheimer's disease. A role for extranigral factors? *Arch Neurol* 1989, 46, 651-657.

Nakayama, M.; Takahashi, K.; Kitamuro, T.; Yasumoto, K.; Katayose, D.; Shirato, K.; Fujii-Kuriyama, Y.; Shibahara, S. Repression of heme oxygenase-1 by hypoxia in vascular endothelial cells. *Biochem Biophys Res Commun* 2000, 271, 665-671.

Nunomura, A.; Perry, G.; Pappolla, M. A.; Wade, R.; Hirai, K.; Chiba, S.; Smith, M. A. RNA oxidation is a prominent feature of vulnerable neurons in Alzheimer's disease. *J Neurosci* 1999, 19, 1959-1964.

Nunomura, A.; Perry, G.; Aliev, G.; Hirai, K.; Takeda, A.; Balraj, E. K.; Jones, P. K.; Ghanbari, H.; Wataya, T.; Shimohama, S.; Chiba, S.; Atwood, C. S.; Petersen, R. B.; Smith, M. A. Oxidative damage is the earliest event in Alzheimer disease. *J Neuropathol Exp Neurol* 2001, 60, 759-767.

Ogborne, R. M.; Rushworth, S. A.; Charalambos, C. A.; O'Connell, M. A. Haem oxygenase-1: A target for dietary antioxidants. *Biochem Soc Trans* 2004, 32, 1003-1005.

Okinaga, S.; Takahashi, K.; Takeda, K.; Yoshizawa, M.; Fujita, H.; Sasaki, H.; Shibahara, S. Regulation of human heme oxygenase-1 gene expression under thermal stress. *Blood* 1996, 87, 5074-5084.

Panahian, N.; Maines, M. D. Site of injury-directed induction of heme oxygenase-1 and -2 in experimental spinal cord injury: Differential functions in neuronal defense mechanisms? *J Neurochem* 2001, 76, 539-554.

Packer, L.; Tritschler, H. J.; Wessel, K. Neuroprotection by the metabolic antioxidant alpha-lipoic acid. *Free Radic Biol Med* 1997, 22, 359-378.

Parfenova, H.; Basuroy, S.; Bhattacharya, S.; Tcheranova, D.; Qu, Y.; Regan, R. F.; Leffler, C. W. Glutamate induces oxidative stress and apoptosis in cerebral vascular endothelial cells: Contributions of HO-1 and HO-2 to cytoprotection. *Am J Physiol Cell Physiol* 2006, 290, C1399-1410.

Perry, G.; Raina, A. K.; Nunomura, A.; Wataya, T.; Sayre, L. M.; Smith, M. A. How important is oxidative damage? Lessons from Alzheimer's disease. *Free Radic Biol Med* 2000, 28, 831-834.

Rego, A. C.; Santos, M. S.; Oliveira, C. R. Glutamate-mediated inhibition of oxidative phosphorylation in cultured retinal cells. *Neurochem Int* 2000, 36, 159-166.

Romanelli, M. F.; Morris, J. C.; Ashkin, K.; Coben, L. A. Advanced Alzheimer's disease is a risk factor for late-onset seizures. *Arch Neurol* 1990, 7, 847-850.

Rubin, E. H.; Morris, J. C.; Storandt, M.; Berg, L. Behavioral changes in patients with mild senile dementia of the Alzheimer's type. *Psychiatry Res* 1987, 21, 55-62.

Ryter, S. W.; Alam, J.; Choi, A. M. Heme oxygenase-1/carbon monoxide: From basic science to therapeutic applications. *Physiol Rev* 2006, 86, 583-650.

Ryter, S. W.; Kim, H. P.; Hoetzel, A.; Park, J. W.; Nakahira, K.; Wang, X.; Choi, A. M. Mechanisms of cell death in oxidative stress. *Antioxid Redox Signal* 2007, 9, 49-89.

Sayre, L. M.; Zelasko, D. A.; Harris, P. L.; Perry, G.; Salomon, R. G.; Smith, M. A. 4-Hydroxynonenal-derived advanced lipid peroxidation end products are increased in Alzheimer's disease. *J Neurochem* 1997, 68, 2092-2097.

Scapagnini, G.; D'Agata, V.; Calabrese, V.; Pascale, A.; Colombrita, C.; Alkon, D.; Cavallaro, S. Gene expression profiles of heme oxygenase isoforms in the rat brain. *Brain Res* 2002, 954, 51-59.

Selkoe, D. J. Alzheimer's disease: Genes, proteins, and therapy. *Physiol Rev* 2001, 81, 741-766.

Shibahara, S. The heme oxygenase dilemma in cellular homeostasis: New insights for the feedback regulation of heme catabolism. *Tohoku J Exp Med* 2003, 200, 167-186.

Sisodia, S. S.; Koo, E. H.; Beyreuther, K.; Unterbeck, A.; Price, D. L. Evidence that beta-amyloid protein in Alzheimer's disease is not derived by normal processing. *Science* 1990, 248, 492-495.

Smith, D. G.; Cappai, R.; Barnham, K. J. The redox chemistry of the Alzheimer's disease amyloid beta peptide. *Biochim Biophys Acta* 2007, 1768, 1976-1990.

Smith, M. A.; Rottkamp, C. A.; Nunomura, A.; Raina, A. K.; Perry, G. Oxidative stress in Alzheimer's disease. *Biochim Biophys Acta* 2000, 1502, 139-144.

Srisook, K.; Kim, C.; Cha, Y. N. Molecular mechanisms involved in enhancing HO-1 expression: De-repression by heme and activation by Nrf2, the "one-two" punch. *Antioxid Redox Signal* 2005, 7, 1674-1687.

Stoothoff, W. H.; Johnson, G. V. Tau phosphorylation: Physiological and pathological consequences. *Biochim Biophys Acta* 2005, 1739, 280-297.

Swearer, J. M.; Drachman, D. A.; O'Donnell, B. F.; Mitchell, A. L. Troublesome and disruptive behaviors in dementia. Relationships to diagnosis and disease severity. *J Am Geriatr Soc* 1988, 36, 784-790.

Takahashi, K.; Nakayama, M.; Takeda, K.; Fujia, H.; Shibahara, S. Suppression of heme oxygenase-1 mRNA expression by interferon-gamma in human glioblastoma cells. *J Neurochem* 1999, 72, 2356-2361.

Takahashi, M.; Doré, S.; Ferris, C. D.; Tomita, T.; Sawa, A.; Wolosker, H.; Borchelt, D. R.; Iwatsubo, T.; Kim, S. H.; Thinakaran, G.; Sisodia, S. S.; Snyder, S. H. Amyloid precursor proteins inhibit heme oxygenase activity and augment neurotoxicity in Alzheimer's disease. *Neuron* 2000, 28, 461-473.

Trakshel, G. M.; Kutty, R. K.; Maines, M. D. Purification and characterization of the major constitutive form of testicular heme oxygenase. The noninducible isoform. *J Biol Chem* 1986, 261, 11131-11137.

Trakshel, G. M.; Kutty, R. K.; Maines, M. D. Resolution of the rat brain heme oxygenase activity: Absence of a detectable amount of the inducible form (HO-1). *Arch Biochem Biophys* 1988, 260, 732-739.

Wang, J.; Zhuang, H.; Doré, S. Heme oxygenase 2 is neuroprotective against intracerebral hemorrhage. *Neurobiol Dis* 2006, 22, 473-476.

Van der Worp, H. B.; Bär, P. R.; Kappelle, L. J.; de Wildt, D. J. Dietary vitamin E levels affect outcome of permanent focal cerebral ischemia in rats. *Stroke* 1998, 29, 1002-1005.

Weber, C. M.; Eke, B. C.; Maines, M. D. Corticosterone regulates heme oxygenase-2 and NO synthase transcription and protein expression in rat brain. *J Neurochem* 1994, 63, 953-962.

Won, S. J.; Kim, D. Y.; Gwag, B. J. Cellular and molecular pathways of ischemic neuronal death. *J Biochem Mol Biol* 2002, 35, 67-86.

Yamanaka, M.; Yamabe, K.; Saitoh, Y.; Katoh-Semba, R.; Semba, R. Immunocytochemical localization of heme oxygenase-2 in the rat cerebellum. *Neurosci Res* 1996, 24, 403-407.

Yoshida, T.; Biro, P.; Cohen, T.; Müller, R. M.; Shibahara, S. Human heme oxygenase cDNA and induction of its mRNA by hemin. *Eur J Biochem* 1988, 171, 457-461.

Zenclussion, A. C.; Joachim, R.; Hagen, E.; Peiser, C.; Klapp, B. F.; Arck, P. C. Heme oxygenase is downregulated in stress-triggered and interleukin-12-mediated murine abortion. *Scand J Immunol* 2002, 55, 560-569.

Zenclussion, A. C.; Lim, E.; Knoeller, S.; Knackstedt, M.; Hertwig, K.; Hagen, E.; Klapp, B. F.; Arck, P. C. Heme oxygenases in pregnancy II: HO-2 is downregulated in human pathologic pregnancies. *Am J Reprod Immunol* 2003, 50, 66-76.

Zhang, Y.; Furuyama, K.; Kaneko, K.; Ding, Y.; Ogawa, K.; Yoshizawa, M.; Kawamura, M.; Takeda, K.; Yoshida, T.; Shibahara, S. Hypoxia reduces the expression of heme oxygenase-2 in various types of human cell lines. A possible strategy for the maintenance of intracellular heme level. *FEBS J* 2006, 273, 3136-3147.

Zhang, Y.; Marcillat, O.; Giulivi, C.; Ernster, L.; Davies, K. J. The oxidative inactivation of mitochondrial electron transport chain components and ATPase. *J Biol Chem* 1990, 265, 16330-16336.

Zipfel, G. J.; Babcock, D. J.; Lee, J. M.; Choi, D. W. Neuronal apoptosis after CNS injury: The roles of glutamate and calcium. *J Neurotrauma* 2000, 17, 857-869.

CHAPTER 2

EMBRACING COMPLEXITY: SEARCHING FOR GENE-GENE AND GENE ENVIRONMENT INTERACTIONS IN GENETIC EPIDEMIOLOGY

ALISON A. MOTSINGER and DAVID M. REIF

CONTENTS

2.1   Introduction 20
2.2   Preliminary Analyses... 23
2.3   Traditional Statistical Approaches.. 26
2.4   Novel Approaches to Detect Epistasis.................................. 32
2.5   Examples of Epistasis Found in Humans 48
2.6   Developing an Analysis Plan.. 49
Keywords 50
References... 51

2.1 INTRODUCTION

The search for susceptibility loci in the study of common, complex diseases is a major challenge in the field of human genetics, and has been less successful than for simple Mendelian disorders (Moore, 2003). It is likely that this is due to many complicating factors such as an increased number of contributing loci and susceptibility alleles, incomplete penetrance, and contributing environmental effects (Templeton, 2000; Cordell, 2002; Culverhouse *et al.*, 2002; Moore and Williams, 2002; Moore, 2003; Sing *et al.*, 2004; Thornton-Wells *et al.*, 2004). Additionally, gene-gene and gene-environment interaction, or epistasis, is an increasingly assumed to play an important role in the underlying etiology of such diseases (Templeton, 2000; Cordell, 2002; Culverhouse *et al.*, 2002; Moore and Williams, 2002; Moore, 2003; Sing *et al.*, 2004; Thornton-Wells *et al.*, 2004). While there are several definitions of the word "epistasis", in its simplest definition, epistasis occurs when the action of one gene is modified by one or more other genetic and/or environmental factors. This phenomenon presents a challenge in the search for disease-risk variants, since if the effect of one locus is altered or masked by effects at another locus, the power to detect the first locus is likely to be reduced and elucidation of the joint effects at the two loci will be hindered by their interaction, unless explicitly examined (Cordell, 2002).

The high dimensionality involved in the evaluation of combinations of many such genetic and environmental variables quickly diminishes the usefulness of traditional, parametric statistical methods, known as "the curse of dimensionality" (Bellman, 1961). As the number of factors increases and the number of possible interactions increases exponentially, many contingency table cells will be left with very few, if any, data points. Due to the hierarchical model-building process typically used by traditional methods for variable selection (Moore and Williams, 2002), they are often limited in their ability to deal with many factors and fail to characterize epistasis models in the absence of main effects. This results in increased type II (false negative) errors and decreased power (Moore, 2004). These challenges are magnified by relatively small sample sizes. The time and expense involved in sample collection can make effective studies cost prohibitive with traditional analytical methods.

Additionally, rapid advances in genotyping technology have increased the number of genetic variants included in genetic studies. Genome-wide association studies including as many as one million single nucleotide polymorphisms (SNPs) are now accessible. This results in an additional level of complexity from an analytical perspective (Moore and Ritchie, 2004). An analytical approach must not only have statistical power to detect significant associations, but must also search through thousands or million(s) of variables and identify those that best predict the outcome/disease of interest. Additionally, such a large number of variables amplify the problem of multiple comparisons. As the number of statistical tests increases, it becomes increasingly likely that one will observe data that satisfy the acceptance criterion (*e.g.* is significant at the p<0.05 level) by chance alone. For example, with a genome-wide association study with one million SNPs, you would expect 50,000 significant SNPs using a p value cutoff of 0.05. Analytical methods must limit false positive results without limiting power. These challenges have prompted the development of novel methodologies to detect gene-gene and gene-environment interactions in a variety of study designs.

In this chapter, we discuss the tools available to a genetic epidemiologist to detect such epistatic interactions. We briefly define epistasis and review the study designs and strategies that can be used for an interaction-oriented analysis plan. Next, we discuss the traditional statistical tools available for epidemiological research and how they can be applied to the detection of epistasis. We then discuss a "data-mining" approach to analysis, highlighting novel computational approaches that have been developed to detect epistasis. Finally, we stress the importance of considering interactions in epidemiology by highlighting the interactions that have been previously detected in statistical genetics studies of human populations.

2.1.1   DEFINING EPISTASIS

Bateson in 1909 is credited with first coining the term "epistatic" to describe an effect where a variant at one locus masks the manifestation of the effect of another (Bateson, 1909). This was originally viewed as an extension of the concept of genetic dominance for allelic variants at a single locus. Bateson's definition of epistasis is often used by biologists or biochemists when investigating biological interactions between proteins, however what is meant by biological interaction is not always well defined (Cordell, 2002). It usually corresponds to a situation in which the qualitative nature of the mechanism of action of a factor is affected by the presence or absence of the other (Neuman and Rice, 1992).

As Bateson is credited with pioneering the term "epistasis", Fisher in 1918 is credited with defining a separate statistical sense of the term (Fisher, 1918). Fisher proposed a mathematical definition of epistasis as a deviation from additivity in the effect of alleles at different loci with respect to their contribution to a quantitative phenotype. Epistasis in this sense is closer to the usual concept of statistical interaction (Norton and Pearson, 1976) as a departure from a specific linear model describing the relationship between predictive factors. While the long-term goal of disease-risk mapping is to gain an understanding of functional consequences of variants at the biological level that translate into clinical progress, at the population level epidemiological approaches must rely on statistical methodologies to target genetic regions. So while there are semantic and theoretical differences between these two definitions, they are certainly not separate ideas. A complete discussion of these definitions is beyond the scope of this chapter, but excellent discussions of this topic can be found in Cordell (2002) and Moore and Williams (2005). For our purposes, we are less concerned with a strict definition of the term and will focus on a practical approach to the detection of epistasis in its broadest sense: the action of one gene is modified by one or more other genetic and/or environmental factors.

2.1.2 APPROPRIATE STUDY DESIGNS

There are two broad types of study design that may be used to identify relationships between human genomic variants and phenotypes of interest: linkage analysis and association analysis (Risch, 2000). Linkage analysis determines whether a chromosomal region is preferentially inherited by offspring with the trait of interest by using genotype and phenotype data from multiple biologically – related family members. Linkage analysis capitalizes on the fact that, as a causative gene(s) segregates through a family kindred, other markers nearby on the same chromosome tend to segregate together (are in linkage) with the causative gene due to the lack of recombination in that region. Association analysis, on the other hand, describes the use of case-control, cohort, or even family data to statistically relate genetic variations to a disease/phenotype. Because association analysis directly examines the effect of a candidate locus, rather than an effect that is diffused across large regions of chromosomes, its greatest applicability is in fine localization and identification of causative loci (Daly and Day, 2001).

Interactive effects can be assessed in any study design (Andrieu and Goldstein, 1998; Goldstein and Andrieu, 1999). Case-control, case-only, prospective cohort, and family-based studies have all been successfully used to detect epistatic interactions. Case-control studies are the most commonly used to search for epistatic interactions, and most novel methods development has been focused on such a design. Case-only designs are more controversial for the study of interactive effects. Case-only designs are extremely powerful for the detection of gene-environment interactions (Andrieu and Goldstein, 1998; Goldstein and Andrieu, 1999), but their utility for the detection of gene-gene interactions has been debated in the literature (Cordell, 2003; Vieland and Huang, 2003). One commonly accepted approach for searching for gene-gene interactions in a case-only design is the generation of "pseudo-controls" prior to analysis (Cordell *et al.*, 2004; Cordell, 2004). Pseudocontrols are generated from the alleles not present in a case subject at each genetic marker. Case-control analytical methods can then be applied. Family-based study designs, involving sibling pairs, affected sibling pairs, trios (parents and affected offspring), or extended family designs, are also very powerful for the detection of epistasis (Goldstein and Andieu, 1999).

Both association and linkage analyses have been used to investigate complex genetic and environmental disease etiologies, and appropriate tools are available for both strategies. Association analysis is by far the most commonly used approach, due to advantages in power and ease of sample collection (Risch, 2000), and most computational methods have been designed for such studies. Because of this, the current review focuses on methodologies for association studies, but it is important to keep in mind that linkage approaches can also be used to detect interactive effects. For example, Ordered Subset Analysis (OSA) (Hauser *et al.*, 2004) identifies genetically more homogeneous subsets of the overall data by ordering families according to covariate trait values in ascending or descending order.

2.2 PRELIMINARY ANALYSES

2.2.1 QUALITY CONTROL

Considering the quality of data is a crucial initial step in searching for epistasis, as quality control issues may contribute to spurious patterns of multilocus association. Data should be thoroughly checked for problems such as batch or study-center effects, or for unusual patterns of missing data. While "data cleaning" is not the focus of this review, this is an important first step in a data analysis plan investigating epistasis, and a few major issues are discussed below.

There are several standard approaches for evaluating the level of error in a given dataset. In a family-based study design, Mendelian inconsistencies can be evaluated to estimate a global/overall error rate. For example, if an allele seen in a child is not found in either parent, an error has occurred in either the genotyping or in the assumed family structure. In a case-control design, however, this error-checking method is not available. In all study designs, sex chromosome markers can be used to test for gender errors, as another way of estimating your global error. For example, if a male study subject has two copies of a nonpseudo-autosomal X-linked marker, an error is indicated. Genotyping efficiency can also be used in any study design to evaluate potential error. Low genotyping efficiency of either a single marker or a single individual may indicate a general problem with the data/sample for either the individual or marker. This should be evaluated prior to any statistical analysis.

Traditionally, testing for Hardy – Weinberg equilibrium (HWE) is used to detect errors in the data, but caution should be taken when considering this approach. There are several reasons for deviations from HWE, including inbreeding, population stratification, selection, or genotyping error (Wigginton *et al.*, 2005; Cox and Kraft, 2006). Additionally, deviation could also indicate disease association (especially if this deviation is seen in cases) (Wigginton *et al.*, 2005; Cox and Kraft, 2006), so eliminating SNPs from an analysis on this criterion could potentially be self-defeating. Testing for HWE in controls may serve as a better indicator of genotyping error than in the total population. Deviations can also be seen in the presence of common deletion polymorphisms because of a mutant polymerase chain reaction (PCR) primer binding site, miscalls during genotyping (Bailey and Eichler, 2006; Conrad *et al.*, 2006), or the presence of a copy number variant (Wigginton *et al.*, 2005; Cox and Kraft, 2006). Before discarding loci based on HWE calculations, it is important to consider all these possibilities. Testing for deviations from HWE is usually performed using a Pearson goodness-of-fit test, based on the chi-square distribution because the test statistic has approximately a chi-square null distribution. In the case of low genotype counts, however, a Fisher exact test should be used instead since it does not rely on the chi-square approximation (Guo and Thompson, 1992).

Missing data is another concern in any genetic analysis (Little and Rubin, 2002). When very little data (typically less than 5%) is missing, and there is no pattern to this missing data, it is a negligible problem. However, increasing amounts of missing data or irregular patterns present analytical challenges. Data imputation is one proposed solution to this problem. Data imputation involves replacing missing genotypes with predicted values based on the observed genotypes at nearby SNPs. In the case of very tightly linked markers, this can be reliable. There are several methods of imputation. First, maximum likelihood estimation can be used to seek a "best" prediction of a missing genotype (a single imputation). Alternatively, a genotype value can randomly be selected from a probability distribution (multiple imputations). "Hot-deck" approaches (Little and Rubin, 2002) can also be used, where a missing genotype is copied form another individual whose genotypes match at surrounding loci. Finally, regression models may be used that are based on the genotypes of all individuals at several neighboring loci (Souverein *et al.*, 2006).

These imputation methods should be used with caution, especially with a case-control study design. Many imputation strategies require two steps: an initial imputing phase (the information needed to determine the two haplotypes underlying a multilocus genotype within a chromosomal segment) followed by imputing missing values based on the initial phase. Additionally, these methods assume that missingness is independent of true genotype and of phenotype.

These assumptions are not always met, leading to serious biases (Clayton *et al.*, 2005). These assumptions should be considered and tested before applying any imputation procedures.

2.2.2   POPULATION STRATIFICATION

Population stratification is a well-documented confounding factor in genetic association studies (Pritchard and Przeworski, 2001; Ardlie *et al.*, 2002; Salisbury *et al.*, 2003), and should be considered prior to any association testing - including interaction analysis. This is of particular concern in a case-control study design, since the realities of sample collection do not often allow for homogeneous samples and complete knowledge of subjects' heritage is usually unknown. Family-based studies are immune to this potential problem.

This confounding occurs when individuals are selected from two genetically different populations in different proportions in cases and controls. Thus, the cases and controls are not matched for their genetic background. This may cause spurious associations, or it may mask true associations (Wang *et al.*, 2006). Most methods designed to deal with population stratification require a minimal number (usually > 100) of widely spaced null SNPs that have been genotyped in both cases and controls specifically for these analyses (Devlin and Roeder, 1999).

Genomic Control (Devlin and Roeder, 1999) is a commonly used approach that computes the Armitage test statistic (λ) at each null SNP. λ is calculated as the empirical median divided by its expectation under a chi-square distribution with one degree of freedom. If $\lambda > 1$, λ is then divided by λ. A $\lambda > 1$ is likely to be due to the effect of population stratification and dividing by λ cancels this effect for candidate SNPs. Genomic control is useful in a variety of scenarios but can be conservative in extreme settings and anticonservative if too few null SNPs are used (Marchini *et al.*, 2004).

Structured association methods are also used to control for population stratification (Pritchard *et al.*, 2000; Satten *et al.*, 2001; Hoggart *et al.*, 2003). These approaches are based on the idea of attributing the genomes of study individuals to hypothetical subpopulations and testing for association that is conditional on the specific allocations. This approach is computationally expensive, and the number of subpopulations to use is unresolved. The true number of subpopulations should ideally be used for analysis, but this number is often unknown.

Another approach to deal with this potential confounding factor is regression analysis that includes population structure as a covariate (Setakis *et al.*, 2006). Null SNPs can mitigate the effects of population structure when included in a regression model. This approach is very computationally efficient, and does not explicitly model the population structure as the structured association methods do. In addition, it is more flexible than genomic control because epistatic and covariate effects can be included.

There are several other methods being developed to diagnose population structure when many null SNPs are available, including principle components analysis (PCA) (Price *et al.*, 2006) and mixed-model approaches (Yu *et al.*, 2006). As mentioned previously, these methods require that null SNPs be genotyped for the purpose of assessing any potential confounding population structure. If this was not included in the genotyping, stratified analyses (stratified on population/ethnicity information) could be used to try to reduce and identify the impact of population stratification.

2.3   TRADITIONAL STATISTICAL APPROACHES

Traditional statistical approaches to detect genetic associations have been successful in identifying single-SNP associations and have had success in detecting interactions when properly applied. In any genetic analysis plan, both single-SNP and epistatic models should be considered. Below, we discuss some general concerns of these methods. Then, we describe the traditional methods most commonly used in genetic epidemiology and their application in the search for epistatic interactions. Table 2.1 summarizes the traditional methods discussed below.

There are several advantages to traditional statistical approaches that must not be overlooked. First, they are easily computed, and most methods are readily available in common statistical software packages. Additionally, the results are easily interpreted since the mathematical implications of most parameters have been extensively evaluated, and there is a long history of model interpretation. Finally, these models are readily accepted in both the biological and statistical communities.

However, there are several disadvantages to traditional methods that much be considered. First, as mentioned above, the curse of dimensionality limits the power of traditional methods to detect interactive effects. In regression analysis, for example, this can result in increased type 1 errors and parameter estimates with very large standard errors (Concato *et al.*, 1993). Additionally, simulation studies have demonstrated that 10 outcome events per independent variable are required for each parameter estimate (Peduzzi *et al.*, 1996). As genome-wide association studies become more common, this is an unrealistic sample size requirement.

Variable selection is another concern with the use of traditional methods. Most classical statistical tests were designed to test a specific, *a priori* hypoth-

TABLE 2.1   Traditional statistical analytical methods.

Method	Study designs	Outcome variables	Input variables	Parametric Genetic	Parametric Statistical
Pearson's Chi-Square	Case-control	discrete	discrete	yes	yes
Fisher's Exact Test	Case-control	discrete	discrete	yes	no
Armitage Cochran	Case-control	binary	discrete	yes	no
McNemar's Chi-square	Case-control	binary	discrete	yes	no
TDT	Trios	binary	discrete	no	yes
Sib-TDT	Dis-cordant sibling pairs	binary	discrete	no	yes
1-TDT	Proband+One parent	binary	discrete	no	yes
PDT	Extended Ped-igrees	binary	discrete	no	yes
FBAT	Extended Ped-igrees	binary	discrete	no	yes
GTDT	Trios	binary	discrete	no	yes
Tmhet	Trios	binary	discrete	no	yes
ccTDT	Mixed designs (Case-control and Family-based)	binary	discrete	no	yes
Linear Regression	Population-based, Cohort, Family-based	continuous	discrete or continuous	depends on encoding	yes
Logistic Regression	Case-control	binary	discrete or continuous	depends on encoding	yes
Cox Pro-portional Hazards Regression	Cohort	binary	discrete or continuous	depends on encoding	yes
Poisson Regression	Cohort	binary	discrete or continuous	depends on encoding	yes
Restricted Cubic Splines Regression	Case-Control, Cohort, Family-based	depends on types	discrete or continuous	depends on encoding	no
Kernel Regression	Population-based, Cohort, Family-based	continuous	discrete or continuous	depends on encoding	no
GLM	Population-based; Family-based	multiple variables; discrete or continuous	discrete or continuous	no	no
T-Test	Population-based; Family-based	continuous	discrete	yes	yes
MANOVA/MANCOVA	Population-based; Family-based	continuous	discrete	no	yes
Wald-Wolfowitz runs test	Population-based	continuous	discrete	no	no
Mann-Whitney U test	Population-based	continuous	discrete	no	no
Kolmo-gorov-Smirnov two-sample test	Population-based	continuous	discrete	no	no
Kruskal-Wallis analysis of ranks	Population-based	continuous	discrete	no	no
Median test	Population-based	continuous	discrete	no	no
Sign test	Population-based; Family-based	continuous	discrete	no	no
Wilcoxon's matched pairs test	Population-based; Family-based	continuous	discrete	no	no
Friedman's two-way ANOVA	Population-based; Family-based	continuous	discrete	no	no
Cochran's Q	Population-based; Family-based	continuous	discrete	no	no
QTDT	Trios	continuous	discrete	no	yes

Contingency table methods (rows: Pearson's Chi-Square through ccTDT)

Generalized linear models (rows: Linear Regression through GLM)

Analysis of variance (rows: T-Test through QTDT)

esis: the association between a prespecified variable(s) and an outcome of interest. They were not designed to identify which variables are the most important in predicting that outcome. Variable selection approaches, such as stepwise selection and best subset selection, are often applied as "wrappers" around traditional methods to try to address this challenge. These "wrappers" have been most often applied in a regression framework (discussed below), where well-known criteria like Mallows' C_p (Mallows, 1973), the Akaike information criterion (AIC) (Akaike and Hirotugu, 1981) and the Bayesian information criterion (BIC) (McQuarrie and Tsai, 1998) are often used to penalize the number of nonzero parameters. Shrinkage estimation approaches have also been employed to achieve better prediction and reduce the variances of estimators, such as ridge regression (Frank and Friedman, 1993).

While these procedures can be extremely useful in certain situations, they may not be appropriate for the detection of gene-gene and gene-environment interactions. Step-wise regression, for example, is a well-known and widely used form of variable selection within a regression framework (Mantel, 1970). There are several important limitations with such an approach. Especially in the case of small sample sizes, this approach can yield biased r-squared values (Copas, 1983; Derksen and Keselman, 1992), confidence intervals for effects and predicted values that are falsely narrow (Altman and Andersen 1989), p-values that do not have proper meaning (Hurvich and Tsai, 1990), biased regression coefficients that need shrinkage (Tibshirana, 1996). Further, it is based on methods (*i.e.* F tests for nested models) that were intended to test prespecified hypotheses (Hurvich and Tsai, 1990). Additionally, most of these methods rely on some criteria for hierarchical model building. In a genetic context, this means that they would be dependent on marginal main effects to even begin to build interactive models.

Another important consideration with high-dimensional studies is the risk of false discovery due to multiple testing, especially if traditional methods are used to individually test loci or multilocus combinations. As the number of loci increases so do the number of statistical tests typically performed. This problem may become overwhelming as the field embraces genome-wide studies. For example, Chi-square testing of a whole genome association dataset of 1,000,000 SNPs may yield 50,000 chance associations at p<0.05, and 100 at p<0.0001. Traditional approaches such as the Bonferroni correction adjust for type I error, but are extremely conservative (Van *et al.*, 2005) and thus may reject true associations. For genomic studies, these correction procedures may demand unrealistically small significance levels, and often ignore issues of between-test dependence (Van *et al.*, 2005) due to linkage between markers. It may be more appropriate to correct for multiple testing with the false discovery rate method, which considers the expected number of false rejections divided by the total number of rejections (Benjamini, 1995). The false discovery rate method is less conservative than the Bonferroni correction (Benjamini, 1995), but still may be too conservative for very large numbers of variables. Permutation testing is also used to decrease the impact of multiple comparisons through empirical estimates of significance. Permutation testing is a commonly used nonparametric statistical procedure. Rather than make specific distributional assumptions, a permutation test randomly permutes the data many times to actually construct the distribution of the test statistic under the null hypothesis. If the value of the test statistic based on the original samples is extreme relative to this distribution (i.e. if it falls far into the tail of the distribution), then, the null hypothesis is rejected (Good, 2000). The validity of a permutation test relies only on the data maintaining the property of exchangeability under the null hypothesis - so permutation testing makes no statistical or genetic assumptions and produces an unbiased p-value (Good, 2000; Mukherjee *et al.*, 2003; Neuhauser, 2005) Permutation testing can partially control for multiple comparison by significance testing only the best/final model as opposed to all individual tests. The chief drawback of this method is that it is computationally expensive, and for extremely large datasets, this limitation may make this type of significance testing prohibitive.

2.3.1   CONTINGENCY TABLE ANALYSES

One of the simplest sets of methods for detecting associations in genetic epidemiological studies is contingency table analysis. For case-control data, this includes the Pearson chi-square and Fisher's exact test. It is important to consider both genotypic and allelic versions of these analyses. The two degree of freedom (df) Pearson and Fisher tests (with genotypic encodings) are generally powerful, but in the case of an additive genetic model, an allelic encoding (with one df for each test) is most powerful. While this encoding may be more powerful to detect an additive association, it is important to remember that these tests assume HWE (Sasieni, 1997), which might not always hold true. The Armitage-Cochran test (Armitage, 1955), which is similar to the allele-count test, may be better suited. It tends to be conservative and does not assume HWE.

Contingency table analyses have been extended to family-based study designs as well. Possibly the most commonly used example of this is the Transmission Disequilibrium Test (TDT) (Terwilliger and Ott, 1992; Terwilliger *et al.*, 1992; Spielman *et al.*, 1993). This test compares observed versus expected transmitted alleles in trio data, and follows a Chi-square distribution. Additionally, there are a number of versions of this test to extend it to additional study designs, including sib-TDT (for discordant sib pair data) (Spielman and Ewens, 1998), 1-TDT (for a proband and one parent designs) (Sun *et al.*, 1999), Pedigree Disequilibrium Test (PDT) (Martin *et al.*, 2000) and Family Based Association Test (FBAT) (Horvath *et al.*, 2001) (for extended pedigrees), generalized TDT (GTDT) (Gordon *et al.*, 2004) and Tmhet (Kaplan *et al.*, 1997) (for multiallelic markers), and case-controlTDT (ccTDT) (Ruiz-Narvaez and Campos, 2004) (for combined case-control and family data). One additional important extension - which is technically an analysis of variance method - is quantitative TDT (QTDT) (Rabinowitz, 1997) for quantitative traits. These tests are most powerful to detect single-locus associations, but can be extended to detect multilocus associations. For low-dimensional data with large numbers of samples this approach can be successful, but the dimensionality of exploring interaction effects quickly diminishes the usefulness of simple contingency table analyses.

2.3.2 GENERALIZED LINEAR MODELS

Generalized linear models (GLMs) are a broad class of methods that can be thought of as an extended multivariate regression framework (Dobson, 2001). The general purpose of multivariate regression is to quantify the relationship between several independent variables and a single dependent variable. Generalized linear models extend this concept to multiple dependent variables. Just as in multivariate regression analysis, interactions between independent variables can be modeled and tested.

The most familiar example of a GLM is regression analysis (Fox, 1997), which remains the accepted standard for association studies in genetic epidemiology. Regression analysis models the relationship between response variables (dependent variables) and predictors (independent variables). Multivariate regression describes models that include more than one predictor variable. This allows the modeling of any genetic or environmental covariates and interactions (discrete or continuous) (Mardia *et al.*, 1979).

There are many types of regression analyses, making this approach applicable to any study design. Linear regression models a continuous response variable using a linear equation (for population-based studies). Logistic regression models a binary response variable using the logit function (for a case-control design). Cox Proportional Hazards regression is appropriate for cohort study designs with survival data. Less often used forms of regression analysis include Poisson regression, supervised learning, and unit-weighted regression. The above mentioned forms of regression are all statistically parametric (Fox, 1997). Nonparametric forms of regression analysis exist also (Fox, 2000), and include kernel regression and restricted cubic splines. Regression analysis is also appropriate for family data, since inter- and intraclass correlations can be adjusted for in the analysis.

In addition to limitations in variable selection and power, regression techniques are also limited in their ability to deal with nonindependent predictor variables (Hosmer and Lemeshow, 2000), which is very often the case in genetic epidemiology as SNPs are often in linkage disequilibrium. Another important consideration with regression methods is their limitations in dealing with situations of genetic heterogeneity. Regression approaches model the relation between predictors and risk of disease of all individuals in the data and do not consider subgroups (Hosmer and Lemeshow, 2000).

The general linear model goes a step beyond the multivariate regression model by allowing for linear transformations or linear combinations of multiple dependent variables. This extension gives the general linear model important advantages over the multiple and the multivariate regression models, both of which are inherently univariate in regards to the number of outcome variables evaluated. One advantage is that multivariate tests of significance can be employed when responses on multiple dependent variables are correlated. Separate univariate tests of significance for correlated dependent variables are not independent and may not be appropriate. Multivariate tests of significance of independent linear combinations of multiple dependent variables also can give insight into which dimensions of the response variables are, and are not, related to the predictor variables. Another advantage is the ability to analyze effects of repeated measure factors (which are normally analyzed using ANOVA techniques). Linear combinations of responses reflecting a repeated measure effect (for example, the difference of responses on a measure under differing conditions) can be constructed and tested for significance using either the univariate or multivariate approach to analyzing repeated measures in the general linear model. A second important advantage of the general linear model is its ability to handle nonindependent predictor variables. This is of particular utility in genetics, since genetic variants are often in linkage disequilibrium so they are nonindependent (Dobson, 2001). The general linear model is frequently applied to analyze any analysis of variance (discussed below) design with categorical or continuous predictor variables, as well as any multiple or multivariate regression design with continuous predictor variables (Dobson, 2001).

2.3.3 ANALYSIS OF VARIANCE

Analysis of variance (ANOVA) methods are a popular family of methods for association studies with continuous outcomes (Cobb, 1998). In general, the purpose of analysis of variance (ANOVA) is to test for significant differences between group means. If only two means are compared, then ANOVA will give the same results as the *t* test for independent samples (if comparing two different groups of cases or observations) or the *t* test for dependent samples (if comparing two variables in one set of cases or observations). ANOVA analyses are also applicable to numerous, and increasingly complex, study designs. Between groups, repeated measures, and nested study designs can be analyzed using ANOVA methods (Cobb, 1998). ANOVA methods have a distinct advantage over a simple *t* test in that interaction effects can be evaluated. Multivariate ANOVA (MANOVA) and multivariate analysis of covariance (MANCOVA) methods can evaluate more than one dependent variable and/or covariates (Cobb, 1998).

The ANOVA methods mentioned above are all statistically parametric, but there are nonparametric versions of the same tests if the statistical assumptions are not met. Differences in independent groups can be tested using the Wald-Wolfowitz runs test, the Mann-Whitney U test, the Kolmogorov-Smirnov two-sample test, the Kruskal-Wallis analysis of ranks, and the Median test. For differences between dependent groups, the sign test, Wilcoxon's matched-pairs test, Friedman's two-way analysis of variance, and Cochran's Q test are all appropriate (Cobb, 1998). ANOVA methods are most readily applicable to population-based study designs, but can also handle family data. This family of methods is very flexible as far as the hypothesis asked and very powerful for detecting interactions.

2.4 NOVEL APPROACHES TO DETECT EPISTASIS

Because of the limitations of traditional methodologies, particularly the curse of dimensionality and the variable selection problem, the development of novel methods to detect epistasis is a booming subdiscipline of human genetics. Recent methods have been developed for both association and linkage studies, in both population-based and family-based datasets. In the current section, we briefly discuss some of the basic tools and strategies used by these methods that take a "data mining" approach to detecting and characterizing epistasis. While the methods covered here do not represent a comprehensive list, the most commonly used methods are covered. Table 2.2 summarizes the methods discussed below.

TABLE 2.2 Data mining methods to detect epistasis.

	Method	Study design	Outcome	Input	Needs main effect	Parametric Genetic	Parametric Statistic
Tree-based	Classification and Regression Trees (CART)	Case-control	binary	discrete or continuous	yes	no	no
	Random Forests (RF)	Population-based; Case-control	discrete or continuous	discrete or continuous	yes	no	no
	Mutivariate Adaptive Regression Splines (MARS)	Population-based; Case-control	discrete or continuous; multiple outcome variables	discrete or continuous	yes	no	no
Combinatorial	Combinatorial Partitioning Method (CPM)	Population-based	continuous	discrete	no	no	no
	Restricted Partition Method (RPM)	Population-based	continuous	discrete	no	no	no
	Multifactor Dimensionality Reduction (MDR)	Case-control	binary	discrete	no	no	no
	MDR-PDT	Family-based	binary	discrete	no	no	no
	Generalized MDR	Case-control or population-based	discrete or continuous	discrete or continuous	no	no	no
	Patterning and Recursive Partitioning (PRP)	Case-control	discrete or continuous	discrete or continuous	no	no	no
	Detection of Informative Combined Effect (DICE)	Case-control	discrete or continuous	discrete or continuous	yes	yes	no
Neural Networks	Parameter Decreasing Method (PDM)	Case-control	binary	discrete or continuous	yes	no	no
	Genetic Programming NN (GPNN)	Case-control	binary	discrete or continuous	no	no	no
	Grammatical Evolution NN (GENN)	Case-control	binary	discrete or continuous	no	no	no
Clustering algorithms	CLADHC	Case-control or Family-based	binary	discrete	yes	no	yes or no
	HapMiner	Case-control or Family-based	discrete or continuous	discrete	N/A	no	yes
	k-means clustering	Population-based or Family-based	continuous	discrete or continuous	yes	no	no
	EM clustering	Population-based or Family-based	continuous	discrete or continuous	yes	no	no
Two-Step	Set Association	Case-control	binary	discrete	no	no	no
	Focused Interaction Testing Framework (FITF)	Case-control	binary	discrete	no	no	no
	Principle Components analysis (PCA)	Population-based or Family-based	continuous	discrete or continuous	yes	no	no

2.4.1 DATA MINING METHODS

Data mining is an analytical process designed to explore large amounts of data (like large-scale genetic studies) in search of consistent patterns and/or systematic relationships between variables, and then to validate the findings by applying the detected patterns to new subsets of data. Data mining is often considered "a blend of statistics, AI [artificial intelligence], and database research" (Pregibon, 1997). Data mining has sometimes received a tepid reception from traditionalists, even considered by some "a dirty word in Statistics" (Pregicon, 1997). However, as the practical importance and success of this approach is increasingly recognized, and the scale of genetic studies exponentially expands, this sort of approach is gaining acceptance.

The ultimate goal of any data mining approach is usually prediction - in the case of genetic epidemiology this prediction is in the form of disease-risk loci (Weiss and Indurkhya, 1997). As opposed to traditional hypothesis testing designed to verify *a priori* hypotheses about relations between variables, data mining typically falls under an exploratory data analysis framework. It is used to identify relations between variables when there are no, or incomplete, *a priori* expectations as to the nature of those relations.

There are three general stages to any data mining application (Witten and Frank, 2000). First there is data exploration. In genetic epidemiology, this may include simply the preliminary analysis discussed above, or a filter step in the analysis, where a certain number of independent variables are selected based on a criterion of choice. Filters that have been used in genetic applications include ReliefF (Kira and Rendell, 1992), genetic algorithms (GA) (Crosby, 1973), and genetic main effects as measured by Chi-square or regression (Evans *et al.*, 2006). The second step in any data mining process is model building and internal validation. It is this step that differs greatly from method to method. The third step is deployment, which involves using the model selected as best in the previous stage and applying it to new data to estimate its predictive ability. Many data mining approaches combine steps two and three by using a data resampling technique, such as bagging, boosting, cross-validation, jackknifing, or bootstrapping to simultaneously build and test a model. An excellent discussion of resampling and internal model validation techniques can be found in (Hastie *et al.*, 2001).

There are two general, broad categories of data mining methods: pattern recognition (Theodoridis and Koutroumbas, 2006) and data reduction (Bevington and Robinson, 1991). The pattern recognition family of methods considers the full dimensionality of the data, and aims to classify based on information extracted from the patterns. Tree- based methods, neural networks (NN), and clustering algorithms are all included in this family of methods (Theodoridid and Koutroumbas, 2006). The term data reduction in the context of data mining is usually applied to projects where the goal is to aggregate or amalgamate the information contained in large datasets into manageable (smaller) information nuggets. Data reduction methods can include simple tabulation, aggregation (computing descriptive statistics), or more sophisticated techniques like principal components analysis, etc (Bevington and Robinson, 1991). The combinatorial approaches discussed below fall into this category.

2.4.2   TREE-BASED APPROACHES

Arguably the simplest group of data mining approaches used in genetic epidemiology is tree-building algorithms. Also referred to as recursive partitioning methods, this group of tools determines a set of if-then logical (split) conditions that permit accurate prediction or classification of cases.

As with any analysis strategy, there are important advantages and disadvantages to any choice of algorithm. There are several important advantages to the tree-based algorithms that make them particularly useful in the context of genetic epidemiology. First, they can handle a large number of input variables, which is important as the scale of genetic studies increases. Also, learning is fast and computation time is modest even for very large datasets (Robnik-Sikonja, 2004). Additionally, tree methods are suited to dealing with certain types of genetic heterogeneity (roughly, where different variants can lead to the same disease), since splits near the root node define separate model subsets in the data. Also, trees-based algorithms produce an easily interpretable final model that is essentially a set of if-then rules (an example of a "white box" solution representation). Finally, these algorithms can uncover interactions among factors that do not exhibit strong marginal effects, without demanding a prespecified model (McKinney *et al.*, 2006). One important limitation of these methods to consider when looking for interactions is that they are dependent on slight marginal effects to model epistasis. If marginal main effects are not present, these methods will likely fail to characterize the interaction.

A major issue that arises when applying tree-based methods is a decision to stop splitting in the tree-building process. Model "over-fitting" is a concern with this group of methods. For example, in a data set with 10 cases, with 9 splits every single case could be perfectly classified. In general, with enough splits, a model could be found that perfectly describes any given dataset. Unfortunately this is not useful since such complex results most often fail to replicate in a sample of new observations. Therefore internal model validation approaches (Hastie *et al.*, 2001) are generally used with a resampling technique to avoid overfitting. Once a tree-building algorithm has stopped, it is always useful to further evaluate the quality of the prediction of the current tree in samples of observations that did not participate in the original computations. These methods are used to "prune back" the tree, *i.e.*, to eventually select a simpler tree than the one obtained when the tree-building algorithm stopped, but one that is equally as accurate for predicting or classifying 'unseen' observations (Brieman *et al.*, 1984).

Classification and Regression Trees (CART) (Brieman *et al.*, 1984) is a highly successful analytic procedure for predicting the values of a continuous response variable or categorical response variable from continuous or categorical predictors. The technique is referred to as "Classification Trees" when the dependent or response variable of interest is categorical in nature and as "Regression Trees" when the response variable of interest is continuous in nature. The goal of CART analysis is generally to find a tree where the terminal tree nodes are relatively "pure", *i.e.*, contain observations that (almost) all belong to the same category or class; for regression tree problems, node purity is usually defined in terms of the sums-of-squares deviation within each node.

Another popular tree-building algorithm is Random Forests (RF) (Brieman, 2001). Random Forests (RF) builds a forest of classification trees (similar to those built in CART) wherein each member of the forest is a tree is grown from a bootstrap sample of the data, and the variable at each tree node is selected from a random subset of all variables in the data (Brieman, 2001). Final classification of an individual is determined by voting over all trees in the forest. The importance of particular variables is determined by randomly permuting the values of that variable and testing whether these permutations adversely affect the predictive ability of trees in unseen (out-of-bag) samples. If randomly permuting the values of a particular variable drastically impairs the ability of trees to correctly predict the class of out-of-bag samples, then the importance score of that variable will be high. If randomly permuting values of a particular variable does not affect the predictive ability of trees on out-of-bag samples, that variable is assigned a low importance score. By running out-of-bag samples down entire trees during the permutation procedure, interactions are taken into account when calculating importance scores, since class is assigned in the context of other variable nodes in the tree.

Another tree-based method for detecting genetic associations is Multivariate Adaptive Regression Splines (MARS) (Hastie *et al.*, 2001). MARS is a nonparametric regression procedure that makes no assumption about the underlying functional relationship between the dependent and independent variables. Instead, MARS constructs this relation from a set of coefficients and basic functions that are entirely "driven" from the regression data. In a sense, the method can be thought to use a "divide and conquer" strategy, by partitioning the input space into regions, each with its own regression equation. This strategy makes

MARS particularly suitable for problems with higher dimensions. The step of the MARS algorithm proceeds as follows. First, the simplest model involving only the constant basis function is evaluated for each variable and for all possible spline knots, such that the space of basic functions is exhaustively searched. The model is grown by adding variables that maximize a certain measure of goodness of fit (minimized prediction error). The addition of variables to the developing model is recursively performed until a model of predetermined maximum complexity is derived. Finally, a pruning procedure is applied where the variables and corresponding basis functions are removed that contribute least to the overall goodness of fit. MARS can be thought of as a generalization of regression trees, where the "hard" binary splits are replaced by "smooth" basis functions. The MARS algorithm can be used with either discrete or continuous outcome variables, and can analyze discrete or continuous predictor/input variables, making it a useful tool for detecting both gene-gene and gene-environment interactions in a wide range of study designs. Additionally, MARS can be applied to multiple outcome variables of interest, similarly to generalized linear models. The flexibility of this method makes it useful for detecting interactions in a variety of study designs.

2.4.3   COMBINATORIAL APPROACHES

Another important group of methods used in genetic epidemiology are the combinatorial approaches. The defining feature of these methods is that they search over all possible variable combinations to find the combination(s) that best predict the outcome of interest. This exhaustive search approach is ideal for detecting interactions, including high-order interactions, since no marginal main effects are needed for variable selection during the training/model-building stage. While this is an important theoretical advantage for these methods, the computation time required grows exponentially with the number of markers evaluated. Certainly for genome-wide association studies, and even for some large-scale candidate gene studies, computational time may limit the ability to explore high-order interactions with these methods.

Three closely related methods that fall under this category include: the Combinatorial Partitioning Method (CPM) (Nelson et al., 2001), Restricted Partition Method (RPM) (Culverhouse et al., 2004) and Multifactor Dimensionality Reduction (MDR) (Ritchie et al., 2001). CPM and RPM are designed to detect interactions in quantitative phenotypes of interest, while MDR was originally designed for a binary outcome. Each of these methods uses cross-validation to avoid overfitting and assess the predictive performance of each model as discussed above in the context of tree-based methods. Patterning and Recursive

Partitioning (PRP) (Bastone *et al.*, 2004) combines a combinatorial approach with a tree-building method. All four methods rely on a form of permutation testing to ascribe statistical significance to a final model. Permutation testing has the advantages of being an assumption-free approach to significance testing but is a heavy computational burden, especially for these already computation-heavy methods.

As mentioned above, CPM evaluates multifactor combinations that predict quantitative phenotypes. In step one, the method searches the state space (the fitness landscape) by first evaluating all possible loci combinations and then evaluating the amount of phenotype variability explained by partitions of multi-locus genotypes. These multilocus genotypes are divided into sets of genotypic partitions. This first step is repeated, and those sets of genotypic partitions that explain a significant amount of phenotypic variability are retained for use in the second step. In step two, those genotypic partitions that are retained are validated using ten-fold cross-validation. Cross-validation is used to assess the predictive ability of each partition, so that step three involves the selection of the "best" sets of partitions. This selection is based on two criteria—the proportion of phenotypic variability explained by a set and the number of individuals in each set. The predictive ability of these final sets is then compared to a null distribution generated by permutation testing to ascribe statistical significance to a final model (Nelson *et al.*, 2001).

RPM was developed to improve computation time as compared to CPM. Where CPM searches over all possible combinations, RPM restricts its search to avoid fully evaluating genotype partitions that will not explain much of the variation in the phenotype. This restriction is performed in three steps. First, using a multiple comparison test, the difference between mean values of genotype groups (sets) is tested for significance. Second, from all nonsignificant pairs of genotype groups, those with the smallest differences between their mean values are combined into a new group. This step reduces the number of genotype groups to be evaluated. Finally, step two is repeated until all differences between pairs of genotype groups are significantly different. This approach works since a group consisting of genotypes for which the difference between their mean values is large (thus having a large within-group variance) will not explain much of the total variance (Culverhouse *et al.*, 2004). After this reduction in the number of sets to evaluate, RPM proceeds with the steps described above for CPM.

As mentioned above, MDR was originally designed for studies with a binary outcome variable and only discrete predictor variables. These predictor variables and their multifactor classes are divided in *n*-dimensional space. Then the ratio of cases to controls is calculated within each multifactor class. Each multi-factor cell class is then labeled "high risk" or "low risk" based on the ratio calculated, therefore reducing the n-dimensional space to one dimension with two levels. The collection of these multifactor classes composes the MDR model for a particular combination of factors. For each possible model size (one-locus, two-locus, etc.) a single MDR model is chosen that has the lowest number of misclassified individuals. To evaluate the predictive ability of the model, prediction error is calculated on the testing sets (from cross-validation). The result is a set of models, one for each model size considered. From these models, a final model is chosen based on minimization of prediction error and maximization of cross-validation consistency (number of times a particular set of factors is identified across the cross-validation subsets) (Ritchie *et al.*, 2001). MDR was designed for case-control studies, but has also been applied to discordant sibling pair data and trio data by generating pseudo-controls from untransmitted alleles (Motsinger and Ritchie, 2006). Recently, the PDT statistic has been incorporated into MDR (MDR-PDT) for application to extended pedigree data (Martin *et al.*, 2006). Other fitness functions are also being evaluated in the application of the MDR approach, such as calculating an odds ratio instead of a classification accuracy (Chung *et al.*, 2007). Most recently, the MDR algorithm has been extended to examine both continuous and discrete outcome variables (Lou *et al.*, 2007).

As mentioned above, the Patterning and Recursive Partitioning (PRP) (Bastone *et al.*, 2004) method combines a tree-based and combinatorial approach. PRP is an extension of the CART method discussed previously. In the first step of PRP, individuals within a dataset are assigned to genotype groups based on their multilocus genotypes and the resulting classification is used as a predictor variable in a recursive partitioning framework. This recursive partitioning framework evaluates the impurity of a model using a decision tree approach (Bastone *et al.*, 2004). The PRP extends a decision tree approach to be able to capture purely epistatic models by not limiting the split in the tree-building process to a single variable at a time.

The Detection of Informative Combined Effect (DICE) (Tahri-Daizadeh *et al.*, 2003) is another example of a combinatorial approach. Briefly, the DICE algorithm exhaustively explores all combinations of independent variables (either discrete or continuous), first assuming an additive model and then assuming an interactive model. Akaike's information criterion (AIC) is used to evaluate the fitness of each combination. Models of increasing complexity are successively fitted to the data and the difference (Δ_s) of the Akaike's information criterion between models indicates whether the fit is substantially improved. That is, when Δ_s exceeds a predetermined threshold, a model is considered improved over another. The algorithm stops when no model leads to a Δ_s higher than the fixed threshold (Tahri-Daizadeh *et al.*, 2003). This model-building approach is more similar to a traditional logistic regression framework than the other combinatoric methods.

2.4.4 NEURAL NETWORKS

Unlike the data reduction approaches discussed above, Neural Networks (NN) is a class of pattern recognition techniques. NN are modeled after the (hypothesized) processes of learning in the cognitive system and the neurological functions of the brain. They are capable of predicting new observations (on specific variables) from other observations (on the same or other variables) after executing a process of so-called "learning" from existing data (Anderson, 1995). Neural networks are a type of directed graph consisting of nodes that represent the processing elements (or neurons), arcs that represent the connections of the nodes (or synaptic connections), and directionality on the arcs that represent the flow of information (Skapura, 1995). The nodes (processing elements) are arranged in layers such that the input layer receives the external pattern vector that is to be processed by the network. Each node in the input layer is then connected to one or more nodes in a hidden layer, and these are in turn connected to nodes in additional hidden layers or to each output node. Each connection in the network has a weight (a_i), or coefficient, associated with it. The signal is conducted from the input layer through the hidden layers to the output layer. The output layer, which often consists of a single node, generates an output signal that is then used to classify the input pattern. In the context of genetic epidemiology, this output can be case/control status or a quantitative trait.

There are several features of NN that make them appealing for genetic epidemiology: they are able to handle large quantities of data, they are universal function approximators and therefore should be able to approximate any genetic penetrance function, and they are genetic model free, meaning that no assumptions of the genetic model need be made. Also, because of their parallel nature, NNs are well suited to model interactions. Additionally, NN in general may have an advantage in situations with intercorrelated variables over other statistical and machine-learning approaches. NNs are somewhat protected against the problems caused by multicollinearity due to their parallel nature (Smith, 1996). This is important since many genetic variables are correlated (are in linkage disequilibrium). Also, unlike many traditional statistical methods, NN do not assume independence of either individuals in the dataset or input variables. Adjustments of weights between network connections are assumed to correct for variable intercorrelation (De Veaux and Ungar, 1994). This makes NN useful tools for looking at family-based and population-based data.

One important disadvantage of traditional NN approaches is that the selection of inputs, arrangement of nodes and connections (known as the architecture of the NN), and weights are important decisions in a NN analysis that can drastically alter the results. Because of this, several approaches for constructing appropriate NN architecture have been developed. Most often a "trial and error" approach is taken for choosing the architecture, and a back-propagation strategy is used to optimize the weights. This approach has met with mixed success in genetic epidemiology. Novel methods have been developed to incorporate variable selection into a NN analysis, including the Parameter Decreasing Method (PDM) (Tomita *et al.*, 2004), Genetic Programming Neural Networks (GPNN) (Ritchie *et al.*, 2003), and most recently Grammatical Evolution Neural Networks (Motsinger *et al.*, 2006). Each of these methods use cross-validation to prevent overfitting, and permutation testing to ascribe statistical significance to final models. All of these methods consider the full dimensionality of the data, making them appropriate for finding gene-gene and gene-environment interactions. Additionally, NN are flexible tools for combining both discrete and continuous predictor variables.

PDM was designed to perform variable selection while utilizing a user-specified NN architecture. In step one, one predictor variable is deleted from the total number and a model containing all remaining variables is constructed. This process is repeated for each variable such that in turn each variable is deleted from the total number and a model is constructed with the remaining. From all of the constructed models, the model with the lowest number of misclassified subjects in both the training and evaluation set is selected (as mentioned above, cross-validation is used to prevent overfitting). This process is repeated until one variable remains. For each selected model, a measure of prediction accuracy is calculated by the sum of true predicted cases and controls divided by the total number in the evaluation sample. The prediction accuracy is calculated for each evaluation set created by multifold cross-validation and the sum of the prediction accuracies divided by the number of evaluation sets gives the average prediction accuracy (Tomita *et al.*, 2004). This accuracy can be compared to a permutation distribution to ascribe statistical significance.

While PDM does perform variable selection, the NN architecture must be established *a priori*, and the proper set up may vary for each individual dataset. To try to solve this problem, evolutionary computation algorithms have been utilized to optimize the architecture of the NN while simultaneously performing variable selection. GPNN (Ritchie *et al.*, 2003) and GENN (Motsinger *et al.*, 2006) use different evolutionary computation algorithms to perform these tasks, but the general steps of each method are the same.

First, GPNN/GENN has a set of parameters that must be initialized before beginning the evolution of NN models. These parameters determine aspects of the evolutionary process such as population size, mutation rate, and crossover rate. Second, the data are divided for cross-validation. Third, model training begins by generating an initial random population of potential solutions. Each solution is a NN. Fourth, each potential solution (NN) is evaluated on the training set and its fitness (a classification error) recorded. Fifth, the best solutions are selected for crossover and reproduction using a selection technique, where lower classification error represents higher relative fitness in the population. The more fit a solution, the more likely that NN is to be passed onto the next generation. A predefined proportion of the best solutions will be directly copied (reproduced) into the new generation. Another proportion of the solutions will be used for crossover with other best solutions. The new generation, which is equal in size to the original population, begins the cycle again. This continues until some criterion is met at which point the evolutionary process stops. The stopping criteria are a classification error of zero, or a prespecified number of generations. Sixth, this best GPNN model is tested on the testing data to estimate the prediction error of the model. Steps two through six are repeated for each cross-validation interval. The results of a GPNN/GENN analysis include 10 models, one for each split of the data. A classification error and prediction error is recorded for each of the models. A cross-validation consistency can be measured to determine those variables that have a strong signal and should be included in the final model. Cross-validation consistency is the number of times a particular combination of variables are present in the ten cross-validation data splits. Thus, a high cross-validation consistency would indicate a strong signal, whereas a low cross-validation consistency would indicate a weak signal and a potentially false positive result. The loci combination with the highest cross-validation consistency is chosen as the final model. A prediction error is then determined for the final model and compared to a permutation distribution.

2.4.5 CLUSTERING ALGORITHMS

Clustering algorithms are a subgroup of the pattern recognition family. Cluster analysis aims to sort different objects into groups such that the degree of association between two objects is maximal if they belong to the same group and minimal otherwise. Given the above, cluster analysis can be used to discover structures in data without providing an explanation/interpretation. In other words, cluster analysis simply discovers structures in data without explaining why they exist. Clustering approaches have been previously applied in medicine for clustering diseases, cures for diseases, or symptoms of diseases to generate useful taxonomies. Their application to the identification of disease-susceptibility genes is more novel, and is often used in concert with more traditional measures of association. This class of methods also considers the full dimensionality of the data, so they are also able the cluster according to interactive models. These methods are particularly appealing for cases of genetic heterogeneity, as they can detect "clusters" of individuals whose phenotype variation is explained by different genetic models (Thornton-Wells *et al.*, 2006). Recently developed methods like HapMiner (Li and Jiang, 2005) and CLADHC (Bardel *et al.*, 2005) capitalize on these advantages. Clustering methods have also been used in concert with other computational methodologies to better define phenotypes for analysis or pick-apart genetic heterogeneity (Thornton-Wells *et al.*, 2006). There are several broad classes of clustering techniques available to an epidemiologist: joining (tree clustering), two-way joining, k-means (Hartigan, 1975), and expectation maximization (EM) clustering (Witten and Frank, 2000).

The joining or tree clustering method uses the dissimilarities (similarities) or distances between objects when forming the clusters (Witten and Frank, 2000). Similarities are a set of rules that serve as criteria for grouping or separating items. These distances (similarities) can be based on a single dimension or multiple dimensions, with each dimension representing a rule or condition for grouping objects. The most straightforward way of computing distances between objects in a multidimensional space is to compute Euclidean distances, but other distance measures are also applicable, such as squared Euclidean, City-block (Manhattan), Chebychev, power, and % disagreement distances (Witten and Frank, 2000). In the first step of joining clustering, when each object represents its own cluster, the distances between those objects are defined by the chosen distance measure. Linkage or amalgamation rules are then used to determine when two clusters are sufficiently similar to be linked together. Numerous rules have been proposed including single linkage (nearest neighbor), complete linkage (furthest neighbor), unweighted/weighted pair-group averages, and weighted/unweighted pair-group centroids. Two-way joining clustering approaches extend the concept of the joining approaches to simultaneously cluster based on both predictor and outcome variables (Witten and Frank, 2000). In genetic epidemiology, this sort of application could be used to dissect both phenotypic and genetic heterogeneity (Thornton-Wells *et al.*, 2006).

HapMiner (Li and Jiang, 2005) is one such joining clustering method developed for association testing. This method is designed to cluster haplotypes (either phased or inferred) for family-based or case-control study designs. For case-control data, the disease status of each individual is used to label both of its haplotypes. For case-parent data, transmitted haplotypes can be labeled as case haplotypes and untransmitted haplotypes as controls. The algorithm scans each marker one by one. For each marker position, a haplotype segment with certain length centered at the position will be considered. The segment length is determined based on marker interval distances. Clusters are identified based on a unique distance measure via a density-based clustering algorithm. The Pearson phi^2 statistic or Z-score, based on a contingency table derived from the numbers of case haplotypes and control haplotypes in a cluster can be used as an indicator of the degree of association between the cluster and disease. Both measures can then be used to test association. A significance threshold is chosen by the user and all findings that exceed the threshold are reported. Currently, this algorithm performs single-locus tests of association, but could theoretically be extended for interactive effects (Li and Jiang, 2005).

CLADHC (Bardel *et al.*, 2005) is a joining clustering method for case-control studies. This method uses haplotypes composed of a combination of SNPs, where each haplotype is labeled either as case or control depending on the phenotype of the individual. After building a phylogenetic tree of these different haplotypes, a series of nested homogeneity tests are performed to detect differences in the distribution of cases and controls in the different clades. Briefly, at each level of the tree, homogeneity in the distribution of cases and controls is tested among all the *n* clades defined at this level. If the test is significant, an association is detected and the analysis ends. If the test is not significant, one homogeneity test is performed between all the subclades descending from the *n* clades. Once an association is detected, a new character is defined according to the proportion of cases carrying a haplotype. Then, it is optimized on the tree and the sites that significantly mutate with this new character (similar in concept to correlation) are putative susceptibility sites for the disease of interest. This method has not been directly tested on epistatic models, but it performs well on noninteractive multifactorial models (Bardel *et al.*, 2005).

K-means clustering (Hartigan, 1975) is actually very different from the joining clustering methods. In general, the *k*-means method will produce exactly *k* different clusters of greatest possible distinction. The best number of clusters *k* leading to the greatest separation (distance) is not normally known *a priori* and is often computed from the data. This method can be thought of as ANOVA "in reverse". The algorithm will start with *k* random clusters, and then move objects between those clusters with the goal to 1) minimize variability within clusters and 2) maximize variability between clusters. The similarity rules will apply maximally to the members of one cluster and minimally to members belonging to the rest of the clusters. This is analogous to "ANOVA in reverse" in the sense that the significance test in ANOVA evaluates the between group variability against the within-group variability when computing the significance test for the hypothesis that the means in the groups are different from each other. In *k*-means clustering, the program tries to move individuals in and out of groups (clusters) to get the most significant ANOVA results (Hartigan, 1975).

The EM clustering algorithm (Witten and Frank, 2000) is an extension of the *k*-means approach. There are two important differences that distinguish EM clustering. First, instead of assigning individuals to clusters to maximize the differences in means for continuous variables, the EM clustering algorithm computes probabilities of cluster memberships based on one or more probability distributions. The goal of the clustering algorithm then is to maximize the overall probability or likelihood of the data, given the (final) clusters. Second, unlike the classic implementation of *k*-means clustering, the general EM algorithm can be applied to both continuous and categorical variables (note that the classic *k*-means algorithm can also be modified to accommodate categorical variables) (Witten and Frank, 2000).

As mentioned above, the number of clusters that should be used in k-means or EM clustering is not known *a priori*. Usually, this number is determined using cross-validation. Most commonly, the *v*-fold cross-validation algorithm is used. The general idea of this method is to divide the overall sample into a number of *v* folds. The same type of analysis is then successively applied to the observations belonging to the *v*-1 folds (training sample), and the results of the analyses are applied to sample *v* (the sample or fold that was not used to estimate the parameters, build the tree, determine the clusters, etc.; this is the testing sample) to compute some index of predictive validity. The results for the *v* replications are aggregated (averaged) to yield a single measure of the stability of the respective model, i.e., the validity of the model for predicting new observations (Witten and Frank, 2000).

2.4.6   TWO-STEP APPROACHES

Several novel methods have taken a two stage approach to detecting genetic associations by first determining a small number of potentially interesting markers, and then modeling interactions between those potential predictors. Focusing on gene-gene and gene-environment interactions, it is crucial that the first step of these approaches considered not just single markers, but sets of markers that could potentially interact. If only markers with strong main effects are considered in the first step, strictly epistatic models will be missed. This multistep approach is not unique to these methods, but is a defining feature. Set Association (Hoh *et al.*, 2001) and Focused Interaction Testing Framework (FITF) (Millstein *et al.*, 2006) are two popular methods designed specifically to detect interactions with this framework. These methods can be considered data reduction methods because they address the dimensionality problem by reducing the number of variables examined, and try to estimate global levels of significance.

Set Association (Hoh *et al.*, 2001) was designed for a binary outcome variable (such as case/control status), and can examine both discrete and continuous predictor variables. This makes it especially useful in looking for gene-environment interactions. In step one, a test statistic is calculated for each marker separately. The test-statistic used is a product of two test statistics, where the first measures the association of a marker with disease outcome and the second measures deviation of a marker from the null-hypothesis of HWE. Any measure of association can be used for the first statistic but a Chi-square statistic calculated from the contingency table of alleles (or genotypes) with disease status is traditionally used. Chi-square values for deviations from HWE are calculated in the cases only for the second statistic. Large deviations from HWE are assumed to indicate an association between the marker and the disease. HWE is also tested in controls to check and control for genotyping errors. The markers are then ordered based on their value for the overall test-statistic. The marker with the largest overall test statistic is selected and sum statistics are calculated by sequentially adding the most important marker from the group of unselected markers. Increasing sums of markers are formed and the number of markers in the sums ranges from 1 to a predefined maximum number of M markers. Permutation testing is used to assess the significance of each sum. The sum with the lowest significance level (smallest p-value) is selected as the best set of markers. This first p-value is then used as a test statistic and is evaluated by a second permutation test testing the null-hypothesis of no association of the selected markers with the disease outcome to give an overall p-value. The main disadvantage of the set association approach is that genetic interactions are only tested for the markers that are selected in the sum. Thus, important interactions with weak main effects will be missed (Hoh *et al.*, 2001).

FITF (Millstein *et al.*, 2006) is a modification of the Interaction Testing Framework (ITF) method that prescreens all possible gene sets to focus on those that potentially are the most informative. In the ITF method, a series of logistic regression analyses is performed in incremental stages, where the highest-order interaction parameter considered increases at each subsequent stage. In stage 1, the main effect of each genetic variant is considered, in stage two, all pair-wise combinations are tested, in stage three all three-way interactions are tested, etc. If a variant or multilocus combination is declared significant in an earlier stage, those variants are not retested in subsequent stages to avoid retesting the same effects. This means that significant main effects are removed before testing for potential interactive models. The overall type one error is controlled by dividing the overall desired alpha level by the number of stages and allocating this adjusted alpha level to each stage. Within each stage, the significance threshold is adjusted by controlling the False Discovery Rate (FDR) (Benjamini, 1995). This approach allows the method to detect interactions in the absence of any marginal main effects. The FITF algorithm modifies the ITF approach to reduce the overall number of variants tested with an initial filter process. A chi-square goodness-of-fit statistic that compares the observed with the expected Bayesian distribution of multilocus genotype combinations in a combined case-control population, referred to as the Chi-square subset (CSS), is used in a prescreening initial stage (Millstein *et al.*, 2006). All markers that pass the initial filter stage enter the ITF process.

2.4.7   PRINCIPLE COMPONENTS ANALYSIS

Principal Components Analysis (PCA) (Pearson, 1901; Fukunaga and Keino-suke, 1990) is a method that reduces data dimensionality by performing a cova-riance analysis between factors. PCA uses eigen analysis to transform a number of potentially correlated variables into a smaller number of uncorrelated vari-ables called principle components. The first principle component accounts for as much variability in the data as possible and then each succeeding component amounts for as much of the remaining variation as possible. The main objectives of this method are to reduce the dimensionality of the data set, and to identify new meaningful underlying variables, such as interactive predictors (Pearson, 1901; Fukunaga and Keinosuke, 1990). Because of this, it is readily applied to large-scale studies in human genetics. PCA has been used for microarray analysis, population stratification analysis, and is appropriate for data mining applications in genetics studies with continuous outcome variables.

2.4.8   METHODS TO INTERPRET EPISTATIC MODELS

As well as methods to detect interactive models, there are also bioinformatics approaches designed to help interpret and understand epistatic models. Such approaches combine aspects of expert knowledge, visualization, and reexamina-tion of model solution content. Some novel tools will aid in literature searches to help characterize the biological interactions of statistical methods. Chilibot (Chen and Sharp, 2004) is a web-based text mining (using natural language pro-cessing) application that extracts term-term relationships from MEDLINE ab-stracts. The results of a Chilibot search are returned in a graphical format, where connections between genetic or proteomic variables represent interactions, and these connections are color-coded according to the type of interaction hypoth-esized. A variety of tools for knowledge-based visualization of results, including K-graph (Kelly *et al.*, 2006), EVA (Reif *et al.*, 2005), and the Onto-Tools suite (Draghici *et al.*, 2003), can help identify important patterns postanalysis. Other tools designed to help interpret interactive models are based on information gain and mutual information criteria, such as interaction dendrograms (Seo *et al.*, 2002). These tools can be extremely helpful in better understanding complex models. While this is in no way an exhaustive list of tools, we wanted to make the reader aware of such methods in order to encourage their use, and further development.

2.5   EXAMPLES OF EPISTASIS FOUND IN HUMANS

With a myriad of both traditional and novel tools available to a genetic epidemiologist, it is important to get an empirical understanding of the importance of using them in one's own work. The field of human genetics has experienced a paradigm shift as common diseases are now assumed to be due to the complex interactions among numerous genetic and environmental factors (Moore, 2003). Even diseases once thought to be exclusively Mendelian in etiology (like sickle cell anemia and cystic fibrosis) are now known to much more complex that previously assumed. Excellent reviews of modifier genes impacting the clinical manifestation of these disorders can be found in (Steinberg and Adewoye, 2006) and (Knowles, 2006) respectively. The dissection of the epistatic nature of the Hirshsprung's disease on both a statistical and functional level (Carrasquillo *et al.*, 2002) also demonstrates this shift.

Possibly the most convincing argument for incorporating an exploration of interactions in any analytical plan is the numerous interactions already found in the study of human phenotypes using statistical/computational methods. In a PubMed search performed in January 2007, over 250 examples of epistatic interactions found in genetic epidemiological studies were found. While space limitations prevent listing all examples found, some interesting patterns were seen in this list. Of the interactions found, ~92% were found using an association study approach, while only ~8% used a linkage strategy. Over 73% used a population-based study design, while ~27% used a family-based approach. A population-based case-control design is by far the most used study design, with ~53% of interactions detected in this design. Case-only designs were used in less than 1% of studies, and a mixed (both population and family-based) design was used in ~3%. This demonstrates the flexible nature of the analytical tools to find interactions in any study design. More than 77% of interactions were discovered with traditional methods (used in a straightforward, classical approach or with stratification), while ~23% of studies used novel techniques.

Figure 2.1 shows the number of interactions discovered per year, starting in 1984. As the histogram illustrates, the number of interactions found per year is growing exponentially as the field recognizes their importance and more readily applies novel methodologies or extends traditional ones to account for interactions. As the enthusiasm for this type of search grows, we are also starting to see validated epistatic models. Studies in asthma provide a good example. An interactive effect between the IL-13 and IL4Ralpha genes that predicted asthma was first seen by Howard *et al.* (2002) in a case-control study of Dutch individuals. In 2006, this interaction was replicated in a completely different human population of Chinese cases and controls using both traditional and novel analytical methods (Chan *et al.*, 2006). This is encouraging since epistasis has been proposed as an explanation for lack of replication in studies of complex human diseases. As investigators continue to explore interactions, we will surely see more replication of epistatic interactions.

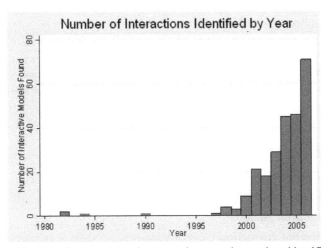

FIGURE 2.1   Number of gene-gene and gene-environment interactions identified by year.

Unfortunately, as in any human population-based study, the risk of false-positives is just as applicable in the search for epistatic interactions. While the reported number of interactions is growing at an exciting pace, it is important to remember that an unknown number of such reports represent false positive findings. As with any linkage or association results, replication, and validation are key to revealing true results.

2.6   DEVELOPING AN ANALYSIS PLAN

When developing an analysis plan, an investigator has broad options. Particularly for large-scale or genome-wide studies, investigators may consider using a combination of several of the tools discussed above. For example, the first stage of analysis could involve a filter method, the second stage could involve a novel tool that performs both variable selection and modeling (such as MDR), and as a final step a traditional method like logistic regression could be used to put the model in a more interpretable or familiar framework. Another option could involve a nonexhaustive combinatorial search to perform variable selection over the entire dataset followed by knowledge-based interpretation of the results. The combinations of choices are effectively infinite, and Table 2.1 and 2.2 are presented as a launching point for identifying appropriate methods.

The choice comes down to the details of the particular study, and the investigator should carefully consider these details. Are there well-characterized mechanisms or candidate genes in the literature? Is the etiology likely to involve accumulation of minor epistatic effects or one large main effect with modifiers? What is the scale of the study in the number of variables and sample size? Is a validation cohort available?

Continued methods development with aid an investigator is making these choices, and will hopefully encourage the search for interactions in even more studies. Better curation of a web of knowledge about certain diseases, navigability of knowledge databases, and standards for high-throughput data (genome-wide studies, etc.) will all aid in this pursuit.

KEYWORDS

- **Analysis of Variance**
- **Armitage test statistics**
- **Association analysis**
- **Bonferroni correction**
- **Chi-square statistics**
- **Classification and regression trees**
- **Clustering algorithms**
- **Combinatorial partitioning method**
- **Contingency table**
- **Data mining**
- **Epistasis**
- **Fisher exact test**
- **Focused interaction testing framework**
- **Gene environment interactions**
- **Gene-gene interactions**
- **Genetic epidemiology**
- **Genome-wide association**
- **Genotyping technology**

- **Hardy-Weinberg equilibrium**
- **Interaction testing framework**
- **Linkage analysis**
- **Multifactor dimensionality reduction**
- **Multivariate adaptive regression splines**
- **Neural networks**
- **Ordered subset analysis**
- **Parametric statistical methods**
- **Population stratification**
- **Shrinkage estimation**
- **Single nucleotide polymorphisms**
- **Step-wise regression**
- **Tree-building algorithms**

REFERENCES

Akaike, H.; Hirotugu S. Likelihood of a model and information criteria. *J Econ* 1981, 16, 3-14.

Altman, D. G.; Andersen, P. K. Bootstrap investigation of the stability of a Cox regression model. *Stat Med* 1989, 8, 771-783.

Anderson, J. *An Introduction to Neural Networks*, MIT Press: Cambridge, Massachusetts, 1995.

Andrieu, N.; Goldstein, A. M. Epidemiologic and genetic approaches in the study of gene-environment interaction: An overview of available methods. *Epidemiol Rev* 1998, 20, 137-147.

Ardlie, K. G.; Kruglyak, L.; Seielstad, M. Patterns of linkage disequilibrium in the human genome. *Nat Rev Genet* 2002, 3, 299-309.

Armitage, P. Tests for linear trends in proportions and frequencies. *Biometrics* 1955, 11, 375-386.

Bailey, J. A.; Eichler, E. E. Primate segmental duplications: Crucibles of evolution, diversity and disease. *Nat Rev Genet* 2006, 7, 552-564.

Bardel, C.; Danjean, V.; Hugot, J. P.; Darlu, P.; Genin, E. On the use of haplotype phylogeny to detect disease susceptibility loci. *BMC Genet* 2005, 6, 24.

Bastone, L.; Reilly, M.; Rader, D. J.; Foulkes, A. S. MDR and PRP: A Comparison of methods for high-order genotype-phenotype associations. *Hum Hered* 2004, 58, 82-92.

Bateson, W. *Mendel's Principles of Heredity*, Cambridge University Press: Cambridge, 1909.

Bellman, R. *Adaptive Control Processes*, Princeton University Press: Princeton, 1961

Benjamini, Y.; Hochburg, Y. Controlling the false discovery rate: A practical and powerful approach to multiple testing. *J Roy Stat Soc-Series B* (*Methodological*) 1995, 57, 289-300.

Bevington, P. R.; Robinson, D. *Data reduction and error analysis for the physical sciences*, McGraw-Hill: New York, 1991.

Breiman, L.; Friedman, J. H.; Olshen, R. A.; Stone, C. J. *Classification and regression trees*, Wadsworth and Brooks/Cole Advanced Books and Software: Pacific Grove, CA, USA, 1984.

Brieman, L. Random Forests. *Machine Learning* 2001, 45, 5-32.

Carrasquillo, M. M.; McCallion, A. S.; Puffenberger, E. G.; Kashuk, C. S.; Nouri, N.; Chakravarti, A. Genome-wide association study and mouse model identify interaction between RET and EDNRB pathways in Hirschsprung disease. *Nat Genet* 2002, 32, 237-244.

Chan, I. H.; Leung, T. F.; Tang, N. L.; Li, C. Y.; Sung, Y. M.; Wong, G. W. K.; Wong, C. K.; Lam, C. W. K. Gene-gene interactions for asthma and plasma total IgE concentration in Chinese children. *J Allergy Clin Immunol* 2006, 117, 127-133.

Chen, H.; Sharp, B. M. Content-rich biological network constructed by mining PubMed abstracts. *BMC Bioinformatics* 2004, 5, 147.

Chung, Y.; Lee, S. Y.; Elston, R. C.; Park, T. Odds ratio based multifactor-dimensionality reduction method for detecting gene-gene interactions. *Bioinformatics* 2007, 23, 71-76.

Clayton, D. G.; Walker, N. M.; Smyth, D. J.; Pask, R.; Cooper, J. D.; Maier, L. M.; Sminck, L. J.; Lam, A. C.; Ovington, N. R.; Stevens, S. E.; Nutland, S.; Howson, J. M. M.; Faham, M.; Moorhead, M.; Jones, H. B.; Falkowski, M.; Hardenbol, P.; Willis, T. D.; Todd, J. A. Population structure, differential bias and genomic control in a large-scale, case-control association study. *Nat Genet* 2005, 37, 1243-1246.

Cobb, G. W. *Introduction to Design and Analysis of Experiments*, Springer: New York, 1998.

Concato, J.; Feinstein, A. R.; Holford, T. R. The risk of determining risk with multivariable models. *Ann Intern Med* 1993, 118, 201-210.

Conrad, D. F.; Andrews, T. D.; Carter, N. P.; Hurles, M. E.; Pritchard, J. K. A high-resolution survey of deletion polymorphism in the human genome. *Nature Gen* 2006, 38, 75-81.

Copas, J. B. Regression, prediction and shrinkage (with discussion). *J Roy Stat Soc-Series B (Methodological)* 1983, 45, 311-354.

Cordell, H. J. Properties of case/pseudocontrol analysis for genetic association studies: Effects of recombination, ascertainment, and multiple affected offspring. *Genet Epidemiol* 2004, 26, 186-205.

Cordell, H. J. Epistasis: What it means, what it doesn't mean, and statistical methods to detect it in humans. *Hum Mol Genet* 2002, 11, 2463-2468.

Cordell, H. J. Affected-sib-pair data can be used to distinguish two-locus heterogeneity from two-locus epistasis. *Am J Hum Genet* 2003, 73, 1468-1471.

Cordell, H. J.; Barratt, B. J.; Clayton, D. J. Case/pseudocontrol analysis in genetic association studies: A unified framework for detection of genotype and haplotype associations, gene-gene and gene-environment interactions, and parent-of-origin effects. *Genet Epidemiol* 2004, 26, 167-185.

Cox, D. G.; Kraft, P. Quantification of the power of Hardy-Weinberg equilibrium testing to detect genotyping error. *Hum Hered* 2006, 61, 10-14.

Crosby, J. L. *Computer Simulation in Genetics*, John Wiley and Sons: London, 1973.

Culverhouse, R.; Klein, T.; Shannon, W. Detecting epistatic interactions contributing to quantitative traits. *Genet Epidemiol* 2004, 27, 141-152.

Culverhouse, R.; Suarez, B. K.; Lin, J.; Reich, T. A perspective on epistasis: Limits of models displaying no main effect. *Am J Hum Genet* 2002, 70, 461-471.

Daly, A. K.; Day, C. P. Candidate gene case-control association studies: Advantages and potential pitfalls. *Br J Clin Pharmacol* 2001, 52, 489-499.

De Veaux, R. D.; Ungar, L. H. Multicollinearity: A Tale of two non-parametric regressions. In *Selecting Models from Data: AI and Statistics IV*, 1994. pp. 293-302.

Derksen, S.; Keselman, H. J. Backward, forward and stepwise automated subset selection algorithms: Frequency of obtaining authentic and noise variables. *Br J Math Stat Psych* 1992, 45, 265-282.

Devlin, B.; Roeder, K. Genomic control for association studies. *Biometrics* 1999, 55, 997-1004.

Dobson, A. J. *An Introduction to Generalized Linear Models*, 2nd Ed.; Chapman and Hall/CRC: London, 2001.

Draghici, S.; Khatri, P.; Bhavsar, P.; Shah, A.; Krawetz, S. A.; Tainsky, M. A. Onto-Tools, the toolkit of the modern biologist: Onto-Express, Onto-Compare, Onto-Design and Onto-Translate. *Nucleic Acids Res* 2003, 31, 3775-3781.

Evans, D. M.; Marchini, J.; Morris, A. P.; Cardon, L. R. Two-stage two-locus models in genome-wide association. *PLoS Genet* 2006, 2, e157.

Fisher, R. A. The correlation between relatives on the supposition of Mendelian inheritance. *Trans R Soc Edin* 1918, 52, 399-433.

Fox, J. *Applied Regression Analysis, Linear Models and Related Methods*, SAGE Publications: Thousand Oaks, CA, USA, 1997.

Fox, J. *Multiple and Generalized Nonparametric Regression*, SAGE Publications: Thousand Oaks, CA, USA, 2000.

Frank, I. E.; Friedman, J. H. A statistical view of some chemometrics regression tools. *Technometrics* 1993, 35, 109-148.

Fukunaga, K. *Introduction to statistical pattern recognition.* Elsevier, New York, 1990.

Goldstein, A.; Andrieu, N. Detection of interaction involving identified genes: Available study designs. *J Nat Canc Inst Monographs* 1999, 49-54.

Good, P. *Permutation Tests: A Practical Guide to Resampling Methods for Testing Hypotheses*, Springer-Verlag: New York, 2000.

Gordon, D.; Haynes, C.; Johnnidis, C.; Patel, S. B.; Bowcock, A. M.; Ott, J. A transmission disequilibrium test for general pedigrees that is robust to the presence of random genotyping errors and any number of untyped parents. *Eur J Hum Genet* 2004, 12, 752-761.

Guo, S. W.; Thompson, E. A. Performing the exact test of Hardy-Weinberg proportion for multiple alleles. *Biometrics* 1992, 48, 361-372.

Hartigan, J. A. *Clustering Algorithms (Probability and Mathematical Statistics).* John Wiley and Sons: London, 1975.

Hastie, T.; Tibshirani, R.; Friedman, J. H. *The elements of statistical learning - Springer Series in Statistics.* Springer Verlag: Basel, 2001.

Hauser, E. R.; Watanabe, R. M.; Duren, W. L.; Bass, M. P.; Langefeld, C. D.; Boehnk, M. Ordered subset analysis in genetic linkage mapping of complex traits. *Genet Epidemiol* 2004, 27, 53-63.

Hoggart, C. J.; Parra, E. J.; Shriver, M. D.; Bonilla, C.; Kittles, R. A.; Clayton, D. G.; McKeigue, P. M. Control of confounding of genetic associations in stratified populations. *Am J Hum Genet* 2003, 72, 1492-1504.

Hoh, J.; Wille, A.; Ott, J. Trimming, weighting, and grouping SNPs in human Case-control association studies. *Genome Res* 2001, 1, 2115-2119.

Horvath, S.; Xu, X.; Laird, N. M. The family based association test method: Strategies for studying general genotype-phenotype associations. *Eur J Hum Genet* 2001, 9, 301-306.

Hosmer, D. W.; Lemeshow, S. *Applied Logistic Regression.* John Wiley and Sons Inc.: New York, 2000.

Howard, T. D.; Koppelman, G. H.; Xu, J.; Zheng, S. L.; Postma, D. S.; Meyers, D. A.; Bleeker, E. R. Gene-gene interaction in asthma: IL4RA and IL13 in a Dutch population with asthma. *Am J Hum Genet* 2002, 70, 230-236.

Hurvich, C. M.; Tsai, C. L. The impact of model selection on inference in linear regression. *Am Stat* 1990, 44, 214-217.

Kaplan, N. L.; Martin, E. R.; Weir, B. S. Power studies for the transmission/disequilibrium tests with multiple alleles. *Am J Hum Genet* 1997, 60, 691-702.

Kelly, R. J.; Jacobsen, D. M.; Sun, Y. V.; Smith, J. A.; Kardia, S. L. KGraph: A system for visualizing and evaluating complex genetic associations. *Bioinformatics* 2007, 23, 249-251.

Kira, K.; Rendell, I. A. The feature selection problem: Traditional methods and a new algorithm. *Proc. 10th National Conference on Artificial Intelligence, MIT Press* 1992, 1, 129-132.

Knowles, M. R. Gene modifiers of lung disease. *Curr Opin Pulm Med* 2006, 12, 416-421.

Li, J.; Jiang. Haplotype-based linkage disequilibrium mapping via direct data mining. *Bioinformatics* 2005, 21, 4384-4393.

Little, R. J. A.; Rubin, D. B. *Statistical analysis with missing data.* Wiley: New York, 2002.

Lou, X. Y.; Guo-Bo, C.; Yan, L.; Ma, J. Z.; Zhu, J.; Elston, R. C.; Li, M. D. A generalized combinatorial approach for detecting gene-by-gene and gene-by-environment interactions with application to nicotine dependence. *Am J Hum Gen* 2007, 80, 1125-1137.

Mallows, C. L. Some comments on CP. *Technometrics* 1973, 15, 661-675.

Mantel, N. Why stepdown procedures in variable selection. *Technometrics* 1970, 12, 621-625.

Marchini, J.; Cardon, L. R.; Phillips, M. S.; Donnelly, P. The effects of human population structure on large genetic association studies. *Nat Genet* 2004, 36, 512-517.

Mardia, K. V.; Kent, J. T.; Bibby, J. M. *Multivariate Analysis.* Academic Press: London, 1979.

Martin, E. R.; Monks, S. A.; Warren, L. L.; Kaplan, N. L. A test for linkage and association in general pedigrees: The pedigree disequilibrium test. *Am J Hum Gen* 2000, 67, 146-154.

Martin, E. R.; Ritchie, M. D.; Hahn, L. W.; Kang, S.; Moore, J. H. A novel method to identify gene-gene effects in nuclear families: The MDR-PDT. *Genet Epidemiol* 2006, 30, 111-123.

McKinney, B. A.; Reif, D. M.; Ritchie, M. D.; Moore, J. H. Machine learning for detecting gene-gene interactions: A Review. *Appl Bioinformatics* 2006, 5, 77-88.

McQuarrie, A. D. R.; Tsai, C. *Regression and Time Series Model Selection.* World Scientific: Singapore, 1998.

Millstein, J.; Conti, D. V.; Gilliland, F. D.; Gauderman, W. J. A testing framework for identifying susceptibility genes in the presence of epistasis. *Am J Hum Genet* 2006, 78, 15-27.

Moore, J. H. The ubiquitous nature of epistasis in determining susceptibility to common human diseases. *Hum Hered* 2003, 56, 73-82.

Moore, J. H. Computational analysis of gene-gene interactions using multifactor dimensionality reduction. *Expert Rev Mol Diagn* 2004, 4, 795-803.

Moore, J. H.; Ritchie, M. D. STUDENTJAMA. The challenges of whole-genome approaches to common diseases. *JAMA* 2004, 291, 1642-1643.

Moore, J. H.; Williams, S. M. New strategies for identifying gene-gene interactions in hypertension. *Ann Med* 2002, 34, 88-95.

Moore, J. H.; Williams, S. M. Traversing the conceptual divide between biological and statistical epistasis: Systems biology and a more modern synthesis. *Bioessays* 2005, 27, 637-646.

Motsinger, A. A.; Dudek, S. M.; Hahn, L. W.; Ritchie, M. D. Comparison of neural network optimization approaches for studies of human genetics. *Lec Notes in Comp Sci* 2006, 3907, 103-114.

Motsinger, A. A.; Ritchie, M. D. Multifactor dimensionality reduction: An analysis strategy for modeling and detecting gene-gene interactions in human genetics and pharmacogenomics studies. *Hum Genomics* 2006, 2, 318-328.

Mukherjee, S.; Golland, P.; Panchenko, D. *Permutation tests for classification. AI Memo #2003-019.* Massachusetts Institure of Technology: MA, Cambridge, 2003.

Nelson, M. R.; Kardia, S. L.; Ferrell, R. E.; Sing, C. F. A combinatorial partitioning method to identify multilocus genotypic partitions that predict quantitative trait variation. *Genome Res* 2001, 11, 458-470.

Neuhauser, M. Efficiency comparisons of rank and permutation tests. *Stat Med* 2005, 24, 1777-1778.

Neuman, R. J.; Rice, J. P. Two-locus models of disease. *Genet Epidemiol* 1992, 9, 347-365.

Norton, B.; Pearson, E. S. A note on the background to and refereeing of R.A. Fisher's 1918 paper 'The correlation between relatives on the supposition of Mendelian inheritance'. *Notes Rec R Soc Lond* 1976, 31, 151-162.

Pearson, K. On lines and planes of closest fit to systems of points in space. *Philosophical Mag* 1901, 2, 559-572.

Peduzzi, P.; Concato, J.; Kemper, E.; Holford, T. R.; Feinstein, A. R. A simulation study of the number of events per variable in logistic regression analysis. *J Clin Epidemiol* 1996, 49, 1373-1379.

Pregibon, D. *Data Mining.* Statistical Computing and Graphics, 1997.

Price, A. L.; Patterson, N. J.; Plenge, R. M.; Weinblatt, M. E.; Shadick, N. A.; Reich D. Principal components analysis corrects for stratification in genome-wide association studies. *Nat Genet* 2006, 38, 904-909.

Pritchard, J. K.; Przeworski, M. Linkage disequilibrium in humans: Models and data. *Am J Hum Genet* 2001, 69, 1-14.

Pritchard, J. K.; Stephens, M.; Rosenberg, N. A.; Donnelly, P. Association mapping in structured populations. *Am J Hum Genet* 2000, 67, 170-181.

Rabinowitz, D. A transmission disequilibrium test for quantitative trait loci. *Hum Hered* 1997, 47, 342-350.

Reif, D. M.; Dudek, S. M.; Shaffer, C. M.; Wang, J.; Moore, J. H. Exploratory visual analysis of pharmacogenomic results. *Pac Symp Biocomput* 2005, 1, 296-307.

Risch, N. J. Searching for genetic determinants in the new millennium. *Nature* 2000, 405, 847-856.

Ritchie, M. D.; Hahn, L. W.; Roodi, N.; Bailey, L. R.; Dupont, W. D.; Moore, J. H. Multifactor-dimensionality reduction reveals high-order interactions among estrogen-metabolism genes in sporadic breast cancer. *Am J Hum Genet* 2001, 69, 138-147.

Ritchie, M. D.; White, B. C.; Parker, J. S.; Hahn, L. W.; Moore, J. H. Optimization of neural network architecture using genetic programming improves detection and modeling of gene-gene interactions in studies of human diseases. *BMC Bioinformatics* 2003, 4, 28.

Robnik-Sikonja, M. Improving random forests. *Machine Learning: Ecml 2004, Proceedings,* 3201, 2004, 359-370.

Ruiz-Narvaez, E. A.; Campos, H. Transmission disequilibrium test (TDT) for case-control studies. *Eur J Hum Genet* 2004, 12, 105-114.

Salisbury, B. A.; Pungliya, M.; Choi, J. Y.; Jiang, R.; Sun, X. J.; Stephens, J. C. SNP and haplotype variation in the human genome. *Mutat Res* 2003, 526, 53-61.

Sasieni, P. D. From genotypes to genes: Doubling the sample size. *Biometrics* 1997, 53, 1253-1261.

Satten, G. A.; Flanders, W. D.; Yang, Q. Accounting for unmeasured population substructure in case-control studies of genetic association using a novel latent-class model. *Am J Hum Gene* 2001, 68, 466-477.

Seo, J.; Shneiderman, B.; Hoffman, E. Interactively exploring hierarchical clustering results. *IEEE Computer* 2002, 35, 80-86.

Setakis, E.; Stirnadel, H.; Balding, D. J. Logistic regression protects against population structure in genetic association studies. *Genome Res* 2006, 16, 290-296.

Sing, C. F.; Stengard, J. H.; Kardia, S. L. Dynamic relationships between the genome and exposures to environments as causes of common human diseases. *World Rev Nutr Diet* 2004, 93, 77-91.

Skapura, D. *Building neural networks.* ACM Press: New York, 1995.

Smith, M. *Neural Networks for Statistical Modeling.* International Thomson Computer Press: Boston, 1996.

Souverein, O. W.; Zwinderman, A. H.; Tanck, M. W. Multiple imputation of missing genotype data for unrelated individuals. *Ann Hum Genet* 2006, 70, 372-381.

Spielman, R. S.; Ewens, W. J. A sibship test for linkage in the presence of association: The sib transmission/disequilibrium test. *Am J Hum Genet* 1998, 62, 450-458.

Spielman, R.; McGinnis, S.; Ewens, W. J. Transmission test for linkage disequilibrium: The insulin gene region and insulin-dependent diabetes mellitus (IDDM). *Am J Hum Genet* 1993, 52, 506-516.

Steinberg, M. H.; Adewoye, A. H. Modifier genes and sickle cell anemia. *Curr Opin Hematol* 2006, 13, 131-136.

Sun, F.; Flanders, W. D.; Yang, Q.; Khoury, M. J. Transmission disequilibrium test (TDT) when only one parent is available: the 1-TDT. *Am J Epidemiol* 1999, 150, 97-104.

Tahri-Daizadeh, N.; Tregouet, D. A.; Nicaud, V.; Manuel, N.; Cambien, F.; Tiret, L. Automated detection of informative combined effects in genetic association studies of complex traits. *Genome Res* 2003, 13, 1952-1960.

Templeton, A. Epistasis and complex traits. In *Epistasis and the Evolutionary Process*, Wade, M.; Broadie, B.; Wolf, J. Ed., Oxford University Press: Oxford, 2000; pp. 41-57.

Terwilliger, J. D.; Ding, Y.; Ott, J. On the relative importance of marker heterozygosity and inter-marker distance in gene mapping. *Genomics* 1992, 13, 951-956.

Terwilliger, J. D.; Ott, J. A haplotype-based 'haplotype relative risk' approach to detecting allelic associations. *Hum Hered* 1992, 42, 337-346.

Theodoridis, S.; Koutroumbas, K. *Pattern Recognition, 3rd Edn.*, Academic Press: London, 2006.

Thornton-Wells, T. A.; Moore, J. H.; Haines, J. L. Genetics, statistics and human disease: Analytical retooling for complexity. *Trends Genet* 2004, 20, 640-647.

Thornton-Wells, T. A.; Moore, J. H.; Haines, J. L. Dissecting trait heterogeneity: A comparison of three clustering methods applied to genotypic data. *BMC Bioinformatics* 2006, 7, 204.

Tibshirani, R. Regression shrinkage and selection via the lasso. *J Roy Stat Soc-Series B* (*Methodological*) 1996, 58, 267-288.

Tomita, Y.; Tomida, S.; Hasegawa, Y.; Suzuki, Y.; Shirakawa, T.; Kobayashi, T.; Honda, H. Artificial neural network approach for selection of susceptible single nucleotide polymorphisms and construction of prediction model on childhood allergic asthma. *BMC Bioinformatics* 2004, 5, 120.

Van, S. K.; McQueen, M. B.; Herbert, A.; Raby, B.; Lyon, H.; Demeo, D. L.; Murphy, A.; Su, J.; Datta, S.; Rosenow, C.; Christman, M.; Silverman, E. K.; Laird, N. M.; Weiss, S. T.; Lange, C.

Genomic screening and replication using the same data set in family-based association testing. *Nat Genet* 2005, 37, 683-691.

Vieland, V. J.; Huang, J. Two-locus heterogeneity cannot be distinguished from two-locus epistasis on the basis of affected-sib-pair data. *Am J Hum Genet* 2003, 73, 223-232.

Wang, Y.; Localio, R.; Rebbeck, T. R. Evaluating bias due to population stratification in epidemiologic studies of gene-gene or gene-environment interactions. *Cancer Epidemiol Biomarkers Prev* 2006, 15, 124-132.

Weiss, S. M.; Indurkhya, N. *Predictive Data Mining: A practical guide.* Morgan-Kaufman: New York, 1997.

Wigginton, J. E.; Cutler, D. J.; Abecasis, G. R. A note on exact tests of Hardy-Weinberg equilibrium. *Am J Hum Genet* 2005, 76, 887-893.

Witten, I.; Frank, E. *Data mining: Practical machine learning tools and techniques.* Morgon Kaufman: New York, 2000.

Yu, J.; Pressoir, G.; Briggs, W. H.; Vroh, B.; Yamasaki, I. M.; Doebley, J. F.; McMullen, M. D.; Gaut, B. S.; Nielson, D. M.; Holland, J. B.; Kresovich, S.; Buckler, E. S. A unified mixed-model method for association mapping that accounts for multiple levels of relatedness. *Nat Genet* 2006, 38, 203-208.

CHAPTER 3

ELUCIDATION OF PROTO-ONCOGENE C-ABL FUNCTION WITH THE USE OF MOUSE MODELS AND THE DISEASE MODEL OF CHRONIC MYELOID LEUKEMIA

JENNY F. L. CHAU and BAOJIE LI

CONTENTS

3.1   Identification of C-ABL and BCR-ABL 60

3.2   Structure and Activation 61

3.3   Activation by Growth Factors 62

3.4   Activation by DNA Damage 63

3.5   Potential Substrates for C-ABL 63

3.6   ABL Function: Cell Biology Approach 64

3.7   ABL Function: Genetic Approach 72

3.8   ABL Function: CML Studies 75

3.9   Conclusion 79

3.10 Future Perspectives 81

Acknowledgements 82

Keywords 82

References 83

3.1   IDENTIFICATION OF C-ABL AND BCR-ABL

In 1969, Abelson and Rabstein isolated a new tumor inducing variant of Moloney Murine Leukemia Virus (M-MuLV), a virus known to induce thymomas in mice (Abelson and Rabstein, 1969). In contrast to the parental strain, this new strain is a lymphosarcoma-producing virus that is characterized by a rapid development of solid lymphoid and massive meningeal tumors, without affecting the thymus, in infected mice. While a polymorphonuclear leukemoid reaction is observed in these mice, there is no lymphocytic invasion of their organs (Abelson and Rabstein, 1970). This new virus variant was then named the Abelson Murine Leukemia Virus (A-MuLV). A-MuLV was demonstrated to have transformation ability. It can transform fibroblasts and myeloid cells *in vitro* (Rabstein *et al.*, 1971; Scher and Siegler, 1975). Later, Witte *et al.* (1978) identified an A-MuLV encoded protein present in A-MuLV transformed cells. This product contains a viral amino-terminal region derived from the Gag gene of M-MuLV and a carboxyl-terminal region from a normal cellular gene. The cellular gene is referred to as the ABL region. The Gag-Abl fusion protein was later referred to as v-Abl. Comparison of the genome sequences of A-MuLV and M-MuLV revealed a DNA fragment that is present only in A-MuLV. This fragment was used to probe the human cDNA library and to pull out a cellular homolog, which was identified as c-Abl (Witte *et al.*, 1978; Witte *et al.*, 1979; Goff *et al.*, 1980).

About 10 years before the identification of A-MuLV, Nowell and Hungerford performed cytogenetic studies on normal and leukemic leukocytes and revealed the presence of a minute chromosome in most of the chronic myelogenous leukemia (CML) cells (Nowell and Hungerford, 1960). This minute chromosome was named the Philadelphia chromosome, and was later demonstrated by Rowley to be a result of a reciprocal translocation between chromosomes 9 and 22 (Rowley, 1973). The exchange of the chromosomal material generates a longer chromosome 9 and a shorter chromosome 22, with the latter referred to as the Philadelphia chromosome. de Klein and colleagues found that c-Abl, a gene normally located on chromosome 9, is present on the Philadelphia chromosome (de Klein *et al.*, 1982), and that the Philadelphia chromosomal breakpoints are clustered within a limited region, termed the "breakpoint cluster region" (BCR), on chromosome 22 (Groffen *et al.*, 1984). By using Abl cDNA as probes, cells from CML patients were found to contain an abnormal Abl mRNA species, consisting of a fusion of BCR sequences to the Abl sequence (Canaani *et al.*, 1984; Collins *et al.*, 1984; Shtivelman *et al.*, 1985). Due to the fusion of c-Abl with different portions of the BCR gene, there are at least three isoforms of the chimerical BCR-ABL proteins: p180, p210, and p230. p210 BCR-ABL is present in more than 90 % cases of CML. Fusion of c-Abl to BCR alters the c-Abl protein structure and results in a constitutively active tyrosine kinase, which is the leading cause of CML (Konopka *et al.*, 1985).

3.2   STRUCTURE AND ACTIVATION

c-Abl is ubiquitously expressed. The c-Abl gene encodes two mRNA transcripts as a result of alternative splicing of the first two exons. They are translated into two ~ 140 kd isoforms, type 1a and type 1b in human, and type I and type IV in mouse. Type 1b and IV differ from type 1a and I respectively by carrying a C_{14} myristoyl fatty acid at the amino terminus. Adjacent to the myristoyl fatty acid is the Cap region, which is present in all different splice variants but is less conserved. This is followed by a SH3, a SH2, and a kinase domain, which are shared by all the Src family members. This conserved region is then followed by a carboxyl terminal, which is not found in other Src family members except c-Abl's homolog, Arg. The carboxyl terminus of c-Abl consists of three nuclear localization signals (NLS) and one nuclear export signal (NES) (Van Etten *et al.*, 1989; Taagepera *et al.*, 1998), which allows the shuttling of c-Abl between the nucleus and the cytoplasm. The carboxyl terminus also carries a DNA binding domain (Kipreos and Wang, 1992), a RNA polymerase II binding site (Baskaran *et al.*, 1996), and a filamentous and globular actin binding domain (Van Etten *et al.*, 1994). These various domains are important for c-Abl's functions, which include the DNA damage response, cell cycle control, cytoskeleton reorganization, and cell spreading and cell mobility.

c-Abl's localization to the cytoplasm, plasma membrane, nucleus, mitochondria, and endoplasmic reticulum depends on the cell type and cellular conditions. The activity of c-Abl is tightly controlled and deregulation of c-Abl kinase activity causes deleterious effects to the cell. c-Abl is mostly inactive in unstimulated cells. In this inactive state, the protein is folded into a structure in which the SH3 and SH2 domains face the distal side of the kinase domain. The SH3 domain lies opposing to the N-terminal lobe while the SH2 domain is connected to the C-terminal lobe of the kinase domain that contains an activation loop with Y412 (Barila and Superti-Furga,1998; Nagar *et al.*, 2003). The SH3 and SH2 domains, with the help of the SH3-SH2 connector and the SH2-PTK linker, clamp the kinase domain into an inactive state. The inactive state is further enforced with the N-terminal Cap segment and the myristoyl group in type Ib/VI c-Abl, as the Cap and the myristol group lock the SH3-SH2 domains onto the distal surface of the kinase domain (Nagar *et al.*, 2003; Nagar *et al.*, 2006). Based on the structures of c-Abl and Src and numerous biochemical studies, it has been proposed that there are three steps in the activation of c-Abl. Firstly, the myristol group and the Cap segment need to be released from the C-lobe of the kinase domain. This is then followed by *"unclamping"*, which results from the dissociation of the SH3-SH2 domains from the kinase domain and exposes Y412 for phosphorylation. The third step is the phosphorylation of Y412, which results in the activation of c-Abl. Phosphorylation of Y245, which is also packed tightly in the inactive form of the kinase domain, further enhances the kinase activity (Harrison, 2003; Li, 2006). While what triggers these changes is not well understood, we do know that c-Abl can be activated by TGF, EGF, PDGF, and DNA damage.

3.3   ACTIVATION BY GROWTH FACTORS

c-Abl can be activated by transforming growth factor β (TGFβ), a growth factor that regulates cell proliferation and differentiation. TGFβ binds to the TGFβ receptors and initiates a cascade of cellular signaling responses in a Smad-dependent or Smad-independent manner. It is reported that c-Abl takes part in TGFβ induced, Smad-independent pathway. In mesenchymal stem cells, TGFβ activated Akt and PAK2 though PI3K. While Akt had no effect on the activation of c-Abl by TGFβ, inhibition of PAK2 completely abolished the activation of c-Abl by TGFβ (Wilkes and Leof, 2006). It is proposed that activation of c-Abl by TGF β promotes cell proliferation.

   c-Abl could also be activated by epidermal growth factor (EGF). EGF binds to EGF receptors and stimulates the intrinsic protein tyrosine kinase activity of the receptor to regulate cell proliferation. It was found that the EGF treatment lead to the binding of Abl SH2 domain to EGF receptor (Zhu *et al.*, 1993). Activation of EGF receptors also activate the Src kinase which mediates the activation of c-Abl (Plattner *et al.*, 1999). However, the consequences of c-Abl's activation by EGF are not clear. Another growth factor that has been shown to activate c-Abl is the platelet-derived growth factor (PDGF). Similar to EGF, PDGF activates c-Abl through the receptor mediated activation of Src. c-Src induces c-Abl activation through the phosphorylation of Y245 and Y412. The activated c-Abl mediates the elevation of c-Myc expression. The activation of c-Abl can mediate the chemotaxis toward PDGF (Plattner *et al.*, 1999; Furstoss *et al.*, 2002). In addition, the membrane fraction of c-Abl can be activated by phospholipase C-γ1 (PLC-γ1), in addition to activation by Src, in response to PDGF. Activated PDGF receptors phosphorylate PLC-γ1, leading to the hydrolysis of phosphatidylinositol-4,5-bisphosphate (PIP2) to inositol-1,4,5-trisphophate (IP3) and diacylglycerol (DAG). It has been demonstrated that PIP2 inhibits c-Abl. Hydrolysis of PIP2 by PLC-γ1 decreases the level of PIP2 and leads to activation of c-Abl. It is also shown that c-Abl can bind to PLC-γ1 and phosphorylates it at Y771 and Y1003. This phosphorylation serves as a negative feedback which leads to the inactivation of PLC-γ1 (Plattner *et al.*, 2003).

3.4 ACTIVATION BY DNA DAMAGE

Upon DNA damage, cells undergo cell cycle arrest to allow themselves to have more time to repair the damaged DNA. However, if the damage is overwhelming, cells will undergo apoptosis to prevent the passing of damaged DNA to the daughter cells. c-Abl is activated by DNA damage caused by ionizing irradiation (IR), cisplatin, mitomycin C, etoposide, doxorubicin, camptothecin, antimetabolite 1-beta-D-arabinofuranosylcytosine, but not by UV radiation (Wang, 2000). Following DNA damage, the PI3K related family members ATM (Ataxia Telangiectasia, mutated), ATR (ATM- and Rad3-Related), and DNA-PKcs (DNA protein kinase catalytic subunit) initiate a cascade of signaling responses (Shafman et al., 1997). c-Abl has been reported to be involved in the ATM and DNA-PK initiated signaling pathways. DNA-PKcs is able to phosphorylate and activate c-Abl, and this activation is defective in DNA-PK deficient cells. In parallel, phosphorylation of DNA-PKcs is also dependent on c-Abl upon IR. Thus, DNA-PKcs and c-Abl regulate each other in response to DNA damage (Kharbanda et al., 1997). In addition, c-Abl binds constitutively to ATM, with its SH3 domain interacting with the DPAPNPPHFP motif of ATM (Shafman et al., 1997). Upon IR, c-Abl is phosphorylated at serine 465 by ATM. As the activation of c-Abl by DNA damage is diminished in ATM deficient cells, this suggests that ATM activates c-Abl upon DNA damage (Baskaran et al., 1997; Shafman et al., 1997).

3.5 POTENTIAL SUBSTRATES FOR C-ABL

There are more than 500 putative protein kinases encoded by the human genome. About 90 of them encode protein tyrosine kinases (Manning et al., 2002), and in which about 1/3 are nonreceptor tyrosine kinases. c-Abl is one of the nonreceptor tyrosine kinases. Protein tyrosine kinases are enzymes that catalyze the transfer of the γ-phosphoryl group from ATP to tyrosine residues in the substrates. Substrates for c-Abl include c-Crk (Ren et al., 1994), CrkL, c-Cbl, Shc, RasGAP (Bose et al., 2006), Shb (Hagerkvist et al., 2007), p73 (Agami et al., 1999), Rad51 (Yuan et al., 1998a), Abi (Dai and Pendergast, 1995), RNA polymerase II (Baskaran et al., 1993), IkappaBalpha (Kawai et al., 2002), SHPTP1 (Kharbanda et al., 1996), PKCδ (Sun et al., 2000b), catalase (Cao et al., 2003b), glutathione peroxidase 1 (Cao et al., 2003a), phospholipid scramblase 1 (PLSCR1) (Sun et al., 2001), Cables (Zukerberg et al., 2000), Dok1 (Woodring et al., 2004), PSTPIP1 (Cong et al., 2000) and others. In response to different stimuli, c-Abl phosphorylates various substrates to regulate cell death, proliferation, differentiation, cell movement, and so on.

3.6   ABL FUNCTION: CELL BIOLOGY APPROACH

3.6.1   THE ROLE OF C-ABL IN CELL DEATH

Apoptosis or programmed cell death, is an important process in the development of multicellular organisms. It involves a series of biochemical events and helps to eliminate the unwanted cells. Defects in apoptosis may be disastrous. Excessive apoptosis may lead to organ hypotrophy while insufficient apoptosis may lead to the accumulation of unwanted or damaged cells, resulting in uncontrolled cell proliferation and cancer. Cellular studies demonstrated that c-Abl is mainly a pro-apoptotic protein. Overexpression of c-Abl results in apoptosis whereas cells with inactive c-Abl show increased resistance to apoptosis (Yuan *et al.*, 1997a; Theis and Roemer, 1998).

3.6.2   OVEREXPRESSION INDUCED APOPTOSIS

Induction of apoptosis by c-Abl requires its NLS and can be mediated by MAPKs. c-Abl has been shown to regulate stress-activated protein kinases, including those activators upstream of p38 and c-Jun N-terminal kinase (JNK). c-Abl's effect on JNK has been contradictory. While some studies show that overexpressed c-Abl does have an effect on JNK, other reports show the opposite (Kharbanda *et al.*, 1995). Thus, the involvement of JNK in c-Abl induced apoptosis is still controversial. The activation of p38 and its upstream activator mitogen-activated protein kinase kinase 6 (MKK6) by overexpressed c-Abl leads to apoptosis (Cong and Goff, 1999). c-Abl-induced apoptosis can be blocked by dominant negative MKK6, suggesting that MKK6 works downstream of c-Abl in this pathway. However, while it is known that c-Abl mediated p38 activation acts through MKK6, inhibition of p38 fails to block c-Abl induced apoptosis, suggesting that c-Abl induced apoptosis is independent of p38 MAPK (Cong and Goff, 1999). One possible way that c-Abl can induce apoptosis may be through p73, a proapoptotic protein and a p53 homolog. It has been shown that inhibition of MKK6 and p38 abolished the accumulation of p73 and the threonine phosphorylation of p73 (adjacent to a proline residue) in the presence of c-Abl overexpression. The presence of p38 can stabilize p73 and is required for the transcriptional activation of p73 by c-Abl (Sanchez-Prieto *et al.*, 2002). Whether p38 is involved in apoptosis induced by c-Abl needs further studies. Unexpectedly, p53 was reported not to play an essential role in apoptosis induced by c-Abl overexpression as c-Abl can induce apoptosis in p53 deficient cells (Theis and Roemer, 1998).

3.6.3 GENOTOXIC STRESS INDUCED APOPTOSIS

Cells deficient for c-Abl or expressing the kinase dead c-Abl are reported to resist apoptosis induced by DNA damage reagents. Numerous studies suggest that c-Abl might regulate DNA damage induced apoptosis through several distinct mechanisms. These include ATM-p53-p73 pathway, translocation of c-Abl to the nucleus, and regulation of caspases directly by c-Abl.

ATM can activate both p53 and p73. p53 is a well known tumor suppressor and p53 loss-of-function is frequently linked to tumorigenesis. p53 has effects on both apoptosis and cell proliferation. p73 belongs to the p53 family, and shares almost an identical architecture. The DNA binding domain and transactivation domain are basically similar between p53 and p73, and as a result, p73 can transactivate quite a number of p53 target genes. p73 is a substrate of c-Abl, and is phosphorylated at Y99 by c-Abl. c-Abl stabilizes p73 in response to DNA damage and enhances its apoptotic ability. This apoptotic activity is lost under c-Abl deficient conditions, indicating that p73 induced apoptosis requires a functional c-Abl protein (Agami *et al.*, 1999; Gong *et al.*, 1999; Yuan *et al.*, 1999). c-Abl stabilizes p73 through promoting the interaction between Pin1 and p73 (Mantovani *et al.*, 2004). p38 MAPK is likely to participate in this process as well. p38 has been demonstrated to be essential for the transcriptional activation of p73 by c-Abl as overexpressed p38 stabilizes p73 and promotes its binding to Pin1 (Sanchez-Prieto *et al.*, 2002). Binding of Pin1 not only stabilizes p73 but also enhances its transcriptional activity, resulting in the up-regulation of p73 target genes such as p21 and Bax (Mantovani *et al.*, 2004). The elevation of Bax may lead to apoptosis. Bax increases the permeability of the mitochondria membrane, resulting in the release of cytochrome c, activation of caspase 9 and the downstream caspases, and cell death.

Recently, it has been demonstrated that both Bax as well as Bak can be activated by c-Abl through the c-Abl-PKCγ-Rac1-p38 MAPK pathway (Choi *et al.*, 2006). It has been demonstrated that IR activated c-Abl is responsible for the binding, phosphorylation and activation of PKCγ (Yuan *et al.*, 1998b; Choi *et al.*, 2006), which in turn activates p38 through Rac1. The activation of p38 is associated with the activation of Bak and Bax, leading to the dissipation of mitochondrial membrane potential, release of cytochrome c, and finally cell death (Choi *et al.*, 2006). Indeed, c-Abl is also localized in the mitochondria and can directly activate caspases. A portion of c-Abl colocalizes with PKCγ and Bcl-xL at the mitochondria.

c-Abl can also mediate apoptosis in a p73 independent manner, through the cleavage of PARP in response to DNA damage (Lasfer *et al.*, 2006). Loss of mitochondrial transmembrane potential was also demonstrated to be associated with c-Abl, as c-Abl$^{-/-}$ MEFs displayed an attenuated loss of mitochondrial transmembrane potential compared with wild-type MEFs. Treatment with imatinib, an inhibitor of c-Abl, attenuates ara-C-induced caspase-3 activation. Moreover, c-Abl also associates with and phosphorylates caspase 9 at Y153, leading to the activation of caspase 3 and apoptosis under genotoxic stress (Raina *et al.*, 2005). Furthermore, c-Abl can act downstream of caspases in genotoxic stress and TNF induced apoptosis, as the caspase inhibitor suppresses c-Abl activation and the onset of apoptosis (Dan *et al.*, 1999). Caspases cleave c-Abl into fragments with functional domains. This increases the kinase activity of c-Abl, promoting apoptosis (Machuy *et al.*, 2004).

c-Abl is also associated with Rad9 and phosphorylates Rad9 at Y28 in response to genotoxic stress. This induces the binding of Rad9 and Bcl-xL in a c-Abl depending manner, leading to apoptosis (Yoshida *et al.*, 2002). c-Abl has also been shown to interact and phosphorylate Shb. c-Abl augments the apoptotic activity of Shb to various stress stimuli such as genotoxic stress, oxidative stress and ER stress (Hagerkvist *et al.*, 2007).

It is proposed that nuclear localization of c-Abl after DNA damage can lead to apoptosis. The nucleus translocation of c-Abl can be prevented by binding to 14-3-3 protein. This binding requires the phosphorylation of c-Abl at T735, which is independent of DNA damage. Binding of 14-3-3 probably masks the NLS and thus prevents it from importing to the nucleus (Yoshida and Miki, 2005; Yoshida *et al.*, 2005). Upon DNA damage, activated JNK phosphorylates 14-3-3, leading to the dissociation of 14-3-3 from c-Abl and allowing c-Abl to transport into the nucleus. In addition, 14-3-3 also mediates the pro-apoptotic activity of other proteins, such as Bax and Bad, by sequestering them in the cytoplasm (Zha *et al.*, 1996; Tsuruta *et al.*, 2004). Activated JNK can phosphorylate 14-3-3 and release Bax, which then translocates to the mitochondria to induce apoptosis (Tsuruta *et al.*, 2004). Studies on oncoprotein MUC1, however, gave rise to inconsistent results. It was found that while MUC1 attenuated the phosphorylation of c-Abl at T735, and blocked the binding of c-Abl to 14-3-3, it still sequesters c-Abl in the cytoplasm (Raina *et al.*, 2006). More studies are needed to understand these inconsistent results.

3.6.4 OXIDATIVE STRESS INDUCED CELL DEATH

It has been proposed that the nuclear c-Abl is activated by genotoxic stress, while the cytosolic fraction responses to oxidative stress. Oxidants can be generated during normal cellular metabolisms such as oxidative phosphorylation of the mitochondrial electron transport chain, peroxisomal fatty acid metabolism, macrophage phagocytosis, and inflammation. The reactive oxygen species (ROS) produced may then lead to cellular damage such as protein oxidation, membrane lipid peroxidation, and DNA damage. Some of the ROS can react with chromatin associated proteins, pyrimidines and purines, leading to base modification and genomic instability. Oxidative stress has been implicated in atherosclerosis, tissue damages as a result of ischemia/reperfusion, aging, and carcinogenesis.

H_2O_2 treatment can induce the binding of cytosolic c-Abl and PKCδ. This interaction causes the phosphorylation (at Y512) and activation of PKCδ. Activated PKCδ is pro-apoptotic. Moreover, PKCδ also phosphorylates and activates c-Abl as a feedback regulation (Sun *et al.*, 2000b). In addition, it has been shown that ROS induces the translocation of c-Abl to the mitochondria. This process is dependent on the activation of PKCδ, and is associated with ROS induced loss of mitochondrial membrane potential (Kumar *et al.*, 2001). Mitochondrial cytochrome c is released in response to oxidative stress in a c-Abl dependent manner (Sun *et al.*, 2000a). Arg is also involved in oxidative stress induced apoptosis and c-Abl forms heterodimers with Arg in response to ROS (Cao *et al.*, 2003c).

In addition, c-Abl also interacts with some of the antioxidative enzymes to regulate apoptosis. c-Abl has been shown to interact with catalase. Catalase has a crucial role in antioxidant enzymatic activity as it hydrolyzes H_2O_2 to H_2O and O_2. c-Abl can phosphorylate catalase at Y231 and Y386 under mild oxidative stress, but not at high concentration of H_2O_2 (Cao *et al.*, 2003b). It was shown that the activity of catalase can be down-regulated by ubiquitination and finally degraded by 26S proteosome. Y231 and Y386 phosphorylated by c-Abl and Arg are required for its degradation. Thus, c-Abl mediate the degradation of catalase and thus has a pro-apoptotic activity (Cao *et al.*, 2003d).

c-Abl is also associated with glutathione peroxidase 1 and phosphorylates it at Y96. Glutathione peroxidase converts H_2O_2 to the nonreactive H_2O and O_2 by oxidizing glutathione. It was found that c-Abl and Arg stimulate glutathione peroxidase activity when cells are exposed to H_2O_2 at a concentration lower than 0.5mM. However, the c-Abl and glutathione peroxidase 1 complex was disrupted when cells are exposed to a high concentration of H_2O_2 (≥ 0.5mM) (Cao *et al.*, 2003a). While it is surprising to see that c-Abl, which has pro-apoptotic activity, could bind and activate these antioxidative enzymes to protect the cell from ROS-induced cell death, it is suggested by Cao *et al.* (2003a) that there are dual roles for c-Abl and Arg in response to oxidative stress. c-Abl and Arg could act as antiapoptotic or antiapoptotic proteins in response to different doses of H_2O_2.

3.6.5   OTHERS

c-Abl can also fulfill its pro-apoptotic role by negatively regulating survival signals. The transcriptional factor, NFκB, appears to participate in the c-Abl induced apoptotic pathway. NFκB is believed to be essential for cell survival. It is mainly controlled by its inhibitor IκB that binds and prevents it from entering the nucleus. IκBα, a member of the IκB family, is a substrate of c-Abl. c-Abl interacts with and phosphorylates IκBα at Y305, leading to the stabilization of the protein and the nuclear accumulation of IκBα. The retention of IκBα in the nucleus abolishes the transcription ability of NFκB for its target genes. As a consequence, it sensitizes the cell to apoptosis (Kawai et al., 2002). Another survival pathway that is negatively regulated by c-Abl is the phosphatidylinositol-3-kinase (PI3K) pathway. PI3K is mainly activated by growth factor receptor stimulation and transduces survival signals in the cell (Kennedy et al., 1997). c-Abl binds to PI3K and causes the phosphorylation of the p85 subunit of PI3K, inhibiting PI3K. This inhibition may affect the transduction of antiapoptotic signal to downstream targets in the cell (Yuan et al., 1997b).

3.6.6   CELL PROLIFERATION

Overexpression of c-Abl inhibits cell growth and causes cell cycle arrest. This growth suppression requires the kinase activity, nucleus localization sequences, and the SH2 domain of c-Abl while deficiency of c-Abl disrupts cell cycle control (Sawyers et al., 1994). c-Abl induced cell cycle arrest required p53. p53 is a pivotal component of cell response to stress and DNA damage. p53 inhibits cell proliferation by mediating cell cycle arrest, senescence, or apoptosis. While c-Abl contains a DNA binding domain, it does not possess transactivation activity and it may cooperate with other transcription factors such as p53 to regulate transcription. c-Abl was found to bind to the C-terminal domain of p53 and to enhance the transcription of p53 target genes that are involved in cell cycle arrest (Goga et al., 1995; Nie et al., 2000). c-Abl's effect on p53-mediated transcription is highly selective. As c-Abl stabilizes p53 tetrameric conformation, resulting in a more stable p53-DNA complex (Nie et al., 2000), it assists the binding of p53 to the promoter of p53 responsive genes containing a "perfect binding sequence" with all four quarter binding site. As a consequence, p21, which contains a "perfect binding sequence" for p53, is transcriptionally enhanced by the more stable p53 tetramer, while Bax, which contains a less perfect p53 binding sequence, is not (Wei et al., 2005).

The level of p53 can be controlled by its negative regulators and positive regulators, some of which are also under the regulation of c-Abl. The murine double minute (Mdm2) or Hdm2 in human serves as an important negative regulator of p53. Mdm2 acts as an E3 ubiquitin ligase, targeting p53 for degradation through the proteasome. On the contrary, p14Arf serves as a positive regulator of p53. p14Arf interacts with Mdm2 and sequesters Mdm2 in the nucleus, leading to the inhibition of Mdm2 nuclear export and a decrease in p53 turnover. It has been found that c-Abl interacts and phosphorylates Mdm2 or Hdm2. Phosphorylation of Hdm2 at Y394 impairs its effect on p53 degradation, while phosphorylation at Y276 stimulates its interactions with Arf, leading to the nuclear retention of Hdm2 and thus an increase in p53 levels (Goldberg *et al.*, 2002; Dias *et al.*, 2006). Thus, c-Abl could up-regulate p21 through stabilizing p53 (Wei *et al.*, 2005). However, c-Abl mediated growth arrest can also be independent of p21. c-Abl can down regulate the levels of Cdks in p21 deficient cells, suggesting that c-Abl mediated growth arrest can be in a p21 independent manner (Yuan *et al.*, 1996).

There is also some evidence supporting that c-Abl might induce growth arrest through retinoblastoma protein (Rb). Rb plays a crucial role in regulating cell cycle progression. Rb, in its hypophosphorylated form, binds to E2F and inhibits the transactivation of E2F target genes that are required for cell cycle progression. Phosphorylation of Rb results in its release from E2F and allows the activation of E2F target genes and the entry of cells into S phase. While some studies have shown that c-Abl has strong cytostatic effects in normal cells, but not in Rb deficient cells (Wen *et al.*, 1996), other studies using fibroblasts with dysfunctional p53 or Rb showed that c-Abl could mediate growth arrest independent of Rb (Goga *et al.*, 1995). This discrepancy cannot be currently explained. To make the case more complicated, it has been shown that the activity of c-Abl is also affected by Rb. A portion of c-Abl binds to Rb in a cell cycle-dependent manner. Rb can bind to the ATP-binding lobe of c-Abl to inhibit the activation of c-Abl. Phosphorylation of Rb releases c-Abl and activates c-Abl (Welch and Wang, 1993). In addition, another protein that can interact with and inhibit the activity of c-Abl is the proliferation associated gene, PAG. PAG encodes a protein with antioxidative properties that is implicated in cellular response to oxidative stress as well as cellular proliferation and differentiation. Overexpression of PAG can inhibit c-Abl's autophosphorylation and phosphorylation of other substrates, as well as rescue the cytostatic effects induced by c-Abl. It has been suggested that the inhibitory property of PAG on c-Abl is not due to its antioxidative properties, as the truncation mutant of PAG that lacks the portion for antioxidative function still retains its c-Abl inhibitory property (Wen and Van Etten, 1997). Instead, PAG serves as a physiological inhibitor for c-Abl at least in cell cycle regulation.

c-Abl also regulates cell cycle progression through the ubiquitin-proteasome pathway. c-Abl associates and phosphorylates proteosome PSMA ($\alpha4$) subunit at Y153. As a consequence, proteolysis mediated by proteasome was attenuated. In addition, this phosphorylation leads to cell cycle regulation. Cell expressing mutant PSMA at the site of phosphorylation (Y153F) displays G1/S cycle arrest and impaired S/G2 progression (Liu *et al.*, 2006).

3.6.7   CYTOSKELETON AND CELL MOVEMENT

c-Abl plays a very important role in the modulation of cytoskeleton organization. Cytoskeleton is a network of protein fibers that are responsible for cell morphology, cell movement, neurite elongation, endocytosis, phagocytosis, exocytosis, and intracellular trafficking. These processes require the formation of membrane ruffles, filopodia, lamellipodia, and focal adhesions, which are dynamic structures that rely on the ability of the cells to perform actin polymerization and depolymerization. c-Abl consists of an actin binding domain. Together with other associated proteins such as Paxillin, c-Abl could relay the extracellular signals to regulate cytoskeleton reorganization.

Growth factors such as PDGF have been shown to regulate the cytoskeletal reorganization. Membrane ruffling is observed after PDGF treatment. c-Abl is greatly accumulated at the membrane ruffles upon PDGF stimulation (Ting *et al.*, 2001). The number of membrane ruffling is much less in c-Abl deficient cells, suggesting that c-Abl is required for the reorganization of cytoskeleton after PDGF treatment (Plattner *et al.*, 1999). The underlying mechanisms are still not very clear. It is likely that Abi and PAK2 are involved as both of them can interact with c-Abl and are localized in the membrane ruffles induced by PDGF. It is also known that the interaction of Abi (Abl interacting protein) and PAK2 is required for membrane ruffle formation (Machuy *et al.*, 2007). However, the role of c-Abl in this pathway still requires further investigation. Lamellipodia, filopodia and focal adhesions are cytoskeletal structures, which are required for cell migration. Attachment of cells to fibronectin stimulates Abl activity and transiently redistributes the nuclear Abl to the cytoplasm and focal adhesions (Lewis *et al.*, 1996). c-Abl deficient cells display fewer actin microspikes, the precursors of filopodia, than normal cells when spread on fibronectin coated surface (Woodring *et al.*, 2002).

The filopodia of the neurite is also affected by c-Abl. The branching of the cortical embryonic neurons is defected in the absence of c-Abl (Woodring *et al.*, 2002), and overexpressing c-Abl in primary neurons can increase the length of the neurite (Zukerberg *et al.*, 2000). One of the mechanisms by which c-Abl promotes filopodia during cell spreading is probably due to the phosphorylation of

the down-stream of kinase (Dok) family by c-Abl. c-Abl is able to phosphorylate p62 docking protein (Dok1) at Y361 and Dok-R (Master *et al.*, 2003). Phosphorylation of these Dok proteins promotes its association with Nck (Woodring
et al., 2004), which has been shown to trigger actin polymerization (Rivera *et
al.*, 2004). Moreover, the association of Nck with some actin polymerization
proteins, such as N-Wasp (neural Wiskott-Aldrich syndrome protein) (Rivera
et al., 2004), WAVE1 (WASP-family verprolin homologous protein) (Eden *et
al.*, 2002) and PAK (p21-activated kinase) (Zhou *et al.*, 2003), may also mediate this effect. Less filopodia were found in fibroblast lacking c-Abl, Dok1, or
Nck (Woodring *et al.*, 2004). The WASP and WAVE proteins play an essential
role in connecting the membrane to the cytoskeleton (Takenawa and Suetsugu,
2007). They mediate the upstream signals to the activation of the actin-related
protein-2/3 (ARP2/3) complex, which is essential for the nucleation of actin
polymerization. c-Abl interacts with some of the proteins in the pathways that
regulate cytoskeleton reorganization (Hirao *et al.*, 2006; Stuart *et al.*, 2006).
c-Abl can also be recruited to the WAVE2 macromolecular complex by Abi-1,
where it phosphorylates WAVE2 at Y150, leading to the activation of WAVE2
for actin polymerization (Leng *et al.*, 2005; Hirao *et al.*, 2006; Stuart *et al.*,
2006). It has also been shown that c-Abl is involved in the Rap1 guanine nucleotide exchange factor (C3G) mediated cytoskeleton rearrangement and filopodia
formation. C3G is localized in the cytoplasm where it interacts with c-Abl and
localizes c-Abl to cytoplasm. Together with the activity of N-WASP, filopodia
formation is promoted (Radha *et al.*, 2007).

c-Abl is able to mediate the activation of mammalian Ena/VASP family protein, Mena. It has been found that Abi-1 and Abi-2 interact with both c-Abl and
Mena to promote the phosphorylation of Mena by c-Abl at Y296 (Tani *et al.*,
2003; Hirao *et al.*, 2006). Activated Mena then binds to the barbed end of actin
filaments to promote the elongation of the target filament (Bear *et al.*, 2002).

However, it is interesting to find that c-Abl can also negatively regulate cell
migration, despite the fact that it can promote the formation of cytoskeleton
structure. c-Abl is able to phosphorylate Crk at Y221, disrupting the Crk-CAS
(Crk-associated substrate) complexes (Kain and Klemke, 2001). As the association of Crk with CAS induces cell migration (Klemke *et al.*, 1998), inhibition
of Crk-CAS coupling probably leads to prevention of cell migration. Another
proposed mechanism is that c-Abl prolongs the exploratory phase before cell
movement, resulting in slower cell migration (Woodring *et al.*, 2003).

The action of c-Abl on actin cytoskeleton can also be negatively regulated
by actin. c-Abl can promote the formation of the actin cytoskeletal structure,
but the activity of c-Abl can be inhibited by F-actin. c-Abl can bind to F-actin
through its actin binding domain. Mutation of the F-actin binding domain re-

lieves the inhibition (Woodring *et al.*, 2002). c-Abl can be dephosphorylated by a process mediated by an actin associated protein PSTPIP1 (proline, serine, threonine phosphatase interacting protein). PSTPIP1 is a substrate of c-Abl, and is primarily phosphorylated by c-Abl at Y344 (Cong *et al.*, 2000). On the other hand, PSTPIP1 is also a substrate of the PEST-type protein tyrosine phosphatases (PTP) (Spencer *et al.*, 1997). Thus, PSTPIP1 is able to associate PEST-type PTP with c-Abl, leading to the dephosphorylation of c-Abl by PEST-type PTP (Cong *et al.*, 2000). Indeed cells deficient in PEST-type PTP displays signs of increased c-Abl activity, increased tyrosine phosphorylation of PSTPIP, as well defects in motility. Thus, PEST-type PTP may serve as a negative regulator to mediate the effect of c-Abl on cytoskeletal modulation (Angers-Loustau *et al.*, 1999).

3.7   ABL FUNCTION: GENETIC APPROACH

3.7.1   *GENERAL PHENOTYPES OF C-ABL KNOCKOUT MICE*

The expression of c-Abl has been detected throughout mouse embryonic development and in all the mouse tissues tested, with higher expression in the thymus (Muller *et al.*, 1982; Renshaw *et al.*, 1988). In human tissues, higher expression is observed in the hyaline cartilages, adipocytes, and ciliated epithelium in adults, while the strongest expression is observed at the sites of endochondral ossification in fetuses (O'Neill *et al.*, 1997). Two lines of c-Abl knockout mice has been generated (Schwartzberg *et al.*, 1989; Tybulewicz *et al.*, 1991). One line, c-Abl[2], reported by Tybulewicz *et al.* (1991) contains a deletion within the N-terminal part of the tyrosine kinase domain, resulting in a deletion in the DNA binding domain and the ATP binding site. No c-Abl or c-Abl kinase activity could be detected in these mice. The other line, Abl[m1], generated by Schwartzberg *et al.* (1991), expresses a C-terminal truncated c-Abl protein with DNA binding and actin binding domains missing. This truncated c-Abl still has tyrosine kinase activity. The phenotypes of these two lines of mice were quite similar, suggesting that the carboxyl-terminal region is essential for c-Abl's function *in vivo* (Schwartzberg *et al.*, 1991). The homozygous knockout mice showed a higher rate of perinatal lethality, decreased fertility, runtedness, immunodeficiency, spleen and thymus atrophy, and osteoporosis. The timing of eye opening was also affected in the homozygous knockout mice that often develop cataracts or permanent eye damage. Some of the homozygous mutant mice that survived to 3 or 4 months developed megaesophagus and anal prolapse (Schwartzberg *et al.*, 1991; Tybulewicz *et al.*, 1991; Li *et al.*, 2000). The hepatocytes of the knockout mice also displayed fatty vacuoles, and some of these cells showed signs of degeneration. To identify which c-Abl transcript and domains are responsible for the different phenotypes, a few lines of c-Abl transgenic mouse were generated and mated with homozygous c-Abl knockout mice (Hardin *et al.*, 1996a). It was found that the presence of either type I or type IV isoform could rescue the major phenotypes of the c-Abl knockout mice, suggesting that these two isoforms have redundant functions. Yet the kinase defective c-Abl transgene could not rescue the defects found in the homozygous knockout mice (Hardin *et al.*, 1996a). Further studies show that the actin binding domain and NLS are required to rescue the phenotypes of the knockout mice while the DNA binding domain is dispensable (Goff S.P., unpublished data, personal communication).

Mice deficient for both c-Abl and its sole paralog Arg are embryonic lethal at 11 dpc. These mice have defects in neurotube closure (Koleske *et al.*, 1998). Mice deficient for Arg are normal. Massive apoptosis and hemorrhage were observed in the brains of the double knockout embryos (Koleske *et al.*, 1998). These data suggest that c-Abl and its paralog Arg are essential for early embryonic development. More importantly, these results indicate a critical role for c-Abl and Arg in brain development and neuron survival. Note that Arg is solely localized in the cytoplasm and its C-terminal region is less conserved with c-Abl. Therefore, the redundant function between c-Abl and Arg lies in the cytoplasm.

3.7.2 OSTEOPOROSIS- THE FUNCTION OF C-ABL IN BMP SIGNALING AND OSTEOBLAST FUNCTION

Bone remodeling is mainly contributed by two types of bone cells, the osteoblasts and the osteoclasts. Osteoblasts are responsible for bone formation, and the increase in bone mass, while osteoclasts are involved in bone resorption. Osteoporosis occurs if bone resorption is faster than bone formation when osteoclasts are overactive, osteoblasts are less active, or there are an insufficient number of osteoblasts. The c-Abl knockout mice develop an osteoporotic phenotype. Their long bones display thinner cortical bone and reduced trabecular bone volume and reduced bone formation rate. The mineralization of the bone is also decreased in c-Abl knockout mice (Li *et al.*, 2000). It is mainly due to the delay of osteoblast maturation rather than osteoclast dysfunction. Altered bone cell activity and mineral metabolism was observed in some of the patients who were treated with imatinib, a drug used to treat CML by inhibiting BCR-ABL activity (Berman *et al.*, 2006). Cell culture study showed that c-Abl was up regulated during early stages of osteoblast differentiation. Cultured c-Abl[-/-] osteoblast showed delayed maturation and decreased mineral deposition.

Osteoblast maturation or differentiation is stimulated by cytokines and growth factors, such as bone morphogenetic proteins (BMPs). c-Abl osteoblasts also have a compromised response to the stimulation of BMP2. These data suggested that c-Abl is important in osteoblast differentiation and it may participate in the BMP triggered signaling pathway (Li *et al.*, 2000). It has also been demonstrated that the defective of differentiation in c-Abl[-/-] osteoblast could be corrected by p53 deficiency, in which high bone mass and accelerated osteoblast differentiation were found, suggesting that p53 is acting downstream of c-Abl in osteoblast differentiation (Wang *et al.*, 2006). In addition, c-Abl[-/-] osteoblast was hypersensitive to oxidative stress (Li *et al.*, 2004). Under oxidative stress, it was found that although the osteoblasts of c-Abl knockout mice displayed elevated peroxiredoxin I, an antioxidant protein mediated by a basic leucine zipper transcription factor Nrf2 through PKCδ, an increase in the cell death rate was observed. Thus, with such a hypersensitive nature, the c-Abl[-/-] osteoblast may experience more damage even under normal oxidative stress leading to cell death and subsequently reduced number of osteoblasts. c-Abl hence protects the osteoblasts against oxidative stress. Moreover, oxidative stress has been shown to inhibit the differentiation of osteoblasts (Mody *et al.*, 2001). It is likely that it may also contribute to the defects in differentiation observed in c-Abl deficient osteoblasts.

3.7.3 *IMMUNODEFICIENCY-REVEALS THE FUNCTION OF C-ABL IN THE IMMUNE SYSTEM*

c-Abl knockout mice display a variety of immune system defects and they suffer from a wide range of infections. The knockout mice also display spleen abnormality and thymus atrophy with deficiency in thymocytes. They show both T-cell and B-cell lymphopenia (Schwartzberg *et al.*, 1991; Tybulewicz *et al.*, 1991). Later studies also demonstrated that the c-Abl deficient mice experience reduction in pre-B cells and pre-T cells in adult spleen, thymus, bone marrow, and peripheral blood. Furthermore, the bone marrow cells show reduced response to interleukin IL-7, and the spleen B-cell show compromised response to lipopolysaccharide, a mitogen that can stimulate B-cells proliferation and differentiation. Short-term B-cells lymphopoiesis culture showed a reduction of pro-B-cells. Long-term lymphoid bone marrow cultures also demonstrated that B-cell progenitors are more sensitive to apoptotic stimuli, such as glucocorticoid and IL-7 deprivation, in the absence of c-Abl.

The decrease in B-cell numbers is probably due to impaired B-cell development as well as increased apoptosis of B-cell precursors. Moreover, the Ig heavy chain rearrangement, V(D)J rearrangement, was also defective in c-Abl knockout B-cell progenitors (Lam *et al.*, 2007). The antigen activated B-cell expansion and differentiation need to be tightly regulated. Binding of antigen stimulates the B-cell-antigen receptor, which triggers a cascade of signaling responses to initiate the proliferation and differentiation of B cells. The membrane-spanning glycoprotein CD19 has been found to play a key role in this pathway. CD19 acts as a costimulatory molecule to assist the activation of B-cell by B-cell-antigen receptor. It was found that CD19 contains a c-Abl phosphorylation site and two potential c-Abl SH2 binding sites, and that it serves as a substrate and binding partner for c-Abl. The amount of c-Abl is increased upon B-cell-antigen receptor signaling. The B cells isolated from c-Abl knockout mice are hyporesponsive to the proliferative signal triggered by the B-cell-antigen receptor, probably due to the insufficient activation of CD19 as a result of c-Abl deficiency. Thus, c-Abl may regulate the proliferation of B cells through CD19 upon B-cell-antigen receptor activation (Zipfel *et al.*, 2000). On the other hand, B-cell Fc receptors transmit inhibitory signals to regulate the antigen-driven activation and proliferation of lymphocytes, so as to prevent the disastrous effects of antibody overproduction. When the antibody level in an organism reaches a certain level in response to immune stimulation, it forms an immune complex with the antigen. These immune complexes then bind to the Fc receptor to result in either inhibition of the B-cell-antigen receptor triggered signaling pathway, when coligated to the B-cell-antigen receptor, or to trigger apoptosis, when the aggregation is independent of the B-cell-antigen receptor (Pearse *et al.*, 1999). It was found that the Fc receptor can trigger apoptosis in a c-Abl dependent manner that is independent of the B-cell-antigen receptor (Tzeng *et al.*, 2005).

The peripheral blood of c-Abl deficient mice has reduced response to concanavalin A, a T-cell mitogen that can induce cell division and T-cell function (Hardin *et al.*, 1995; Dorsch and Goff, 1996; Hardin *et al.*, 1996b). More interestingly, it has been reported that the Abl kinase participates in T-cell signaling, and mediates T-cell activation possibly by phosphorylating ZAP70 and the transmembrane adaptor linker for the activation of T-cell (LAT). Abl kinase also plays a role in T-cell receptor signaling pathway in IL-2 production and T-cell proliferation (Zipfel *et al.*, 2004). In addition, c-Abl can also activate T cells by stabilizing c-Jun. It has been reported that c-Abl binds to and phosphorylates c-Jun at the tyrosine residue within the PPXY motif. This phosphorylation blocks the ubiquitination of c-Jun by Itch, an E3 ubiquitin ligase, thus stabilizing c-Jun. In c-Abl deficient T cells, c-Jun degradation is accelerated, leading to the suppression of T-cell proliferation (Gao *et al.*, 2006).

3.8 ABL FUNCTION: CML STUDIES

Uncontrolled activation of c-Abl can cause deleterious effects. The fusion protein BCR-ABL is an oncoprotein. Expression of BCR-ABL is found in 95 % of the CML patients, and in 5-10 % of the acute lymphoblastic leukemia patients. CML is a clonal myeloproliferative disease, which is originated from an abnormal pluripotent bone marrow stem cell. Disease progression in CML usually occurs through three phases, a chronic, an accelerated, and a blast phase. In BCR-ABL, much of the N-terminal Cap of c-Abl has been deleted during the fusion. As mentioned earlier, the Cap helps to keep c-Abl in an inactive conformation. Loss of the Cap may weaken the inhibitory effect of the SH3-SH2 domains on the kinase domain. BCR is a serine/tyrosine kinase with multiple domains, including a coiled-coil oligomerization domain, a GTPase exchange factor domain, and a RacGAP domain. The coiled-coil domain of the BCR leads to the oligomerization of the protein, facilitating the transphosphorylation of Y412 and enhancing the activation of the ABL kinase (Harrison, 2003). BCR-ABL is hence constitutively active. While c-Abl is localized in various subcellular compartments, BCR-ABL is localized exclusively in the cytoplasm (Dierov et al., 2004).

The mechanisms by which BCR-ABL causes CML include promoting cell proliferation, preventing apoptosis, and altering cell adhesion. BCR-ABL activates multiple signaling pathways to fulfill these functions, in which the Ras-MAPK pathway and PI3K pathway are the prominent ones. Auto-phosphorylation of BCR at Y177 recruits the adaptor proteins Grb2 and Gab2, resulting in the activation of Ras-MAPK and PI3K pathways (Million and Van Etten, 2000; He et al., 2002). Activation of Ras-MAPK has been implicated in the downregulation of the pro-apoptotic factor Bim and the activation of the antiapoptotic factor Mcl-1 in BCR-ABL positive CML cells (Aichberger et al., 2005a; Aichberger et al., 2005b). Suppression of Ras and MAPK leads to the attenuation of BCR-ABL mediated cell transformation (Peters et al., 2001). Activation of PI3K-Akt-mTor pathway by BCR-ABL in CML cells is important for cell survival, cell cycle progression and cell transformation (Skorski et al., 1997; Mayerhofer et al., 2005). BCR-ABL also promotes leukemogensis through the activation of GTPase Rap1, phosphorylation of Rb and enhanced activity of Stat5. These signaling events lead to the up regulation of Id1, PIM1/2, and Bcl-xL for invasion and survival (Gesbert and Griffin, 2000; Cho et al., 2005; Adam et al., 2006; Nieborowska-Skorska et al., 2006; Nagano et al., 2006).

Cytoskeleton remodeling is a fundamental process that regulates cell shape, cell adhesion and motility. Deregulation of this process is often associated with cellular transformation and tumorigenesis (Rao and Li, 2004). BCR-ABL consists of an actin binding domain, which enables it to interact with actin filaments and associate with cytoskeleton remodeling proteins. BCR-ABL positive CML cells exhibit abnormal cytoskeletal functions, including increased motility and altered adhesion. The underlying mechanism by which BCR-ABL leads to these abnormalities is not very clear. It is proposed that abnormal regulation of Abi1/ WAVE2, CRKL, and Rac by BCR-ABL may contribute to the altered cytoskeletal organization (Hemmeryckx et al., 2001; Sini et al., 2004; Li et al., 2007).

Conventionally, busulfan, hydroxyurea, radiation, bone marrow transplantation, arsenic, IFN-alpha, and cytarabine are used for treatment of CML. Although these therapies could delay the disease progression, the results are not as satisfactory as the drug imatinib, an inhibitor for BCR-ABL. Imatinib is a 2-phenylaminopyrimidine derivative with a higher affinity to Abl than other kinases. It binds to the kinase domain of BCR-ABL, through competitive inhibition at the ATP-binding site, and inhibits the tyrosine phosphorylation of BCR-ABL substrates. Delayed disease progression is observed in CML patients treated with imatinib and it also prolongs the overall survival of CML patients. In early chronic phase CML patients, imatinib achieved 95 % complete hematological response rate. 76 % of the patients achieved a complete cytogenetic response rate and 97 % of the patients survived without progressing to the accelerated or blast phase at 19 months (Baccarani et al., 2006).

Besides, its inhibitory effect on the oncogenic protein BCR-ABL, imatinib can also inactivate c-Abl, PDGFR, c-Kit, Arg, and probably other unknown targets which play important roles in cellular functions (Buchdunger et al., 1996; Heinrich et al., 2000; Okuda et al., 2001). These nonspecific targets of imatinib may cause adverse effects on patients. Imatinib had been administrated to rats, dogs, and monkeys to study its toxicity. Hematological problems were observed in rats, dogs, and monkeys. Rats and dogs also developed lymphoid atrophy and lymphoid depletion as a result of imatinib administration, suggesting that their immune system may be affected. Renal toxicity was also encountered by rats and monkeys, while dogs suffered from hepatic toxicity. Decreased spermatogenesis and enlarged hemorrhagic ovaries were also noted. Imatinib also showed teratogenic effect in pregnant rats, with increased skeletal malformations and anomalies, suggesting that imatinib affect the development of the rat particularly in the skeletal formation. Some of these defects are also observed in the c-Abl deficient mice, such as immune system problems, reproductive defects, skeletal abnormalities and other developmental problems (Schwartzberg et al., 1991; Tybulewicz et al., 1991; Li et al., 2000). This suggests that the inhibitory effect on c-Abl by imatinib may contribute at least partially (if not all) to these side effects (Cohen et al., 2002).

Clinically, the common side effects of imatinib were noted in humans including nausea, vomiting, edema, dyspnea, diarrhea, elevated liver enzymes, muscle cramps, musculoskeletal pain, rash and other skin problems, abdominal pain, fatigue, joint pain, headache, neutropenia, thrombocytopenia, and anemia (Cohen et al., 2002; Druker et al., 2006). Recently, adverse effects of imatinib were also observed in bone and heart (Berman *et al.*, 2006; Kerkela *et al.*, 2006). Serum levels of osteocalcin, a marker of bone formation, was low in patients with either CML or gastrointestinal stromal tumors who were receiving imatinib, suggesting that the drug may affect bone-cell activity and inhibit bone remodeling. Imatinib also leads to hypophosphatemia in a portion of the patients. The bone deficits observed in patients treated with imatinib was probably due to the inhibitory effect of imatinib in c-Abl, as c-Abl is important for skeletal development in mouse (Li *et al.*, 2000; Berman *et al.*, 2006). The inhibitory effect of imatinib on PDGFR may also contribute to this side effect, as PDGFR has been claimed to play a role in skeletal development (Berman *et al.*, 2006). The cardiotoxicity of imatinib has been revealed recently. During the clinical trails, although the cardiotoxicity was not directly examined, it has been observed that some of the patients suffer from edema (~65 %) and dyspnea (~15 %), which are signs of heart failure (Cohen *et al.*, 2002). Druker *et al.* (2006) also reported a case of congestive heart failure in their five-year follow up study that is related to imatinib. Kerkela *et al.* (2006) reported 10 patients who have developed advanced heart failure in receiving imatinib (Druker *et al.*, 2006). These patients suffered from 50 % decrease in the pumping capacity of the heart due to left ventricular dysfunction. Mice treated with imatinib also developed left ventricular contractile dysfunction. In both human and mice administrated with imatinib, the cardiomyocytes showed signs of toxic myopathy, including mitochondrial abnormalities and accumulation of membrane whorls in the sarcoplasmic reticulum. In cultured cardiomyocytes, imatinib activates the stress response in ER through JNK. It was found that the mitochondrial membrane potential was collapsed and cytochrome c was released into cytosol, depleting the cell of ATP, and causing cell death. The cardiotoxicity effect of imatinib is likely to be due to the inactivation of cellular c-Abl as introduction of an imatinib resistant c-Abl mutant prevented imatinib induced cell death (Kerkela *et al.*, 2006).

Although imatinib has been proven as an effective therapy for CML, clinical resistance has been observed while treating CML patients with imatinib. The most common reason of acquired imatinib resistance is the reactivation of the BCR-ABL activity. This may be achieved by BCR-ABL gene amplification and the development of point mutations on BCR-ABL, leading to the loss of affinity for imatinib. The incidence of BCR-ABL mutation is about 10 % within 4 years of imatinib treatment, and more than 40 % with more than 4 years of drug treatment (Branford *et al.*, 2003). It is also proposed that imatinib potentially inhibits the production of differentiated leukemic cells, but does not deplete leukemic stem cells (Michor *et al.*, 2005; Brendel *et al.*, 2007), leading to the relapse of the disease after termination of imatinib treatment. Moreover, the survived leukemic stem cells may accumulate mutations during imatinib therapy, leading to imatinib resistance and disease progression.

New therapies are needed for patients who develop imatinib-resistance. Firstly, increase in the dosage of imatinib may help to overcome imatinib resistance due to BCR-ABL gene amplification and mutations in the BCR-ABL kinase domain, which lead to inefficient binding of imatinib. This also overcomes drug resistance due to up-regulation of α-acid glycoprotein, which binds imatinib, and P-glycoprotein, which causes the efflux of imatinib from the cell. The use of greater potency inhibitors, such as BMS354825, AMN107, and INNO406, may also help in the treatment. Secondly, a combination of imatinib therapy with conventional therapy can be used. It has been found to be more effective when imatinib was co-administered with IFN-alpha, cytarabine, or arsenic trioxide (Wong and Witte, 2004). Thirdly, the use of alternative BCR-ABL inhibitors, such as PD180970, PD173955, ON012380, AMN107, BMS344825, PD166326, AP23464, SKI606, INNO406, may help to battle the imatinib-sensitive-mutant BCR-ABL. Fourthly, the downstream signaling molecules of BCR-ABL, for example, Ras, PI3K, and Stat5 pathways can also be targeted to diminish BCR-ABL signaling (Martinelli et al., 2005).

CML patients treated with imatinib are also more susceptible to Varicella-Zoster virus infections, an infection that normally happens in immunosuppressed patients, possibly due to the decline of CD4 positive T-cells (Mattiuzzi et al., 2003). This corresponds to lymphoid atrophy and lymphoid depletion in rats and dogs which are treated with imatinib (Cohen et al., 2002). The cardiotoxicity that has been observed in both patients and mice treated with imatinib is speculated to be due to the inactivation of cellular c-Abl (Kerkela et al., 2006). While it is surprising that heart related problems have not been reported in the c-Abl knockout mice, this may be due to the fact that the hearts of c-Abl knockout mice have not received enough detailed study.

3.9 CONCLUSION

To date, the results on c-Abl's role in cell death are inconsistent and controversial. On one hand, the biochemical studies demonstrated that the overexpression of c-Abl causes apoptosis, and that DNA damage induced apoptosis is compromised in MEFs deficient for c-Abl, suggesting a pro-apoptotic role for c-Abl. On the other hand, neuronal cells deficient for c-Abl and Arg show massive apoptosis during early development. MEFs deficient in c-Abl and Arg show increased cell death in response to oxidative stress, and osteoblasts deficient in c-Abl also show hypersensitivity to oxidative stress. Similarly, pre-B and pro-B cells isolated form c-Abl knockout mice are also hypersensitive to IL-7 deprivation induced apoptosis. These results, together with the fact that BCR-ABL is highly antiapoptotic, support a pro-survival role for c-Abl under most of the conditions. This contradicts with studies showing that c-Abl regulates proteins, e.g. caspases that are directly involved in apoptosis. It is possible that c-Abl's role in cell death can be cell type specific and stimuli specific. For example, DNA damage mainly activates nuclear c-Abl, and overexpression of c-Abl leads to the accumulation of c-Abl in the nucleus, resulting in apoptosis. On the contrary, activation of cytoplasmic c-Abl, such as BCR-ABL, might help the cell to survive against stress.

A role for c-Abl in proliferation has always been an attractive concept since c-Abl is a proto-oncogene. Activated Abl kinases, such as BCR-ABL and v-Abl, are capable of promoting cell proliferation and transformation of cells. However, overexpression of c-Abl was found to repress cell growth and cells are arrested at G1/G0 phase. This requires that c-Abl could enter the nucleus. In consistent with this, several studies show that c-Abl deficient MEFs show defects in cell cycle checkpoint in response to double stranded DNA breaks. Another possibility is that c-Abl plays an important role in stem cell renewal. BCR-ABL can transform myeloid stem cells. In the absence of c-Abl, osteoblasts, pre-B and pro-B cell numbers are reduced.

Although many c-Abl interacting proteins as well as substrates have been identified, most of them have not been confirmed *in vivo*. Correlation of c-Abl to these interacting partners or putative substrates in terms of cellular functions is lacking. Moreover, no efforts have been made on comparing the phenotypes of c-Abl knockout mice with those of the mice deficient for the putative substrates or interacting proteins. Nevertheless, various studies suggest that c-Abl plays an important role in DNA damage response, as well as in receptor mediated signaling pathways including, TGF, BMP, PDGF and TCR. In most of the cases, c-Abl appears to act on the receptors or signaling molecules directly downstream of the receptors (Figure 3.1). This function might be carried out by the membrane attached type IV c-Abl that is attached to the plasma membrane by myristoylation. However the molecular mechanisms by which c-Abl regulates these signaling pathways are not clear.

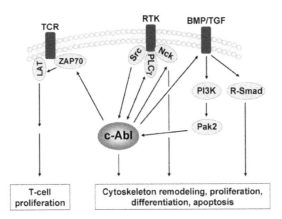

FIGURE 3.1   Participation of c-Abl in T cell receptor, receptor tyrosine kinase, and BMP/ TGFβ signaling.

While the cytoplasmic c-Abl may regulate signaling pathways mediated by the receptors, nuclear c-Abl appears to affect the signaling pathway triggered by DNA damage, especially double stranded DNA breaks (Figure 3.2). Around 100 proteins are assembled at DSBs, forming a DNA repair center and a signaling center. ATM is among the earliest signaling molecules that are activated by DSBs. It then interacts with, phosphorylates, and activates c-Abl. Further downstream of the signaling pathway are some of c-Abl substrates that are critical in DNA damage induced cell cycle checkpoint control, apoptosis, and DNA repair. One of these proteins is p73, which is stabilized by activated c-Abl and promotes apoptosis. Another protein is Arf, which interacts with Mdm2 to regulate the protein stability of p53. Without c-Abl, DNA damage induced p53 accumulation is compromised. This might directly affect cell cycle progression and apoptosis. Finally, the protein Rad51 can also be phoshorylated by c-Abl to facilitate Rad51 mediated DNA repair.

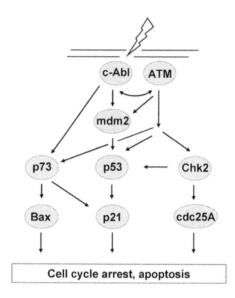

FIGURE 3.2   Participation of c-Abl in cell response to DNA damage.

3.10   FUTURE PERSPECTIVES

The problem facing the c-Abl research community is a lack of agreement between the *in vitro* studies and the *in vivo* studies. While great efforts have been made on studying c-Abl's role in the genotoxic/oxidative stress response, there has been no *in vivo* evidence to back these *in vitro* results. This is partially because c-Abl knockout mice are hard to get, due to perinatal lethality. Another problem is that it is difficult to find a common ground for c-Abl and BCR-ABL. Overexpression of c-Abl did not reproduce the effects of expression of BCR-ABL, and BCR-ABL's expression did not reproduce phenotypes that are opposite of c-Abl deficiency. This could be due to the complexity of c-Abl activation and localization, and the fact that differentially localized c-Abl might have distinct functions. However mouse genetic studies and CML patient studies have revealed that c-Abl deficiency and inhibition of c-Abl/BCR-ABL do have something in common. These include alterations in bone remodeling, immune function, and heart function. Future research should be directed at using these common features to dissect the signaling pathways that are affected by c-Abl deficiency or inhibition, and to integrate the biochemical studies with the *in vivo* studies to understand the true function of c-Abl.

ACKNOWLEDGEMENTS

We thank Yue Wan and Wai Fook Leong for helpful discussion. The work in the laboratory of B. Li is supported by the Agency for Science, Technology and Research of the Republic of Singapore. B. Li is an adjunct member of Department of Medicine of the National University of Singapore.

KEYWORDS

- **Abl function**
- **Antimetabolite**
- **Apoptosis**
- **BCR-ABL proteins**
- **c-Abl gene**
- **Camptothecin**
- **Cap region**
- **Cell cycle arrest**
- **Cell cycle progression**
- **Cell death**
- **Cell movement**
- **Cell proliferation**
- **Chronic myeloid leukemia**
- **Cisplatin**

- **C-terminal lobe**
- **DNA binding domain**
- **DNA damage**
- **Doxorubicin**
- **Endocytosis**
- **Etoposide**
- **Exocytosis**
- **Genotoxic stress induced apoptosis**
- **Growth factors**
- **Human cDNA library**
- **Intracellular trafficking**
- **Ionizing irradiation**
- **Knockout mice**
- **MAPK pathway**
- **Mitomycin C**
- **Moloney Murine Leukemia Virus**
- **Mouse models**
- **Neurite elongation**
- **Osteoporosis**
- **Overexpression induced apoptosis**
- **Oxidative stress**
- **Phagocytosis**
- **Phosphorylation**
- **Protein tyrosine kinases**
- **Proto-oncogene**
- **Thymomas**
- **Transforming growth factor**

REFERENCES

Abelson, H. T.; Rabstein, L. S. A new tumor inducing variant of Moloney Leukemia Virus. *Proc Am Assoc Cancer Res* 1969, 10, 1.

Abelson, H. T.; Rabstein, L. S. Lymphosarcoma: Virus-induced thymic-independent disease in mice. *Cancer Res* 1970, 30, 2213-2222.

Adam, M.; Pogacic, V.; Bendit, M.; Chappuis, R.; Nawijn, M. C.; Duyster, J.; Fox, C. J.; Thompson, C. B.; Cools, J.; Schwaller, J. Targeting PIM kinases impairs survival of hematopoietic cells transformed by kinase inhibitor-sensitive and kinase inhibitor-resistant forms of Fms-like tyrosine kinase 3 and BCR/ABL. *Cancer Res* 2006, 66, 3828-3835.

Agami, R.; Blandino, G.; Oren, M.; Shaul, Y. Interaction of c-Abl and p73alpha and their collaboration to induce apoptosis. *Nature* 1999, 399, 809-813.

Aichberger, K. J.; Mayerhofer, M.; Krauth, M. T.; Skvara, H.; Florian, S.; Sonneck, K.; Akgul, C.; Derdak, S.; Pickl, W. F.; Wacheck, V.; Selzer, E.; Monia, B. P.; Moriggl, R.; Valent, P.; Sillaber, C. Identification of mcl-1 as a BCR/ABL-dependent target in chronic myeloid leukemia (CML): Evidence for cooperative antileukemic effects of imatinib and mcl-1 antisense oligonucleotides. *Blood* 2005a, 105, 3303-3311.

Aichberger, K. J.; Mayerhofer, M.; Krauth, M. T.; Vales, A.; Kondo, R.; Derdak, S.; Pickl, W. F.; Selzer, E.; Deininger, M.; Druker, B. J.; Sillaber, C.; Esterbauer, H.; Valent, P. Low-level expression of proapoptotic Bcl-2-interacting mediator in leukemic cells in patients with chronic myeloid leukemia: Role of BCR/ABL, characterization of underlying signaling pathways, and reexpression by novel pharmacologic compounds. *Cancer Res* 2005b, 65, 9436-9444.

Angers-Loustau, A.; Cote, J. F.; Charest, A.; Dowbenko, D.; Spencer, S.; Lasky, L. A.; Tremblay, M. L. Protein tyrosine phosphatase-PEST regulates focal adhesion disassembly, migration, and cytokinesis in fibroblasts. *J Cell Biol* 1999, 144, 1019-1031.

Baccarani, M.; Saglio, G.; Goldman, J.; Hochhaus, A.; Simonsson, B.; Appelbaum, F.; Apperley, J.; Cervantes, F.; Cortes, J.; Deininger, M.; Gratwohl, A.; Guilhot, F.; Horowitz, M.; Hughes, T.; Kantarjian, H.; Larson, R.; Niederwieser, D.; Silver, R.; Hehlmann, R. Evolving concepts in the management of chronic myeloid leukemia: Recommendations from an expert panel on behalf of the European LeukemiaNet. *Blood* 2006, 108, 1809-1820.

Barila, D.; Superti-Furga, G. An intramolecular SH3-domain interaction regulates c-Abl activity. *Nat Genet* 1998, 18, 280-282.

Baskaran, R.; Chiang, G. G.; Wang, J. Y. Identification of a binding site in c-Abl tyrosine kinase for the C-terminal repeated domain of RNA polymerase II. *Mol Cell Biol* 1996, 16, 3361-3369.

Baskaran, R.; Dahmus, M. E.; Wang, J. Y. Tyrosine phosphorylation of mammalian RNA polymerase II carboxyl-terminal domain. *Proc Natl Acad Sci USA* 1993, 90, 11167-11171.

Baskaran, R.; Wood, L. D.; Whitaker, L. L.; Canman, C. E.; Morgan, S. E.; Xu, Y.; Barlow, C.; Baltimore, D.; Wynshaw-Boris, A.; Kastan, M. B.; Wang, J. Y. Ataxia telangiectasia mutant protein activates c-Abl tyrosine kinase in response to ionizing radiation. *Nature* 1997, 387, 516-519.

Bear, J. E.; Svitkina, T. M.; Krause, M.; Schafer, D. A.; Loureiro, J. J.; Strasser, G. A.; Maly, I. V.; Chaga, O. Y.; Cooper, J. A.; Borisy, G. G.; Gertler, F. B. Antagonism between Ena/VASP proteins and actin filament capping regulates fibroblast motility. *Cell* 2002, 109, 509-521.

Berman, E.; Nicolaides, M.; Maki, R. G.; Fleisher, M.; Chanel, S.; Scheu, K.; Wilson, B. A.; Heller, G.; Sauter, N. P. Altered bone and mineral metabolism in patients receiving imatinib mesylate. *N Engl J Med* 2006, 354, 2006-2013.

Bose, R.; Holbert, M. A.; Pickin, K. A.; Cole, P. A. Protein tyrosine kinase-substrate interactions. *Curr Opin Struct Biol* 2006, 16, 668-675.

Branford, S.; Rudzki, Z.; Walsh, S.; Parkinson, I.; Grigg, A.; Szer, J.; Taylor, K.; Herrmann, R.; Seymour, J. F.; Arthur, C.; Joske, D.; Lynch, K.; Hughes, T. Detection of BCR-ABL mutations in patients with CML treated with imatinib is virtually always accompanied by clinical resistance, and mutations in the ATP phosphate-binding loop (P-loop) are associated with a poor prognosis. *Blood* 2003, 102, 276-283.

Brendel, C.; Scharenberg, C.; Dohse, M.; Robey, R. W.; Bates, S. E.; Shukla, S.; Ambudkar, S. V.; Wang, Y.; Wennemuth, G.; Burchert, A.; Boudriot, U.; Neubauer, A. Imatinib mesylate and nilotinib (AMN107) exhibit high-affinity interaction with ABCG2 on primitive hematopoietic stem cells. *Leukemia* 2007, 21, 1267-1275.

Buchdunger, E.; Zimmermann, J.; Mett, H.; Meyer, T.; Muller, M.; Druker, B. J.; Lydon, N. B. Inhibition of the Abl protein-tyrosine kinase *in vitro* and *in vivo* by a 2-phenylaminopyrimidine derivative. *Cancer Res* 1996, 56, 100-104.

Canaani, E.; Gale, R. P.; Steiner-Saltz, D.; Berrebi, A.; Aghai, E.; Januszewicz, E. Altered transcription of an oncogene in chronic myeloid leukaemia. *Lancet* 1984, 1, 593-595.

Cao, C.; Leng, Y.; Huang, W.; Liu, X.; Kufe, D. Glutathione peroxidase 1 is regulated by the c-Abl and Arg tyrosine kinases. *J Biol Chem* 2003a, 278, 39609-39614.

Cao, C.; Leng, Y.; Kufe, D. Catalase activity is regulated by c-Abl and Arg in the oxidative stress response. *J Biol Chem* 2003b, 278, 29667-29675.

Cao, C.; Leng, Y.; Li, C.; Kufe, D. Functional interaction between the c-Abl and Arg protein-tyrosine kinases in the oxidative stress response. *J Biol Chem* 2003c, 278, 12961-12967.

Cao, C.; Leng, Y.; Liu, X.; Yi, Y.; Li, P.; Kufe, D. Catalase is regulated by ubiquitination and proteosomal degradation. Role of the c-Abl and Arg tyrosine kinases. *Biochemistry* 2003d, 42, 10348-10353.

Cho, Y. J.; Hemmeryckx, B.; Groffen, J.; Heisterkamp, N. Interaction of Bcr/Abl with C3G, an exchange factor for the small GTPase Rap1, through the adapter protein Crkl. *Biochem Biophys Res Commun* 2005, 333, 1276-1283.

Choi, S. Y.; Kim, M. J.; Kang, C. M.; Bae, S.; Cho, C. K.; Soh, J. W.; Kim, J. H.; Kang, S.; Chung, H. Y.; Lee, Y. S, Lee, S. J. Activation of Bak and Bax through c-abl-protein kinase Cdelta-p38 MAPK signaling in response to ionizing radiation in human non-small cell lung cancer cells. *J Biol Chem* 2006, 281, 7049-7059.

Cohen, M. H.; Williams, G.; Johnson, J. R.; Duan, J.; Gobburu, J.; Rahman, A.; Benson, K.; Leighton, J.; Kim, S. K.; Wood, R.; Rothmann, M.; Chen, G.; Km U.; Staten, A. M.; Pazdur, R. Approval summary for imatinib mesylate capsules in the treatment of chronic myelogenous leukemia. *Clin Cancer Res* 2002, 8, 935-942.

Collins, S. J.; Kubonishi, I.; Miyoshi, I.; Groudine, M. T. Altered transcription of the c-abl oncogene in K-562 and other chronic myelogenous leukemia cells. *Science* 1984, 225, 72-74.

Cong, F.; Goff, S. P. c-Abl-induced apoptosis, but not cell cycle arrest, requires mitogen-activated protein kinase kinase 6 activation. *Proc Natl Acad Sci USA* 1999, 96, 13819-13824.

Cong, F.; Spencer, S.; Cote, J. F.; Wu, Y.; Tremblay, M. L.; Lasky, L. A.; Goff, S. P. Cytoskeletal protein PSTPIP1 directs the PEST-type protein tyrosine phosphatase to the c-Abl kinase to mediate Abl dephosphorylation. *Mol Cell* 2000, 6, 1413-1423.

Dai, Z.; Pendergast, A. M. Abi-2, a novel SH3-containing protein interacts with the c-Abl tyrosine kinase and modulates c-Abl transforming activity. *Genes Dev* 1995, 9, 2569-2582.

Dan, S.; Naito, M.; Seimiya, H.; Kizaki, A.; Mashima, T.; Tsuruo, T. Activation of c-Abl tyrosine kinase requires caspase activation and is not involved in JNK/SAPK activation during apoptosis of human monocytic leukemia U937 cells. *Oncogene* 1999, 18, 1277-1283.

de Klein, A.; van Kessel, A. G.; Grosveld, G.; Bartram, C. R.; Hagemeijer, A.; Bootsma, D.; Spurr, N. K.; Heisterkamp, N.; Groffen, J.; Stephenson, J. R. A cellular oncogene is translocated to the Philadelphia chromosome in chronic myelocytic leukaemia. *Nature* 1982, 300, 765-767.

Dias, S. S.; Milne, D. M.; Meek, D. W. c-Abl phosphorylates Hdm2 at tyrosine 276 in response to DNA damage and regulates interaction with ARF. *Oncogene* 2006, 25, 6666-6671.

Dierov, J.; Dierova, R.; Carroll, M. BCR/ABL translocates to the nucleus and disrupts an ATR-dependent intra-S phase checkpoint. *Cancer Cell* 2004, 5, 275-285.

Dorsch, M.; Goff, S. P. Increased sensitivity to apoptotic stimuli in c-abl-deficient progenitor B-cell lines. *Proc Natl Acad Sci USA* 1996, 93, 13131-13136.

Druker, B. J.; Guilhot, F.; O'Brien, S. G.; Gathmann, I.; Kantarjian, H.; Gattermann, N.; Deininger, M. W.; Silver, R. T.; Goldman, J. M.; Stone, R. M.; Cervantes, F.; Hochhaus, A.; Powell, B. L.; Gabrilove, J. L.; Rousselot, P.; Reiffers, J.; Cornelissen, J. J.; Hughes, T.; Agis, H.; Fischer, T.; Verhoef, G.; Shepherd, J.; Saglio, G.; Gratwohl, A.; Nielsen, J. L.; Radich, J. P.; Simonsson, B.; Taylor, K.; Baccarani, M.; So, C.; Letvak, L.; Larson, R. A. Five-year follow-up of patients receiving imatinib for chronic myeloid leukemia. *N Engl J Med* 2006, 355, 2408-2417.

Eden, S.; Rohatgi, R.; Podtelejnikov, A. V.; Mann, M.; Kirschner, M. W. Mechanism of regulation of WAVE1-induced actin nucleation by Rac1 and Nck. *Nature* 2002, 418, 790-793.

Furstoss, O.; Dorey, K.; Simon, V.; Barila, D.; Superti-Furga, G.; Roche, S. c-Abl is an effector of Src for growth factor-induced c-myc expression and DNA synthesis. *EMBO J* 2002, 21, 514-524.

Gao, B.; Lee, S. M.; Fang, D. The tyrosine kinase c-Abl protects c-Jun from ubiquitination-mediated degradation in T cells. *J Biol Chem* 2006, 281, 29711-29718.

Gesbert, F.; Griffin, J. D. Bcr/Abl activates transcription of the Bcl-X gene through STAT5. *Blood* 2000, 96, 2269-2276.

Goff, S. P.; Gilboa, E.; Witte, O. N.; Baltimore, D. Structure of the Abelson murine leukemia virus genome and the homologous cellular gene: Studies with cloned viral DNA. *Cell* 1980, 22, 777-785.

Goga, A.; Liu, X.; Hambuch, T. M.; Senechal, K.; Major, E.; Berk, A. J.; Witte, O. N.; Sawyers, C. L. p53 dependent growth suppression by the c-Abl nuclear tyrosine kinase. *Oncogene* 1995, 11, 791-799.

Goldberg, Z.; Vogt, S. R.; Berger, M.; Zwang, Y.; Perets, R.; Van Etten, R. A.; Oren, M.; Taya, Y.; Haupt, Y. Tyrosine phosphorylation of Mdm2 by c-Abl: Implications for p53 regulation. *EMBO J* 2002, 21, 3715-3727.

Gong, J. G.; Costanzo, A.; Yang, H. Q.; Melino, G.; Kaelin, W. G.; Jr. Levrero, M.; Wang, J. Y. The tyrosine kinase c-Abl regulates p73 in apoptotic response to cisplatin-induced DNA damage. *Nature* 1999, 399, 806-809.

Groffen, J.; Stephenson, J. R.; Heisterkamp, N.; de Klein, A.; Bartram, C. R.; Grosveld, G. Philadelphia chromosomal breakpoints are clustered within a limited region, bcr, on chromosome 22. *Cell* 1984, 36, 93-99.

Hagerkvist, R.; Mokhtari, D.; Lindholm, C.; Farnebo, F.; Mostoslavsky, G.; Mulligan, R. C.; Welsh, N.; Welsh, M. Consequences of Shb and c-Abl interactions for cell death in response to various stress stimuli. *Exp Cell Res* 2007, 313, 284-291.

Hardin, J. D.; Boast, S.; Mendelsohn, M.; de los, S. K.; Goff, S. P. Transgenes encoding both type I and type IV c-abl proteins rescue the lethality of c-abl mutant mice. *Oncogene* 1996a, 12, 2669-2677.

Hardin, J. D.; Boast, S.; Schwartzberg, P. L.; Lee, G.; Alt, F. W.; Stall, A. M.; Goff, S. P. Bone marrow B lymphocyte development in c-abl-deficient mice. *Cell Immunol* 1995, 165, 44-54.

Hardin, J. D.; Boast, S.; Schwartzberg, P. L.; Lee, G.; Alt, F. W.; Stall, A. M.; Goff, S. P. Abnormal peripheral lymphocyte function in c-abl mutant mice. *Cell Immunol* 1996b, 172, 100-107.

Harrison, S. C. Variation on an Src-like theme. *Cell* 2003, 112, 737-740.

He, Y.; Wertheim, J. A.; Xu, L.; Miller, J. P.; Karnell, F. G.; Choi, J. K.; Ren, R.; Pear, W. S. The coiled-coil domain and Tyr177 of bcr are required to induce a murine chronic myelogenous leukemia-like disease by bcr/abl. *Blood* 2002, 99, 2957-2968.

Heinrich, M. C.; Griffith, D. J.; Druker, B. J.; Wait, C. L.; Ott, K. A.; Zigler, A. J. Inhibition of c-kit receptor tyrosine kinase activity by STI 571, a selective tyrosine kinase inhibitor. *Blood* 2000, 96, 925-932.

Hemmeryckx, B.; van Wijk, A.; Reichert, A.; Kaartinen, V.; de Jong, R.; Pattengale, P. K.; Gonzalez-Gomez, I.; Groffen, J.; Heisterkamp, N. Crkl enhances leukemogenesis in BCR/ABL P190 transgenic mice. *Cancer Res* 2001, 61:1398-1405.

Hirao, N.; Sato, S.; Gotoh, T.; Maruoka, M.; Suzuki, J.; Matsuda, S.; Shishido, T.; Tani, K. NESH (Abi-3) is present in the Abi/WAVE complex but does not promote c-Abl-mediated phosphorylation. *FEBS Lett* 2006, 580, 6464-6470.

Kain, K. H.; Klemke, R. L. Inhibition of cell migration by Abl family tyrosine kinases through uncoupling of Crk-CAS complexes. *J Biol Chem* 2001, 276, 16185-16192.

Kawai, H.; Nie, L.; Yuan, Z. M. Inactivation of NF-kappaB-dependent cell survival, a novel mechanism for the proapoptotic function of c-Abl. *Mol Cell Biol* 2002, 22, 6079-6088.

Kennedy, S. G.; Wagner, A. J.; Conzen, S. D.; Jordan, J.; Bellacosa, A.; Tsichlis, P. N.; Hay, N. The PI 3-kinase/Akt signaling pathway delivers an anti-apoptotic signal. *Genes Dev* 1997, 11, 701-713.

Kerkela, R.; Grazette, L.; Yacobi, R.; Iliescu, C.; Patten, R.; Beahm, C.; Walters, B.; Shevtsov, S.; Pesant, S.; Clubb, F. J.; Rosenzweig, A.; Salomon, R. N.; Van Etten, R. A.; Alroy, J.; Durand, J. B.; Force, T. Cardiotoxicity of the cancer therapeutic agent imatinib mesylate. *Nat Med* 2006, 12, 908-916.

Kharbanda, S.; Bharti, A.; Pei, D.; Wang, J.; Pandey, P.; Ren, R.; Weichselbaum, R.; Walsh, C. T.; Kufe, D. The stress response to ionizing radiation involoves c-Abl-dependent phosphorylation of SHPTP1. *Proc Natl Acad Sci USA* 1996, 93, 6898-6901.

Kharbanda, S.; Pandey, P.; Jin, S.; Inoue, S.; Bharti, A.; Yuan, Z. M.; Weichselbaum, R.; Weaver, D.; Kufe, D. Functional interaction between DNA-PK and c-Abl in response to DNA damage. *Nature* 1997, 386, 732-735.

Kharbanda, S.; Pandey, P.; Ren, R.; Mayer, B.; Zon, L.; Kufe, D. c-Abl activation regulates induction of the SEK1/stress-activated protein kinase pathway in the cellular response to 1-beta-D-arabinofuranosylcytosine. *J Biol Chem* 1995, 270, 30278-30281.

Kipreos, E. T.; Wang, J. Y. Cell cycle-regulated binding of c-Abl tyrosine kinase to DNA. *Science* 1992, 256, 382-385.

Klemke, R. L.; Leng, J.; Molander, R.; Brooks, P. C.; Vuori, K.; Cheresh, D. A. CAS/Crk coupling serves as a "molecular switch" for induction of cell migration. *J Cell Biol* 1998, 140, 961-972.

Koleske, A. J.; Gifford, A. M.; Scott, M. L.; Nee, M.; Bronson, R. T.; Miczek, K. A.; Baltimore, D. Essential roles for the Abl and Arg tyrosine kinases in neurulation. *Neuron* 1998, 21, 1259-1272.

Konopka, J. B.; Watanabe, S. M.; Singer, J. W.; Collins, S. J.; Witte, O. N. Cell lines and clinical isolates derived from Ph1-positive chronic myelogenous leukemia patients express c-abl proteins with a common structural alteration. *Proc Natl Acad Sci USA* 1985, 82, 1810-1814.

Kumar, S.; Bharti, A.; Mishra, N. C.; Raina, D.; Kharbanda, S.; Saxena, S.; Kufe, D. Targeting of the c-Abl tyrosine kinase to mitochondria in the necrotic cell death response to oxidative stress. *J Biol Chem* 2001, 276, 17281-17285.

Lam, Q. L.; Lo, C. K.; Zheng, B. J.; Ko, K. H.; Osmond, D. G.; Wu, G. E.; Rottapel, R.; Lu, L. Impaired V(D)J recombination and increased apoptosis among B cell precursors in the bone marrow of c-Abl-deficient mice. *Int Immunol* 2007, 19, 267-276.

Lasfer, M.; Davenne, L.; Vadrot, N.; Alexia, C.; Sadji-Ouatas, Z.; Bringuier, A. F.; Feldmann, G.; Pessayre, D.; Reyl-Desmars, F. Protein kinase PKC delta and c-Abl are required for mitochondrial apoptosis induction by genotoxic stress in the absence of p53, p73 and Fas receptor. *FEBS Lett* 2006, 580, 2547-2552.

Leng, Y.; Zhang, J.; Badour, K.; Arpaia, E.; Freeman, S.; Cheung, P.; Siu, M.; Siminovitch, K. Abelson-interactor-1 promotes WAVE2 membrane translocation and Abelson-mediated tyrosine phosphorylation required for WAVE2 activation. *Proc Natl Acad Sci USA* 2005, 102, 1098-1103.

Lewis, J. M.; Baskaran, R.; Taagepera, S.; Schwartz, M. A.; Wang, J. Y. Integrin regulation of c-Abl tyrosine kinase activity and cytoplasmic-nuclear transport. *Proc Natl Acad Sci USA* 1996, 93, 15174-15179.

Li, B. Inhibition of Abl kinase by intramolecular and intermolecular interactions: a lesson from struction studies and CML therapy. *Current Enzyme Inhibition* 2006, 2, 135-146.

Li, B.; Boast, S.; de los, S. K.; Schieren, I.; Quiroz, M.; Teitelbaum, S. L.; Tondravi, M. M.; Goff, S. P. Mice deficient in Abl are osteoporotic and have defects in osteoblast maturation. *Nat Genet* 2000, 24, 304-308.

Li, B.; Wang, X.; Rasheed, N.; Hu, Y.; Boast, S.; Ishii, T.; Nakayama, K.; Nakayama, K. I.; Goff, S. P. Distinct roles of c-Abl and Atm in oxidative stress response are mediated by protein kinase C delta. *Genes Dev* 2004, 18, 1824-1837.

Li, Y.; Clough, N.; Sun, X.; Yu, W.; Abbott, B. L.; Hogan, C. J.; Dai, Z. Bcr-Abl induces abnormal cytoskeleton remodeling, beta1 integrin clustering and increased cell adhesion to fibronectin through the Abl interactor 1 pathway. *J Cell Sci* 2007, 120, 1436-1446.

Liu, X.; Huang, W.; Li, C.; Li, P.; Yuan, J.; Li, X.; Qiu, X. B. Ma, Q.; Cao, C. Interaction between c-Abl and Arg tyrosine kinases and proteasome subunit PSMA7 regulates proteasome degradation. *Mol Cell* 2006, 22, 317-327.

Machuy, N.; Campa, F.; Thieck, O.; Rudel, T. c-Abl-binding protein interacts with p21-activated kinase 2 (PAK-2) to regulate PDGF-induced membrane ruffles. *J Mol Biol* 2007, 370, 620-632.

Machuy, N.; Rajalingam, K.; Rudel, T. Requirement of caspase-mediated cleavage of c-Abl during stress-induced apoptosis. *Cell Death Differ* 2004, 11, 290-300.

Manning, G.; Whyte, D. B.; Martinez, R.; Hunter, T.; Sudarsanam, S. The protein kinase complement of the human genome. *Science* 2002, 298, 1912-1934.

Mantovani, F.; Piazza, S.; Gostissa, M.; Strano, S.; Zacchi, P.; Mantovani, R.; Blandino, G.; Del Sal, G. Pin1 links the activities of c-Abl and p300 in regulating p73 function. *Mol Cell* 2004, 14, 625-636.

Martinelli, G.; Soverini, S.; Rosti, G.; Cilloni, D.; Baccarani, M. New tyrosine kinase inhibitors in chronic myeloid leukemia. *Haematologica* 2005, 90, 534-541.

Master, Z.; Tran, J.; Bishnoi, A.; Chen, S. H.; Ebos, J. M.; Van Slyke, P.; Kerbel, R. S.; Dumont, D. J. Dok-R binds c-Abl and regulates Abl kinase activity and mediates cytoskeletal reorganization. *J Biol Chem* 2003, 278, 30170-30179.

Mattiuzzi, G. N.; Cortes, J. E.; Talpaz, M.; Reuben, J.; Rios, M. B.; Shan, J.; Kontoyiannis, D.; Giles, F. J.; Raad, I.; Verstovsek, S.; Ferrajoli, A.; Kantarjian, H. M. Development of Varicella-Zoster virus infection in patients with chronic myelogenous leukemia treated with imatinib mesylate. *Clin Cancer Res* 2003, 9, 976-980.

Mayerhofer, M.; Aichberger, K. J.; Florian, S.; Krauth, M. T.; Hauswirth, A. W.; Derdak, S.; Sperr, W. R.; Esterbauer, H.; Wagner, O.; Marosi, C.; Pickl, W. F.; Deininger, M.; Weisberg, E.; Druker, B. J.; Griffin ,J. D.; Sillaber, C.; Valent, P. Identification of mTOR as a novel bifunctional target in chronic myeloid leukemia: Dissection of growth-inhibitory and VEGF-suppressive effects of rapamycin in leukemic cells. *FASEB J* 2005, 19, 960-962.

Michor, F.; Hughes, T. P.; Iwasa, Y.; Branford, S.; Shah, N. P.; Sawyers, C. L.; Nowak, M. A. Dynamics of chronic myeloid leukaemia. *Nature* 2005, 435, 1267-1270.

Million, R. P.; Van Etten, R. A. The Grb2 binding site is required for the induction of chronic myeloid leukemia-like disease in mice by the Bcr/Abl tyrosine kinase. *Blood* 2000, 96, 664-670.

Mody, N.; Parhami, F.; Sarafian, T. A.; Demer, L. L. Oxidative stress modulates osteoblastic differentiation of vascular and bone cells. *Free Radic Biol Med* 2001, 31, 509-519.

Muller, R.; Slamon, D. J.; Tremblay, J. M.; Cline, M. J.; Verma, I. M. Differential expression of cellular oncogenes during pre- and postnatal development of the mouse. *Nature* 1982, 299, 640-644.

Nagano, K.; Itagaki, C.; Izumi, T.; Nunomura, K.; Soda, Y.; Tani, K.; Takahashi, N.; Takenawa, T.; Isobe, T. Rb plays a role in survival of Abl-dependent human tumor cells as a downstream effector of Abl tyrosine kinase. *Oncogene* 2006, 25, 493-502.

Nagar, B.; Hantschel, O.; Seeliger, M.; Davies, J. M.; Weis, W. I.; Superti-Furga, G.; Kuriyan, J. Organization of the SH3-SH2 unit in active and inactive forms of the c-Abl tyrosine kinase. *Mol Cell* 2006, 21, 787-798.

Nagar, B.; Hantschel, O.; Young, M. A.; Scheffzek, K.; Veach, D.; Bornmann, W.; Clarkson, B.; Superti-Furga, G.; Kuriyan, J. Structural basis for the autoinhibition of c-Abl tyrosine kinase. *Cell* 2003, 112, 859-871.

Nie, Y.; Li, H. H.; Bula, C. M.; Liu, X. Stimulation of p53 DNA binding by c-Abl requires the p53 C terminus and tetramerization. *Mol Cell Biol* 2000, 20, 741-748.

Nieborowska-Skorska, M.; Hoser, G.; Rink, L.; Malecki, M.; Kossev, P.; Wasik, M. A.; Skorski, T. Id1 transcription inhibitor-matrix metalloproteinase 9 axis enhances invasiveness of the breakpoint cluster region/abelson tyrosine kinase-transformed leukemia cells. *Cancer Res* 2006, 66, 4108-4116.

Nowell, P. C.; Hungerford, D. A. Chromosome studies on normal and leukemic human leukocytes. *J Natl Cancer Inst* 1960, 25, 85-109.

O'Neill, A. J.; Cotter, T. G.; Russell, J. M.; Gaffney, E. F. Abl expression in human fetal and adult tissues, tumours, and tumour microvessels. *J Pathol* 1997, 183, 325-329.

Okuda, K.; Weisberg, E.; Gilliland, D. G.; Griffin, J. D. ARG tyrosine kinase activity is inhibited by STI571. *Blood* 2001, 97, 2440-2448.

Pearse, R. N.; Kawabe, T.; Bolland, S.; Guinamard, R.; Kurosaki, T.; Ravetch, J. V. SHIP recruitment attenuates Fc gamma RIIB-induced B cell apoptosis. *Immunity* 1999, 10, 753-760.

Peters, D. G.; Hoover, R. R.; Gerlach, M. J.; Koh, E. Y.; Zhang, H.; Choe, K.; Kirschmeier, P.; Bishop, W. R.; Daley, G. Q. Activity of the farnesyl protein transferase inhibitor SCH66336 against BCR/ABL-induced murine leukemia and primary cells from patients with chronic myeloid leukemia. *Blood* 2001, 97, 1404-1412.

Plattner, R.; Irvin, B. J.; Guo, S.; Blackburn, K.; Kazlauskas, A.; Abraham, R. T.; York, J. D.; Pendergast, A. M. A new link between the c-Abl tyrosine kinase and phosphoinositide signalling through PLC-gamma1. *Nat Cell Biol* 2003, 5, 309-319.

Plattner, R.; Kadlec, L.; DeMali, K. A.; Kazlauskas, A.; Pendergast, A. M. c-Abl is activated by growth factors and Src family kinases and has a role in the cellular response to PDGF. *Genes Dev* 1999, 13, 2400-2411.

Rabstein, L. S.; Gazdar, A. F.; Chopra, H. C.; Abelson, H. T. Early morphological changes associated with infection by a murine nonthymic lymphatic tumor virus. *J Natl Cancer Inst* 1971, 46, 481-491.

Radha, V.; Rajanna, A.; Mitra, A.; Rangaraj, N.; Swarup, G. C3G is required for c-Abl-induced filopodia and its overexpression promotes filopodia formation. *Exp Cell Res* 2007, 313, 2476-2492.

Raina, D.; Ahmad, R.; Kumar, S.; Ren, J.; Yoshida, K.; Kharbanda, S.; Kufe, D. MUC1 oncoprotein blocks nuclear targeting of c-Abl in the apoptotic response to DNA damage. *EMBO J* 2006, 25, 3774-3783.

Raina, D.; Pandey, P.; Ahmad, R.; Bharti, A.; Ren, J.; Kharbanda, S.; Weichselbaum, R.; Kufe, D. c-Abl tyrosine kinase regulates caspase-9 autocleavage in the apoptotic response to DNA damage. *J Biol Chem* 2005, 280, 11147-11151.

Rao, J.; Li, N. Microfilament actin remodeling as a potential target for cancer drug development. *Curr Cancer Drug Targets* 2004, 4, 345-354.

Ren, R.; Ye, Z. S.; Baltimore, D. Abl protein-tyrosine kinase selects the Crk adapter as a substrate using SH3-binding sites. *Genes Dev* 1994, 8, 783-795.

Renshaw, M. W.; Capozza, M. A.; Wang, J. Y. Differential expression of type-specific c-abl mRNAs in mouse tissues and cell lines. *Mol Cell Biol* 1988, 8, 4547-4551.

Rivera, G. M.; Briceno, C. A.; Takeshima, F.; Snapper, S. B.; Mayer, B. J. Inducible clustering of membrane-targeted SH3 domains of the adaptor protein Nck triggers localized actin polymerization. *Curr Biol* 2004, 14, 11-22.

Rowley, J. D. Letter: A new consistent chromosomal abnormality in chronic myelogenous leukaemia identified by quinacrine fluorescence and Giemsa staining. *Nature* 1973, 243, 290-293.

Sanchez-Prieto, R.; Sanchez-Arevalo, V. J.; Servitja, J. M.; Gutkind, J. S. Regulation of p73 by c-Abl through the p38 MAP kinase pathway. *Oncogene* 2002, 21, 974-979.

Sawyers, C. L.; McLaughlin, J.; Goga, A.; Havlik, M.; Witte, O. The nuclear tyrosine kinase c-Abl negatively regulates cell growth. *Cell* 1994, 77, 121-131.

Scher, C. D.; Siegler, R. Direct transformation of 3T3 cells by Abelson murine leukaemia virus. *Nature* 1975, 253, 729-731.

Schwartzberg, P. L.; Goff, S. P.; Robertson, E. J. Germ-line transmission of a c-abl mutation produced by targeted gene disruption in ES cells. *Science* 1989, 246, 799-803.

Schwartzberg, P. L.; Stall, A. M.; Hardin, J. D.; Bowdish, K. S.; Humaran, T.; Boast, S.; Harbison, M. L.; Robertson, E. J.; Goff, S. P. Mice homozygous for the ablm1 mutation show poor viability and depletion of selected B and T cell populations. *Cell* 1991, 65, 1165-1175.

Shafman, T.; Khanna, K. K.; Kedar, P.; Spring, K.; Kozlov, S.; Yen, T.; Hobson, K.; Gatei, M.; Zhang, N.; Watters, D.; Egerton, M.; Shiloh, Y.; Kharbanda, S.; Kufe, D.; Lavin, M. F. Interaction between ATM protein and c-Abl in response to DNA damage. *Nature* 1997, 387, 520-523.

Shtivelman, E.; Lifshitz, B.; Gale, R. P.; Canaani, E. Fused transcript of abl and bcr genes in chronic myelogenous leukaemia. *Nature* 1985, 315, 550-554.

Sini, P.; Cannas, A.; Koleske, A. J.; Di Fiore, P. P.; Scita, G. Abl-dependent tyrosine phosphorylation of Sos-1 mediates growth-factor-induced Rac activation. *Nat Cell Biol* 2004, 6, 268-274.

Skorski, T.; Bellacosa, A.; Nieborowska-Skorska, M.; Majewski, M.; Martinez, R.; Choi, J. K.; Trotta, R.; Wlodarski, P.; Perrotti, D.; Chan, T. O.; Wasik, M. A.; Tsichlis, P. N.; Calabretta, B. Transformation of hematopoietic cells by BCR/ABL requires activation of a PI-3k/Akt-dependent pathway. *EMBO J* 1997, 16, 6151-6161.

Spencer, S.; Dowbenko, D.; Cheng, J.; Li, W.; Brush, J.; Utzig, S.; Simanis, V.; Lasky, L. A. PSTPIP: A tyrosine phosphorylated cleavage furrow-associated protein that is a substrate for a PEST tyrosine phosphatase. *J Cell Biol* 1997, 138, 845-860.

Stuart, J. R.; Gonzalez, F. H.; Kawai, H.; Yuan, Z. M. c-Abl interacts with the WAVE2 signaling complex to induce membrane ruffling and cell spreading. *J Biol Chem* 2006, 281, 31290-31297.

Sun, J.; Zhao, J.; Schwartz, M. A.; Wang, J. Y.; Wiedmer, T.; Sims, P. J. c-Abl tyrosine kinase binds and phosphorylates phospholipid scramblase 1. *J Biol Chem* 2001, 276, 28984-28990.

Sun, X.; Majumder, P.; Shioya, H.; Wu, F.; Kumar, S.; Weichselbaum, R.; Kharbanda, S.; Kufe, D. Activation of the cytoplasmic c-Abl tyrosine kinase by reactive oxygen species. *J Biol Chem* 2000a, 275, 17237-17240.

Sun, X.; Wu, F.; Datta, R.; Kharbanda, S.; Kufe, D. Interaction between protein kinase C delta and the c-Abl tyrosine kinase in the cellular response to oxidative stress. *J Biol Chem* 2000b, 275, 7470-7473.

Taagepera, S.; McDonald, D.; Loeb, J. E.; Whitaker, L. L.; McElroy, A. K.; Wang, J. Y.; Hope, T. J. Nuclear-cytoplasmic shuttling of C-ABL tyrosine kinase. *Proc Natl Acad Sci USA* 1998, 95, 7457-7462.

Takenawa, T.; Suetsugu, S. The WASP-WAVE protein network: Connecting the membrane to the cytoskeleton. *Nat Rev Mol Cell Biol* 2007, 8, 37-48.

Tani, K.; Sato, S.; Sukezane, T.; Kojima, H.; Hirose, H.; Hanafusa, H.; Shishido, T. Abl interactor 1 promotes tyrosine 296 phosphorylation of mammalian enabled (Mena) by c-Abl kinase. *J Biol Chem* 2003, 278, 21685-21692.

Theis, S.; Roemer, K. c-Abl tyrosine kinase can mediate tumor cell apoptosis independently of the Rb and p53 tumor suppressors. *Oncogene* 1998, 17, 557-564.

Ting, A. Y.; Kain, K. H.; Klemke, R. L.; Tsien, R. Y. Genetically encoded fluorescent reporters of protein tyrosine kinase activities in living cells. *Proc Natl Acad Sci USA* 2001, 98, 15003-15008.

Tsuruta, F.; Sunayama, J.; Mori, Y.; Hattori, S.; Shimizu, S.; Tsujimoto, Y.; Yoshioka, K.; Masuyama, N.; Gotoh, Y. JNK promotes Bax translocation to mitochondria through phosphorylation of 14-3-3 proteins. *EMBO J* 2004, 23, 1889-1899.

Tybulewicz, V. L.; Crawford, C. E.; Jackson, P. K.; Bronson, R. T.; Mulligan, R. C. Neonatal lethality and lymphopenia in mice with a homozygous disruption of the c-abl proto-oncogene. *Cell* 1991, 65, 1153-1163.

Tzeng, S. J.; Bolland, S.; Inabe, K.; Kurosaki, T.; Pierce, S. K. The B cell inhibitory Fc receptor triggers apoptosis by a novel c-Abl family kinase-dependent pathway. *J Biol Chem* 2005, 280, 35247-35254.

Van Etten, R. A.; Jackson, P.; Baltimore, D. The mouse type IV c-abl gene product is a nuclear protein, and activation of transforming ability is associated with cytoplasmic localization. *Cell* 1989, 58, 669-678.

Van Etten, R. A.; Jackson, P. K.; Baltimore, D.; Sanders, M. C.; Matsudaira, P. T.; Janmey, P. A. The COOH terminus of the c-Abl tyrosine kinase contains distinct F- and G-actin binding domains with bundling activity. *J Cell Biol* 1994, 124, 325-340.

Wang, J. Y. Regulation of cell death by the Abl tyrosine kinase. *Oncogene* 2000, 19, 5643-5650.

Wang, X.; Kua, H. Y.; Hu, Y.; Guo, K.; Zeng, Q.; Wu, Q.; Ng, H. H.; Karsenty, G.; de Crombrugghe, B.; Yeh, J.; Li, B. p53 functions as a negative regulator of osteoblastogenesis, osteoblast-dependent osteoclastogenesis, and bone remodeling. *J Cell Biol* 2006, 172, 115-125.

Wei, G.; Li, A. G.; Liu, X. Insights into selective activation of p53 DNA binding by c-Abl. *J Biol Chem* 2005, 280, 12271-12278.

Welch, P. J.; Wang, J. Y. A C-terminal protein-binding domain in the retinoblastoma protein regulates nuclear c-Abl tyrosine kinase in the cell cycle. *Cell* 1993, 75, 779-790.

Wen, S. T.; Jackson, P. K.; Van Etten, R. A. The cytostatic function of c-Abl is controlled by multiple nuclear localization signals and requires the p53 and Rb tumor suppressor gene products. *EMBO J* 1996, 15, 1583-1595.

Wen, S. T.; Van Etten, R. A. The PAG gene product, a stress-induced protein with antioxidant properties, is an Abl SH3-binding protein and a physiological inhibitor of c-Abl tyrosine kinase activity. *Genes Dev* 1997, 11, 2456-2467.

Wilkes, M. C.; Leof, E. B. Transforming growth factor beta activation of c-Abl is independent of receptor internalization and regulated by phosphatidylinositol 3-kinase and PAK2 in mesenchymal cultures. *J Biol Chem* 2006, 281, 27846-27854.

Witte, O. N.; Rosenberg, N.; Paskind, M.; Shields, A.; Baltimore, D. Identification of an Abelson murine leukemia virus-encoded protein present in transformed fibroblast and lymphoid cells. *Proc Natl Acad Sci USA* 1978, 75, 2488-2492.

Witte, O. N.; Rosenberg, N. E.; Baltimore, D. A normal cell protein cross-reactive to the major Abelson murine leukaemia virus gene product. *Nature* 1979, 281, 396-398.

Wong, S.; Witte, O. N. The BCR-ABL story: bench to bedside and back. *Annu Rev Immunol* 2004, 22, 247-306.

Woodring, P. J.; Hunter, T.; Wang, J. Y. Regulation of F-actin-dependent processes by the Abl family of tyrosine kinases. *J Cell Sci* 2003, 116, 2613-2626.

Woodring, P. J.; Litwack, E. D.; O'Leary, D. D.; Lucero, G. R.; Wang, J. Y.; Hunter, T. Modulation of the F-actin cytoskeleton by c-Abl tyrosine kinase in cell spreading and neurite extension. *J Cell Biol* 2002, 156, 879-892.

Woodring, P. J.; Meisenhelder, J.; Johnson, S. A.; Zhou, G. L.; Field, J.; Shah, K.; Bladt, F.; Pawson, T.; Niki, M.; Pandolfi, P. P.; Wang, J. Y.; Hunter, T. c-Abl phosphorylates Dok1 to promote filopodia during cell spreading. *J Cell Biol* 2004, 165, 493-503.

Yoshida, K.; Komatsu, K.; Wang, H. G.; Kufe, D. c-Abl tyrosine kinase regulates the human Rad9 checkpoint protein in response to DNA damage. *Mol Cell Biol* 2002, 22, 3292-3300.

Yoshida, K.; Miki, Y. Enabling death by the Abl tyrosine kinase: mechanisms for nuclear shuttling of c-Abl in response to DNA damage. *Cell Cycle* 2005, 4, 777-779.

Yoshida, K.; Yamaguchi, T.; Natsume, T.; Kufe, D.; Miki, Y. JNK phosphorylation of 14-3-3 proteins regulates nuclear targeting of c-Abl in the apoptotic response to DNA damage. *Nat Cell Biol* 2005, 7, 278-285.

Yuan, Z. M.; Huang, Y.; Ishiko, T.; Kharbanda, S.; Weichselbaum, R.; Kufe, D. Regulation of DNA damage-induced apoptosis by the c-Abl tyrosine kinase. *Proc Natl Acad Sci USA* 1997a, 94, 1437-1440.

Yuan, Z. M.; Huang, Y.; Ishiko, T.; Nakada, S.; Utsugisawa, T.; Kharbanda, S.; Wang, R.; Sung, P.; Shinohara, A.; Weichselbaum, R.; Kufe, D. Regulation of Rad51 function by c-Abl in response to DNA damage. *J Biol Chem* 1998a, 273, 3799-3802.

Yuan, Z. M.; Huang, Y.; Whang, Y.; Sawyers, C.; Weichselbaum, R.; Kharbanda, S.; Kufe, D. Role for c-Abl tyrosine kinase in growth arrest response to DNA damage. *Nature* 1996, 382, 272-274.

Yuan, Z. M.; Shioya, H.; Ishiko, T.; Sun, X.; Gu, J.; Huang, Y. Y.; Lu, H.; Kharbanda, S.; Weichselbaum, R.; Kufe, D. p73 is regulated by tyrosine kinase c-Abl in the apoptotic response to DNA damage. *Nature* 1999, 399, 814-817.

Yuan, Z. M.; Utsugisawa, T.; Huang, Y.; Ishiko, T.; Nakada, S.; Kharbanda, S.; Weichselbaum, R.; Kufe, D. Inhibition of phosphatidylinositol 3-kinase by c-Abl in the genotoxic stress response. *J Biol Chem* 1997b, 272, 23485-23488.

Yuan, Z. M.; Utsugisawa, T.; Ishiko, T.; Nakada, S.; Huang, Y.; Kharbanda, S.; Weichselbaum, R.; Kufe, D. Activation of protein kinase C delta by the c-Abl tyrosine kinase in response to ionizing radiation. *Oncogene* 1998b, 16, 1643-1648.

Zha, J.; Harada, H.; Yang, E.; Jockel, J.; Korsmeyer, S. J. Serine phosphorylation of death agonist BAD in response to survival factor results in binding to 14-3-3 not BCL-X(L). *Cell* 1996, 87, 619-628.

Zhou, G. L.; Zhuo, Y.; King, C. C.; Fryer, B. H.; Bokoch, G. M.; Field, J. Akt phosphorylation of serine 21 on Pak1 modulates Nck binding and cell migration. *Mol Cell Biol* 2003, 23, 8058-8069.

Zhu, G.; Decker, S. J.; Mayer, B. J.; Saltiel, A. R. Direct analysis of the binding of the abl Src homology 2 domain to the activated epidermal growth factor receptor. *J Biol Chem* 1993, 268, 1775-1779.

Zipfel, P. A.; Grove, M.; Blackburn, K.; Fujimoto, M.; Tedder, T. F.; Pendergast, A. M. The c-Abl tyrosine kinase is regulated downstream of the B cell antigen receptor and interacts with CD19. *J Immunol* 2000, 165, 6872-6879.

Zipfel, P. A.; Zhang, W.; Quiroz, M.; Pendergast, A. M. Requirement for Abl kinases in T cell receptor signaling. *Curr Biol* 2004, 14, 1222-1231.

Zukerberg, L. R.; Patrick, G. N.; Nikolic, M.; Humbert, S.; Wu, C. L.; Lanier, L. M.; Gertler, F. B.; Vidal, M.; Van Etten, R. A.; Tsai, L. H. Cables links Cdk5 and c-Abl and facilitates Cdk5 tyrosine phosphorylation, kinase upregulation, and neurite outgrowth. *Neuron* 2000, 26, 633-646.

CHAPTER 4

A THEORY ON THE MOLECULAR NATURE OF POST-ZYGOTIC REPRODUCTIVE ISOLATION

FRANCISCO PROSDOCIMI

CONTENTS

4.1   Introduction 96
4.2   The Hypothesis.. 97
4.3   Embryogenesis and Impaired Fecundity 100
4.4   Evolutionary Origin of Incompatibilities 101
4.5   Considerations about Mosaic Evolution................................110
4.6   Conclusion..112
Acknowledgements..115
Keywords 115
References..116

4.1 INTRODUCTION

The molecular nature of reproductive isolation (RI) is classically studied in hybrids of *Drosophila* species. Trying to understand the causes hybrid sterility and/or inviability, researchers have evidenced specific molecular incompatibilities between genes that avoid reproduction with fertile offspring between hybrids of different fly species. The present paper presents a new model on the origin and evolution of RI based on the evolution of Dobzhansky-Muller heteromers. It is suggested that RI is achieved when alleles from genes forming heteromeric protein complexes, mainly those acting during embryogenesis or gametogenesis, accumulate mutations in such a way that some heteromeric proteins become unable to interact with others in the multimeric complex. The failure in the production of a functional heteromer interrupts the developmental process and produces inviable or sterile offspring, even inside sexual-reproducing biological species. This is one of the most remarkable derivations of the Dobzhansky-Muller allele evolution model presented here: RI is common even inside populational interbreeding groups and so its molecular nature is suggested to be studied, from now on, also inside these groups. However, evidences of intrapopulation RI are today scarce since it seems that no one has suggested this hypothesis earlier and it is so that the data available are not formatted for the ones considering it. Even though, humans living in a monogamous society have shown to be a good source of data for the study of intrapopulation RI. Reproductive medicine data about a so-called "impaired fecundity" fertility problem in humans has shown to be a strong example of the theory here presented. At last, the evolution of heteromeric Dobzhansky-Muller-like allele incompatibilities may substance a gradual theory of speciation and provides a glimpse about the molecular basis of mosaic evolution, outbreeding depression and sympatric speciation.

4.1.1 INFERTILITY AND IMPAIRED FECUNDITY

In humans, a status of infertility is frequently characterized as a failure to achieve pregnancy after 12 months or more of frequent and unprotected sexual intercourse (CDC, 2005; Jose-Miller *et al.*, 2007). According to the American Academy of Family Physicians, 10-15 % of couples in the United States are infertile (AAFP, 2007). Consequently, after this unsuccessful reproductive year, physicians diagnose the couple as infertile and suggest an evaluation of male and/or female fertility in order to verify the specific causes of the problem. In the last years, infertility has been extensively studied and many factors bringing to male or female genetic infertility have been discovered. Studying spermatogenesis in knockout mice, researchers have identified more than 300 genes necessary to the normal production of spermatozoids (Miki *et al.*, 2004; Kumar, 2005; Kuo *et al.*, 2005; Escalier, 2006). Malfunction of these genes may cause azoospermia or oligospermia in men. Moreover, deletions in three different spermatogenesis loci in the Y chromosome–known as "azoospermia factors": AZFa, b, and c–cause severe testiculopathy, producing male infertility (Najmabadi *et al.*, 1996; Martinez *et al.*, 2000; Maurer and Simoni, 2000; Foresta *et al.*, 2001). Considering women, infertility causes are mainly related to obstructions in the fallopian tubes (Palagiano, 2005), endometriosis (Honore, 1997; Alpay *et al.*, 2006) or ovarian dysfunction (Hull *et al.*, 1985; Jose-Miller *et al.*, 2007). Moreover, gene knockout studies in mice have also identified about 80 genes related to female fertility. Alterations in these genes may influence many important processes, such like ovarian function, oocyte fertilization, embryonic development, implantation, and difficulties to delivery (Naz and Rajesh, 2005). Although dysfunctions in parents' genes is clearly relevant when considering infertility, the present work focus in the molecular aspects of a *couple* being able or not to produce babies due to a *molecular (in)compatibility among parents*. It also tries to generalize the observations when considering other sexual-reproducing species.

A couple is considered fecund when both the woman and her husband (or cohabiting partner) have no known barrier to having a child (CDC, 2005). It is interesting to realize that many of these *fecund couples* may be diagnosed as infertile, characterizing a fertility problem known as *impaired fecundity* (IF) (Chandra and Stephen, 1998). An analysis of the US women population in 2002 have identified IF in 12 % of them (~7.3 million women according to CDC (2005). Another interesting data reveals that IF levels are higher among married women, representing about 15 % of all married ones (~4.3 million women) (CDC, 2005), suggesting incompatibility between partners to achieve pregnancy. The precise causes of IF are unknown and they are frequently though to be a result of some woman specific physical barrier to getting pregnant or carry a baby to term (CDC, 2005). But, an alternative hypothetical explanation that takes on account of the molecular (in)compatibility among partners may also be proposed to explain a number of IF cases.

4.2   THE HYPOTHESIS

The present work aims to propose a gradualistic evolutionary hypothesis to explain, at least in part, the partner incompatibility that characterizes IF, suggesting a model to describe *how reproductive isolation (RI) may be originated* molecularly inside a population of sexual-reproducing organisms. So, the problem stressed here is: which genomic modifications may turn two specific individuals within the same population molecularly incompatible in a way to avoid the production of fertile offspring by them? Can these levels of RI (evidenced by the existence of IF) rise in the population? Is there any chance that this growing in RI levels may turn a great part of population incompatible? Can these levels grow enough to achieve complete RI and, therefore, speciation? Basically, the theory claims that RI is a common feature inside biological populations and some degree of isolation resulting from molecular incompatibilities does exist in any population.

In fact, evolving sexual-reproducing species must change their genomic content in a coordinated fashion to achieve molecular compatibility. An individual from a sexual-reproducing species that has been target of a massive modification in its genome presents a small chance to be molecular compatible with other individuals in the same population and probably it will not be able to reproduce. However, small random mutations such as nucleotide base changes are always happening in the genome of organisms and just the individuals presenting compatible mutations (when compared with others organisms inside the same population) will be able to reproduce with fertile offspring. But which kind of mutations may frequently turn individuals molecularly incompatible?

Once random mutations take place in genes whose proteins act as heterodimers, they may produce incompatible versions of one and another protein molecules that interact to form these heterodimers. Thus, two individuals presenting *3D incompatible isoforms of protein molecules in certain heteromers* may not be able to produce either viable or fertile offspring. The interaction failure to produce a functional heteromer may not allow embryogenesis or sexual-gonad development processes to occur. Hence, it is not every male-female couple in some sexual-reproducing dioecious species that will be able to reproduce with fertile offspring. If this hypothesis is true, then a number of human couples would be diagnosed as IF; and they are.

Independently, Dobzhansky (1937) and Muller (1940) have proposed a similar model for the evolution of incompatibilities to explain RI among *different species*. Although this model was used as basis for the present work, it is here suggested specific molecular mechanisms to achieve RI. Another difference between the present model and the classical one is: here, it is suggested that some RI does exist inside biological species, avoiding the production of fertile offspring by a pair of specific and molecular incompatible individuals. Finally, a generalization is made from heterodimers to heteromers. Therefore, this hypothesis suggests that IF is not necessarily a problem in the woman reproductive apparatus, but it may be caused by evolutionary consequences of DNA mutations into the genome of a constant evolving biological species; in case, ourselves. *Homo sapiens* seems to be the best model organism to study the origins of postzygotic RI, since reproductive medicine data is reasonably well-described and abundant. Direct consequences on the present theory will also bring light into processes involved in speciation, mosaic evolution, outbreeding depression and even the Dobzhansky-Mayr's biological concept of species. However, before going deep into this hypothesis, it is necessary to understand the current status on the study of RI in natural species.

4.2.1 CURRENT STATUS ON THE STUDY OF MOLECULAR MECHANISMS OF REPRODUCTIVE ISOLATION

The molecular basis of RI is barely known, although biologists have been thinking about it since a very long time. The effective understanding of RI process is direct related to the concept of biological species since most of species' concepts used today still define species in regard of the capability to produce fertile offspring indefinitely (Dobzhansky, 1937; May, 1963; Coyne and Orr, 2004; de Queiroz, 2005). Once the paradigmatic belief consider that RI is only achieved when species are already completely separated, contemporary researches in the field are always trying to explain the molecular causes of inviability and/or sterility in *hybrids* of *Drosophila* species (Coyne, 1986; Coyne and Charlesworth, 1986; Coyne, 1989; Coyne and Charlesworth, 1989; Coyne *et al.*, 1991; Coyne and Berry, 1994; Coyne, 1996; Coyne *et al.*, 1998; Coyne *et al.*, 2002; Coyne *et al.*, 2005; Llopart *et al.*, 2005; Moehring *et al.*, 2006).

As already mentioned, the molecular basis of postzygotic RI was firstly proposed by Theodosius Dobzhansky, suggesting that hybrid failure could be explained when considering a pair of genes whose proteins interact with each other (forming a heterodimer) evolving along different paths after some population split (Dobzhansky, 1937). Independently, the Nobel laureate Hermann Joseph Muller proposes a very similar thesis (Muller, 1940) and the so-called Dobzhansky-Muller (DM) model gained wide acceptance in the following years (Pennisi, 2006). This model stated that RI would be achieved by the result of incompatibilities between gene variants arising independently in populations, and these variants would be deleterious in different genetic backgrounds, bringing to inviability or sterility of hybrids. Since then, many studies have identified genes that might be responsible for RI in *species hybrids*. At least three putative pars of DM genes have shown evidences to be related to the process of RI in *Drosophila* species (Orr, 2005). Firstly, a presumed transcription factor named *Odysseus*-Homeobox gene (*OdsH*) was found after mapping a locus causing sterility of *D. simulans*-*D. mauritiana* hybrid males (Coyne and Charlesworth, 1986; Perez *et al.*, 1993). After a while, a gene responsible to regulate the trafficking of RNA and proteins in and out of nucleus, *Nucleoporin96* (*Nup96*), have been revealed to cause inviability of *D. melanogaster*-*D. simulans* hybrids (Presgraves *et al.*, 2003). These genes were supposed to interact with genes from other unknown loci to produce sterility in *Drosophila* hybrids, in agreement with DM model (Orr, 2005). However, the best known example discovered for DM model of speciation was described last year (Brideau *et al.*, 2006). *Hybrid male rescue* (*Hmr*) gene was known to encode a transcriptional factor putative related to the MYB family and it was initially found to cause inviability in hybrids of *Drosophila* species (Hutter *et al.*, 1990; Barbash *et al.*, 2003). In a wonderful work published by Brideau *et al.* (2006) have shown, for the first time, the precise molecular interactions between heterodimers formed by *Hmr* and *Lhr* (*Lethal hybrid rescue*) gene alleles leading to RI in species from the *Drosophila* genus. All genes related to RI already identified have shown to evolve rapidly, probably driven by positive Darwinian selection (Presgraves *et al.*, 2003; Orr, 2005; Brideau *et al.*, 2006).

Therefore, modern researches on RI are done in order to explain why *species hybrids* are inviable or infertile (Charlesworth *et al.*, 1993; Orr, 1995; Orr *et al.*, 1997; Orr, 1999; Orr and Irving, 2000; Orr and Irving, 2001; Presgraves *et al.*, 2003; Masly *et al.*, 2006). The present hypothesis, once taken seriously, opens a new scientific research program (Lakatos, 1977) in the sense that *it suggests the study of RI shall be done inside biological populations*. So, scientists trying to discover why some fecund couples are infertile will be able to identify the series of small steps producing molecular incompatibilities that will cause failure in reproduction with fertile offspring. Moreover, the study of these small intra-population molecular incompatibilities may help to understand how speciation (both sympatric and allopatric) may arise in biological populations by gradual modification processes.

Both Dobzhansky and Muller have published their models before the molecular nature of genes have been effectively discovered by Watson and Crick (1953). Therefore, the present model tries to bring DM model into a molecular evolutionary perspective in the light of current genomic research. In a very influential paper in the modern RI field, Coyne and Orr have suggested that 'Curiously, there have been few theoretical studies of the Dobzhansky-Muller model' (Coyne and Orr, 1998). The present work may be seen as an initial tentative to fill this theoretical gap.

4.3   EMBRYOGENESIS AND IMPAIRED FECUNDITY

In order to explain inviability cases characterizing IF, we shall suppose the action of many heteromeric DM genes during *embryogenesis*. However, the study of molecular mechanisms underlying the gene interaction networks during embryo development is still incipient. In the last years, molecular developmental mechanisms have been extensively studied in the model organism *Caenorhab-ditis elegans* (Chen and Meister, 2005; Gunsalus *et al.*, 2005; Ge *et al.*, 2006; Updike and Mango, 2006). Although *C. elegans* seems to be the most used or-ganism for developmental studies in the molecular level, some molecular de-velopmental processes had been also studied in many other organisms (Zwijsen *et al.*, 2001; Ang and Constam, 2004), including humans (Breitwieser *et al.*, 1996; Vortkamp, 2001; McCarthy and Argraves, 2003; Olson, 2006). Therefore, although it is still not possible to define precise candidate genes (producing a database of genes for RI) on which proteins may fail to interact producing inviable offspring and find single nucleotide polymorphisms (SNPs) in these genes, it is clear that the precise interaction of many proteins is essential to de-velopment. Once human gene networks acting in embryogenesis will be better known, precise genotyping experiments will be able to be performed in order to verify precisely *which alleles from an infertile couple are, in fact, incompatible.* Further *yeast two-hybrid* experiments as well as other proteomic techniques to verify protein interactions shall be performed to determine how the precise in-teractions between alleles of DM protein-coding genes have been affected by specific DNA mutations. Moreover, even if this chapter emphasizes the causes of inviability characterizing IF problems in humans, it is clear that the fail on the interaction of DM heteromers working to promote sexual maturation is also involved in RI. This is why some hybrids (and any individual inside a given population) may be viable, even if not fertile. Thus, the present theory may also help to explain the molecular causes of some infertility cases and dysfunctions in heteromers may cause a number of well-known phenotypes observed in in-fertile individuals.

4.4   EVOLUTIONARY ORIGIN OF INCOMPATIBILITIES

Two models for the origin of molecular incompatibility on DM gene alleles are presented here: a sympatric (Figure 4.1a) and an allopatric model (Figure 4.1b). These figures show how incompatible alleles may appear gradually in populations and produce a complex pattern of allelic compatibility (Figure 4.2). Both models in Figure 4.1 consider biological populations evolving along time and accumulating mutations in heteromers that will turn some DM gene alleles incompatible. The *origin of allele incompatibility* inside a sympatric population may be explained using combined effects of well-known genetic mechanisms, such as inbreeding and genetic drift. Instead of considering only two incompat-ible DM gene alleles per locus, it is suggested a more likely and intricate pos-sibility of allele combination (Figure 4.2) in order to promote a network of allele compatibility and allow new alleles originated by mutations to interact with old wild ones. The dynamic of new allele formation in heteromers and the risen of incompatibilities is clearly disposed in Figure 4.1, evidencing the *coevolution* of DM gene alleles. It is precisely the coevolution of new DM gene alleles originated by random DNA mutations that will guide us into the understanding of the origin of RI at the first moment; and speciation, at the second.

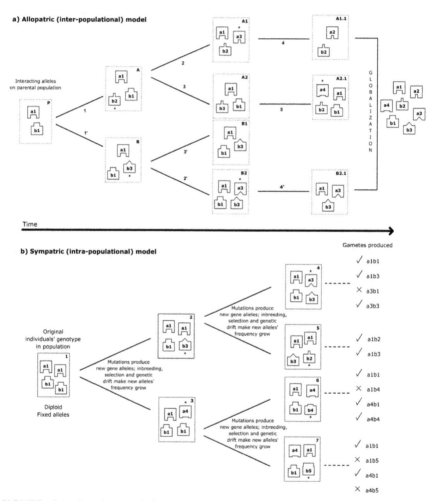

FIGURE 4.1 Putative evolutionary scenarios producing RI. In both models, two heterodimeric DM genes (*a* and *b*) evolving along time have been considered. The models are independent and they explain, using different mechanisms, how genes initially presenting fixed alleles (in the past, left side of figure) may undergo modification and mutations along time to produce incompatible alleles (present, right side of figure). Alleles of *a* and *b* genes were identified by numbers and a schematic proposed conformation of the interaction site

FIGURE 4.1   *(Caption continued)*

between proteins codified by them was also provided. Although intuitive, the allowed allele interactions capable to produce a functional heterodimer in this model must be described. They are: *a1-b1, a1-b2, a1-b3, a2-b2, a3-b3,* and *a4-b1.* When a new allele is produced for the first time in some population, it is identified by a "*". In the *allopatric model* (a) alleles inside boxes are polymorphic in populations. The initial parental population (P) presents fixed alleles for *a* and *b*, that is: *a1* and *b1.* Other populations are named as shown at the top-right side of the squares; squares are made of dashed lines since some individuals of a population may sometimes reproduce with individuals living in any other population existing at that time. All numbered lines represent divisions of populations and the populations at the right side of lines represent new formed populations from the left-sided ancestral ones, generated by migration. In all the splitting processes represented by lines, a number of genetic processes are happening (migration, genetic drift, founder effect, and mutation). At the end (present time), all different alleles originated in populations are brought together in human populations by the effect of globalization. In the *sympatric model* (b) squares now represent diploid individuals sampled from the population and containing the particular set of alleles shown. The entire scheme represents a single and undivided population evolving along time. At the very left side of the picture we can see an individual presenting the most common genotype inside an initial considered population: *a1a1b1b1.* Along time, mutations occur into individuals, producing new alleles; and some of them become more frequent in population due to inbreeding, natural selection, and genetic drift. Other alleles are probably being originated and vanished without growing their frequencies by effects of genetic drift (these alleles are not represented in the picture). Solid lines linking individuals represent a consider amount of time on which individuals in population are intercrossing, and the alleles originated are becoming more frequent, while mutations bringing to new alleles are also happening. Broken lines at the right side of picture link some represented individuals to the alleles present in their haploid gametes. Some of these gametes present incompatible alleles and these gametes were identified by "X"s. When individuals presenting gametes with incompatible alleles try to reproduce, they will only be successful in the production of offspring whether the gametes of their partners are able to complement each other.

In both models described for the evolution of allele incompatibility (intra and interpopulational), it is described the gradual emergence of incompatible allele sets of a single pair of heteromeric DM genes: *a* and *b.* The emergence of these incompatible pairs of DM genes acting on embryogenesis will produce offspring inviability inside biological populations. Although it is clear that many heteromers may act in embryogenesis (including house-keeping genes), a simplistic model on the emergence of incompatibility in a single pair of DM genes is presented (Figure 4.1). The model shall be generalized in the sense that allele modifications are happening constantly at organisms' genomes and the coevolution of DM gene alleles will depend on the susceptibility to mutation of each locus.

Both models are based in the fact that ancestral alleles of DM genes (*a1* and *b1*) will accumulate mutations over the time, producing new alleles that will be able (or not) to interact in a heteromer. Whether they interact, a specific molecular modification will be produced in the cell bearing them; whether they do not interact, the cell will not develop properly. Therefore, I suppose that this modification produced by the correct interaction of *a* and *b* alleles will be the trigger of a cascade of other effects and it will allow the embryogenesis process to go ahead. The absence of both *a-b* interaction and further cellular modification that this interaction would produce (such like generation of a chemical intermediate on a biochemical pathway, the transcription of a given RNA molecule, some kind of polarization characteristic of developmental process, etc.) will culminate in an inviable offspring.

The schematic representation in the pictures (Figure 4.1 and 4.2) present DM genes *a* and *b* followed by a number that represents each specific allele. All alleles, when translated by cellular machinery will produce a protein with a specific three-dimensional structure. In the diagrams, the active site where DM gene pairs *a* and *b* interact were represented in a way that will be somewhat intuitive to the reader to understand which alleles interact to each other (Figure 4.2). Although clearly occurring, mutations in DM gene pairs that do not alter the conformation of the site of protein interaction in the heteromer will not be relevant for the model and they have not been considered. Finally, three main processes are strikingly relevant to the model suggested and must be kept in mind: (1) the origin of new alleles by random mutations; (2) the growth in the percentage of new alleles in population due to genetic mechanisms (such like migration, founder effect, genetic drift and inbreeding) and; (3) the production of gametes containing an incompatible set of DM gene alleles.

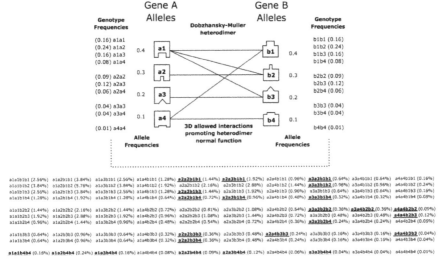

a1a1b1b1 (2.56%) a1a2b1b1 (3.84%) a1a3b1b1 (2.56%) a1a4b1b1 (1.28%) **a2a2b1b1** (1.44%) **a2a3b1b1** (1.92%) a2a4b1b1 (0.96%) **a3a3b1b1** (0.64%) a3a4b1b1 (0.64%) a4a4b1b1 (0.16%)
a1a1b1b2 (3.84%) a1a2b1b2 (5.76%) a1a3b1b2 (3.84%) a1a4b1b2 (1.92%) a2a2b1b2 (2.16%) a2a3b1b2 (2.88%) a2a4b1b2 (1.44%) **a3a3b1b2** (0.96%) a3a4b1b2 (0.96%) a4a4b1b2 (0.24%)
a1a1b1b3 (2.56%) a1a2b1b3 (3.84%) a1a3b1b3 (2.56%) a1a4b1b3 (1.28%) **a2a2b1b3** (1.44%) a2a3b1b3 (1.92%) a2a4b1b3 (0.96%) a3a3b1b3 (0.64%) a3a4b1b3 (0.64%) a4a4b1b3 (0.16%)
a1a1b1b4 (1.28%) a1a2b1b4 (1.92%) a1a3b1b4 (1.28%) a1a4b1b4 (0.64%) **a2a2b1b4** (0.72%) **a2a3b1b4** (0.96%) a2a4b1b4 (0.48%) **a3a3b1b4** (0.32%) a3a4b1b4 (0.32%) a4a4b1b4 (0.08%)

a1a1b2b2 (1.44%) a1a2b2b2 (2.16%) a1a3b2b2 (1.44%) a1a4b2b2 (0.72%) a2a2b2b2 (0.81%) a2a3b2b2 (1.08%) a2a4b2b2 (0.54%) **a3a3b2b2** (0.36%) **a3a4b2b2** (0.36%) **a4a4b2b2** (0.09%)
a1a1b2b3 (1.92%) a1a2b2b3 (2.88%) a1a3b2b3 (1.92%) a1a4b2b3 (0.96%) a2a2b2b3 (1.08%) a2a3b2b3 (1.44%) a2a4b2b3 (0.72%) a3a3b2b3 (0.48%) **a3a4b2b3** (0.48%) **a4a4b2b3** (0.12%)
a1a1b2b4 (0.96%) a1a2b2b4 (1.44%) a1a3b2b4 (0.96%) a1a4b2b4 (0.48%) a2a2b2b4 (0.54%) a2a3b2b4 (0.72%) a2a4b2b4 (0.36%) **a3a3b2b4** (0.24%) a3a4b2b4 (0.24%) a4a4b2b4 (0.06%)

a1a1b3b3 (0.64%) a1a2b3b3 (0.96%) a1a3b3b3 (0.64%) a1a4b3b3 (0.32%) **a2a2b3b3** (0.36%) a2a3b3b3 (0.48%) **a2a4b3b3** (0.24%) a3a3b3b3 (0.16%) a3a4b3b3 (0.16%) **a4a4b3b3** (0.04%)
a1a1b3b4 (0.64%) a1a2b3b4 (0.96%) a1a3b3b4 (0.64%) a1a4b3b4 (0.32%) **a2a2b3b4** (0.36%) a2a3b3b4 (0.48%) a2a4b3b4 (0.24%) a3a3b3b4 (0.16%) a3a4b3b4 (0.16%) a4a4b3b4 (0.04%)

a1a1b4b4 (0.16%) **a1a2b4b4** (0.24%) **a1a3b4b4** (0.16%) a1a4b4b4 (0.08%) **a2a2b4b4** (0.09%) a2a3b4b4 (0.12%) a2a4b4b4 (0.06%) a3a3b4b4 (0.04%) a3a4b4b4 (0.04%) a4a4b4b4 (0.01%)

FIGURE 4.2 Population genetics model in a scenario on which incompatible heterodimeric DM alleles have evolved. The model consider the existence of 4 different polymorphic alleles for each gene on the DM heterodimer with their respectively frequencies in a putative population. Allowed interactions of *a* and *b* protein isoforms (derived from different alleles) that allow heterodimer molecular function process to occur are represented by lines linking alleles. Genotypic frequencies are also shown. Using a population genetics model, there were produced all possible genotypes allowed in this population and individuals presenting molecular incompatibilities were shown in bold, underlined. In this example, unless there is another redundant gene network to produce the function done by *a-b* interaction, 11.40% of gamete encounters will produce IF considering this heterodimer as necessary for embryogenesis.

4.4.1 ALLOPATRIC OR INTERPOPULATIONAL MODEL

Allopatric or interpopulation model for the *emergence of allele incompatibility* consider humans migrating in populations on which new alleles are produced by random DNA mutations in groups geographically dispersed. In Figure 4.1a, numbered lines shall be thought as human migrations and it is assumed that migrations occur concomitantly with genetic drift, founder effect, inbreeding, and the putative occurrence of mutations producing new alleles. This interpopulational model shows the evolution of a single DM gene pair on which the alleles are coevolving over time after many migration events. It does not matter whether mutations are adaptive or just neutral, although new alleles presenting higher fitness will probably spread faster in population. At the very right side of Figure 4.1a, we see that all gene alleles produced during evolution (some incompatible) are put together through the effect of globalization. In natural populations, it would not be expected to occur as a general pattern, although particular populations from the same species may hybridize after a considered amount of time separated from a common ancestral population. It has been shown that some individuals of these populations are not capable to intercross properly and this phenomenon particularly relevant for conservation biology purposes is frequently called outbreeding depression (Storfer, 1999; Edmands, 2002). The allele incompatibility model here exposed may also explain molecularly why these reproductive problems happen.

4.4.2   SYMPATRIC OR INTRAPOPULATIONAL MODEL

Sympatric or intrapopulation hypothesis for allele incompatibility origin consider humans (or any other diploid sexual-reproducing natural species) intercrossing inside a single small population on which new alleles have been produced by DNA mutations. It must be considered that DNA mutations are always happening in populations and the allelic content of the individuals are continuously changing along time (Kimura, 1969). Thus, these new alleles originated increase their frequencies due to one or more of the following processes: genetic drift, natural selection and inbreeding. Figure 4.1b evidences how gametes presenting incompatible versions of DM gene pairs may be originated and gives a first clue about how IF works molecularly and how some degree of RI may arise sympatrically. In this figure, when we look into the gametes produced by individuals 4 and 7 (the number of each individual is shown in the top-right side of the squares), for example, we realize that they will not produce viable offspring at 6.25 % of conceptions (25 % times 25 %), i.e., when a zygote is formed by the gametes *a3b1* (produced in 25 % of cases) from individual 4 and gamete *a4b5* (produced in 25 % of cases) from individual 7. This conception failure happens because neither *a3* is capable to interact with *b1* or *b5*, nor *a4* is capable to interact with *b3* or *b5* and, consequently, the DM heterodimer will not function properly. Since we are supposing this heterodimer as necessary to embryogenesis, there will not be produced a viable offspring. When considering the evolution of RI, failure in the production of functional reproductive apparatus will also be relevant and DM gene pairs involved in the offspring fertility shall be taken on account. However, for all other interactions between gametes from individuals 4 and 7 considering these loci, embryogenesis would occur properly. Thus, in order to understand better how DM allele incompatibility may produce a common IF phenotype (such as sympatric speciation, species isolation mechanisms and the existence of mosaic evolution) we shall consider a population genetics model.

4.4.3 A POPULATION GENETICS MODEL TO UNDERSTAND IMPAIRED FECUNDITY AND THE ORIGIN OF REPRODUCTIVE ISOLATION

Although IF must consider the *genotype of parents*, a broad population genetics model may help us to understand how a set of incompatible alleles may be spread in a given population. Once incompatible alleles of DM genes pairs have been originated in sympatry or allopatry (Figure 4.1), there will be the existence of an incompatible group of alleles inside a given population. In order to understand how this process will produce inviable (or infertile) offspring, it is necessary to introduce the concept of an *incompatible gamete*. An incompatible gamete may be considered as a gamete presenting an incompatible set of gene alleles. It must be kept in mind we are considering evolution in diploid species and, in Figure 4.2, incompatible alleles can be observed when the gamete present any cluster of incompatible alleles of DM genes *a* and *b*. Genotypes of incompatible gametes in this example (Figure 4.2) are: *a1b4*, *a2b1*, *a2b3*, *a2b4*, *a3b1*, *a3b2*, *a3b4*, *a4b2*, *a4b3* (compatibility was here defined by step-wised evolutionary events shown in Figure 4.1). Whether two incompatible gametes found each other to produce a zygote, two events may happen: (1) the gamete coming from a parent may complement the incompatibility of the one coming from the other parent, restoring the viability of the zygote (in the case of a gamete with genotype *a2b3* fecund another gamete *a4b2* genotype, the *a2-b2* interaction will be restored and the zygote will be viable); (2) the gametes may not be allelic complementary and the zygote will not be viable (in case of a gamete with genotype *a2b4* fecund another gamete with *a2b3* genotype, neither *a2-b3* nor *a2-b4* interactions are compatible; and the precise protein interaction will not be restored). It is suggested that a number of IF cases fall in this category. Moreover, any encounter of a compatible gamete with other gametes (compatible or incompatible) will allow viability (concerning the interaction of the DM genes under consideration), unless dominance or imprinting have been considered. So, the interaction of alleles in a compatible gamete is enough to allow the molecular modifications produced by heterodimer function necessary for embryogenesis. For example, a gamete *a2b2* will produce viable offspring no matter which other gamete it encounter to form a zygote, since the interactions between *a2* and *b2* gene alleles are enough to produce viable offspring (considering only the effect of this heterodimer).

Due to allele incompatibility in a single pair of heterodimeric DM genes, Figure 4.2 suggests a general case on which 11.40 % of theoretical allele interactions in a given population would not produce viable offspring. In actual cases, however, two *viable* organisms trying to reproduce must present compatible alleles in their genomes and it is so that no organism shall present a genotype *a2a3b1b4*, since otherwise it would be inviable. Thus, at least one out four possible parental gametes for these two loci must be a viable compatible gamete–the one that allow the parent being viable. If this was the only factor to be considered, it would be expected that IF would never happen, since only 6.25 % (25 % times 25 %) of conceptions would not occur properly. However, if the frequency of incompatible gametes encounter for one single heteromer may be 6.25 %, this percentage would heavily rise whether we consider the existence of a number of independent heteromers acting in embryogenesis. Since it is likely that a high number of heteromers acts to promote development, we shall suppose that the frequency of compatible alleles in nature must be high to allow the existence of sexual-reproducing organisms. Otherwise, we shall suppose the existence of a high-level redundancy in gene function.

When thinking in a wide-range model of population genetics to deal with RI in a species clade, it is possible to predict some factors that will clearly culminate in the production of inviable (or infertile) offspring based in the processes already described. The main factors to be considered in order to produce reproductive fitted offspring seems to be: (1) the number of heteromers necessary to allow viability and reproductive fitness in a given species; (2) the number of polymorphic incompatible alleles of these heteromers originated by events like the ones described in Figure 4.1; (3) the compatibility relationship between the alleles of each heteromer; (4) the frequency of the heteromeric alleles in the population. Using (2), (3) and (4) rules, the scheme shown in the below part of Figure 4.2 was produced, evidencing the theoretical number of incompatible gamete matches in a population, given the data of a single heterodimeric DM gene pair. Rule number (1), however, is presently unknown and we can only speculate about it. When considering this population genetics model, it must be kept in mind that the incompatibility status will depend on the parents' genotype and, therefore, this broad and theoretical model is just illustrative and must be seen with caution.

4.4.4   REDUNDANCY IN GENE FUNCTION

It is clear that the interaction of many genes is responsible to allow embryogenesis (in the case of IF) and embryogenesis plus fertility (in the case of a broader postzygotic RI molecular genetics model). Moreover, although the example described in the bottom part of Figure 4.2 produces a number of ~12 % of incompatible offspring (similar to the value truly observed in present human population according to CDC), it is clear that a tremendously intricate network of genes interacting is involved in the process of reproductive fitness and also that there are many mechanisms of gene network redundancy to achieve a given cellular phenotype. Once the mechanisms underlying embryogenesis, for example, have been more extensively understood, we will probably be able to find viable living organisms presenting incompatible alleles of some heteromer. These organisms may probably be viable due to gene redundancy mechanisms, on which an alternative cellular pathway may complement the nonfunctional status of another pathway that may be interrupted by allele incompatibility causing heteromer dysfunction. It will be interesting to evaluate how far it is going to be the extension of these gene redundancy networks. Moreover, consistent incompatibilities among parental alleles of redundant gene networks may be able to explain why after a long time trying to have babies; a human couple may finally be succeeded. It may be the result of a specific arrange of alleles for these redundant gene networks randomly disposed in their gametes. In these cases, a factor concerning only the random arrangement of gene alleles into parental gametes may be enough to finally produce a viable and fertile offspring, without the help of medical intervention.

Although we know about the existence of some redundancy in gene action, i.e., two different genes or gene networks producing the same effect during embryogenesis (Onda *et al.*, 2004), we should not suppose that all molecular processes will present redundancy. Moreover, these evolutionary processes described here may happen in both redundant gene pathways together and, in fact, the origin and coevolution of new gene alleles in DM gene pairs probably keeps happening in many and different gene interaction pathways in biological organisms. Considering this last observation, our problem may be observed upside down and it might be difficult to explain how a high number of organisms always accumulating mutations, evolving, and modifying the allelic content of their heteromers would be able to reproduce and maintain their species alive for so long time. Here, it becomes clear the rule of *natural selection* in order to keep stable the genetic pool of some species. As already noted, sexual-reproducing organisms accumulating too much mutation in their genes will probably be negative selected, since their gene pools will not fit the ones from other individuals in the same population. Once more, there would probably exist a more stringent limit on the rate of DNA changing in sexual than asexual-reproducing species since the evolution of individuals belonging to sexual-reproducing ones depends on the evolution of others sharing the same gene pool. Moreover, individuals in asexual-reproducing species may activate some SOS system that promotes high-level mutation rates to allow rapid adaptation to environmental conditions. Thus, sexual-reproducing species must change (it is a random process, of course) their compatible DNA content along time in a coordinated fashion. And it is so that sexual-reproducing species *will be observed* to evolve in the direction of *self-homeostasis*; maybe even more than evolving into environmental adaptation.

4.4.5 FROM INTRAPOPULATION RI TO SPECIATION

Whether this hypothesis about the evolution of allele incompatibility is not false, the concept of reproduction with fertile offspring leaves to be a characteristic shared by any organisms inside a given population or species and it turns to be a characteristic of a *pair of individuals* sharing compatible genomes. This clearly explains IF as a common case of intrapopulation RI and it may also explain some cases of infertility in humans. The gradual origin of molecular mechanisms producing RI in natural populations is still unknown. Wu presents a putative scenario for the evolution of RI beginning with a population in the Stage I of differentiation on which there would be no apparent RI (Wu, 2001). The author advances RI stages until the Stage IV, passing from races to subspecies until completely separated species, on which a complete RI between individuals in two populations have been achieved (Wu and Ting, 2004). The present hypothesis considers RI in any natural species such as Wu consider a stage II of species separation and suggests that "not apparent" RI may not exist in biological populations. A considerable amount of RI is intrinsic to the continuous process of DNA mutations happening in heteromeric alleles and it must be kept in mind that 7.3 million American women have been diagnosed with IF in 2002 (CDC, 2005). Although the pathway from intrapopulation RI to speciation is long, it seems reasonable to understand it as an *initial step to speciation*. In regard of human population, recent globalization processes have been sharing human molecular biodiversity overall the planet. According to this observation, recent analyses using SNP data have shown that human biodiversity is high (Weir *et al.*, 2005; Frazer *et al.*, 2007) and it may be supposed that human heterozygosity has never been as high as nowadays. And it is so that human speciation is very unlikely to occur.

4.5 CONSIDERATIONS ABOUT MOSAIC EVOLUTION

Additionally, many cases of reproductive isolated species morphologically indistinguishable (cryptic species) are well-known as well as other cases on which highly differentiated morphological individuals are capable to reproduce with fertile offspring (Sonneborn, 1975; Stebbins, 1983). Such cases of mosaic evolution have been classically described and they suggest the well-known fact that evolutionary rate is not homogeneous in nature. The present theory, however, evidences that the question about mosaic evolution is *not necessarily* a question about evolutionary rate. Even if two recently separated groups of individuals have been accumulating mutations at a slow rate, it may be supposed that random mutational processes act exactly in some alleles of DM heteromers, avoiding reproduction and, therefore, causing speciation. However, it is also clear that higher the rate of evolutionary change, higher the chance of alleles for heteromers being affected and modified by mutations. Therefore, it is possible to say that mosaic evolution is often associated with higher rates of evolutionary changing, although sometimes it may not be the case.

It is interesting to consider some examples about the emergence of putative mosaic evolution scenarios. Let us suppose a population (1) on which *a2* and *b2* alleles (Figure 4.2) have been fixed by events like natural selection, genetic drift and inbreeding; and another population (2) on which *a3* and *b3* alleles are the ones that have been fixed by any genetic processes. In this case, all individuals from populations (1) and (2) will be completely reproductive isolated, even if all alleles for all other genes have been present in each population in an identical percentage. This is an example of a phenomenon molecular explained that produces cryptic but reproductive isolated species. On the other hand, it is also possible to suppose a situation on which many alleles bringing to different phenotypes will present a high amount of variation between two populations, while most alleles for heteromers relevant for RI will be compatible among the individuals from both populations. In this last example, there would be produced highly differentiated organisms capable to reproduce with fertile offspring. Consequences of this observation suggest that biodiversity may be better explained looking into molecular variation among different individuals and it may happen that a single reproductive continuous species present more molecular biodiversity than a group of closely related species.

Thus, it is possible to conclude that the so-called "speciation genes" are, in fact, DM genes forming heteromers. The different accumulation of mutations in the alleles of these heteromers, avoiding protein interaction and function in embryogenesis or sexual fitness causes RI (and further, speciation), despite of the mutations happening in the rest of organisms genomes. Therefore, allele incompatibility processes are capable to explain clearly the relationship between mosaic evolution and RI mechanisms.

4.5.1   A FOOTNOTE INTO THE BIOLOGICAL CONCEPT OF SPECIES

At last, whether the evolution of allele incompatibility in DM genes producing RI inside populations has been accepted by community, even the Biological Concept of Species (BCS) will need to be slightly restructured. Therefore, the present theory may help us to understand better the long-term debate in biology about the species problem, i.e., the problem to define precisely a clear concept for the term "species" (Burma and May, 1949; Haldane, 1956; Beaudry, 1960). The BCS consider species as "groups of actually or potentially interbreeding natural populations, which are reproductively isolated from other such groups" (Mayr, 1963). However the present chapter suggests that the capability to inter-breed (actually or potentially) is a characteristic of two specific individuals pre-senting a precise combination of compatible DM gene alleles in their genomes and *it is not a characteristic of an entire natural population*. In all populations, there should be pairs of individuals presenting an incompatible set of hetero-meric gene alleles that will not be able reproduce, although they will clearly belong to what we are used to call a "natural species". These pairs of *impaired fecund individuals* will probably be able to produce fertile offspring with other genomically compatible individuals in the same population. The question rises: how does it affect the BCS?

A single modification in the BCS is enough to keep it valid once it has been proven that RI is common inside a species' group. The present theory agrees with both BCS and the modern view of species in a sense that species must be identified and understood as a function of RI (Orr, 1995; Coyne and Orr, 1998; Mallet, 2005; Mallet, 2007). Opposite to the current view, it is suggested here that organisms shall be considered as different species not if they are not able to reproduce with fertile offspring, since RI is frequent inside any biological popu-lations. Therefore, two organisms must be considered from different species if their genomes cannot be mixed in future generations by vertical ascendance. Or-ganisms belonging to different species do not share a continuous gene flow or; if organisms participate on the same gene flow (vertically speaking), they shall be considered from the same biological species. Thus, different sexual-reproducing species shall be understood as groups of individuals on which their gene flow have been completely separated by vertical lineages so that the evolution of one and another gene flow happens in separate.

4.6   CONCLUSION

In this chapter, a molecular evolutionary model to explain the unknown causes of IF in humans is proposed. Although preliminary and speculative, this model allows us to understand putative processes producing IF when considering the presence of incompatible versions of gene alleles acting in heteromers that a couple may contain in their genomes. Moreover, the coevolution of heteromeric alleles is a clear generalization of DM speciation theory for molecular incom-patibility. Although today speculative, this model may be incorporated by hard science in a close future, since many research areas of interest are growing, such as the molecular basis of RI, embryogenesis, and infertility. Moreover, future population genomics studies as well as more extensive studies with SNPs will probably be able to identify the overall variability of polymorphic gene alleles in human population. The interaction and compatibility relation of known human polymorphic alleles may also be investigated using yeast two-hybrids experiments, protein arrays and other proteomic techniques.

It is also interesting to discuss briefly some epistemological aspects of the present theory. Differently from other RI studies currently done in already differentiated species, the present work question the paradigmatic belief (Kuhn, 1962) shared by biological community that all individuals inside a single population (or species) can potentially interbreed originating fertile offspring. Here, the break of this dogma is suggested due to a preliminary evidence of intrapopulational RI based on medical statistical data about IF (CDC, 2005); a hypothesis that shall be further confirmed by other methodological approaches. The scientific crisis in speciation research field has been anticipated by some (Mallet, 2001) and well-known researchers on the field have attested for the absence of new theoretical approaches in the study of molecular models of speciation (Coyne and Orr, 1998). The present hypothesis takes gradualism and population genetics in account to explain the origin of RI and speciation. Therefore it fits better Darwinism than other speciation theories, avoiding mystical and saltacionistic aspects sometimes related to this unknown process. Moreover, the nature of the present theory is clearly scientific in the Popperian sense (Popper, 1959), since it is testable (falseable) and it may be proven false whether further crucial experiments fail to verify the principle of intrapopulation incompatibilities. Once a new theoretical basis of speciation is here explained, further studies and experimental techniques may be developed in model organisms trying to evaluate its pertinence in a broader context, evidencing a progressive research program in evolutionary biology (Lakatos, 1977).

Although RI inside populations have recently been taken on account by some researchers (Wu, 2001; Edmands, 2002; Wu and Ting, 2004), it seems that the relationship between a gradualistic view of speciation and intrapopulational RI have not been completely understood. This last observation is corroborated by the fact that RI keeps being studied in species hybrids (Brideau *et al.*, 2006; Mallet, 2006; Masly *et al.*, 2006; Moehring *et al.*, 2006; Pennisi, 2006; Rogers and Bernatchez, 2006; Russell and Magurran, 2006; Mallet, 2007) instead of trying to investigate it intrapopulationally, such as studying the molecular genetics of IF in humans. It seems that researchers avoid the usage of the term "reproductive isolation" in humans since they might be afraid to be bad interpreted. However, it shall be better evaluated whether IF is actually a case of intrapopulation RI, as it seems to be considering the rationale presented here. Although ethical question will certainly arise, in the light of the present theory it will be possible in future to test the sexual-compatibility of couple by genotyping a number of loci before the marriage.

Additionally, new mathematical models in population genetics may be created considering the knowledge about incompatible gene alleles to evaluate: (1) the average chance of a sexual couple (male-female) chosen randomly present IF inside a given biological population; this value may represent a chance of speciation in the sense that higher the IF chance, higher the chance of the entire groups suffer future speciation; (2) the chance of reproduction in organisms suffering outbreeding (or inbreeding) depression, a relevant factor for conservation biology purposes; if we are able to evaluate the compatible alleles of organisms, we can choose only the compatible male-female pair to reproduce; (3) the putative existence of a maximum number of heteromers in a species to allow its viability; (4) the relationship between the number of heteromers in a given species and its speciation rate (it may happen that cryptic species present higher number of proteins interacting to perform biological functions), and many other ones. Some other important variables may also affect these models: such like linkage disequilibrium between genes participating in heteromers, putative occurrence of meiotic drift (leading to nonequal production of alleles in gametes), genetic imprinting and different compatibility network relationships between alleles of two or more genes. Figure 4.1a and 4.1b have also helped us to understand unsolved problems regarding the molecular basis of, respectively, outbreeding depression and sympatric speciation.

This chapter evidences how academic issues are becoming each time more significant in medicine, something already pointed out by some (Goldstein and Chikhi, 2002), but it also shows the other way round. CDC medical statistics have shown, in this chapter, its relevance to the study of academic issues in respect to evolution of gene alleles and natural species (CDC, 2005). The present theory was too much corroborated by the existence of published data about IF; in fact, the absence of these data would make this theory much more speculative. Moreover, although many cases of IF may happen due to the molecular causes here delineated, a number of other unknown processes may be responsible for a couple being unable to have babies. It also must be noted that IF data are still difficult to obtain and, since researchers have not understood the molecular causes of IF and directed their experimental methods to study it molecularly, there is no much data available about this subject. This might also explain why a theory like this one was never proposed before. Furthermore, when working with animal behavior, the failure in producing fertile offspring is easily solved by breeders changing the couples of animals to reproduce. In fact, probably many breeders have not realized which specific animals are not capable to reproduce with some other specific ones. Humans living in a monogamous society have shown to be the very best substrates of study to allow us the understanding of the molecular basis of RI. IF would be hardly identified in animals or even in humans living in a nonmonogamous society. Social variables, therefore, have shown here their relevance for the development of scientific theories.

Finally, further studies in medical reproductive genetics may reveal whether allopatric or sympatric mechanisms of allele differentiation have been more frequent during the evolution of human populations. Tests still to be performed may answer the question whether IF would be generally more frequent inside some specific populations or among different populations of humans; or any other natural species. The frequency of molecular incompatibility inside and among natural populations may help in the debate about the frequency of allopatric and sympatric models of speciation.

ACKNOWLEDGEMENTS

I would like to thank FAPEMIG for financial support. I also thank Dr. Sávio Farias, Dr. James Mallet, Dr. Mauricio Mudado, Rodrigo Loyola and mainly Dr. Francisco Lobo for helpful discussion on some ideas enclosed here. I am in debt with Dr. Rodrigo Veras too for the critical reviewing of an initial version of this chapter and with Dr. Mônica Bucciarelli Rodriguez for criticizing its final version.

KEYWORDS

- **Allopatric model**
- **Biological concept of species**
- **Dobzhansky-Muller heteromers**
- **Gene function**
- **Gene network redundancy**
- **Impaired fecundity**
- **Infertility**
- **Mosaic evolution**
- **Population genetics model**
- **Reproductive isolation**
- **Speciation**
- **Sympatric model**

REFERENCES

Alpay, Z.; Saed, G. M.; Diamond, M. P. Female infertility and free radicals: Potential role in adhesions and endometriosis. *J Soc Gynecol Investig* 2006, 13, 390-398.

American Academy of Family Physicians. Information from your family doctor. Infertility: What you should know. *Am Fam Physician* 2007, 75, 857-858.

Ang, S. L.; Constam, D. B. A gene network establishing polarity in the early mouse embryo. *Semin Cell Dev Biol* 2004, 15, 555-561.

Barbash, D. A.; Siino, D. F.; Tarone, A. M.; Roote, J. A rapidly evolving MYB-related protein causes species isolation in Drosophila. *Proc Natl Acad Sci USA* 2003, 100, 5302-5307.

Beaudry, J. R. The species concept: Its evolution and present status. *Rev Can Biol* 1960, 19, 219-240.

Breitwieser, W.; Markussen, F. H.; Horstmann, H.; Ephrussi, A. Oskar protein interaction with Vasa represents an essential step in polar granule assembly. *Genes Dev* 1996, 10, 2179-2188.

Brideau, N. J.; Flores, H. A.; Wang, J.; Maheshwari, S.; Wang, X.; Barbash, D. A. Two Dobzhansky-Muller genes interact to cause hybrid lethality in Drosophila. *Science* 2006, 314, 1292-1295.

Burma, B. H.; Mayr, E. The species concept. Evolution. *Int J Org Evolution* 1949, 3, 369-373.

CDC. Fertility, Family Planning, and Reproductive Health of U.S. Women: Data from the 2002 National Survey of Family Growth, Vital and Health statistics. Centers for Disease Control and Prevention: Hyattsville, Maryland, 2005, 21-24.

Chandra, A.; Stephen, E. H. Impaired fecundity in the United States: 1982-1995. *Fam Plann Perspect* 1998, 30, 34-42.

Charlesworth, B.; Coyne, J. A.; Orr, H. A. Meiotic drive and unisexual hybrid sterility: A comment. *Genetics* 1993, 133, 421-432.

Chen, P. Y.; Meister, G. microRNA-guided posttranscriptional gene regulation. *Biol Chem* 2005, 386, 1205-1218.

Coyne, J. A. Meiotic segregation and male recombination in interspecific hybrids of Drosophila. *Genetics* 1986, 114, 485-494.

Coyne, J. A. Genetics of sexual isolation between two sibling species, *Drosophila simulans* and *Drosophila mauritiana. Proc Natl Acad Sci USA* 1986, 86, 5464-5468.

Coyne, J. A. Genetics of sexual isolation in male hybrids of *Drosophila simulans* and *D. mauritiana. Genet Res* 1996, 68, 211-220.

Coyne, J. A.; Berry, A. Effects of the fourth chromosome on the sterility of hybrids between *Drosophila simulans* and its relatives. *J Hered* 1994, 85, 224-227.

Coyne, J. A.; Charlesworth, B. Location of an X-linked factor causing sterility in male hybrids of *Drosophila simulans* and *D. mauritiana. Heredity* 1986, 57, 243-246.

Coyne, J. A.; Charlesworth, B. Genetic analysis of X-linked sterility in hybrids between three sibling species of Drosophila. *Heredity* 1989, 62, 97-106.

Coyne, J. A.; Elwyn, S.; Rolan-Alvarez, E. Impact of experimental design on Drosophila sexual isolation studies: Direct effects and comparison to field hybridization data. *Int J Org Evolution* 2005, 59, 2588-2601.

Coyne, J. A.; Kim, S. Y.; Chang, A. S.; Lachaise, D.; Elwyn, S. Sexual isolation between two sibling species with overlapping ranges: *Drosophila santomea* and *Drosophila yakuba. Int J Org Evolution* 2002, 56, 2424-2434.

Coyne, J. A.; Orr, H. A. The evolutionary genetics of speciation. *Philos Trans R Soc Lond B Biol Sci* 1998, 353, 287-305.

Coyne, J. A.; Orr, H. A. *Speciation*. Sinauer Associates, Inc.: Sunderland, Massachusetts, USA, 2004.

Coyne, J. A.; Rux, J.; David, J. R. Genetics of morphological differences and hybrid sterility between *Drosophila sechellia* and its relatives. *Genet Res* 1991, 57, 113-122.

Coyne, J. A.; Simeonidis, S.; Rooney, P. Relative paucity of genes causing inviability in hybrids between *Drosophila melanogaster* and *D. simulans. Genetics* 1998, 150, 1091-1103.

de Queiroz, K. Ernst Mayr and the modern concept of species. *Proc Natl Acad Sci USA* 2005, 102, 6600-6607.

Dobzhansky, T. *Genetics and the origin of species*. Columbia University Press: New York, 1937.

Edmands, S. Does parental divergence predict reproductive compatibility? *Trends Ecol Evol* 2002, 17, 520-527.

Escalier, D. Animal models: Candidate genes for human male infertility. *Gynecol Obstet Fertil* 2006, 34, 827-830.

Foresta, C.; Moro, E.; Ferlin, A. Y chromosome microdeletions and alterations of spermatogenesis. *Endocr Rev* 2001, 22, 226-239.

Frazer, K. A.; Ballinger, D. G.; Cox, D. R.; Hinds, D. A.; Stuve, L. L.; Gibbs, R. A.; Belmont, J. W.; Boudreau, A.; Hardenbol, P.; Leal, S. M.; Pasternak, S.; Wheeler, D. A.; Willis, T. D.; Yu, F.; Yang, H.; Zeng, C.; Gao, Y.; Hu, H.; Hu, W.; Li, C.; Lin, W.; Liu, S.; Pan, H.; Tang, X.; Wang, J.; Wang, W.; Yu, J.; Zhang, B.; Zhang, Q.; Zhao, H.; Zhou, J.; Gabriel, S. B.; Barry, R.; Blumenstiel, B.; Camargo, A.; Defelice, M.; Faggart, M.; Goyette, M.; Gupta, S.; Moore, J.; Nguyen, H.; Onofrio, R. C.; Parkin, M.; Roy, J.; Stahl, E.; Winchester, E.; Ziaugra, L.; Altshuler, D.; Shen, Y.; Yao, Z.; Huang, W.; Chu, X.; He, Y.; Jin, L.; Liu, Y.; Sun, W.; Wang, H.; Wang, Y.; Xiong, X.; Xu, L.; Waye, M. M.; Tsui, S. K.; Xue, H.; Wong J. T.; Galver, L. M.; Fan, J. B.; Gunderson, K.; Murray, S. S.; Oliphant, A. R.; Chee, M. S.; Montpetit, A.; Chagnon, F.; Ferretti, V.; Leboeuf, M.; Olivier, J. F.; Phillips, M. S.; Roumy, S.; Sallee, C.; Verner, A.; Hudson, T. J.; Kwok, P. Y.; Cai, D.; Koboldt, D. C.; Miller, R. D.; Pawlikowska, L.; Taillon-Miller, P.; Xiao, M.; Tsui, L. C.; Mak, W.; Song, Y. Q.; Tam, P. K.; Nakamura, Y.; Kawaguchi, T.; Kitamoto, T.; Morizono, T.; Nagashima, A.; Ohnishi, Y.; Sekine, A.; Tanaka, T.; Tsunoda, T.; Deloukas, P.; Bird, C. P.; Delgado, M.; Dermitzakis, E. T.; Gwilliam, R.; Hunt, S.; Morrison, J.; Powell, D.; Stranger, B. E.; Whittaker, P.; Bentley, D. R.; Daly, M. J.; de Bakker, P. I.; Barrett, J.; Chretien, Y. R.; Maller, J.; McCarroll, S.; Patterson, N.; Pèer, I.; Price, A.; Purcell, S.; Richter, D. J.; Sabeti, P.; Saxena, R.; Schaffner, S. F.; Sham, P. C.; Varilly, P.; Stein, L. D.; Krishnan, L.; Smith, A. V.; Tello-Ruiz, M. K.; Thorisson, G. A.; Chakravarti, A.; Chen, P. E.; Cutler, D. J.; Kashuk, C. S.; Lin, S.; Abecasis, G. R.; Guan, W.; Li, Y.; Munro, H. M.; Qin, Z. S.; Thomas, D. J.; McVean, G.; Auton, A.; Bottolo, L.; Cardin, N.; Eyheramendy, S.; Freeman, C.; Marchini, J.; Myers, S.; Spencer, C.; Stephens, M.; Donnelly, P.; Cardon, L. R.; Clarke, G.; Evans, D. M.; Morris, A. P.; Weir, B. S.; Mullikin, J. C.; Sherry, S. T.; Feolo, M.; Skol, A.; Zhang, H.; Matsuda, I.; Fukushima, Y.; Macer, D. R.; Suda, E.; Rotimi, C. N.; Adebamowo, C. A.; Ajayi, I.; Aniagwu, T.; Marshall, P. A.; Nkwodimmah, C.; Royal, C. D.; Leppert, M. F.; Dixon, M.; Peiffer, A.; Qiu, R.; Kent, A.; Kato, K.; Niikawa, N.; Adewole, I. F.; Knoppers, B. M.; Foster, M. W.; Clayton, E. W.; Watkin, J.; Muzny, D.; Nazareth, L.; Sodergren, E.; Weinstock, G. M.; Yakub, I.; Birren, B. W.; Wilson, R. K.; Fulton, L. L.; Rogers, J.; Burton, J.; Carter, N. P.; Clee, C. M.; Griffiths, M.; Jones, M. C.; McLay, K.; Plumb, R. W.; Ross, M. T.; Sims, S. K.; Willey, D. L.; Chen, Z.; Han, H.; Kang, L.; Godbout, M.; Wallenburg, J. C.; L'Archeveque, P.; Bellemare, G.; Saeki, K.; An, D.; Fu, H.; Li, Q.; Wang, Z.; Wang, R.; Holden, A. L.; Brooks, L. D.; McEwen, J. E.; Guyer, M. S.; Wang, V. O.; Peterson, J. L.; Shi, M.; Spiegel, J.; Sung, L. M.; Zacharia, L. F.; Collins, F. S.; Kennedy, K.; Jamieson, R.; Stewart, J. A. Second generation human haplotype map of over 3.1 million SNPs. *Nature* 2007, 449, 851-861.

Ge, H.; Player, C. M.; Zou, L. Toward a global picture of development: Lessons from genome-scale analysis in *Caenorhabditis elegans* embryonic development. *Dev Dyn* 2006, 235, 2009-2017.

Goldstein, D. B.; Chikhi, L. Human migrations and population structure: What we know and why it matters. *Annu Rev Genomics Hum Genet* 2002, 3, 129-152.

Gunsalus, K. C.; Ge, H.; Schetter, A. J.; Goldberg, D. S.; Han, J. D.; Hao, T.; Berriz, G. F.; Bertin, N.; Huang J.; Chuang, L. S.; Li, N.; Mani, R.; Hyman, A. A.; Sonnichsen, B.; Echeverri, C. J.; Roth, F. P.; Vidal, M.; Piano, F. Predictive models of molecular machines involved in *Caenorhabditis elegans* early embryogenesis. *Nature* 2005, 436, 861-865.

Haldane, J. Sex ratio and unisexual sterility in hybrid animals. *J Genet* 1922, 12, 101-109.

Haldane, J. Can a species concept be justified? In *The species concept in paleontology*, Sylvester-Bradley, P., ed., Systematics Association: London, UK, 1956; pp. 95-96.

Honore, L. H. Pathology of female infertility. *Curr Opin Obstet Gynecol* 1997, 9, 37-43.

Hull, M. G.; Glazener, C. M.; Kelly, N. J.; Conway, D. I.; Foster, P. A.; Hinton, R. A.; Coulson, C.; Lambert P. A .; Watt, E. M .; Desai, K. M. Population study of causes, treatment, and outcome of infertility. *Br Med J* 1985, 291, 1693-1697.

Hutter, P.; Roote, J.; Ashburner, M. A genetic basis for the inviability of hybrids between sibling species of Drosophila. *Genetics* 1990, 124, 909-920.

Jose-Miller, A. B.; Boyden, J. W.; Frey, K. A. Infertility. *Am Fam Physician* 2007, 75, 849-856.

Kimura, M. The rate of molecular evolution considered from the standpoint of population genetics. *Proc Natl Acad Sci USA* 1969, 63, 1181-1188.

Kuhn T. *The Structure of Scientific Revolutions.* University of Chicago Press: Chicago, USA, 1962.

Kumar, T. R. What have we learned about gonadotropin function from gonadotropin subunit and receptor knockout mice? *Reproduction* 2005, 130, 293-302.

Kuo, Y. M.; Duncan, J. L.; Westaway, S. K.; Yang, H.; Nune, G.; Xu, E. Y.; Hayflick, S. J.; Gitschier, J. Deficiency of pantothenate kinase 2 (Pank2) in mice leads to retinal degeneration and azoospermia. *Hum Mol Genet* 2005, 14, 49-57.

Lakatos, I. *The Methodology of Scientific Research Programmes: Philosophical Papers, Volume 1.* Cambridge University Press: Cambridge, UK, 1977.

Llopart, A.; Lachaise, D.; Coyne, J. A. An anomalous hybrid zone in Drosophila. *Int J Org Evolution* 2005, 59, 2602-2607.

Mallet, J. The speciation revolution. *J Evol Biol* 2001, 14, 887-888.

Mallet, J. Speciation in the 21st century. *Heredity* 2005, 95, 105-109.

Mallet, J. What does Drosophila genetics tell us about speciation? *Trends Ecol Evol* 2006, 21, 386-393.

Mallet, J. Hybrid speciation. *Nature* 2007, 446, 279-283.

Martinez, M. C.; Bernabe, M. J.; Gomez, E.; Ballesteros, A.; Landeras, J.; Glover, G.; Gil-Salom, M.; Remohi, J.; Pellicer, A. Screening for AZF deletion in a large series of severely impaired spermatogenesis patients. *J Androl* 2000, 21, 651-655.

Masly, J. P.; Jones, C. D.; Noor, M. A.; Locke, J.; Orr, H. A. Gene transposition as a cause of hybrid sterility in Drosophila. *Science* 2006, 313, 1448-1450.

Maurer, B.; Simoni, M. Y. Chromosome microdeletion screening in infertile men. *J Endocrinol Invest* 2000, 23, 664-670.

Mayr, E. *Animal Species and Evolution.* Harvard University Press: Cambridge, Massachusetts, 1963.

McCarthy, R. A.; Argraves, W. S. Megalin and the neurodevelopmental biology of sonic hedgehog and retinol. *J Cell Sci* 2003, 116, 955-960.

Miki, K.; Qu, W.; Goulding, E. H.; Willis, W. D.; Bunch, D. O.; Strader, L. F.; Perreault, S. D.; Eddy, E. M.; O'Brien, D. A. Glyceraldehyde 3-phosphate dehydrogenase-S, a sperm-specific glycolytic enzyme, is required for sperm motility and male fertility. *Proc Natl Acad Sci USA* 2004, 101, 16501-16506.

Moehring, A. J.; Llopart, A.; Elwyn, S.; Coyne, J. A.; Mackay, T. F. The genetic basis of postzygotic reproductive isolation between *Drosophila santomea* and *D. yakuba* due to hybrid male sterility. *Genetics* 2006, 173, 225-233.

Muller, H. Bearings of the Drosophila work on systematics. In *The new systematics*, Huxley, J. ed., Clarendon Press: Oxford, UK, 1940; pp. 185-268.

Najmabadi, H.; Huang, V.; Yen, P.; Subbarao, M. N.; Bhasin, D.; Banaag, L.; Naseeruddin, S.; de Kretser, D. M.; Baker, H. W.; McLachlan, R. I.; Loveland, K. A.; Bhasin, S. Substantial prevalence of microdeletions of the Y-chromosome in infertile men with idiopathic azoospermia and oligozoospermia detected using a sequence-tagged site-based mapping strategy. *J Clin Endocrinol Metab* 1996, 81, 1347-1352.

Naz, R. K.; Rajesh, C. Gene knockouts that cause female infertility: Search for novel contraceptive targets. *Front Biosci* 2005, 10, 2447-2459.

Olson, E. N. Gene regulatory networks in the evolution and development of the heart. *Science* 2006, 313, 1922-1927.

Onda, M.; Ota, K.; Chiba, T.; Sakaki, Y.; Ito, T. Analysis of gene network regulating yeast multidrug resistance by artificial activation of transcription factors: Involvement of Pdr3 in salt tolerance. *Gene* 2004, 332, 51-59.

Orr, H. A. The population genetics of speciation: The evolution of hybrid incompatibilities. *Genetics* 1995, 139, 1805-1813.

Orr, H. A. Does hybrid lethality depend on sex or genotype? *Genetics* 1999, 152, 1767-1769.

Orr, H. A. The genetic basis of reproductive isolation: Insights from Drosophila. *Proc Natl Acad Sci USA* 2005, 102, S6522-6526.

Orr, H. A.; Irving, S. Genetic analysis of the hybrid male rescue locus of Drosophila. *Genetics* 2000, 155, 225-231.

Orr, H. A.; Irving, S. Complex epistasis and the genetic basis of hybrid sterility in the *Drosophila pseudoobscura* Bogota-USA hybridization. *Genetics* 2001, 158, 1089-1100.

Orr, H. A.; Madden, L. D.; Coyne, J. A.; Goodwin, R.; Hawley, R. S. The developmental genetics of hybrid inviability: A mitotic defect in Drosophila hybrids. *Genetics* 1997, 145, 1031-1040.

Palagiano, A. Female infertility: The tubal factor. *Minerva Ginecol* 2005, 57, 537-543.

Pennisi, E. Evolution. Two rapidly evolving genes spell trouble for hybrids. *Science* 2006, 314, 1238-1239.

Perez, D. E.; Wu, C. I.; Johnson, N. A.; Wu, M. L. Genetics of reproductive isolation in the *Drosophila simulans* clade: DNA marker-assisted mapping and characterization of a hybrid-male sterility gene, Odysseus (Ods). *Genetics* 1993, 13, 261-275.

Popper, K. *The logic of scientific discovery*. Hutchinson and Co.: London, UK, 1959.

Presgraves, D. C.; Balagopalan, L.; Abmayr, S. M.; Orr, H. A. Adaptive evolution drives divergence of a hybrid inviability gene between two species of Drosophila. *Nature* 2003, 423, 715-719.

Rogers, S. M.; Bernatchez, L. The genetic basis of intrinsic and extrinsic post-zygotic reproductive isolation jointly promoting speciation in the lake whitefish species complex (*Coregonus clupeaformis*). *J Evol Biol* 2006, 19, 1979-1994.

Russell, S. T.; Magurran, A. E. Intrinsic reproductive isolation between Trinidadian populations of the guppy, *Poecilia reticulata*. *J Evol Biol* 2006, 19, 1294-1303.

Sonneborn, T. The Paramecium aurelia complex of fourteen sibling species. *Trans Am Microsc Soc* 1975, 94, 155-178.

Stebbins, G. L. Mosaic evolution: An integrating principle for the modern synthesis. *Experientia* 1983, 39, 823-834.

Storfer, A. Gene flow and endangered species translocations: A topic revisited. *Biol Conserv* 1999, 87, 173-180.

Updike, D. L.; Mango, S. E. Temporal regulation of foregut development by HTZ-1/H2A.Z and PHA-4/FoxA. *PLoS Genet* 2006, 2, e161.

Vortkamp, A. Interaction of growth factors regulating chondrocyte differentiation in the developing embryo. *Osteoarthritis Cartilage* 2001, 9, S109-117.

Watson, J. D.; Crick, F. H. Molecular structure of nucleic acids; a structure for deoxyribose nucleic acid. *Nature* 1953, 171, 737-738.

Weir, B. S.; Cardon, L. R.; Anderson, A. D.; Nielsen, D. M.; Hill, W. G. Measures of human population structure show heterogeneity among genomic regions. *Genome Res* 2005, 15, 1468-1476.

Wu, C. I. The genic view of the process of speciation. *J Evol Biol* 2001, 14, 851-865.

Wu, C. I.; Ting, C. T. Genes and speciation. *Nat Rev Genet* 2004, 5, 114-122.

Zwijsen, A.; van Grunsven, L. A.; Bosman, E. A.; Collart, C.; Nelles, L.; Umans, L.; Van de Putte, T.; Wuytens, G.; Huylebroeck, D.; Verschueren, K. Transforming growth factor beta signalling *in vitro* and *in vivo*: Activin ligand-receptor interaction, Smad5 in vasculogenesis, and repression of target genes by the deltaEF1/ZEB-related SIP1 in the vertebrate embryo. *Mol Cell Endocrinol* 2001, 180, 13-24.

CHAPTER 5

NEXT-GENERATION SEQUENCING AND MICROBIOME EVALUATION: MOLECULAR MICROBIOLOGY AND ITS IMPACT ON HUMAN HEALTH

A. FOCA, M. C. LIBERTO, and N. MARASCIO

CONTENTS

5.1   Introduction 122

5.2   Next-Generation Sequencing Technology – Past, Present and Future.....124

5.3   Next-Generation Sequencing and Human Microbiome 130

5.4   Next-Generation Sequencing in Clinical Microbiology....................... 132

5.5   Conclusion.. 135

Keywords 136

References.. 136

5.1   INTRODUCTION

We live among millions of microorganisms, whose ubiquitous communities have a profound impact on our health. A large variety of microbial populations (microbiota) and their genetic apparatus (microbiome) playing a substantial role in the maintenance of health and in the onset of disease. While the human genome is basically stable (Weinstock, 2011), the microbiome – the genetic material pertaining to all the microbes that live within the human body and, therefore, represents a sort of *second human genome* (according to Weinstock GM) – undergoes significant changes, varying not only between individuals but also within the same person. Viruses, bacteria, and eukaryotic microorganisms coexist in complex communities, and can interact with the micro- and macro-environments in which they live, influencing the development, metabolism, and functions of higher organisms – namely us (Relman, 2011). At any one time, we can play host to up to three types of microorganisms. The first type, considered native or resident, comprises a population arising from the first colonization after the beginning of extra-uterine life. We can also be colonized by transient populations, which invade through a lumen without causing major changes to the resident bacterial population, as well as contaminant or frankly pathogenic microorganisms, which differ from the former by causing dismicrobism and the onset of infectious disease. Each such microbial population thrives in different "environmental" conditions (pH, temperature, etc.), and reacts differently to the action of drugs and/or exogenous chemicals (antimicrobials, enzymes, etc.). They also differ in terms of developmental changes to the genome, which affect the so-called *quasispecies* as a whole, and must compete for their microbial and nutritional needs, etc.

An individual host contains thousands of symbiotic species, of which an estimated 90 % have not yet been cultivated in the laboratory. To expand our knowledge, we also need to further investigate the role of individual components of the microbiota in a large variety of physiological conditions and disease states. To undertake such a monumental task, the sharing of protocols, methods, understanding, and results are essential, particularly with regard to innovative tools and the isolation/sequencing of uncultured microorganisms. In this context, recent advances in sequencing technologies (Next Generation Sequencing, NGS) and the widespread use of mathematics, bioinformatics, and genomics databases have produced great leaps forward in terms of our understanding of the pathophysiological mechanisms at work. In particular, by exploiting such databases, the Human Microbiome Project (HMP) (Human Microbiome Project Consortium, 2012), alongside other studies, have provided scientific demonstrations that have revolutionized how we view the relationships between ourselves and the complex microbial populations that exist within us. Such collaborative studies on the human microbiome comprise a useful metagenomic analysis that is helping unravel the complex mechanisms between microbial community structure and function and the human body (Gilbert and Dupont, 2011). As stated by the HMP Consortium, *"Collectively the data represent a treasure trove that can be mined to identify new organisms, gene functions, and metabolic and regulatory networks, as well as correlations between microbial community structure and health and disease"* (Human Microbiome Project Consortium, 2012).

Indeed, next-generation sequencing provides us with a unique opportunity to study the microbial composition of the human body, and the interactions between these microbes and their host (Xia *et al.*, 2014). In the light of the emerging evidence, long-used terms such as indigenous residents and transitional pathogens are being redefined, and many new or more precise definitions of terms such as microbiota, microbiome, virome, bacteriome, metabolome, proteome, phenotype, genotype, prototype, transcriptome, and metagenome are appearing. Metagenomics exploits the remarkable potentials of computational biology, using databases for the synthesis and understanding of the most important features of the genomes of individual microbial (bacterial, viral, micro eukaryotic genomes) loads (raw form) and their groupings in the different host environments. The persistent and transient microbial communities that live in the various tissues of the human body feature different compositions and therefore express different metagenomes.

The composition of microbial populations residing in the gut, in particular, present considerable variability, even within the same individual, with dietary change and other environmental/host factors playing a key role in their alteration. Such changes in the resident microbial community can cause pathophysiological imbalances in distant districts of the body, as well as in the intestinal populations themselves. For instance, *Streptococcus sanguinis* resides in the buccal cavity, requiring manganese and ribonucleotide reductase RNR 1b expression for growth and virulence, yet is the causative agent of infectious endocarditis and damage to heart valves. The composition and metabolic products of the gut microbiota also affect the modulation of immune reactivity and inflammatory responses, thereby affecting the course (and perhaps even onset) of inflammatory diseases such as asthma (Segata *et al.*, 2012). In this two-way relationship, the immune system also affects the gut microbial flora, and a deficiency in adaptive/innate immunity, such as the lack of immune response to pathogens or gene polymorphisms involved in innate immunity, has a dramatic effect on the composition of the intestinal microbiota (Maslowski and Mackay, 2011).

This delicate balance is especially importance in the newborn, whose gut already plays host to as many as a thousand species of indigenous bacteria (up to 100 trillion organisms), which are crucial to neonatal development and health (Jost *et al.*, 2012). Studies on early microbial colonization soon after birth, focusing in particular on the relationships between the host transcriptome, microbiome (genomic and transcriptomic data), and dietary substrates, indicate that differences in diet can even affect host gene expression – particularly in relation to the innate immune system (Schwartz *et al.*, 2012). Moreover, gut microbial communities heavily influence metabolic processes and environmental factors, and are implicated in the onset of metabolism-related diseases of later life, such as obesity, insulin resistance, diabetes, and cardiovascular disease (Tremaroli and Bäckhed, 2012).

It is clear then that functional metagenomic and metabolomic analysis of the gut microbiota, with a view to elucidating the complex interactions with prebiotic and probiotic foods, will have strong repercussions on our understanding of human health (Moore *et al.*, 2011). Furthermore, research has also shown that studying the different compositions of the resident bacterial community, in particular vaginal microbiota in pregnant and nonpregnant women, can provide us with potential diagnostic, prognostic, and therapeutic information (Romero *et al.*, 2014). In such studies, computational biology and next-generation sequencing (NGS) are the key, helping to unlock the secrets of the synthesis and major features of microbial genomes, and shed light on how they interact with the different environments within us.

5.2   NEXT-GENERATION SEQUENCING TECHNOLOGY – PAST, PRESENT AND FUTURE

Our understanding of biodiversity, and indeed our success in managing diseases, is directly linked to our ability to acquire genomic information. Take, for example, the great leaps forward in terms of the diagnosis and prevention of infection directly ascribable to the introduction of molecular assays, which far outstripped the culture techniques that we had hitherto relied upon. Despite such advances, understanding the pathogenic mechanisms related to the microorganisms associated with human health and disease still represents a major challenge. To this end, sequencing is crucial method that enables us not only to quantify microbial communities, but also to identify novel genomes, new taxa, and microorganisms that have never even been cultured (Sharpton, 2014). Moreover, sequencing provides us with a means of analyzing the molecular evolution and circulation of microbial pathogens in specific areas of the world (Marascio *et al.*, 2014).

An ideal sequencing technology would be fast, accurate, and inexpensive, and these are the goals that their developers are moving toward. The first sequencing method, called Sanger chemistry, or first-generation sequencing (FGS), required a specific primer, and sequencing had to be started at a specific location along the DNA template. Furthermore, it remains unable to detect molecular mutants present at low frequencies. Pyrosequencing, on the other hand, is able to detect some minor sequence variants (lamivudine resistant in heterogeneous Hepatitis B virus populations) (Lindström *et al.*, 2004). This, along with other limitations of Sanger sequencing, fueled the search for next-generation sequencing (NGS) methods, which, since their advent in 2005, have become the preferred way to study genomic data.

NGS is essentially a suite of new sequencing technologies that can produce massive amounts of DNA sequence data, detecting pathogens without the need for specific PCR primers. With these methods, DNA is broken into small pieces and read randomly along the entire genome, and the raw throughput for the mass parallel sequencing used in NGS is significantly greater than that handled by electrophoresis (FGS) (Margulies *et al.*, 2005). Briefly, there are three fundamental steps in NGS systems, and all three, namely preparation of the library of nucleic acids, clonal amplification of the libraries and massive parallel sequencing of DNA fragments, can be carried out in a single experiment. Among the new platforms on the market, the Roche GS-FLX 454 Genome Sequencer (originally 454 sequencing) (http://www.454.com/) (Bialasiewicz *et al.*, 2014; Knapp *et al.*, 2014), the Illumina Genome Analyzer (originally Solexa Technology) (http://www.illumina.com/) (Monecke *et al.*, 2013; Idris *et al.*, 2014) and the ABI SOLiD analyzer (http://solid.appliedbiosystems.com) (Umemura *et al.*, 2013) stand out.

These sequencing platforms generally rely on "tailor-made" nucleic acid libraries that require conversion beforehand (van Dijk *et al.*, 2014). The HiSeq platform (Illumina/Solexa), for example, generates DNA libraries on the surface of lanes in a flow cell (Yegnasubramanian, 2013). Several types of library can be created, depending on their ultimate purpose: DNA-seq (genomic DNA sequencing), RNA-seq (genomic RNA sequencing), mRNA-seq (mRNA sequencing), smallRNA-seq (small RNA sequencing), CHIP-seq (Chromatin Immunoprecipitation Sequencing), MeDip-seq (methylated DNA sequencing), BS-seq (Bisulfite sequencing to measure the methylation state of a whole genome), and RIP-seq (RNA immunoprecipitation) (Chen *et al.*, 2013; Ford *et al.*, 2014; Head *et al.*, 2014) comprise those developed so far.

Once the libraries have been prepared, the second step in NGS, clonal amplification, also tends to be system-specific, with Roche 454 and Applied Biosystems SOLiD systems in particular employing emulsion PCR to amplify template

DNA molecules clonally. In this process individual templates are sequestered, along with PCR reagents, within an aqueous droplet. This droplet is surrounded by a hydrophobic shell within an oil-in-water emulsion. In the Roche 454 system, the droplets are then deposited in picoliter plate-wells, while in the Applied Biosystems method, the amplified DNA molecules are modified and randomly attached to the surface of a glass slide, which is subsequently loaded onto the instrument (Yegnasubramanian, 2013).

In the third step, high-throughput sequencing (HTS), each DNA fragment is sequenced individually. This is performed by either paired-end or mate-pair sequencing (Padmanabhan *et al.*, 2013). The optimal insert size of the DNA fragments is determined by the specific sequencing application (Head *et al.*, 2014). Commercial platforms use different chemistries to sequence template DNA molecules: pyrosequencing (i.e., Roche system), sequence-by-synthesis (i.e., Roche, Illumina), or sequence-by-ligation (i.e., Applied Biosystems). Such technologies have thereby enabled advances such as the identification and quantification of a host of transcripts. Furthermore, in 2014 Quin *et al.*, showed through transcriptome analysis that resveratrol would disturb the expression of genes related to quorum sensing, surface, and secreted proteins, and capsular polysaccharides. They also showed that resveratrol and ursolic acid could be useful adjunct therapies for the treatment of methicillin-resistant *Staphylococcus aureus* (MRSA) biofilm-involved infections (Quin *et al.*, 2014). Nevertheless, second-generation systems (SGS), while improving throughput with respect to first NGS techniques, suffer from the same major limitation, namely short reads (Zhang *et al.*, 2011).

The latest generation of sequencers, appropriately termed "next-next" or third-generation sequencing (TGS), are, however, capable of generating longer sequence reads in a shorter time. This is because PCR is not needed before sequencing, and signals can be captured in real time (Liu *et al.*, 2012). TGS technologies in use today include, among others, true single-molecule sequencing (tSMS™) which requires no amplification, ligation, or cDNA synthesis (http://www.helicosbio.com/); single-molecule real-time sequencing (SMRT™) (http://www.pacificbiosciences.com/); fluorescence resonance energy transfer (FRET); nonfluorescent sequencing systems using nanopores or nanoedges; and Raman-based methods such as sequencing using surface-enhanced Raman spectroscopy (SERS) (Thudi *et al.*, 2012).

In reality, there is little consensus on the distinction between second- and third-generation technologies. For instance, the Ion Torrent sequencer (Life Technologies) could be considered as either; by eliminating the need for light and simplifying the overall sequencing process, it significantly accelerates the time to result (Schadt *et al.*, 2010), a defining characteristic of TGS technologies. However, despite their increased throughput and the ability to generate reads that can span many thousands of bases, the accuracy of TGS technologies is still significantly lower than that of their predecessors. To resolve this issue, Bashir *et al.*, suggested combining the highly accurate SGS data with the longer read lengths provided by TGS systems. Indeed, by merging data generated from Roche 454/Illumina and SMRT sequencing (discussed below), they were able to identify the complete genome of *Vibrio cholerae* isolates from the Haitian cholera outbreak (Bashir *et al.*, 2012). Accuracy can also be increased, whilst preserving low working times, by sequencing RNA directly; direct RNA sequencing and *de novo* assembly are two of the major innovations that distinguish the latest sequencing technologies. Table 5.1 summarizes the main features of three generations (first NGS, SGS, and TGS) of sequencing technologies created to date.

TABLE 5.1   Main features of NGS, SGS, and TGS platforms.

HTS Technology	Read Length	Throughput	Results Turn-around	Accuracy	Data Analysis
NGS: SBS or Pyrose-quencing	Short	Moderate	Days	High	Complex
SGS: Wash-and-scan SBS	Short	High	Days	High	Complex
TGS: SBS, Degradation, direct physical inspection	Long	Moderate	Hours	Moderate	Complex

Sequence alignment is also an indispensible feature of new sequencing technologies, as correct genome assembly is essential for the study of bacteria, viruses, fungi, and protozoa. Traditionally, the BLAST algorithm (http://blast.ncbi.nlm.nih.gov/Blast.cgi) has been used to classify different pathogens (Altschul *et al.*, 1990), but the challenge in new technologies is to process massive amounts of raw DNA sequences, which is still a bottleneck in terms of understanding genomes. Indeed, the various bioinformatics programmes required to handle sequence alignment and assembly, and processing of the data produced by the various NGS platforms, which differ in terms of both format and length of reads, can be difficult to use without informatics expertize. The more the technology advances, the more data the software has to handle, and consequently the more complex the entire operation becomes. Genome assembly, for example, is more difficult to carry out using short-read sequencing data than it is using Sanger data. Likewise, as TGS generates much longer reads than SGS, raw read error rates are a drawback in most third-generation technologies.

To overcome these difficulties, and the accuracy problems mentioned above, in 2012, Au *et al.*, proposed the longest common sequence (LCS) method for the combined analysis of data from second- and third-generation sequencers (Au *et al.*, 2012). Moreover, since the early 2000s, online public server resources have been enhanced to improve data sharing and analysis, such as that found on the NCBI genome page (http://www.ncbi.nlm.nih.gov/genome). A variety of software tools are available online for data analysis, including JR-Assembler (http://jrassembler.iis.sinica.edu.tw/) (Chu *et al.*, 2013), ABySS (http://www.bcgsc.ca/platform/bioinfo/software/abyss) (Simpson *et al.*, 2009), Edena (http://www.genomic.ch/edena) (Hernandez *et al.*, 2008), SOAPdenovo (http://soap.genomics.org.cn/soapdenovo.html) (Li *et al.*, 2010), Taipan (http://taipan.sourceforge.net) (Schmidt *et al.*, 2009), Velvet (http://www.ebi.ac.uk/~zerbino/velvet) (Zerbino and Birney, 2008) and GAM-NGS (http://github.com/vice87/gam-ngs) (Vicedomini *et al.*, 2013), and a list of relevant software can be found on the Seqanswer home page http://seqanswers.com/. The first freely available platform for use in microbiology research, *Orione* (http://orione.crs4.it) is a bioinformatics tool based on Galaxy (https://main.g2.bx.psu.edu/) (Goecks *et al.*, 2010). It enables preprocessing, that is, quality control of reads and their trimming; reads mapping (for instance aligners used SOAP); *de novo* assembly (i.e., Velvet, ABySS); scaffolding; and postassembly through contigs statistics, (multi) aligning and variant calling. It is also able to identify meaningful biological information from sequences, and carry out bacterial RNA-Seq analysis, as well as metagenomics and metatranscriptomics operations (Cuccuru *et al.*, 2014).

The driving force behind this profusion of dedicated software is the need to analyze results accurately and in a short time. To this end, Naccache *et al.*, have described a sequence-based ultra-rapid pathogen identification (SURPI) pipeline capable of identifying pathogens in complex metagenomic data generated from clinical samples, thereby contributing to real-time microbial diagnosis in acutely ill patients (Naccache *et al.*, 2014). The parallel development of software and algorithms capable of handling the massive amount of data generated by high-throughput sequencing (HTS) platforms also expedites the implementation of molecular epidemiology studies. The primary tool in molecular epidemiology is phylogenetic analysis, and several dedicated computational methods are available for processing metagenomic samples. Unfortunately, however, these methods are generally unable to distinguish between different bacterial strains. An exception to this rule, MetaID, was, however, introduced in 2013 by Srinivasan *et al.*, MetaID is an alignment-free "*n*-gram" (referring to nucleotide sequence of fixed length (*n*) approach that can accurately identify microorganisms at the strain level and estimate the abundance of each bacterial organism in a sample (Srinivasan and Guda, 2013).

Specific software can also be used to identify new viruses, as well as confirming the presence and quantifying or studying sequence variations of known exemplars. As concluded by a recent international bioinformatics workshop on phylogenetic analysis, NGS data analysis require specific practical means of analyzing large sets of data to study the evolution and molecular epidemiology of viruses (http://regaweb.med.kuleuven.be/veme-workshop/2014/index.php). To this end, a new tool, *Virus Finder*, was developed by Wang *et al.*, This enables not only characterization of viruses but also that of their integration sites in the human genome (Wang *et al.*, 2013). To aid diffusion of information, many algorithms are distributed in binary for Windows, MacOS X and Linux (Magi *et al.*, 2010). The workflow of high-throughput sequencing can be summarized as in Figure 5.1.

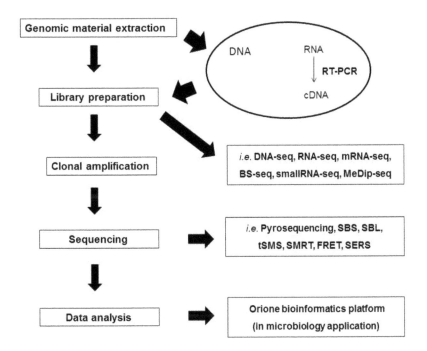

FIGURE 5.1 Workflow of high-throughput sequencing.

Next-, second- and third-generation sequencing methods, by accessing the entire genome of microorganisms and obtaining DNA/RNA from mixed genomes in natural environments or clinical samples, continue to make important contributions to microbial genome research. In the future, when processing such information becomes more practical, it is foreseeable that diagnostic workups will comprise whole genome screening. In the next section, we report the practical applications of next-generation sequencing already in use.

5.3   NEXT-GENERATION SEQUENCING AND HUMAN MICROBIOME

NGS methods have helped us begin to understand the intricate biodiversity of the human microbiome made up of the entire complex of microorganisms that colonize every niche of the human body. Before the development of NGS, the main system used to amplify genes from a wide range of organisms was the cloning of full-length 16S rRNA genes. However, this method was superseded by NGS in terms of cost and sampling capacities – a vital step forward when one considers the ultimate goal of human microbiome research, namely its clinical application as a diagnostic, therapeutic, and preventative tool.

Thanks to such high-throughput technology, since 2009 the US Human Microbiome Project (HMP) (NIH HMP Working Group et al., 2009) has reported sequencing data pertaining to more than 10,000 specimens from 18 body sites in healthy adults, providing the first in-depth characterization of the human microbiome. Building on this research, and using additional data derived from a DMM (Dirichlet multinomial mixture)-based approach, Ding and Schloss (2014) came to three important conclusions: (i) there is a strong association between certain features of an individual's life history (whether they were breastfed, their gender, and their level of education) and their microbiome composition at several body sites; (ii) although there are differences in the specific compositions of oral and gut microbiomes within an individual, these are predictive of each other; and (iii) the most stable microbial community types are found in the stools and vagina, while the least stable in the oral cavity. The important points to note are that even though there are considerable intra- and interpersonal differences in the human microbiome, this variation can be split into community types predictive of each other, and a particular type, or changing types, could be used to assess disease risk and to personalize antimicrobial treatment.

The main sites studied so far in terms of their microbial constituents are the gut (Greenblum et al., 2012), urinary tract (Nelson et al., 2010; Wolfe et al., 2012), lower and upper respiratory tract (Willner et al., 2009), vagina (Oakley et al., 2008; Hummelen et al., 2010), skin (Grice et al., 2009; Capone et al., 2011;

Fahlén *et al.*, 2012), gingiva (Lazarevic *et al.*, 2009), and wounds (Dowd *et al.*, 2008) in health and in disease (Pflughoeft and Versalovic, 2012). Studying the gut microbiome has given us a better idea of how to classify disease processes such as inflammatory bowel disease, and sheds light on specific conditions such as Crohn's disease or idiopathic bowel syndrome (Manichanh *et al.*, 2006). It will not be long before we are able to assess a range of diseases in this manner, giving us useful tools for improved clinical management.

Indeed, studying the human microbiome will eventually enable us to monitor its diversity during and in response to therapy, providing information that can be used to both predict prognosis and improve poor clinical outcomes. These advances will also extend far beyond bacterial infection, as the human microbiome, while predominantly made up of bacteria, also contains eukaryotic microbes and viruses (both human viruses and bacteriophages). The human virome is the collection of all viruses that are found in or on humans, and includes both eukaryotic and prokaryotic viruses, viruses causing acute, persistent, or latent infection, and viruses integrated into the human genome, such as endogenous retroviruses (Duerkop and Hooper, 2013; Minot *et al.*, 2013; Virgin, 2014). Recently a pioneering high-throughput metagenomic sequencing study of the skin microbiome in five healthy individuals and one patient with Merkel cell carcinoma revealed the high diversity of the viral cutaneous flora detected on normal-appearing human skin. The eukaryotic DNA viruses exhibited particularly high diversity, including various representatives of *Papillomaviridae*, *Polyomaviridae*, and *Circoviridae*. These findings emphasize the complexities of the viral flora composition on the surface of the skin, whose alteration may eventually be used to diagnose and/or monitor disease, particularly skin disorders and tumors (Foulongne *et al.*, 2012).

In similar studies of the intestinal microbiota, viruses that replicate in eukaryotic cells and in bacteria have been detected. Such studies have also highlighted the presence of viruses that infect plants, presumably introduced along with the diet (Minot *et al.*, 2011). The emerging evidence regarding resident viruses and their interactions with the immune system, in which the intestine plays a key role, suggests that these pathways are very complex. For instance, it has been observed that mice latently infected with either murine gammaherpesvirus 68 or murine cytomegalovirus, which are genetically very similar to the human pathogens Epstein-Barr virus and human cytomegalovirus, are resistant to lethal infection with both *Listeria monocytogenes* and *Yersinia pestis*. This protection has been convincingly ascribed to the increased production of interferon γ (IFN-γ) expression and systemic activation of macrophages triggered by the latent bacterial infection (Barton *et al.*, 2007).

Conversely, and more recently, Handley *et al.*, have demonstrated that enteric viral infections may affect AIDS enteropathy and disease progression. In particular, in nonhuman primates, the enteric virome has been shown to enhance the progression of SIV infection to AIDS by causing damage to the intestinal epithelium and activating the immune system through pathogen release (Handley *et al.*, 2012). It is therefore evident that the release of bacterial, viral, fungal, or other PAMPs and antigens into host tissues and the systemic circulation can have a wide variety of effects in various disease states, and future investigations into the interactions between resident and invading organisms and the immune system will no doubt have huge repercussions on patient management and human health.

5.4   NEXT-GENERATION SEQUENCING IN CLINICAL MICROBIOLOGY

As well as its appeal in the research field, next-generation sequencing (NGS) is finding a growing range of applications in basic and clinical laboratory microbiology. It is proving useful in the diagnosis of emerging infections, molecular epidemiology of pathogens, and drug-resistance testing (Dunne *et al.*, 2012), and enables detection of viruses as well as viral genome sequencing. By these means, NGS is providing us additional information on viral genome variability, molecular epidemiology of viral infections, quasi-species and virus–host interaction, adaptation strategies, and resistance to antiviral molecules, fueling the search for vaccine discovery. In addition, NGS has been applied to a metagenomics-based approach to identify and characterize communities of previously unknown viruses (Bexfield and Kellam, 2011; Radford *et al.*, 2012). These far-reaching techniques are even opening up avenues in terms of their applications in diagnostic virology, not merely replacing conventional or molecular methods.

5.4.1   APPLICATIONS OF NGS IN VIROLOGY

The most influential new application of NGS in virology is whole genome sequencing, and the list of viral genomes now unlocked is growing rapidly. This, in turn, is enabling us to shine a light on viral genome diversity and evolution, for example in HIV-1 (Tebit and Arts, 2011). Since its isolation, different molecular methods applied to a variety of samples have vastly improved knowledge regarding the origins and genetic evolution of HIV-1, in particular in terms of the geographical distribution of subtypes and how their forms are changing over time. Indeed, once the HIV-1 genome had been fully sequenced, we were able define a specific classification of subtypes and recombinants, as well the various *quasispecies* circulating within infected individuals. Current data on new circulating recombinant forms (CRFs), which are continuously emerging worldwide, emphasize the importance of NGS methods in keeping abreast of developments, further to the future control and prevention of HIV (Tebit and Arts, 2011).

Aside from such medium-to-long-term research targets, as reported by Baillie *et al.* (2012), whole-genome analysis, molecular epidemiology and surveillance can also be used to investigate rapidly evolving viruses, such as the influenza virus, during the course of a single epidemic or pandemic, (Baillie *et al.*, 2012). In this way full-length sequencing can not only provide information vital for defining vaccination strategies, but also affect viral fitness and the outcome of antiviral therapies, in HCV for instance, by enabling the identification of pathogenic mutations of the virus genome (Ninomiya *et al.*, 2012).

Ultra-deep targeted sequencing can be used for virus genotyping (e.g., HCV, human papillomaviruses), as well as analysis of viral populations, and identification of *quasispecies*, virulence factors and drug-resistance genes (Barzon *et al.*, 2011; Ninomiya *et al.*, 2012). For instance, Gorzer *et al.* (2010) discovered different HCMV human cytomegalovirus genotypes in clinical samples isolated from lung transplant patients by genotyping three of the most variable genes within the HCMV genome. In this study genotyping of gN, gO, and UL139 genes by ultra-deep pyrosequencing (UDPS) technology revealed a very high assortment of HCMV genotype populations in patients with mixed HCMV infection. This approach also evidenced distinct quantitative distribution pattern, relative to each other, of HCMV strains (Gorzer *et al.*, 2010). Deep RNA sequencing methodologies have also been used to conduct genome-wide transcript (transcriptome) analysis, thereby identifying new viral genes and transcript isoforms of the gammaherpesviruses (γHVs) (Concha *et al.*, 2012). This is particularly important when one considers that the gammaherpesviruses (γHV), such as Epstein-Barr virus (EBV) or Kaposi's sarcoma-associated herpesvirus (KSHV), are correlated with risk of developing several cancers. Although herpesvirus genomes are known to possess a definite genomic organization containing nonoverlapping ORFs with virus-specific unique genes, the great assortment of novel RNA transcripts from coding and "non-coding" regions detected by RNA sequencing methods suggest that their genome map is very complex indeed. Using such methods to unlock the organization, biogenesis, and function of these novel RNA transcripts will doubtless further increase our bank of information on the pathogenesis and oncogenic mechanisms triggered by such viruses.

5.4.2 APPLICATIONS OF NGS IN BACTERIOLOGY

Not only virology will benefit from such in-depth analysis. Whole-genome sequencing of bacterial populations is clarifying various aspects of within-host evolution, transmission history, and population structure, thereby providing insights into the *in vivo* evolution and modification of bacterial pathogens, as well as their pathogenic mechanisms, i.e., the etiology of bacteria-related diseases.

As regards within-host bacterial dynamics, NGS has been applied to several bacterial pathogens, mainly with a view to resequencing individual isolates. Nevertheless, with respect to viruses, very little is known about the evolution of bacteria *in vivo*. This knowledge gap is mainly due to the large size of bacterial genomes, which also demonstrate less genetic variation, and a slow rate of mutation, even in many of the most virulent bacterial pathogens (Achtman, 2008, 2012; Wilson, 2012). That being said, whole genome sequencing has recently been used to investigate the evolution of *S. aureus* in a single asymptomatic long-term carrier during progression to a fatal bloodstream infection. The results of this study showed that protein-shortening mutations in the genome of this invasive bacteria, particularly the substitution of knockout mutations, induced by premature stop codons, could be linked to the transition to invasive disease (Young *et al.*, 2012). *S. aureus* has also been studied by Golubchik *et al.*, who looked at its evolutionary dynamics and population genomics in asymptomatic carriers. Using NGS they were able to detect short-term bacterial change, both within individual hosts and along host-to-host transmission chains (Golubchik *et al.*, 2013). *S. aureus* is a common commensal and highly pathogenic bacteria, and studies such as these can provide major insights into the way it interacts with and evolves with its human hosts. In particular, its links to and behavior during chronic cystic fibrosis infection of the airways has been studied extensively. McAdam *et al.*, (2011), for instance, performed genome quencing of 3 *S. aureus* strains isolated from the sputum of a cystic fibrosis patient over a 26-month period, to investigate its molecular adaptation *in vivo*. They found differences in phage content and isolate-specific polymorphisms in genes with links to antibiotic resistance and virulence regulation. The genetic and phenotypic diversification they detected was highly suggestive of a heterogeneous infecting population arising from a single infecting strain.

Mycobacterium tuberculosis is also known to evolve during the course of infection, and this capacity to acquire chromosomal mutations confers drug resistance. Recent signs of resistance to the preferred drug isoniazid (INH) in latent tuberculosis have therefore prompted whole genome sequencing of the bacterium, and the mutation rate of *Mycobacterium tuberculosis* has thus been analyzed in isolates from cynomolgus macaques during active, latent, and reactivation phases of the disease. The distribution of single-nucleotide polymorphisms (SNPs) observed indicated that mutation rates during latent and active phases are similar, and the polymorphism pattern suggested a link between *in vivo* mutational and oxidative DNA damage (Ford *et al.*, 2011).

In addition, whole-genome sequencing can help us investigate transmission history, detecting transmission events and historical transmission patterns via population genomics. This improves our capacity to assess outbreaks and transmission chains, with clear advantages over traditional approaches. This has been demonstrated in a study on the clonal lineage of methicillin-resistant *Staphylococcus aureus* (MRSA), which tracked its evolution during hospital and intercontinental transmission (Harris *et al.*, 2010). On a more domestic scale, whole genome sequencing has also been used to study the frequency of within-household transmission of a uropathogenic *Escherichia coli* clone, highlighting the effects of host transmission on genetic mutation rates (Reeves *et al.*, 2011). Interestingly, a total of 20 base mutations in this bacterium, which causes persistent urinary tract infection (UTI), were noted in the 14 isolates analyzed from several members of the household, including a dog, over a 3-year period. Overall, 11 genotypes were identified, and the phylogenetic tree of the mutational changes evidenced 6 host transfer events over the monitoring period, with the family dog being infected at least twice. By yielding data on host-to-host transmission, and demonstrating host-specific adaptation during long-term infection, such investigations have major implications on recurrent epidemiology, helping to shed light on its etiology. In this particular case, it demonstrated that recurrent urinary tract infection can be caused by reinfection from family-associated clones, rather than from the normal resident microbiota of the host, or treatment failure.

5.5 CONCLUSION

Population genomics are also enabling us to reconstruct the spread patterns of ancient bacteria, identifying the geographical and temporal origin of historical pandemics and thereby the dominant routes behind global transmission. In particular, the spread of leprosy has been studied by sequencing over 400 isolates of *Mycobacterium leprae*, including strains from Brazil, Thailand, the United States, and skeletal remains from in and around Europe (Monot *et al.*, 2009). By these means, *Mycobacterium leprae* was classified into 16 SNP subtypes, revealing a strong geographical association pointing to leprosy arising in East Africa and then being transmitted along easterly and westerly trade routes (the Silk Road) linking Europe and China. Whole genome sequencing has also enabled us to identify differences between *Mycobacterium leprae* strains collected from patients with different clinical presentations, and to distinguish between virulent and nonvirulent isolates. Virulence is due to production of toxins, adhesins, capsule synthesis and other virulence factors, and a comprehensive knowledge of population structure may improve preventive public health interventions.

Interestingly a study carried out on more than 300 Group A *Streptococcus pyogenes* strains demonstrated that the capability to cause invasive disease is not restricted to specific strains, and closely related bacteria may be invasive or not (Beres *et al.*, 2010; Shea *et al.*, 2011). This observation supports a model in which *in vivo* mutation plays an important role in progression to invasive disease. In short, NGS is already helping us to unveil the mysteries of the microorganisms that surround us in intricate detail, and as its limitations (data processing, accuracy, practicability) are overcome, its applications in health and disease, whether past, present or future, are seemingly endless.

KEYWORDS

- **Accuracy**
- **Bacteriome**
- ***De novo* assembly**
- **Direct RNA sequencing**
- **Microbiome**
- **Next-generation sequencing technology**
- **Second-generation sequencing technology**
- **Third-generation sequencing technology**
- **Virome**

REFERENCES

Achtman, M. Evolution, population structure, and phylogeography of genetically monomorphic bacterial pathogens. *Annu Rev Microbiol* 2008, 62, 53–70.

Achtman, M. Insights from genomic comparisons of genetically monomorphic bacterial pathogens. *Phil Trans R Soc Lond B Biol Sci* 2012, 367, 860–867.

Altschul, S. F.; Gish, W.; Miller, W.; Myers, E. W.; Lipman, D. J. Basic local alignment search tool. *J Mol Biol* 1990, 215(3), 403–410.

Au, K. F.; Underwood, J. G.; Lee, L.; Wong, W. H. Improving PacBio long read accuracy by short read alignment. *PLoS One* 2012, 7(10), e46679.

Baillie, G. J.; Galiano, M.; Agapow, P. M.; Myers, R.; Chiam, R.; Gall, A.; Palser, A. L.; Watson, S. J.; Hedge, J.; Underwood, A.; Platt, S.; McLean, E.; Pebody, R. G.; Rambaut, A.; Green, J.; Daniels, R.; Pybus, O. G.; Kellam, P.; Zambon, M. Evolutionary dynamics of local pandemic

H1N1/2009 influenza virus lineages revealed by whole-genome analysis. *J Virol* 2012, 86(1), 11–18.

Barton, E. S.; White, D. W.; Cathelyn, J. S.; Brett-McClellan, K. A.; Engle, M.; Diamond, M. S.; Miller, V. L.; Virgin, H. W. 4th. Herpesvirus latency confers symbiotic protection from bacterial infection. *Nature* 2007, 447(7142), 326–329.

Barzon, L.; Militello, V.; Lavezzo, E.; Franchin, E.; Peta, E.; Squarzon, L.; Trevisan, M.; Pagni, S.; Dal Bello, F.; Toppo, S.; Palù, G. Human papillomavirus genotyping by 454 next generation sequencing technology. *J Clin Virol* 2011, 52(2), 93–97.

Bashir, A.; Klammer, A. A.; Robins, W. P.; Chin, C. S.; Webster, D.; Paxinos, E.; Hsu, D.; Ashby, M.; Wang, S.; Peluso, P.; Sebra, R.; Sorenson, J.; Bullard, J.; Yen, J.; Valdovino, M.; Mollova, E.; Luong, K.; Lin, S.; La May, B.; Joshi, A.; Rowe, L.; Frace, M.; Tarr, C. L.; Turnsek, M.; Davis, B. M.; Kasarskis, A.; Mekalanos, J. J.; Waldor, M. K.; Schadt, E. E. A hybrid approach for the automated finishing of bacterial genomes. *Nat Biotechnol* 2012, 30(7), 701–707.

Beres, S. B.; Carroll, R. K.; Shea, P. R.; Sitkiewicz, I.; Martinez-Gutierrez, J. C.; Low, D. E.; McGeer, A.; Willey, B. M.; Green, K.; Tyrrell, G. J.; Goldman, T. D.; Feldgarden, M.; Birren, B. W.; Fofanov, Y.; Boos, J.; Wheaton, W. D.; Honisch, C.; Musser, J. M. Molecular complexity of successive bacterial epidemics deconvoluted by comparative pathogenomics. *Proc Natl Acad Sci USA* 2010, 107(9), 4371–4376.

Bexfield, N.; Kellam, P. Metagenomics and the molecular identification of novel viruses. *Vet J* 2011, 190(2), 191–198.

Bialasiewicz, S.; McVernon, J.; Nolan, T.; Lambert, S. B.; Zhao, G.; Wang, D.; Nissen, M. D.; Sloots, T. P. Detection of a divergent Parainfluenza 4 virus in an adult patient with influenza like illness using next-generation sequencing. *BMC Infect Dis* 2014, 14(1), 275.

Capone, K. A.; Dowd, S. E.; Stamatas, G. N.; Nikolovski, J. Diversity of the human skin microbiome early in life. *J Invest Dermatol* 2011, 131, 2026–2032.

Chen, G. Q.; Zhuang, Q. Y.; Wang, K. C.; Liu, S.; Shao, J. Z.; Jiang, W. M.; Hou, G. Y.; Li, J. P.; Yu, J. M.; Li, Y. P.; Chen, J. M. Identification and survey of a novel avian coronavirus in ducks. *PLoS One* 2013, 8(8), e72918.

Chu, T. C.; Lu, C. H.; Liu, T.; Lee, G. C.; Li, W. H.; Shih, A. C. Assembler for *de novo* assembly of large genomes. *Proc Natl Acad Sci USA* 2013, 110(36), E3417–3424.

Concha, M.; Wang, X.; Cao, S.; Baddoo, M.; Fewell, C.; Lin, Z.; Hulme, W.; Hedges, D.; McBride, J.; Flemington, E. K. Identification of new viral genes and transcript isoforms during Epstein-Barr virus reactivation using RNA-Seq. *J Virol* 2012, 86(3), 1458–1467.

Cuccuru, G.; Orsini, M.; Pinna, A.; Sbardellati, A.; Soranzo, N.; Travaglione, A.; Uva, P.; Zanetti, G.; Fotia, G. Orione, a web-based framework for NGS analysis in microbiology. *Bioinformatics* 2014, 30(13), 1928–1929.

Ding, T.; Schloss, P. D. Dynamics and associations of microbial community types across the human body. *Nature* 2014, 509(7500), 357–360.

Dowd, S. E.; Sun, Y.; Secor, P. R.; Rhoads, D. D.; Wolcott, B. M.; James, G. A.; Wolcott, R. D. Survey of bacterial diversity in chronic wounds using pyrosequencing, DGGE, and full ribosome shotgun sequencing. *BMC Microbiol* 2008, 8, 43.

Duerkop, B. A; Hooper, L. V. Resident viruses and their interactions with the immune system. *Nat Immunol* 2013, 14(7), 654–659.

Dunne, W. M. Jr.; Westblade, L. F.; Ford, B. Next-generation and whole-genome sequencing in the diagnostic clinical microbiology laboratory. *Eur J Clin Microbiol Infect Dis* 2012, 31(8), 1719–1726.

Fahlén, A.; Engstrand, L.; Baker, B. S.; Powles, A.; Fry, L. Comparison of bacterial microbiota in skin biopsies from normal and psoriatic skin. *Arch Dermatol Res* 2012, 304, 15–22.

Ford, C. B.; Lin, P. L.; Chase, M. R.; Shah, R. R.; Iartchouk, O.; Galagan, J.; Mohaideen, N.; Ioerger, T. R.; Sacchettini, J. C.; Lipsitch, M.; Flynn, J. L.; Fortune, S. M. Use of whole genome sequencing to estimate the mutation rate of *Mycobacterium tuberculosis* during latent infection. *Nat Genet* 2011, 43(5), 482–486.

Ford, E.; Nikopoulou, C.; Kokkalis, A.; Thanos, D. A method for generating highly multiplexed ChIP-seq libraries. *BMC Res Notes* 2014, 7(1), 312.

Foulongne, V.; Sauvage, V.; Hebert, C.; Dereure, O.; Cheval, J.; Gouilh, M. A.; Pariente, K.; Segondy, M.; Burguière, A.; Manuguerra, J. C.; Caro, V.; Eloit, M. Human skin microbiota: high diversity of DNA viruses identified on the human skin by high throughput sequencing. *PLoS One* 2012, 7(6), e38499.

Gilbert, J. A.; Dupont, C. L. Microbial metagenomics: beyond the genome. *Ann Rev Mar Sci* 2011, 3, 347–371.

Goecks, J.; Nekrutenko, A.; Taylor, J.; Galaxy Team. Galaxy: a comprehensive approach for supporting accessible, reproducible, and transparent computational research in the life sciences. *Genome Biol* 2010, 11(8), R86.

Golubchik, T.; Batty, E. M.; Miller, R. R.; Farr, H.; Young, B. C.; Larner-Svensson, H.; Fung, R.; Godwin, H.; Knox, K.; Votintseva, A.; Everitt, R. G.; Street, T.; Cule, M.; Ip, C. L.; Didelot, X.; Peto, T. E.; Harding, R. M.; Wilson, D. J.; Crook, D. W.; Bowden, R. Within-host evolution of *Staphylococcus aureus* during asymptomatic carriage. *PLoS One* 2013, 8(5), e61319.

Gorzer, I.; Guelly, C.; Trajanoski, S.; Puchhammer-Stockl, E. Deep sequencing reveals highly complex dynamics of human cytomegalovirus genotypes in transplant patients over time. *J Virol* 2010, 84, 7195–7203.

Greenblum, S.; Turnbaugh, P. J.; Borenstein, E. Metagenomic systems biology of the human gut microbiome reveals topological shifts associated with obesity and inflammatory bowel disease. *Proc Natl Acad Sci USA* 2012, 109, 594–599.

Grice, E. A.; Kong, H. H.; Conlan, S.; Deming, C. B.; Davis, J.; Young, A. C.; NISC Comparative Sequencing Program; Bouffard, G. G.; Blakesley, R. W.; Murray, P. R.; Green, E. D.; Turner, M. L.; Segre, J. A. Topographical and temporal diversity of the human skin microbiome. *Science* 2009, 324, 1190–1192.

Handley, S. A.; Thackray, L. B.; Zhao, G.; Presti, R.; Miller, A. D.; Droit, L.; Abbink, P.; Maxfield, L. F.; Kambal, A.; Duan, E.; Stanley, K.; Kramer, J.; Macri, S. C.; Permar, S. R.; Schmitz, J. E.; Mansfield, K.; Brenchley, J. M.; Veazey, R. S.; Stappenbeck, T. S.; Wang, D.; Barouch, D. H.; Virgin, H. W. Pathogenic simian immunodeficiency virus infection is associated with expansion of the enteric virome. *Cell* 2012, 151(2), 253–266.

Harris, S. R.; Feil, E. J.; Holden, M. T.; Quail, M. A.; Nickerson, E. K.; Chantratita, N.; Gardete, S.; Tavares, A.; Day, N.; Lindsay, J. A.; Edgeworth, J. D.; de Lencastre, H.; Parkhill, J.; Peacock, S. J.; Bentley, S. D. Evolution of MRSA during hospital transmission and intercontinental spread. *Science* 2010, 327(5964), 469–474.

Head, S. R.; Komori, H. K.; La Mere, S. A.; Whisenant, T.; Van Nieuwerburgh, F.; Salomon, D. R.; Ordoukhanian, P. Library construction for next-generation sequencing: overviews and challenges. *Biotechniques* 2014, 56(2), 61–77.

Hernandez, D.; François, P.; Farinelli, L.; Osterås, M.; Schrenzel, J. *De novo* bacterial genome sequencing: millions of very short reads assembled on a desktop computer. *Genome Res* 2008, 18(5), 802–809.

Human Microbiome Project Consortium. A framework for human microbiome research. *Nature* 2012, 486(7402), 215–221.

Hummelen, R.; Fernandes, A. D.; Macklaim, J. M.; Dickson, R. J.; Changalucha, J.; Gloor, G. B.; Reid, G. Deep sequencing of the vaginal microbiota of women with HIV. *PLoS One* 2010, 5(8), e12078.

Idris, A.; Al-Saleh, M.; Piatek, M. J.; Al-Shahwan, I.; Ali, S.; Brown, J. K. Viral metagenomics: analysis of begomoviruses by illumina high-throughput sequencing. *Viruses* 2014, 6(3), 1219–1236.

Jost, T.; Lacroix, C.; Braegger, C. P.; Chassard, C. New insights in gut microbiota establishment in healthy breast fed neonates. *PLoS One* 2012, 7(8), e44595.

Knapp, D. J.; McGovern, R. A.; Poon, A. F.; Zhong, X.; Chan, D.; Swenson, L. C.; Dong, W.; Harrigan, P. R. "Deep" sequencing accuracy and reproducibility using Roche/454 technology for inferring co-receptor usage in HIV-1. *PLoS One* 2014, 9(6), e99508.

Lazarevic, V.; Whiteson, K.; Huse, S.; Hernandez, D.; Farinelli, L.; Østerås, M.; Schrenzel, J.; François, P. Metagenomic study of the oral microbiota by Illumina high-throughput sequencing. *J Microbiol Methods* 2009, 79, 266–271.

Li, R.; Zhu, H.; Ruan, J.; Qian, W.; Fang, X.; Shi, Z.; Li, Y.; Li, S.; Shan, G.; Kristiansen, K.; Li, S.; Yang, H.; Wang, J.; Wang, J. *De novo* assembly of human genomes with massively parallel short read sequencing. *Genome Res* 2010, 20(2), 265–272.

Lindström, A.; Odeberg, J.; Albert, J. Pyrosequencing for detection of lamivudine-resistant hepatitis B virus. *J Clin Microbiol* 2004, 42(10), 4788–4795.

Liu, L.; Li, Y.; Li, S.; Hu, N.; He, Y.; Pong, R.; Lin, D.; Lu, L.; Law, M. Comparison of next-generation sequencing systems. *J Biomed Biotechnol* 2012, 251364.

Magi, A.; Benelli, M.; Gozzini, A.; Girolami, F.; Torricelli, F.; Brandi, M. L. Bioinformatics for next generation sequencing data. *Genes (Basel)* 2010, 1(2), 294–307.

Manichanh, C.; Rigottier-Gois, L.; Bonnaud, E.; Gloux, K.; Pelletier, E.; Frangeul, L.; Nalin, R.; Jarrin, C.; Chardon, P.; Marteau, P.; Roca, J.; Dore, J. Reduced diversity of faecal microbiota in Crohn's disease revealed by a metagenomic approach. *Gut* 2006, 55, 205–211.

Marascio, N.; Ciccozzi, M.; Equestre, M.; Lo Presti, A.; Costantino, A.; Cella, E.; Bruni, R.; Liberto, M. C.; Pisani, G.; Zicca, E.; Barreca, G. S.; Torti, C.; Focà, A.; Ciccaglione, A. R. Back to the origin of HCV 2c subtype and spreading to the Calabria Region (Southern Italy) over the last two centuries: a phylogenetic study. *Infect Genet Evol* 2014, 26, 352–358.

Margulies, M.; Egholm, M.; Altman, W. E.; Attiya, S.; Bader, J. S.; Bemben, L. A.; Berka, J.; Braverman, M. S.; Chen, Y. J.; Chen, Z.; Dewell, S. B.; Du, L.; Fierro, J. M.; Gomes, X. V.; Godwin, B. C.; He, W.; Helgesen, S.; Ho, C. H.; Irzyk, G. P.; Jando, S. C.; Alenquer, M. L.; Jarvie, T. P.; Jirage, K. B.; Kim, J. B.; Knight, J. R.; Lanza, J. R.; Leamon, J. H.; Lefkowitz, S. M.; Lei, M.; Li, J.; Lohman, K. L.; Lu, H.; Makhijani, V. B.; McDade, K. E.; McKenna, M. P.; Myers, E. W.; Nickerson, E.; Nobile, J. R.; Plant, R.; Puc, B. P.; Ronan, M. T.; Roth, G. T.; Sarkis, G. J.; Simons, J. F.; Simpson, J. W.; Srinivasan, M.; Tartaro, K. R.; Tomasz, A.; Vogt, K. A.; Volkmer, G. A.; Wang, S. H.; Wang, Y.; Weiner, M. P.; Yu, P.; Begley, R. F.; Rothberg, J. M. Genome sequencing in microfabricated high-density picolitre reactors. *Nature* 2005, 437(7057), 376–380.

Maslowski, K. M.; Mackay, C. R. Diet, gut microbiota and immune responses. *Nat Immunol* 2011, 12(1), 5–9.

McAdam, P. R.; Holmes, A.; Templeton, K. E.; Fitzgerald, J. R. Adaptive evolution of *Staphylococcus aureus* during chronic endobronchial infection of a cystic fibrosis patient. *PLoS One* 2011, 6, e24301.

Minot, S.; Bryson, A.; Chehoud, C.; Wu, G. D.; Lewis, J. D.; Bushman, F. D. Rapid evolution of the human gut virome. *Proc Natl Acad Sci USA* 2013, 110(30), 12450–12455.

Minot, S.; Sinha, R.; Chen, J.; Li, H.; Keilbaugh, S. A.; Wu, G. D.; Lewis, J. D.; Bushman, F. D. The human gut virome: inter-individual variation and dynamic response to diet. *Genome Res* 2011, 21(10), 1616–1625.

Monecke, S.; Baier, V.; Coombs, G. W.; Slickers, P.; Ziegler, A.; Ehricht, R. Genome sequencing and molecular characterisation of *Staphylococcus aureus* ST772-MRSA-V, "Bengal Bay Clone". *BMC Res Notes* 2013, 6, 548.

Monot, M.; Honoré, N.; Garnier, T.; Zidane, N.; Sherafi, D.; Paniz-Mondolfi, A.; Matsuoka, M.; Taylor, G. M.; Donoghue, H. D.; Bouwman, A.; Mays, S.; Watson, C.; Lockwood, D.; Khamesipour, A.; Dowlati, Y.; Jianping, S.; Rea, T. H.; Vera-Cabrera, L.; Stefani, M. M.; Banu, S.; Macdonald, M.; Sapkota, B. R.; Spencer, J. S.; Thomas, J.; Harshman, K.; Singh, P.; Busso, P.; Gattiker, A.; Rougemont, J.; Brennan, P. J.; Cole, S. T. Comparative genomic and phylogeographic analysis of *Mycobacterium leprae*. *Nat Genet* 2009, 41(12), 1282–1289.

Moore, A. M.; Munck, C.; Sommer, M. O.; Dantas, G. Functional metagenomic investigations of the human intestinal microbiota. *Front Microbiol* 2011, 2, 188–198.

Naccache, S. N.; Federman, S.; Veeeraraghavan, N.; Zaharia, M.; Lee, D.; Samayoa, E.; Bouquet, J.; Greninger, A. L.; Luk, K. C.; Enge, B.; Wadford, D. A.; Messenger, S. L.; Genrich, G. L.; Pellegrino, K.; Grard, G.; Leroy, E.; Schneider, B. S.; Fair, J. N.; Martínez, M. A.; Isa, P.; Crump, J. A.; De Risi, J. L.; Sittler, T.; Hackett, J. Jr.; Miller, S.; Chiu, C. Y. A cloud-compatible bioinformatics pipeline for ultrarapid pathogen identification from next-generation sequencing of clinical samples. *Genome Res* 2014, 24(7), 1180–1192.

Nelson, D. E.; Van Der Pol, B.; Dong, Q.; Revanna, K. V.; Fan, B.; Easwaran, S.; Sodergren, E.; Weinstock, G. M.; Diao, L.; Fortenberry, J. D. Characteristic male urine microbiomes associate with asymptomatic sexually transmitted infection. *PLoS One* 2010, 5(11), e14116.

NIH HMP Working Group; Peterson, J.; Garges, S.; Giovanni, M.; McInnes, P.; Wang, L.; Schloss, J. A.; Bonazzi, V.; McEwen, J. E.; Wetterstrand, K. A.; Deal, C.; Baker, C. C.; Di Francesco, V.; Howcroft, T. K.; Karp, R. W.; Lunsford, R. D.; Wellington, C. R.; Belachew, T.; Wright, M.; Giblin, C.; David, H.; Mills, M.; Salomon, R.; Mullins, C.; Akolkar, B.; Begg, L.; Davis, C.; Grandison, L.; Humble, M.; Khalsa, J.; Little, A. R.; Peavy, H.; Pontzer, C.; Portnoy, M.; Sayre, M. H.; Starke-Reed, P.; Zakhari, S.; Read, J.; Watson, B.; Guyer, M. The NIH Human Microbiome Project. *Genome Res* 2009, 19(12), 2317–2323.

Ninomiya, M.; Ueno, Y.; Funayama, R.; Nagashima, T.; Nishida, Y.; Kondo, Y.; Inoue, J.; Kakazu, E.; Kimura, O.; Nakayama, K.; Shimosegawa, T. Use of Illumina deep sequencing technology to differentiate hepatitis C virus variants. *J Clin Microbiol* 2012, 50(3), 857–866.

Oakley, B. B.; Fiedler, T. L.; Marrazzo, J. M.; Fredricks, D. N. Diversity of human vaginal bacterial communities and associations with clinically defined bacterial vaginosis. *Appl Environ Microbiol* 2008, 74, 4898–4909.

Padmanabhan, R.; Mishra, A. K.; Raoult, D.; Fournier, P. E. Genomics and metagenomics in medical microbiology. *J Microbiol Methods* 2013, 95(3), 415–424.

Pflughoeft, K. J.; Versalovic, J. Human microbiome in health and disease. *Annu Rev Pathol* 2012, 7, 99–122.

Qin, N.; Tan, X.; Jiao, Y.; Liu, L.; Zhao, W.; Yang, S.; Jia, A. RNA-Seq-based transcriptome analysis of methicillin-resistant *Staphylococcus aureus* biofilm inhibition by ursolic acid and resveratrol. *Sci Rep* 2014, 4, 5467.

Radford, A. D.; Chapman, D.; Dixon, L.; Chantrey, J.; Darby, A. C.; Hall, N. Application of next-generation sequencing technologies in virology. *J Gen Virol* 2012, 93, 1853–1868.

Reeves, P. R.; Liu, B.; Zhou, Z.; Li, D.; Guo, D.; Ren, Y.; Clabots, C.; Lan, R.; Johnson, J. R.; Wang, L. Rates of mutation and host transmission for an *Escherichia coli* clone over 3 years. *PLoS One* 2011, 6(10), e26907.

Relman, D. A. Microbial genomics and infectious diseases. *N Engl J Med* 2011, 365(4), 347–357.

Romero, R.; Hassan, S. S.; Gajer, P.; Tarca, A. L.; Fadrosh, D. W.; Nikita, L.; Galuppi, M.; Lamont, R. F.; Chaemsaithong, P.; Miranda, J.; Chaiworapongsa, T.; Ravel, J. The composition and stability of the vaginal microbiota of normal pregnant women is different from that of non-pregnant women. *Microbiome* 2014, 2(1), 4.

Schadt, E. E.; Turner, S.; Kasarskis, A. A window into third-generation sequencing. *Hum Mol Genet* 2010, 19(R2), R227–240.

Schmidt, B.; Sinha, R.; Beresford-Smith, B.; Puglisi, S. J. A fast hybrid short read fragment assembly algorithm. *Bioinformatics* 2009, 25(17), 2279–2280.

Schwartz, S.; Friedberg, I.; Ivanov, I. V.; Davidson, L. A.; Goldsby, J. S.; Dahl, D. B.; Herman, D.; Wang, M.; Donovan, S. M.; Chapkin, R. S. A metagenomic study of diet-dependent interaction between gut microbiota and host in infants reveals differences in immune response. *Genome Biol* 2012, 13(4), r32.

Segata, N.; Haake, S. K.; Mannon, P.; Lemon, K. P.; Waldron, L.; Gevers, D.; Huttenhower, C.; Izard, J. Composition of the adult digestive tract bacterial microbiome based on seven mouth surfaces, tonsils, throat and stool samples. *Genome Biol* 2012, 13(6), R42.

Sharpton, T. J. An introduction to the analysis of shotgun metagenomic data. *Front Plant Sci* 2014, 5, 209.

Shea, P. R.; Beres, S. B.; Flores, A. R.; Ewbank, A. L.; Gonzalez-Lugo, J. H.; Martagon-Rosado, A. J.; Martinez-Gutierrez, J. C.; Rehman, H. A.; Serrano-Gonzalez, M.; Fittipaldi, N.; Ayers, S. D.; Webb, P.; Willey, B. M.; Low, D. E.; Musser, J. M. Distinct signatures of diversifying selection revealed by genome analysis of respiratory tract and invasive bacterial populations. *Proc Natl Acad Sci USA* 2011, 108(12), 5039–5044.

Simpson, J. T.; Wong, K.; Jackman, S. D.; Schein, J. E.; Jones, S. J.; Birol, I. ABySS: a parallel assembler for short read sequence data. *Genome Res* 2009, 19(6), 1117–1123.

Srinivasan, S. M.; Guda, C. MetaID: a novel method for identification and quantification of metagenomic samples. *BMC Genomics* 2013, 14 (S8), S4.

Tebit, D. M.; Arts, E. J. Tracking a century of global expansion and evolution of HIV to drive understanding and to combat disease. *Lancet Infect Dis* 2011, 11(1), 45–56.

Thudi, M.; Li, Y.; Jackson, S. A.; May, G. D.; Varshney, R. K. Current state-of-art of sequencing technologies for plant genomics research. *Brief Funct Genomics* 2012, 11(1), 3–11.

Tremaroli, V.; Bäckhed, F. Functional interactions between the gut microbiota and host metabolism. *Nature* 2012, 489(7415), 242–249.

Umemura, M.; Koyama, Y.; Takeda, I.; Hagiwara, H.; Ikegami, T.; Koike, H.; Machida, M. Fine *de novo* sequencing of a fungal genome using only SOLiD short read data: verification on *Aspergillus oryzae* RIB40. *PLoS One* 2013, 8(5), e63673.

van Dijk, E. L.; Jaszczyszyn, Y.; Thermes, C. Library preparation methods for next-generation sequencing: tone down the bias. *Exp Cell Res* 2014, 322(1), 12–20.

Vicedomini, R.; Vezzi, F.; Scalabrin, S.; Arvestad, L.; Policriti, A. GAM-NGS: genomic assemblies merger for next generation sequencing. *BMC Bioinformatics* 2013, 14(S7), S6.

Virgin, H. W. The virome in mammalian physiology and disease. *Cell* 2014, 157(1), 142–150.

Wang, Q.; Jia, P.; Zhao, Z. VirusFinder: software for efficient and accurate detection of viruses and their integration sites in host genomes through next generation sequencing data. *PLoS One* 2013, 8(5), e64465.

Weinstock, G. M. The volatile microbiome. *Genome Biol* 2011, 12(5), 114.

Willner, D.; Furlan, M.; Haynes, M; Schmieder, R.; Angly, F. E.; Silva, J.; Tammadoni, S.; Nosrat, B.; Conrad, D.; Rohwer, F. Metagenomic analysis of respiratory tract DNA viral communities in cystic fibrosis and non-cystic fibrosis individuals. *PLoS One* 2009, 4(11), e7370.

Wilson, D. J. Insights from genomics into bacterial pathogen populations. *PLoS Pathog* 2012, 8(9), e1002874.

Wolfe, A. J.; Toh, E.; Shibata, N.; Rong, R.; Kenton, K.; Fitzgerald, M.; Mueller, E. R.; Schreckenberger, P.; Dong, Q.; Nelson, D. E.; Brubaker, L. Evidence of uncultivated bacteria in the adult female bladder. *J Clin Microbiol* 2012, 50, 1376–1383.

Xia, J. H.; Lin, G.; Fu, G. H.; Wan, Z. Y.; Lee, M.; Wang, L.; Liu, X. J.; Yue, G. H. The intestinal microbiome of fish under starvation. *BMC Genomics* 2014, 15, 266–275.

Yegnasubramanian, S. Explanatory chapter: next generation sequencing. *Methods Enzymol* 2013, 529, 201–208.

Young, B. C.; Golubchik, T.; Batty, E. M.; Fung, R.; Larner-Svensson, H.; Votintseva, A. A.; Miller, R. R.; Godwin, H.; Knox, K.; Everitt, R. G.; Iqbal, Z.; Rimmer, A. J.; Cule, M.; Ip, C. L.; Didelot, X.; Harding, R. M.; Donnelly, P.; Peto, T. E.; Crook, D. W.; Bowden, R.; Wilson, D. J. Evolutionary dynamics of *Staphylococcus aureus* during progression from carriage to disease. *Proc Natl Acad Sci USA* 2012, 109(12), 4550–4555.

Zerbino, D. R.; Birney, E. Velvet: algorithms for *de novo* short read assembly using de Bruijn graphs. *Genome Res* 2008, 18(5), 821–829.

Zhang, J.; Chiodini, R.; Badr, A.; Zhang, G. The impact of next-generation sequencing on genomics. *J Genet Genomics* 2011, 38(3), 95–109.

CHAPTER 6

PROTEOMICS AND PROSTATE CANCER

JAE-KYUNG MYUNG and MARIANNE D. SADAR

CONTENTS

6.1   Introduction 144
6.2   Functional Proteomics Strategy in Prostate Cancer 155
6.3   Affinity Purification with Epitope Tagging Techniques: TAP,
       Flag-Tag, GST-Pull Down 158
6.4   Conclusion.. 163
Acknowledgements... 163
Keywords 164
References.. 164

6.1 INTRODUCTION

6.1.1 PROSTATE CANCER

Prostate cancer is the most common cancer in western men and causes the most frequent cancer-related death after lung cancer. The incidence and death rates are gradually increasing and it is estimated that in the USA alone there will be 218,890 new cases and 27,050 deaths in 2007 (Jemal et al., 2007). There are several risk factors for prostate cancer such as age, family history and ethnicity. Prostate cancer occurs mostly in men aged more than 65 years old and African Americans have the highest incidence and mortality rate compared to European and Caucasian American men (Powell, 2007). In addition, a diet that is high in fat content is suspected to increase the likelihood of being diagnosed with prostate cancer (Stacewicz-Sapuntzakis et al., 2008).

Digital rectal examination, transrectal ultrasonography, and measurement of serum levels of prostate-specific antigen (PSA) are useful tools for detection of prostate cancer. PSA screening has resulted in the majority of men being diagnosed with low to intermediate risk for disease-specific mortality and will die with prostate cancer, not from prostate cancer. PSA screening has led to the overtreatment of prostate cancer thereby resulting in men receiving radical prostatectomy or radiation therapies when they may never develop symptoms of the disease in their lifetime (Etzioni et al., 2002). These forms of therapy can produce significant morbidities such as incontinence and impotence. With the extensive adoption of PSA screening test for men over 50, the rate of false positives of prostate cancer has also increased dramatically. The high false positive rate and the existence of less aggressive forms of prostate cancer indicate a demand for additional tests to verify the most effective course of treatment. There is an urgent need to discover prognostic markers to allow for selective intervention to spare those men from receiving unnecessary treatment, but still provide radical curative treatment for men who will develop clinically significant disease.

Current treatment options for prostate cancer include hormone therapy, radical surgery, radiation therapy and chemotherapy based on the grade and stage of the disease, age, general health, and life expectancy of the patient. Generally radical prostatectomy and radiation therapy are effectively used for localized prostate cancer and androgen ablation can be used neoadjuvantly or adjuvantly to reduce tumor burden and improve survival for some patients receiving radiation therapy. Androgen ablation is also used as a palliative therapy for prostate cancer patients with recurrent and/or systemic disease. Unfortunately androgen ablation therapy will eventually fail when the disease progresses to a hormone refractory or androgen independent stage. Thus, new treatment options are urgently required owing to the lack of curative treatment for recurrent and metastatic disease (Nelson *et al.*, 2007).

6.1.2 PROTEOMICS

Sequencing of the human genome has provided the potential to identify the genetic basis of cancer. However it is probable that most cancers are caused by multiple genes that require investigation at the protein level for changes in protein expression, structure, interactions and activities. Proteomics addresses the relative abundance of the protein product, posttranslational modifications, subcellular localization, turnover, interaction with other proteins, and functional aspects; all of which cannot be addressed by RNA analysis. A poor correlation of less than 0.5 has been determined between mRNA and protein levels. This is due to differences in the rates of degradation of individual mRNAs and proteins, and because many proteins are modified after they have been translated. Also, one mRNA transcript can give rise to more than one protein. Posttranslational modification of proteins is highly important for biological processes and the propagation of cellular signaling pathways. Of these modifications, phosphorylation is known to play a critical role. Verification of a gene product by proteomic methods is an important step in genomic annotation providing key information about true levels of expression, posttranslational modifications, and intracellular localization of gene products.

The major challenges for proteomic approaches involve the dynamic range of detection methods and the inability to amplify proteins thereby making detection of low-abundance proteins difficult. In human cells the dynamic range is estimated to be 10^7-10^8 and in plasma this increases to 10^{12} (Anderson and Anderson, 2002). However, methods have been developed to decrease complexity and improve detection of low abundance proteins. Recent advances in proteomic techniques and high throughput analysis gives rise to more comprehensive systematic approaches on molecular components and mechanisms in cell functions. The development of mass spectrometry technologies has accelerated the ability to gather a huge amount of data elucidating a quantitative and qualitative analysis of protein expression, function, interactions, and structure. Application of proteomics in prostate cancer research will lead to delineation of signaling pathways and new therapeutic targets for development of new drugs plus aid in the discovery of novel prognostic biomarkers that can distinguish aggressive tumors from latent tumors.

Cancer proteomics was categorized as "expression proteomics" and "functional proteomics" (Hoeben *et al.*, 2006). The former includes the expression of gene products or active molecules in the cell, tumor and body fluids by protein profiling and comparison of differences between the "normal" and for example a prostate cancer group. Any protein which shows a significantly different change in expression in the prostate cancer group can be tested as a potential biomarker. "Functional proteomics" includes analysis of protein-protein, DNA or RNA interactions and posttranslational modifications to provide insight into the molecular mechanisms and biological processes in cancer cells (Mocellin *et al.*, 2004). Proteomic technology is based on protein separation followed by protein identification using mass spectrometry and bioinformatics tools. Gel-based protein separation such as two-dimensional gel electrophoresis (2-DE) followed by mass spectrometry analysis is the fundamental of current various proteomic techniques. Currently, mass spectrometry technology coupled with a variety of separation methods especially liquid-based methods makes it possible to examine the expression of hundreds of proteins with better sensitivity, resolution and decrease the complexity of the samples. Microarray technology is also efficient for functional proteomics with a wide range of applications. Many advanced new proteomic technologies are introduced more and more to give better analytical capabilities, automated preparation, fractionation methods and bioinformatics tools with greater resolution, sensitivity, accuracy and higher throughput processing. In this review, we describe the current application of proteomic technologies in prostate cancer research with their advantages and limitations.

6.1.3   PROFILING PROTEOMICS STRATEGY IN PROSTATE CANCER

The key goal of a profiling strategy is the identification of biomarkers that have therapeutic or prognostic value by comparison of protein levels between the normal and disease population (Celis and Gromov, 2003). The application of this approach in prostate cancer research has provided a rich opportunity to develop many novel markers. PSA is the most commonly used serum marker for the screening of prostate cancer. However, the sensitivity and specificity of PSA are limited owing to false negatives and false positives (Thompson *et al.*, 2004). Many candidate biomarkers have been discovered and suggested for prostate cancer (Bradford *et al.*, 2006), but require validation studies to ensure sensitivity and specificity. Some candidate biomarkers discovered for prostate cancer are described in Table 6.1. High-throughput analysis with the aid of mass spectrometry technology and development of separation methods has led to profiling of whole proteins expressed in a cell, tissue or biofluids such as serum, plasma and urine. Here various proteomics tools currently being applied to the drug-discovery process and biomarker discovery for prostate cancer are explained.

TABLE 6.1 Candidate biomarkers for detection of prostate cancer.

Marker	Description	References
AMACR	α-methylacyl-CoA racemase; Isomerase overexpressed in all prostate cancers; Peroxisomal and mitochondrial enzyme involved in fat metabolism and expressed in prostate tissues	Rogers *et al.*, 2004
ApoA-II isoform	Belongs to the apolipoprotein A2 family; may stabilize HDL (high density lipoprotein) structure by its association with lipids, and affect the HDL metabolism	Malik *et al.*, 2005
Chromo granin A	Prohormone peptide secreted by neuroendocrine cells in the prostate gland; belongs to the chromogranin/secretogranin protein family; unknown function that may have a paracrine and/or autocrine role related to its calcium-binding properties	Fracalanza *et al.*, 2005; Marszalek *et al.*, 2005
DBP	Vitamin D-binding protein; multifunctional protein found in plasma, ascitic fluid, cerebrospinal fluid, and urine and on the surface of many cell types; associates with membrane-bound immunoglobulin on the surface of B-lymphocytes and with IgG Fc receptor on the membranes of T-lymphocytes	Corder *et al.*, 1993
EPCA	Early prostate cancer antigen; prostate cancer associated nuclear matrix protein; unknown function but possibly involved in early prostate carcinogenesis	Leman *et al.*, 2007; Paul *et al.*, 2005
hK-2	Human glandular kallikrein-2; belongs to a family of serine protease; overexpressed in prostate cancer tissues	Kurek *et al.*, 2004; Stephan *et al.*, 2000; Stephan *et al.*, 2006
IGF family	Insulin-like growth factors; IGF-1 is limited to its association with preclinical stages of prostate cancer, whereas the IGF-binding proteins (IGFBPs) appear to play a direct role in prostate cancer detection and prognosis	Chan *et al.*, 1998; Harman *et al.*, 2000
PSA	Prostate-specific antigen; glycoprotein secreted by prostatic epithelium; serine protease involved in the liquefaction of seminal fluids	Wang *et al.*, 1981
PSCA	Prostate stem cell antigen; prostate specific glycoprotein expressed on the cell surface of prostate basal cells; unknown function but may play a role in progenitor cell function	Reiter *et al.*,1998
PSMA	Prostate-specific membrane antigen; 100-kDa type 2 integral membrane glycoprotein overexpressed by PCA epithelial cells; unknown function but may have carboxypeptidase activity	Troyer *et al.*,1995
SAA-1	Serum amyloid A protein-1; major acute phase reactant; Apolipoprotein of the HDL complex; G-protein-coupled receptor binding	Le *et al.*, 2005

TABLE 6.2   Commonly used affinity tags.

Tags	Size (aa)	Sequence	Binding agent	Elution agent	References
CBP	26	KRRWKKNFIAVS-AANRFKKISSSGAL	Calmodulin	EGTA	Stofko-Hahn et al., 1992
Cellulose-binding domain	27-189	Domains	Cellulose	Family I: guanidine HCl or urea, Family II/III: ethylene glycol	Xu et al., 2002
Chitin-binding domain	51	TNPGVSAWQVNTAY-TAGQLVTYNGKTYK-CLQPHTSLAGWEP-SNVPALWQLQ	Chitin	DTT, β-ME or cysteine	Humphries et al., 2002; Sharma et al., 2005
c-myc	11	EQKLISEEDL	Monoclonal antibody	Low pH	Terpe, 2003
FLAG	8	DYKDDDDK	Anti-FLAG monoclonal antibody	Low pH or EDTA	Einhauer and Jungbauer,2001; Knappik and Pluckthun, 1994; Slootstra et al., 1997
GST	211	Protein	Glutathione	Reduced glutathione	Scheich et al., 2003; Smyth et al., 2003; Purbey et al., 2005
MBP	396	Protein	Amylose	Maltose	Eliseev et al., 2004; Feher et al., 2004
Poly-Arg	5-6	RRRRR	Cation-exchange resin	NaCl	Sassenfele and Brewer,1984
Poly-His	5-10	HHHHHH	Ni^{2+}-NTA, Co^{2+}-CMA	Imidazole	Chaga et al., 1999; Chatterjee et al., 2005; Ratnala et al., 2004
S	15	KETAAAKFERQH-MDS	S-fragment of RNase A	Guanidine thiocyanate, MgCl$_2$	Slootstra et al., 1997
SBP	38	MDEKTTGWRGGHV-VEGLAGELEQLRAR-LEHHPQGQREP	Streptavidin	Biotin	Wilson et al., 2001
Strep-tag II	8	WSHPQFEK	Strep-Tactin (modified streptavidin)	Desthiobiotin	Skerra and Schmidt, 2000; Witte et al., 2004; Junttila et al., 2005

Abbreviations: (CBP) Calmodulin-binding peptide/protein; (GST) Glutathione S-transferase; (MBP) Maltose-binding protein; (DTT) Dithiothreitol; (β-ME) β-mercaptoethanol

6.1.3.1   TWO-DIMENSIONAL GEL ELECTROPHORESIS (2-DE)

2-DE was established by O'Farrell for comprehensive protein analysis by separation of proteins according to their net charge value in one dimension followed by separation in a second dimension by molecular weight (O'Farrell, 1975). 2-DE with immobilized pH gradients followed by mass spectrometry analysis has been one of the most powerful methods for profiling of proteins (Görg *et al.*, 2000). This is due the power of separation by 2-DE and the ability to detect posttranslational modifications. It is estimated that 2-DE can optimally separate approximately 2,000 protein spots with high resolution to provide quantitative data. An individual cell may express more than 6,000 primary translation products (Celis and Gromov, 1999), and the posttranslational and chemical modifications can be extensive. The proteins spots are visualized by staining with either Coomassie blue dye, silver stains, fluorescent dyes or by radiolabeling. Lower limits of detection are approximately one nanogram depending on the method of visualization. Celis and coworkers have shown that 2-DE immunoblotting in combination with enhanced chemiluminescence can detect as little as 100-500 protein molecules in unfractionated cellular extracts (Celis *et al.*, 1995). Low abundance proteins can be detected by metabolic labeling followed by 2-DE (Celis and Gromov, 1999). Alternatively since low-abundance proteins are often missed due to the presence of high-abundance "housekeeping" proteins, which can be present at 10,000 times the concentration of low-abundance proteins, a form of enrichment can be employed. The simplest way to achieve enrichment is by cellular fractionation (e.g., cytoplasm, nuclear extract and plasma membrane extracts).

In an early proteomic study, a comparison of the profiles of proteins obtained from stromal and epithelial prostate cells using tissue obtained by radical prostatectomy were carried out using 2-DE to detect different levels of cytokeratins (Sherwood *et al.*, 1989). Recently, a number of proteins have been identified in seminal fluids (Utleg *et al.*, 2003), urine (M'Koma *et al.*, 2007) and sera (Diaz *et al.*, 2004) from prostate cancer patients by enhancing the profiling capacity of 2-DE technology (Stastná and Slais, 2005). Weaknesses of the approach include sensitivity and the relatively large amount of protein required for detection of resolved spots. There are also technical difficulties in resolving proteins that are very acidic or basic, hydrophobic, or extremely large or extremely small. Advances are being made for improved separation and fractionation, lysis buffers, and methods of detection (Celis and Gromov, 2000).

6.1.3.2 *TWO-DIMENSIONAL DIFFERENCE GEL ELECTROPHORESIS (2D-DIGE)*

Reproducibility and reliability of 2-DE has been improved by the incorporation of fluorescent dyes into samples. Two-dimensional difference gel electrophoresis (2D-DIGE) gives direct comparison of protein changes between groups in one single gel at the same time by labeling with different fluorescent dyes and measuring intensities of each gel at a specific wavelength respectively (Unlü *et al.*, 1997). Numerous components which could be potential targets for development of novel therapeutic agent in prostate cancer disease management were identified using 2D-DIGE (Wright *et al.*, 2003; Martin *et al.*, 2004; Rowland *et al.*, 2007). One of the major technical problems associated with analysis of the plasma or serum proteome is the large dynamic range and complexity of the sample. Through the application of 2D-DIGE with chromato-focusing fractionation a reduction of the complexity of the serum proteome from prostate cancer patients facilitated the identification of low abundance proteins (Qin *et al.*, 2005).

6.1.3.3   LASER CAPTURE MICRODISSECTION (LCM)

The identification of potential biomarkers for early stage prostate cancer is challenging because of the heterogeneity of cells present in a clinical sample. Laser capture microdissection (LCM) was developed for rapid tissue harvesting procedures and collection of enriched cell populations to reduce cellular heterogeneity and capture "normal" cells from malignant cells (Emmert-Buck *et al.*, 1996). The prostate gland is composed of stroma, basal and luminal epithelium, with rare occurrences of neuroendocrine cells, macrophages and blood cells. LCM is a useful tool in proteomic research to enrich populations of malignant prostate epithelial cells from "normal" or stromal cells (Ornstein *et al.*, 2000; Ornstein *et al.*, 2000; Best and Emmert-Buck, 2001; Paweletz *et al.*, 2001; Grubb *et al.*, 2003). Cellular proteomes of "normal" and cancerous cells from clinical samples of human prostate tissue prepared by LCM were separated and analyzed using 2-DE (Ahram *et al.*, 2002) or Surface-Enhanced Laser Desorption/Ionization (SELDI) (Wright *et al.*, 1999; Cazares *et al.*, 2002; Wellmann *et al.*, 2002; Zheng *et al.*, 2003; Cheung *et al.*, 2004) followed by identification of proteins using mass spectrometry. LCM can be very precise to provide a highly enriched cell population. However, this technique is expensive, labor-intensive (Fuller *et al.*, 2003; Hunt and Finkelstein, 2004) and presents a number of challenges due to the requirement for minimal fixation with no staining, and time constraints to ensure negligible degradation of proteins during the procedure (Hoeben *et al.*, 2006).

6.1.3.4   MASS SPECTROMETRY TECHNOLOGY: MALDI, ESI AND SELDI

Mass spectrometry has become the method of choice for analysis of a wide dynamic range of complex protein samples. Prostate cancer research has benefited from this technology by the identification and analysis of various biomarkers for cancer detection. Two ionization techniques, MALDI (Matrix-Assisted Laser

Desorption/Ionization) and ESI (Electro-Spray Ionization) coupled to the mass spectrometer time-of flight (TOF) produced outstanding achievements in protein biochemistry (Aebersold and Mann, 2003).

SELDI-TOF mass spectrometry is commonly used to facilitate protein capture, purification and analysis with high-throughput protein profiling (Petricoin *et al.*, 2002b). SELDI has the capability to identify changes in levels of proteins using solid-phase protein chip-based mass spectrometry approach with the properties of affinity based chromatographic separation of proteins. Based on these features, SELDI has been used extensively in cancer research to identify candidate biomarkers. SELDI has the advantage of rapid and reproducible data generation with a high degree of sensitivity and access to posttranslational modification in a high-throughput process (Cazares *et al.*, 2002). Typically the limit of detection of the ProteinChip surfaces is in the low femtomole range with a linear response over 2-3 orders of magnitude (Xiao *et al.*, 2000; Diamond *et al.*, 2001). The average % coefficients of variation (% CV) observed for peaks across the 2,500-150,000*m/z* range can be better than 25% and peaks in the 10,000-15,000*m/z* range are on average %CV of 20% (Le *et al.*, 2005). Several biomarkers for prostate cancer have been discovered using SELDI that have excellent sensitivity and specificity (Petricoin *et al.*, 2002a; Qu *et al.*, 2002; Le *et al.*, 2005). Mass spectrometry through the use of SELDI offers the ability to identify small proteins, polypeptides, hydrophobic and low-molecular weight proteins that were previously difficult to detect (Rai and Chan, 2004). However, SELDI is not suitable for high molecular weight proteins and has low resolution compared to other mass spectrometry technologies.

6.1.3.5 CHROMATOGRAPHIC TECHNIQUES: MULTIDIMENSIONAL PROTEIN IDENTIFICATION TECHNOLOGY (MUDPIT)

The large-scale analysis of a wide dynamic range of proteins requires multiple separation technologies such as differential centrifugation, multiple chromatography, one/two-dimensional gel electrophoresis, or capillary electrophoresis (Hochstrasser *et al.*, 2002). Multidimensional Protein Identification Technology (MudPIT) was introduced by Yates and coworkers (Washburn *et al.*, 2001) to enable the direct identification of proteins in a large scale analysis (Wolters *et al.*, 2001). Proteins are digested usually with trypsin and then separated by charge and hydrophobicity using a strong cation exchange resin and reverse phase chromatography, respectively. MudPIT is quicker than 2-DE with improved sensitivity, dynamic range and high-throughput analysis as well as enables characterization of proteins which are difficult to identify with a gel, such as hydrophobic proteins, low abundance proteins, very acidic and basic proteins. We recently employed MudPIT to identify 82 peptides coimmunoprecipitated with the androgen receptor from prostate cancer cell treated with androgen (Comuzzi and Sadar, 2006). A power of this relatively unbiased approach was the ability to identify interactions with different pools of the androgen receptor. However, compared to some other methods available MudPIT is still relatively low throughput, time consuming with complications in comparison of different samples because only one sample can be run at a time. Importantly, MudPIT does not provide quantitative data.

6.1.3.6 ISOTOPE-TAGGED PROTEOMICS: O18-LABELING, ICAT, ITRAQTM AND SILAC

A multitude of proteomic approaches have been developed for quantitation of proteins with the modification of specific amino acids with isotope incorporation. The abundance of a peptide in the analyzed samples is not always reflected by the signal intensity of a peptide ion measured in a mass spectrometry analysis. This is because ionization with ESI or MALDI is variable and sometimes other ions in the sample influence the measured ion intensity of a specific peptide ion. Thus, incorporation of stable isotope tags *in vitro* or *in vivo* can be used to normalize quantitative variations among different mass spectrometry measurements (Yan and Chen, 2005). Although proteomics technologies that employ stable isotope tags followed by mass spectrometry can produce highly reliable data for quantification of proteins, any single method is not without some limitations and each of these approaches has merits and disadvantages. Therefore the choice of method depends on the particular biological question to yield meaningful data that will ultimately have to be validated by alternative approaches (Kolkman *et al.*, 2005).

6.1.3.7 O18-LABELING

One approach to chemically introduce stable isotopes specific to the carboxyl termini of peptides is by using an enzymatic reaction. During an in gel digestion with trypsin, proteins from normal and disease samples can be treated exclusively in heavy water (H_2O^{18}) and normal water (H_2O^{16}) to be incorporated into the C-termini of peptides. Proteins are quantified by the ratio of ion intensities of O^{16}- to O^{18}-labeled peptides measured by mass spectrometry (Mirgorodskaya *et al.*, 2000; Stewart *et al.*, 2001; Wang *et al.*, 2001; Yao *et al.*, 2004). A major drawback of this method is the complication of quantitation owing to incomplete incorporation of isotopic labels or possible losses. If the sample is highly complex, all proteins may not be completely labeled. Back exchange of O^{16}

water and O^{18} water with the terminal isotope-labeled hydroxyl groups can oc-cur when the two digests are mixed together. To circumvent this problem, for-mic acid can be added if the samples are not processed quickly (Stewart et al., 2001). This method is particularly useful in quantifying specific subsets of the proteome, such as protein mixtures separated by immunoprecipitation or organ-elle fractionation procedures (Yan and Chen, 2005). To date, application of this method to prostate cancer has not been reported.

6.1.3.8 ISOTOPE-CODED AFFINITY TAG (ICAT)

ICAT technology involves the chemical tagging of proteins on cysteine residues with a "heavy" (deuterated or C_{13}) or "light" (nondeuterated or C_{12}) stable iso-tope label. ICAT reagent has three components that are a thiol-specific reactive group that labels cysteines only, an ethylene glycol linker group that occurs in a "heavy" or "light" state, and a biotin tag (Gygi et al., 1999). The cysteine side chains of proteins are labeled in two different cell states using "heavy" ICAT reagent for one state and "light" ICAT reagent for the other state before combin-ing both labeled mixtures and digesting using proteolytic enyzmes. The digested mixtures are run on a strepavidin column which significantly reduces sample complexity because only peptides containing a cysteine with the biotin tag are retained on the column and subsequently analyzed by nanoscale liquid chroma-tography-tandem mass spectrometry. The result of this reduction in complexity is the improved detection and quantitation of low-abundant proteins. This ap-proach has the power to analyze proteins that are acidic, basic, hydrophobic, membrane proteins, and high molecular weight proteins (Griffin et al., 2001). Unfortunately the approach detects only tractable cysteine-containing peptides and may miss peptides with posttranslational modifications. Strepavidin col-umns may also introduce artifacts from nonspecific and irreversible binding (Moseley, 2001). However, this problem occurred with the original deuterium label and has been rectified in the second generation labels that use C_{13} and C_{12}. ICAT peptides may be isobaric with untagged peptides that have eluted from the strepavidin column thereby leading to erroneous results. Cleavable ICAT (cIC-AT) tag was developed with the benefits of more sensitivity and improvement in the quality of CID fragmentation spectra obtained from modified peptides, especially larger species (Yi et al., 2005). ICAT has been used to quantitatively identify differentially expressed proteins in prostate cancer cells (Wright et al., 2003; Meehan and Sadar, 2004).

6.1.3.9   ISOTOPE TAGS FOR RELATIVE AND ABSOLUTE PROTEIN QUANTITATION (ITRAQ™)

Recently an amine group-based isotope labeling methodology was developed using isobaric tags to provide relative and absolute quantitation (iTRAQ™) (Ross *et al.*, 2004). Peptides are labeled on lysine residues and the N-terminus with a cleavable iTRAQ™ reagent to produce tandem mass spectrometry signature ions with relative peaks corresponding to the proportion of the labeled peptides. The labeling strategy ensures that most proteins will be labeled at least once. The approach is similar to ICAT but has the advantage of allowing four different samples to be analyzed within a single mass spectrometry run to significantly reduce costs. This is possible due to isobaric labels that enable absolute and accurate quantitation from multiplex samples as well as to increase the probability of correct peptide identification (Yan and Chen, 2005). The isobaric tags contain reporter and balancer groups. The reporter group is cleaved during collision-induced dissociation (CID) to yield quantitative data for a single peptide from each sample in the mass spectrometry run at a known mass. The balancer group is to keep each isotopic tag at the exact same mass. Protein variants (e.g., posttranslational modification) also are problematic. It is possible to give false attribution to down-regulation of a parent protein if the peptide has a posttranslational modification such that it is not isobaric. Such modifications may also result in the peptide being isobaric with other peptides to confound interpretation of the data. iTRAQ™ differs from ICAT by not having a step that reduces complexity which may result in limitations of resolution by liquid chromatography and co-elution of peptides. iTRAQ™ is time-consuming because every peptide must be subjected to tandem mass spectrometry analysis. Many studies have shown promising results using iTRAQ™ coupled to mass spectrometry for proteomic analyses of breast cancer (Gagné *et al.*, 2007), chronic myeloid leukemia (Griffiths *et al.*, 2007), and lung cancer (Keshamouni *et al.*, 2006), but currently there are no reports for prostate cancer.

6.1.3.10   STABLE ISOTOPE LABELING BY AMINO ACIDS IN CELL CULTURE (SILAC)

Stable isotope labeling by amino acids in cell culture (SILAC) is the one of the more widely used methods for proteomics and is based on *in vivo* labeling of whole cellular proteomes for relative quantitation by mass spectrometry. Cells are grown in a culture medium where the natural form of an amino acid is replaced with a stable isotope form, such as lysine or arginine (e.g., C_{13} and N_{15}) (Ong *et al.*, 2002). Incorporation of the "heavy" amino acid takes place through cell growth, protein synthesis and turnover with passaging of the cells. The isotope label is usually 100% incorporated into newly synthesized proteins to replace the naturally occurring amino acids after approximately 5 passages (Amanchy *et al.*, 2005). The "normal" behavior of the protein and the growth and morphology of cells are not altered by replacement of the naturally occurring amino acids with the isotope-containing amino acids (Amanchy *et al.*, 2005; Ong and Mann, 2007). Arginine and lysine are usually the heavy isotopes of choice because trypsin cleaves after these residues which allow every peptide ending with an arginine or lysine to be quantitated and compared to the "light state". Incorporation of both arginine and lysine isotopes provides greater protein coverage which increases confidence of identification. SILAC has been successfully used to quantitate relative protein abundance (Everley *et al.*, 2004), protein-protein interactions (Blagoev *et al.*, 2003), and potentially phosphorylation (Ibarrola *et al.*, 2003; Gruhler *et al.*, 2005). This technology revealed 88 out of 440 proteins have significant changes of levels of expression in metastatic prostate cancer cells (Everley *et al.*, 2004). SILAC is useful for quantitative proteomic analysis but is limited to cells that can be grown in culture and cannot be used for *in vivo* samples such as tissue or body fluids.

6.2 FUNCTIONAL PROTEOMICS STRATEGY IN PROSTATE CANCER

Proteomics technologies for protein profiling have been successful in identifying potential biomarkers and cataloging proteins expressed in prostate cancer samples. However, the large dynamic range of levels of protein expression in biological systems still presents a caveat that requires more precise mass spectrometry-based quantitation followed by extensive validation. Functional proteomics focuses on the common involvement in a cellular function performed by the association of proteins that act together like a molecular machine (Alberts, 1998). Many different functional proteomic technologies are currently used and being developed. For example some of these technologies are based on affinity purification methods and measurement of the associated protein partners such as chip technologies (Zhu and Snyder, 2003), various two-hybrid systems and computational prediction methods based on known three-dimensional (3D) structures and binding motifs (Marcotte *et al.*, 1999; Aloy and Russell, 2006).

Initially prostate cancer is a hormone-responsive tumor that can be treated by blocking the synthesis and action of androgens. The effects of androgens are mediated through the androgen receptor which is ligand-activated transcription factor. To provide the identification of novel molecular targets of prostate cancer and their application into clinical practice, much research has focused on the complex molecular pathways involved in the pathogenesis of malignant transformation and hormonal progression of prostate cancer to the terminal androgen independent stage. Potential mechanisms of hormonal progression in prostate cancer may involve amplification of the androgen receptor gene, mutations that may alter the sensitivity of androgen receptor to low levels of androgen (ligand) or enable transactivation by related steroids, or ligand-independent activation of the androgen receptor (Wang and Sadar, 2006). Many of these mechanisms rely on the transactivation of the androgen receptor which involves co-activator and corepressor proteins. Failure of androgen ablation treatment may be caused by the alteration in the balance of the coregulatory proteins required by the receptor to carry out its function. Unfortunately, very little proteomics work has been done with the androgen receptor. Studies have demonstrated that androgen receptor has three isoforms (Xia et al., 2000) and different phosphoisoforms (Wong et al., 2004) which may have significant functionally importance. There is considerable evidence supporting that some growth factors, such as interleukin-6, insulin like growth factor, keratinocyte growth factor, and epidermal growth factor (Culig et al., 1994) as well as HER-2/neu which activates mitogen-activated protein kinase (Craft et al., 1999) can activate androgen receptor in the absence of ligand. The AKT signaling cascade (Vivanco and Sawyers, 2002; Ghosh et al., 2003) is also known to contribute tumorigenesis with antiapoptotic activity caused by phosphorylation and inactivation of several pro-apoptotic proteins (Feldman and Feldman, 2001). Besides the investigation of complicated signaling cascades that are correlated with cell growth, migration, invasion and resistance to apoptosis, it is important to know how multiple pathways are altered in between normal and disease groups because most proteins are not isolated but often involved in one or more cellular pathways. Thus screening of the function of proteins that includes multiple alterations in protein-protein and protein-nucleic acid interactions would be of great advantage to identify molecular targets as well as its interacting molecules that may account for androgen independent prostate cancer (Soares et al., 2004). Functional approaches are relatively diverse. For example the most established yeast two-hybrid (Y2H) system (Uetz et al., 2000), affinity purification with epitope tagging techniques, phage display, and protein microarray (Mac Beath and Schreiber, 2000; Zhu et al., 2001) can offer a platform to rapidly assess the function of expressed proteins. We discuss here different strategies for experimental functional proteomics that have been applied to prostate cancer with their general advantages and disadvantages.

6.2.1　TWO-HYBRID SYSTEMS

The yeast two-hybrid (Y2H) system is one of the most broadly applied high-throughput screening methods to yield a comprehensive map of protein-protein interactions that occur in the cell (Fields and Song, 1989; Ito *et al.*, 2001). The Y2H system has been used in prostate cancer to reveal interactions between the androgen receptor and coregulators. Examples include ARA70 (Yeh and Chang, 1996), ARA55 (Fujimoto *et al.*, 1998), small nuclear RING finger protein (Moilanen *et al.*, 1998), and JMJD1C (Wolf *et al.*, 2007). The Y2H system is based upon the modular structure of transcription factors containing a DNA-binding domain (DBD) and a transcription activation domain (TAD). The assay requires a given protein ("bait" protein) to be fused to the DBD of GAL4 and a library of unknown partners ("prey" proteins) be fused to the TAD. Interaction is identified when a functional transcription factor is generated (Chien *et al.*, 1991). Isolation of positive clones of yeast is achieved using nutritional markers and enzymatic reporters. The success of this system comes from its simplicity, low cost, and the ability to scale-up for high-throughput analyses. However, investigation of weak protein-protein interactions may require a strong TAD that can lead to increased false positives. Selection of the strain of yeast requires careful consideration to achieve success. Increased copies of the upstream activation sequence in the reporter are another way to increase the sensitivity of the assay with the downside of increasing false positives. False positives arise from "sticky" proteins and proteins that interact with the reporter (Stephens and Banting, 2000). Overexpression of some proteins in yeast may alter permeability and toxicity. There are estimates of 47-91% false-positive rates using genome-wide screens (Mrowka *et al.*, 2001). To reduce false positives, a third selection marker can be used or strategies that include simultaneous use of different reporter genes and multiple strains of yeast. "True" interactions can be isolated by classical transformation followed by a second screening with a yeast genetic mating technique (Tyagi *et al.*, 2000). The original Y2H system is based in the nucleus with transcriptional activation of reporter genes. Thus this method cannot be applied to proteins that may be localized in other intracellular compartments. Some proteins may require posttranslational modifications or cleavage for interactions to occur that are cell-specific or may be absent in yeast (Colland and Daviet, 2004). A number of variations of the Y2H system have been developed including the mammalian two-hybrid system (Toby and Golemis, 2001; Lee and Lee, 2004).

6.2.2 PHAGE DISPLAY

Phage display screening system is a selection method which displays peptides or proteins of interest using recombinant DNA technology. The peptides or proteins of interest are fused to a capsid or coat protein from a relevant library from the C-terminus of phage-coat protein on the surface of bacteriophage (Parmley and Smith, 1989; Scott and Smith, 1990). Phage display has been used for high-throughput screening of protein interactions and libraries to identify protein domains, ligands for receptors and epitopes for monoclonal antibodies. Applications also include analysis of transcription factors which are not applicable to the Y2H system and identification of signaling molecules in the epidermal growth factor receptor signal transduction pathway (Zozulya *et al.*, 1999). Specific applications in prostate cancer research include the identification of peptides that bind to PSA and modulate the activity of this enzyme (Wu *et al.*, 2000). Ligands with applications for imaging prostate cancer have also been identified using phage display. The peptide FRPNRAQDYNTN has specific binding to prostate carcinoma cells but low binding affinity to nontumor cells showing potential as a diagnostic tracer for this disease (Zitzmann *et al.*, 2005). Phage display has identified peptides to prostate cancer cell-specific receptors that affect attachment and invasion (Romanov *et al.*, 2001).

6.3 AFFINITY PURIFICATION WITH EPITOPE TAGGING TECHNIQUES: TAP, FLAG-TAG, GST-PULL DOWN

A complementary approach to the two-hybrid system to identify protein-protein interaction is affinity purification of tagged protein complexes in conjunction with tandem mass spectrometry (Dziembowski and Séraphin, 2004). Affinity purification with epitope tagging technique is an excellent method to purify and identify interactions involving a large number of protein complexes. It greatly reduces the time and costs of purification by selective binding to a fused affinity tag (Fritze and Anderson, 2000). There are many different affinity purification tags available (Table 6.2), but many have low affinity. Technical difficulties include the identification of transient protein interactions and low abundance proteins that are generally recovered in low yield. These affinity based purification methods have contributed to clarify molecular mechanisms and identify novel interacting partners involved in prostate cancer.

6.3.1 TANDEM AFFINITY PURIFICATION (TAP)

One of the best known methods to purify protein complexes is the Tandem Affinity Purification (TAP) method and a detailed protocol can be found at http://www-db.embl-heidelberg.de/. Basically the approach requires transfection of an expression vector for a fusion protein of the target of interest. The fusion protein and associated proteins are immunoprecipitated from cell lysates using an antibody to the tag. The immunoprecipated proteins can be resolved by SDS-PAGE for subsequent mass spectrometry analysis and identification. Alternatively the proteins in the complex can be proteolytically cleaved and analyzed by multidimensional liquid chromatography – tandem mass spectrometry or TAP-MudPIT (Graumann et al., 2004). TAP-technique enhances the specificity of the purification procedure and reduces nonspecific binding as compared to other affinity purification methods. However the identification of low-abundance binding partners requires a relatively large amount of starting sample for purification and with increasing purification steps comes increased cost. A transient or weak protein-protein interaction may be lost during the series of purification steps (Bauch and Superti-Furga, 2006) and low abundance proteins may not be detected due to the detection limit of the mass spectrometry instruments. In addition, this approach has false positive and false negative errors (Edwards et al., 2002; Kemmeren et al., 2002) and depends on transfection which may result in aberrant localization and nonphysiological levels of expression. In spite of these limitations, extensive efforts have achieved reliable high-throughput protein interaction data from yeast (Bader and Hogue, 2002). To date, there is no study using the TAP-method for prostate cancer research.

6.3.2   FLAG-TAG

Sometimes single tags are used to purify protein complexes from a variety of sources by affinity purification because it may increase the yield of low-affinity binding molecules. Commonly used tags for this application are Flag-tag, Myc-tag, LAP-tag, or HA-tag (Terpe, 2003). Flag-tag is recognized by a monoclonal antibody (Einhauer and Jungbauer, 2001). The Flag-tag system has a short, hydrophilic 8-amino-acid peptide, DYKDDDDK which binds to the antibody M1 or M2 and M5 as additional targets with different recognition and binding characteristics (Knappik and Plückthun, 1994). Flag-tag can be fused to the C- or N-terminus of the protein and has been used in bacteria, yeast and mammalian cells (Blanar and Rutter, 1992; Kunz et al., 1992; Schuster et al., 2000). The pitfalls of the system are poor specificity, levels of contamination, and stability of the monoclonal antibody during purification as compared to nickel-nitrilotriacetic acid (NTA). Protein interactions and molecular mechanisms including signaling pathways in prostate cancer have employed this system. Examples include the regulation of stability of the helix-loop-helix Id-1 protein by tumor necrosis factor-alpha through the ubiquitin/proteasome degradation pathway (Ling et al., 2006) and the forkhead transcription factor, FOXO1, by androgen through a proteolytic mechanism (Huang et al., 2004). In addition, the role of mitogen activated protein kinase in apoptosis of prostate cancer cells induced by phenethyl isothiocyanate has been reported (Xiao et al., 2005).

6.3.2.1   GST PULL-DOWN

The glutathione *S*-transferase (GST) pull-down assay is an *in vitro* approach for identifying protein-protein interactions. This assay can be used for the purification of recombinant proteins for the confirmation of suspected interactions from results obtained with other approaches such as the Y2H assay or for identifying novel interactions with a known protein. Both bait and prey may be recombinant fusion proteins. Alternatively the prey may be overexpressed in cells. The bait is expressed as a GST fusion protein usually prepared from *E. coli* and then immobilized on a glutathione-agarose resin or column. To identify unknown interacting proteins, cell extracts are run over the column that has the immobilized bait protein. Proteins from the extract that interact with the GST-fusion protein can be eluted and identified with the aid of mass spectrometry (Dziembowski *et al.*, 2004). Success with this assay relies on the selection and formation of the bait protein and if the interaction under study is stable or transient. There are several weaknesses of the assay that should be realized.

This approach does not consider that the proteins may not colocalize in the cell and never come into proximity with one another under physiological conditions. Mixing together high concentrations of recombinant proteins may yield aberrant interactions from "sticky" proteins. Generally there is a false-positive rate of 61% and false-negative rate of 38% (Edwards *et al.*, 2002). High background levels arise from nonspecific binding to the column. Elution of the complex from the column usually employs SDS-PAGE sample buffer which destroys the 3D conformation and denaturizes protein and the protein complex. It is important that proteins maintain a correctly folded conformation to allow access for interactions that represent that obtained *in vivo*. Conditions of the assay may result in loss of correct folding and lead to aberrant interactions. One relevant example for prostate cancer is the N-terminal domain (NTD) of the androgen receptor that has a high degree of intrinsic disorder with limited structure and is incorrectly folded under *in vitro* conditions (Reid *et al.*, 2002). The GST-pulldown assay has been used to identify specific interactions between the androgen receptor with coregulators (Yeh *et al.*, 1998; He *et al.*, 2002).

6.3.2.2   FLUORESCENCE RESONANCE ENERGY TRANSFER (FRET)

Fluorescence (Förster) resonance energy transfer (FRET) is a powerful method to validate interactions between proteins and can be applied to high-throughput screening of drugs that may alter such interactions. FRET is based on the energy transfer from a donor to an acceptor molecule by the overlap of the emission spectrum with the absorption spectrum of fluorophores of the donor and acceptor, respectively, when they are sufficiently close together. FRET was first described in the late 1940s (Forster, 1948) and has provided a method to measure the dynamic conformational changes by inter- and/or intra-molecular interactions of proteins such as transcription factors (Day, 1998). The most powerful feature of FRET is the ability to measure protein interactions in the living cell. Unfortunately there can be a relatively high background of cellular autofluorescence (Piehler, 2005). FRET has been used to examine the androgen receptor and revealed different recruitment of LXXLL and FXXLF peptides between the wild-type and T877A mutant androgen receptor (Ozers *et al.*, 2007), and intramolecular and intermolecular amino-carboxy interactions of this transcription factor (Schaufele *et al.*, 2005).

6.3.2.3   *PROTEIN ARRAYS*

One of the emerging proteomic strategies in prostate cancer is the protein array. This approach has been useful to screen the expression of proteins and examine protein interactions involved in signaling pathways and other functional states (Liotta *et al.*, 2003). The principle underlying the protein array is similar in concept to the cDNA array and involves immobilizing a variety of proteins onto a membrane or surface-coated glass slide. The molecules that bind to these proteins are detected by direct labeling or by labeled secondary antibodies. Antibodies to proteins can also be immobilized onto a membrane or surface-coated glass slide and biological samples subsequently applied (Bañez *et al.*, 2005). This approach revealed that immunoreactivity against alpha-methylacyl-coenzyme A racemase (AMACR) was significantly higher in sera from prostate cancer patients (Sreekumar *et al.*, 2004). Protein arrays still have substantial limitations but great potential. High-throughput *in vitro* screening for discovery of new drug targets, protein identification, quantification and activity studies are all possible but limited to detection of proteins targeted by the probes and many different kinds of probes are required to be synthesized. Perhaps one of the biggest problems has been the lack of specificity of most available antibodies. Cross reactivity of probes and possible loss of binding of proteins with posttranslational modifications also should be kept in mind. The protein array requires enhanced development in sample handling and detection before it will meet its expected potential to combine the identification of binding partners by screening and characterization of the mechanism of interaction.

6.3.2.4   *3D STRUCTURAL APPROACHES*

An important component of functional proteomics requires investigation of the three dimensional (3-D) structure of proteins. The 3-D structure can aid in virtual docking for drug design and screening and facilitate prediction of function of uncharacterized proteins. 3-D structure together with bioinformatic analyses can save time, labor and cost compared to traditional methods for drug discovery. In prostate cancer, one validated molecular target in successful drug screening is the androgen receptor that has distinct functional domains that include the ligand-binding domain (LBD), a DNA-binding domain (DBD), and an NTD that contains one or more transcriptional activation domains. Studies providing valuable information for rational drug design based on the structure of androgen receptor LBD have been reported (Söderholm *et al.*, 2005; Tamura *et al.*, 2006). Although 3-D structural proteomics generate essential information that can be used to predict protein-protein interactions and drug development, these studies are based on the ability to crystallize the protein which may not be feasible. One relevant example is the NTD of the androgen receptor which has recently been shown to be a potential drug target for advanced prostate cancer (Quayle *et al.*, 2007). It is also possible that the crystal structure of a protein does not reflect its structure in solution.

6.3.2.5   POSTTRANSLATIONAL MODIFICATIONS – PHOSPHORYLATION

The most remarkable advantage of proteomics is the study of posttranslational modifications of proteins. If we draw on the androgen receptor as an example, it is a ligand-dependent transcription factor and regulates the expression genes by translocating to the nucleus and interacting with coregulators to form a transcriptional complex (Gelmann, 2002). Posttranslational modifications of the androgen receptor include phosphorylation, acetylation, and sumoylation (Poukka *et al.*, 2000; Gioeli *et al.*, 2002; Fu *et al.*, 2004; Popov *et al.*, 2007; Wu and Mo, 2007). These modifications can alter protein function, turnover, cellular localization, and protein-protein interactions. There are more than 170 proteins that have been reported to interact with the androgen receptor (Heemers and Tindall, 2007) many of which are also posttranslationally modified to alter function. One example of an important protein that interacts with the androgen receptor is steroid receptor coactivator-1 (SRC-1) which is covalently modified by sumoylation, acetylation and phosphorylation (Li and Shang, 2007). Phosphorylation of SRC-1 facilitates its interaction with p300/CBP. This interaction elicits optimal activation of both ligand-dependent and ligand-independent transcription of the androgen receptor (Ueda *et al.*, 2002a). This is potentially an important mechanism that could be involved in advanced prostate cancer.

Functional proteomic strategies are powerful approaches to monitor global molecular responses to the activation of signaling pathways. A large scale profiling of posttranslational modifications can be completed using the isotope-tagged proteomic strategies coupled to a preenrichment step to purify the peptides of interest. In the case of phosphorylated proteins or peptides they can be enriched by immobilized metal affinity chromatography (IMAC) or titanium dioxide (TiO_2) column. Protein phosphorylation is very rapid and reversible that occurs on at least one third of all proteins in a mammalian cell (Cohen, 1999). Identification of different phosphoprotein isoforms and their phosphorylation sites provides information into signaling pathways. Phosphorylation sites on the androgen receptor have been demonstrated using *in vitro* assays (Zhou *et al.*, 1995) and *in vivo* samples (Gioeli *et al.*, 2002). In the absence of cognate ligand, the androgen receptor can be activated in prostate cells by molecules which stimulate the protein kinase A pathway, interleukin-6 and growth factors (Culig *et al.*, 1994; Nazareth and Weigel, 1996; Sadar, 1999; Sadar and Gleave, 2000; Ueda *et al.*, 2002b). Studies examining changes in the phosphorylation sites of the androgen receptor have not been thoroughly examined in response to many of these signal transduction pathways. Androgen also stimulates rapid signaling involving kinases through nongenotropic mechanisms, but to date there are no reports of large scale analyses of the phosphoproteome.

6.4   CONCLUSION

A significant contribution of several proteomic approaches has been to characterize the dynamic range of protein expression, localization, posttranslational modification, and interactions. These advances have played an important role in the identification of potential therapeutic targets and novel biomarkers. Recent improvements and advanced progress of high-throughput techniques with various identification and separation methods have accelerated the identification of biomarkers and aided in functional studies in prostate cancer. Protein profiling has been successfully performed using gel-based systems, mass spectrometry-based methods with chromatographic separation systems and isotope-tagged proteomic approaches. Profiling and comparative proteomics of prostate cancer samples will continue to contribute to the discovery of key proteins and pathways to enhance our understanding of the disease and uncover new therapeutic approaches.

ACKNOWLEDGEMENTS

Funding provided by the National Institutes of Health grant CA105304.

KEYWORDS

- **Epitope tagging techniques**
- **Expression proteomics**
- **Flag-Tag**
- **Fluorescence resonance energy transfer**
- **Functional proteomics**
- **Glutathione S-transferase pull-down**
- **Isotope tags for relative and absolute protein quantitation**
- **Isotope-coded affinity tag**
- **Isotope-tagged proteomics**
- **Laser capture microdissection**
- **Mass spectrometry**
- **Multidimensional protein identification technology**
- **Phage display**
- **Prostate cancer**
- **Prostate-specific antigen**
- **Protein arrays**
- **Stable isotope labeling by amino acids in cell culture**
- **Tandem affinity purification**
- **Two-dimensional difference gel electrophoresis**
- **Two-dimensional gel electrophoresis**
- **Yeast two-hybrid system**

REFERENCES

Aebersold, R.; Mann, M. Mass spectrometry-based proteomics. *Nature* 2003, 422, 198-207.

Ahram, M.; Best, C. J.; Flaig, M. J.; Gillespie, J. W.; Leiva, I. M.; Chuaqui, R. F.; Zhou, G.; Shu, H.; Duray, P. H.; Linehan, W. M.; Raffeld, M.; Ornstein, D. K.; Zhao, Y.; Petricoin, E. F. 3rd; Emmert-Buck, M. R. Proteomic analysis of human prostate cancer. *Mol Carcinog* 2002, 33, 9-15.

Alberts, B. The cell as a collection of protein machines: Preparing the next generation of molecular biologists. *Cell* 1998, 92, 291-294.

Aloy, P.; Russell, R. B. Structural systems biology: Modelling protein interactions. *Nat Rev Mol Cell Biol* 2006, 7, 188-197.

Amanchy, R.; Kalume, D. E.; Pandey, A. Stable isotope labeling with amino acids in cell culture (SILAC) for studying dynamics of protein abundance and posttranslational modifications. *Sci STKE* 2005, 267, 12.

Anderson, N. L.; Anderson, N. G. The human plasma proteome: History, character, and diagnostic prospects. *Mol Cell Proteomics* 2002, 1, 845-867.

Bader, G. D.; Hogue, C. W. Analyzing yeast protein-protein interaction data obtained from different sources. *Nat. Biotechnol* 2002, 20, 991-997.

Bañez, L. L.; Srivastava, S.; Moul, J. W. Proteomics in prostate cancer. *Curr Opin Urol* 2005, 15, 151-156.

Bauch, A.; Superti-Furga, G. Charting protein complexes, signaling pathways, and networks in the immune system. *Immunol Rev* 2006, 210, 187-207.

Best, C. J.; Emmert-Buck, M. R. Molecular profiling of tissue samples using laser capture microdissection. *Expert Rev Mol Diagn* 2001, 1, 53-60.

Blagoev, B.; Kratchmarova, I.; Ong, S. E.; Nielsen, M.; Foster, L. J.; Mann, M. A proteomics strategy to elucidate functional protein-protein interactions applied to EGF signaling. *Nat Biotechnol* 2003, 21, 315-318.

Blanar, M. A.; Rutter, W. J. Interaction cloning: Identification of a helix-loop-helix zipper protein that interacts with c-fos. *Science* 1992, 256, 1014-1018.

Bradford, T. J.; Tomlins, S. A.; Wang, X.; Chinnaiyan, A. M. Molecular markers of prostate cancer. *Urol Oncol* 2006, 24, 538-551.

Cazares, L. H.; Adam, B. L.; Ward, M. D.; Nasim, S.; Schellhammer, P. F.; Semmes, O. J.; Wright, G. L. Jr. Normal, benign, preneoplastic, and malignant prostate cells have distinct protein expression profiles resolved by surface enhanced laser desorption/ionization mass spectrometry. *Clin Cancer Res* 2002, 8, 2541-2552.

Celis, J. E.; Gromov, P. 2D protein electrophoresis: Can it be perfected? *Current Opinion in Biotechnology* 1999, 10, 16-21.

Celis, J. E.; Gromov, P. High-resolution two-dimensional gel electrophoresis and protein identification using western blotting and ECL detection. *Exs* 2000, 88, 55-67.

Celis, J. E.; Gromov, P. Proteomics in translational cancer research: Toward an integrated approach. *Cancer Cell* 2003, 3, 9-15.

Celis, J. E.; Rasmussen, H. H.; Gromov, P.; Olsen, E.; Madsen, P.; Leffers, H.; Honore, B.; Dejgaard, K.; Vorum, H.; Kristensen, D. B.; Østergaard, M.; Haunsø, A.; Jensen, N. A.; Celis, A.; Basse, B.; Lauridsen, J. B.; Ratz, G. P.; Andersen, A. H.; Walbum, E.; Kjærgaard, I.; Andersen, I.; Puype, M.; Damme, J.; Vandekerckhove, J. The human keratinocyte two-dimensional gel protein database: Mapping components of signal transduction pathways. *Electrophoresis* 1995, 16, 2177-2240.

Chaga, G.; Bochkariov, D. E.; Jokhadze, G. G.; Hopp, J.; Nelson, P. Natural poly-histidine affinity tag for purification of recombinant proteins on cobalt(II)-carboxymethylaspartate crosslinked agarose. *J Chromatogr A* 1999, 864, 247-256.

Chan, J. M.; Stampfer, M. J.; Giovannucci, E.; Gann, P. H.; Ma, J.; Wilkinson, P.; Hennekens, C. H.; Pollak, M. Plasma insulin-like growth factor-I and prostate cancer risk: A prospective study. *Science* 1998, 279, 563-566.

Chatterjee, S.; Schoepe, J.; Lohmer, S.; Schomburg, D. High level expression and single-step purification of hexahistidine-tagged l-2-hydroxyisocaproate dehydrogenase making use of a versatile expression vector set. *Protein Expr Purif* 2005, 39, 137-143.

Cheung, P. K.; Woolcock, B.; Adomat, H.; Sutcliffe, M.; Bainbridge, T. C.; Jones, E. C.; Webber, D.; Kinahan, T.; Sadar, M.; Gleave, M. E.; Vielkind, J. Protein profiling of microdissected prostate tissue links growth differentiation factor 15 to prostate carcinogenesis. *Cancer Res* 2004, 64, 5929-5933.

Chien, C. T.; Bartel, P. L.; Sternglanz, R.; Fields, S. The two-hybrid system: A method to identify and clone genes for proteins that interact with a protein of interest. *Proc Natl Acad Sci USA* 1991, 88, 9578-9582.

Cohen, P. The Croonian Lecture 1998. Identification of a protein kinase cascade of major importance in insulin signal transduction. *Philos Trans R Soc Lond B Biol Sci* 1999, 354, 485-495.

Colland, F.; Daviet, L. Integrating a functional proteomic approach into the target discovery process. *Biochimie* 2004, 86, 625-632.

Comuzzi, B.; Sadar, M. D. Proteomic analyses to identify novel therapeutic targets for the treatment of advanced prostate cancer. *Cell Science* 2006, 3, 61-81.

Corder, E. H.; Guess, H. A.; Hulka, B. S.; Friedman, G. D.; Sadler, M.; Vollmer, R. T.; Lobaugh, B.; Drezner, M. K.; Vogelman, J. H.; Orentreich, N. Vitamin D and prostate cancer: A prediagnostic study with stored sera. *Cancer Epidemiol. Biomarkers Prev* 1993, 2, 467-472.

Craft, N.; Shostak, Y.; Carey, M.; Sawyers, C. L. A mechanism for hormone-independent prostate cancer through modulation of androgen receptor signaling by the HER-2/neu tyrosine kinase. *Nat Med* 1999, 5, 280-285.

Culig, Z.; Hobisch, A.; Cronauer, M. V.; Radmayr, C.; Trapman, J.; Hittmair, A.; Bartsch, G.; Klok-ker, H. Androgen receptor activation in prostatic tumor cell lines by insulin-like growth factor-I, keratinocyte growth factor and epidermal growth factor. *Cancer Res* 1994, 54, 5474-5478.

Day, R. N. Visualization of Pit-1 transcription factor interactions in the living cell nucleus by fluo-rescence resonance energy transfer microscopy. *Mol Endocrinol* 1998, 12, 1410-1419.

Diaz, J. I.; Cazares, L. H.; Corica, A.; John Semmes, O. Selective capture of prostatic basal cells and secretory epithelial cells for proteomic and genomic analysis. *Urol Oncol* 2004, 22, 329-336.

Diamond, D. L.; Kimball, J. R.; Krisanaprakornkit, S.; Ganz, T.; Dale, B. A. Detection of beta-defensins secreted by human oral epithelial cells. *J Immunological Meth* 2001, 256, 65-76.

Dziembowski, A.; Séraphin, B. Recent developments in the analysis of protein complexes. *FEBS Lett* 2004, 556, 1-6.

Dziembowski, A.; Ventura, A. P.; Rutz, B.; Caspary, F.; Faux, C.; Halgand, F.; Laprévote, O.; Séraphin, B. Proteomic analysis identifies a new complex required for nuclear pre-mRNA reten-tion and splicing. *EMBO J* 2004, 23, 4847-4856.

Edwards, A. M.; Kus, B.; Jansen, R.; Greenbaum, D.; Greenblatt, J.; Gerstein, M. Bridging structural biology and genomics: Assessing protein interaction data with known complexes. *Trends Genet* 2002, 18, 529-536.

Einhauer, A.; Jungbauer, A. The FLAG peptide, a versatile fusion tag for the purification of recom-binant proteins. *J. Biochem. Biophys. Methods* 2001,49, 455-465.

Eliseev, R.; Alexandrov, A.; Gunter, T. High-yield expression and purification of p18 form of Bax as an MBP-fusion protein. *Protein Expr Purif* 2004, 35, 206-209.

Emmert-Buck, M. R.; Bonner, R. F.; Smith, P. D.; Chuaqui, R. F.; Zhuang, Z.; Goldstein, S. R.; Weiss, R. A.; Liotta, L. A. Laser capture microdissection. *Science* 1996, 274, 998-1001.

Etzioni, R.; Penson, D. F.; Legler, J. M.; di Tommaso, D.; Boer, R.; Gann, P. H.; Feuer, E. J.Overdiagnosis due to prostate-specific antigen screening: Lessons from U.S. prostate cancer incidence trends. *J Natl Cancer Inst* 2002, 94, 981-990.

Everley, P. A.; Krijgsveld, J.; Zetter, B. R.; Gygi, S. P. Quantitative cancer proteomics: Stable iso-tope labeling with amino acids in cell culture (SILAC) as a tool for prostate cancer research. *Mol Cell Proteomics* 2004, 3, 729-735.

Feher, A.; Boross, P.; Sperka, T.; Oroszlan, S.; Tozser, J. Expression of the murine leukemia virus protease in fusion with maltose-binding protein in *Escherichia coli. Protein Expr Purif* 2004, 35, 62-68.

Feldman, B. J.; Feldman, D. The development of androgen-independent prostate cancer. *Nat. Rev Cancer* 2001, 1, 34-45.

Fields, S.; Song, O. A novel genetic system to detect protein-protein interactions. *Nature* 1989, 340, 245-246.

Forster, T. Intermolecular energy migration and fluorescence. *Ann Physik* 1948, 2, 55-75.

Fracalanza, S.; Prayer-Galetti, T.; Pinto, F.; Navaglia, F.; Sacco, E.; Ciaccia, M.; Plebani, M.; Paga-no, F.; Basso, D. Plasma chromogranin A in patients with prostate cancer improves the diagnostic efficacy of free/total prostate-specific antigen determination. *Urol Int* 2005, 75, 57-61.

Fritze, C. E.; Anderson, T. R. Epitope tagging: General method for tracking recombinant proteins. *Methods Enzymol* 2000, 327, 3-16.

Fu, M.; Wang, C.; Zhang, X.; Pestell, R. G. Acetylation of nuclear receptors in cellular growth and apoptosis. *Biochem Pharmacol* 2004, 68, 1199-1208.

Fujimoto, N.; Yeh, S.; Kang, H. Y.; Inui, S.; Chang, H. C.; Mizokami, A.; Chang, C. Cloning and characterization of androgen receptor coactivator, ARA55, in human prostate. *J Biol Chem* 1999, 274:8316-8321.

Fuller, A. P.; Palmer-Toy, D.; Erlander, M. G.; Sgroi, D. C. Laser capture microdissection and advanced molecular analysis of human breast cancer. *J Mammary Gland Biol Neoplasia* 2003, 8, 335-345.

Gagné, J. P.; Ethier, C.; Gagné, P.; Mercier, G.; Bonicalzi, M. E.; Mes-Masson, A. M.; Droit, A.; Winstall, E.; Isabelle, M.; Poirier, G. G. Comparative proteome analysis of human epithelial ovarian cancer. *Proteome Sci* 2007, 5, 16.

Gelmann, E. P. Molecular biology of the androgen receptor. *J Clin Oncol* 2002, 20, 3001-3015.

Ghosh, P. M.; Malik, S.; Bedolla, R.; Kreisberg, J. I. Akt in prostate cancer: Possible role in androgen-independence. *Curr Drug Metab* 2003, 4, 487-496.

Gioeli, D.; Ficarro, S. B.; Kwiek, J. J.; Aaronson, D.; Hancock, M.; Catling, A. D.; White, F. M.; Christian, R. E.; Settlage, R. E.; Shabanowitz, J.; Hunt, D. F.; Weber, M. J. Androgen receptor phosphorylation. Regulation and identification of the phosphorylation sites. *J Biol Chem* 2002, 277, 29304-29314.

Görg, A.; Obermaier, C.; Boguth, G.; Harder, A.; Scheibe, B.; Wildgruber, R.; Weiss, W. The current state of two-dimensional electrophoresis with immobilized pH gradients. *Electrophoresis* 2000, 21, 1037-1053.

Graumann, J.; Dunipace, L. A.; Seol, J. H.; McDonald, W. H.; Yates, J. R. 3rd, Wold, B. J.; Deshaies, R. J. Applicability of tandem affinity purification MudPIT to pathway proteomics in yeast. *Mol Cell Proteomics* 2004, 3, 226-237.

Griffiths, S. D.; Burthem, J.; Unwin, R. D.; Holyoake, T. L.; Melo, J. V.; Lucas, G. S.; Whetton, A. D. The use of isobaric tag peptide labeling (iTRAQ) and mass spectrometry to examine rare, primitive hematopoietic cells from patients with chronic myeloid leukemia. *Mol Biotechnol* 2007, 36, 81-89.

Griffin, T. J.; Han, D. K.; Gygi, S. P.; Rist, B.; Lee, H.; Aebersold, R.; Parker, K. C. Toward a high-throughput approach to quantitative proteomic analysis: Expression-dependent protein identification by mass spectrometry. *Journal of the American Society for Mass Spectrometry* 2001, 12, 1238-1246.

Grubb, R. L.; Calvert, V. S.; Wulkuhle, J. D.; Paweletz, C. P.; Linehan, W. M.; Phillips, J. L.; Chuaqui, R.; Valasco, A.; Gillespie, J.; Emmert-Buck, M.; Liotta, L. A.; Petricoin, E. F. Signal pathway profiling of prostate cancer using reverse phase protein arrays. *Proteomics* 2003, 3, 2142-2146.

Gruhler, A.; Olsen, J. V.; Mohammed, S.; Mortensen, P.; Faergeman, N. F.; Mann, M.; Jensen, O. N. Quantitative phosphoproteomics applied to the yeast pheromone signaling pathway. *Mol Cell Proteomics* 2005, 4, 310-327.

Gygi, S. P.; Rist, B.; Gerber, S. A.; Turecek, F.; Gelb, M. H.; Aebersold, R. Quantitative analysis of complex protein mixtures using isotope-coded affinity tags. *Nature Biotechnology* 1999, 17, 994-999.

Harman, S. M.; Metter, E. J.; Landis, P. K.; Carter, H. B. Baltimore Longitudinal Study on Aging. Serum levels of insulin-like growth factor I (IGF-I), IGF-II, IGF-binding protein-3, and prostate-specific antigen as predictors of clinical prostate cancer. *J Clin Endocrinol Metab* 2000, 85, 4258-4265.

He, B.; Minges, J. T.; Lee, L. W.; Wilson, E. M. The FXXLF motif mediates androgen receptor-specific interactions with coregulators. *J Biol Chem* 2002, 277, 10226-10235.

Heemers, H. V.; Tindall, D. J. Androgen receptor (AR) coregulators: A diversity of functions converging on and regulating the AR transcriptional complex. *Endocr Rev* 2007, 28, 778-808.

Hochstrasser, D. F.; Sanchez, J. C.; Appel, R. D. Proteomics and its trends facing nature's complexity. *Proteomics* 2002, 2, 807-812.

Hoeben, A.; Landuyt, B.; Botrus, G.; De Boeck, G.; Guetens, G.; Highly, M.; van Oosterom, A. T.; de Bruijn, E. A. Proteomics in cancer research: Methods and application of array-based protein profiling technologies. *Anal Chim Acta* 2006, 564, 19-33.

Huang, H.; Muddiman, D. C.; Tindall, D. J. Androgens negatively regulate forkhead transcription factor FKHR (FOXO1) through a proteolytic mechanism in prostate cancer cells. *J Biol Chem* 2004, 279, 138, 66-13877.

Humphries, H. E.; Christodoulides, M.; Heckels, J. E. Expression of the class 1 outer-membrane protein of *Neisseria meningitidis* in *Escherichia coli* and purification using a self-cleavable affinity tag. *Protein Expr Purif* 2002, 26, 243-248.

Hunt, J. L.; Finkelstein, S. D. Microdissection techniques for molecular testing in surgical pathology. *Arch Pathol Lab Med* 2004, 128, 1372-1378.

Ibarrola, N.; Kalume, D. E.; Gronborg, M.; Iwahori, A.; Pandey, A. A proteomic approach for quantitation of phosphorylation using stable isotope labeling in cell culture. *Anal Chem* 2003, 75, 6043-6049.

Ito, T.; Chiba, T.; Ozawa, R.; Yoshida, M.; Hattori, M.; Sakaki, Y. A comprehensive two-hybrid analysis to explore the yeast protein interactome. *Proc Natl Acad Sci USA* 2001, 98, 4569-4574.

Jemal, A.; Siegel, R.; Ward, E.; Murray, T.; Xu, J.; Thun, M. J. Cancer statistics, 2007. *CA Cancer J Clin* 2007, 57, 43-66.

Junttila, M. R.; Saarinen, S.; Schmidt, T.; Kast, J.; Westermarck, J. Single-step Strep-tag purification for the isolation and identification of protein complexes from mammalian cells. *Proteomics* 2005, 5, 1199-1203.

Kemmeren, P.; van Berkum, N. L.; Vilo, J.; Bijma, T.; Donders, R.; Brazma, A.; Holstege, F. C. Protein interaction verification and functional annotation by integrated analysis of genome-scale data. *Mol Cell* 2002, 9, 1133-1143.

Keshamouni, V. G.; Michailidis, G.; Grasso, C. S.; Anthwal, S.; Strahler, J. R.; Walker, A.; Arenberg, D. A.; Reddy, R. C.; Akulapalli, S.; Thannickal, V. J.; Standiford, T. J.; Andrews, P. C.; Omenn, G. S. Differential protein expression profiling by iTRAQ-2DLC-MS/MS of lung cancer cells undergoing epithelial-mesenchymal transition reveals a migratory/invasive phenotype. *J Proteome Res* 2006, 5, 1143-1154.

Knappik, A.; Plückthun, A. An improved affinity tag based on the FLAG peptide for the detection and purification of recombinant antibody fragments. *Biotechniques* 1994, 17, 754-761.

Kolkman, A.; Dirksen, E. H.; Slijper, M.; Heck, A. J. Double standards in quantitative proteomics: Direct comparative assessment of difference in gel electrophoresis and metabolic stable isotope labeling. *Mol Cell Proteomics* 2005, 4, 255-266.

Kunz, D.; Gerad, N. P.; Gerad, C. The human leukocyte platelet activating factor receptor. *J Biol Chem* 1992, 267, 9101-9106.

Kurek, R.; Nunez, G.; Tselis, N.; Konrad, L.; Martin, T.; Roeddiger, S.; Aumüller, G.; Zamboglou, N.; Lin, D. W.; Tunn, U. W.; Renneberg, H. Prognostic value of combined "triple"-reverse transcription-PCR analysis for prostate-specific antigen, human kallikrein 2, and prostate-specific membrane antigen mRNA in peripheral blood and lymph nodes of prostate cancer patients. *Clin Cancer Res* 2004, 10, 5808-5814.

Le, L.; Chi, K.; Tyldesley, S.; Flibotte, S.; Diamond, D. L.; Kuzyk, M. A.; Sadar, M. D. Identification of serum amyloid A as a biomarker to distinguish prostate cancer patients with bone lesions. *Clin Chem* 2005, 51, 695-707.

Lee, J. W.; Lee, S. K. Mammalian two-hybrid assay for detecting protein-protein interactions *in vivo*. *Methods Mol Biol* 2004, 261, 327-336.

Leman, E. S.; Cannon, G. W.; Trock, B. J.; Sokoll, L. J.; Chan, D. W.; Mangold, L.; Partin, A. W.; Getzenberg, R. H. EPCA-2: A highly specific serum marker for prostate cancer. *Urology* 2007, 69, 714-720.

Li, S.; Shang, Y. Regulation of SRC family coactivators by post-translational modifications. *Cell Signal* 2007, 19, 1101-1112.

Ling, M. T.; Kwok, W. K.; Fung, M. K.; Xianghong, W.; Wong, Y. C. Proteasome mediated degradation of Id-1 is associated with TNFalpha-induced apoptosis in prostate cancer cells. *Carcinogenesis* 2006, 27, 205-215.

Liotta, L. A.; Espina, V.; Mehta, A. I.; Calvert, V.; Rosenblatt, K.; Geho, D.; Munson, P. J.; Young, L.; Wulfkuhle, J.; Petricoin, E. F. 3rd. Protein microarrays: Meeting analytical challenges for clinical applications. *Cancer Cell* 2003, 3, 317-325.

MacBeath, G.; Schreiber, S. L. Printing proteins as microarrays for high-throughput function determination. *Science* 2000, 289, 1760-1763.

Malik, G.; Ward, M. D.; Gupta, S. K.; Trosset, M. W.; Grizzle, W. E.; Adam, B. L.; Diaz, J. I.; Semmes, O. J. Serum levels of an isoform of apolipoprotein A-II as a potential marker for prostate cancer. *Clin Cancer Res* 2005, 11, 1073-1085.

Marcotte, E. M.; Pellegrini, M.; Ng, H. L.; Rice, D. W.; Yeates, T. O.; Eisenberg, D. Detecting protein function and protein-protein interactions from genome sequences. *Science* 1999, 285, 751-753.

Marszalek, M.; Wachter, J.; Ponholzer, A.; Leitha, T.; Rauchenwald, M.; Madersbacher, S. Insulin-like growth factor 1, chromogranin A and prostate specific antigen serum levels in prostate cancer patients and controls. *Eur Urol* 2005, 48, 34-39.

Martin, D. B.; Gifford, D. R.; Wright, M. E.; Keller, A.; Yi, E.; Goodlett, D. R.; Aebersold, R.; Nelson, P. S. Quantitative proteomic analysis of proteins released by neoplastic prostate epithelium. *Cancer Res* 2004, 64, 347-355.

Meehan, K. L.; Sadar, M. D. Quantitative profiling of LNCaP prostate cancer cells using isotope-coded affinity tags and mass spectrometry. *Proteomics* 2004, 4, 1116-1134.

Mirgorodskaya, O. A.; Kozmin, Y. P.; Titov, M. I.; Körner, R.; Sönksen, C. P.; Roepstorff, P. Quantitation of peptides and proteins by matrix-assisted laser desorption/ionization mass spectrometry using (18)O-labeled internal standards. *Rapid Commun. Mass Spectrom* 2000, 14, 1226-1232.

M'Koma, A. E.; Blum, D. L.; Norris, J. L.; Koyama, T.; Billheimer, D.; Motley, S.; Ghiassi, M.; Ferdowsi, N.; Bhowmick, I.; Chang, S. S.; Fowke, J. H.; Caprioli, R. M.; Bhowmick, N. A. Detection of pre-neoplastic and neoplastic prostate disease by MALDI profiling of urine. *Biochem Biophys Res Commun* 2007, 353, 829-834.

Mocellin, S.; Rossi, C. R.; Traldi, P.; Nitti, D.; Lise, M. Molecular oncology in the post-genomic era: The challenge of proteomics. *Trends Mol Med* 2004, 10, 24-32.

Moilanen, A. M.; Poukka, H.; Karvonen, U.; Häkli, M.; Jänne, O. A.; Palvimo, J. J. Identification of a novel RING finger protein as a coregulator in steroid receptor-mediated gene transcription. *Mol Cell Biol* 1998, 18, 5128-5139.

Moseley, M. A. Current trends in differential expression proteomics: Isotopically coded tags. *Trends in Biotechnology* 2001, 19, S10-16.

Mrowka, R.; Patzak, A.; Herzel, H. Is there a bias in proteome research? *Genome Res* 2001, 11, 1971-1973.

Nazareth, L. V.; Weigel, N. L. Activation of the human androgen receptor through a protein kinase A signaling pathway. *J Biol Chem* 1996, 271, 19900-19907.

Nelson, E. C.; Cambio, A. J.; Yang, J. C.; Lara, P. N. J.; Evans, C. P. Biologic agents as adjunctive therapy for prostate cancer: A rationale for use with androgen deprivation. *Nat Clin Pract Urol* 2007, 4, 82-94.

O'Farrell, P. H. High resolution two-dimensional electrophoresis of proteins. *J Biol Chem* 1975, 250, 4007-4021.

Ong, S. E.; Blagoev, B.; Kratchmarova, I.; Kristensen, D. B.; Steen, H.; Pandey, A.; Mann, M. Stable isotope labeling by amino acids in cell culture, SILAC, as a simple and accurate approach to expression proteomics. *Mol Cell Proteomics* 2002, 1, 376-386.

Ong, S. E.; Mann, M. Stable isotope labeling by amino acids in cell culture for quantitative proteomics. *Methods Mol Biol* 2007, 359, 37-52.

Ornstein, D. K.; Englert, C.; Gillespie, J. W.; Paweletz, C. P.; Linehan, W. M.; Emmert-Buck, M. R.; Petricoin, E. F. 3rd. Characterization of intracellular prostate-specific antigen from laser capture microdissected benign and malignant prostatic epithelium. *Clin Cancer Res* 2000, 6, 353-356.

Ornstein, D. K.; Gillespie, J. W.; Paweletz, C. P.; Duray, P. H.; Herring, J.; Vocke, C. D.; Topalian, S. L.; Bostwick, D. G.; Linehan, W. M.; Petricoin, E. F. 3rd.; Emmert-Buck, M. R. Proteomic analysis of laser capture microdissected human prostate cancer and *in vitro* prostate cell lines. *Electrophoresis* 2000, 21, 2235-2242.

Ozers, M. S.; Marks, B. D.; Gowda, K.; Kupcho, K. R.; Ervin, K. M.; De Rosier, T.; Qadir, N.; Eliason, H. C.; Riddle, S. M.; Shekhani, M. S. The androgen receptor T877A mutant recruits LXXLL and FXXLF peptides differently than wild-type androgen receptor in a time-resolved fluorescence resonance energy transfer assay. *Biochemistry* 2007, 46, 683-695.

Parmley, S. F.; Smith, G. P. Filamentous fusion phage cloning vectors for the study of epitopes and design of vaccines. *Adv Exp Med Biol* 1989, 251, 215-218.

Paul, B.; Dhir, R.; Landsittel, D.; Hitchens, M. R.; Getzenberg, R. H. Detection of prostate cancer with a blood-based assay for early prostate cancer antigen. *Cancer Res* 2005, 65, 4097-4100.

Paweletz, C. P.; Liotta, L. A.; Petricoin, E. F. 3rd. New technologies for biomarker analysis of prostate cancer progression: Laser capture microdissection and tissue proteomics. *Urology* 2001, 57, 160-163.

Petricoin, E. F. 3rd.; Ornstein, D. K.; Paweletz, C. P.; Ardekani, A.; Hackett, P. S.; Hitt, B. A.; Velassco, A.; Trucco, C.; Wiegand, L.; Wood, K.; Simone, C. B.; Levine, P. J.; Linehan, W. M.; Emmert-Buck, M. R.; Steinberg, S. M.; Kohn, E. C.; Liotta, L. A. Serum proteomic patterns for detection of prostate cancer. *J Natl Cancer Inst* 2002a, 94, 1576-1578.

Petricoin, E. F.; Ardekani, A. M.; Hitt, B. A.; Levine, P. J.; Fusaro, V. A.; Steinberg, S. M.; Mills, G. B.; Simone, C.; Fishman, D. A.; Kohn, E. C.; Liotta, L. A. Use of proteomic patterns in serum to identify ovarian cancer. *Lancet* 2002b, 359, 572-577.

Piehler, J. New methodologies for measuring protein interactions *in vivo* and *in vitro*. *Curr Opin Struct Biol* 2005, 15, 4-14.

Popov, V. M.; Wang, C.; Shirley, L. A.; Rosenberg, A.; Li, S.; Nevalainen, M.; Fu, M.; Pestell, R. G. The functional significance of nuclear receptor acetylation. *Steroids* 2007, 72, 221-230.

Poukka, H.; Karvonen, U.; Janne, O. A.; Palvimo, J. J. Covalent modification of the androgen receptor by small ubiquitin-like modifier 1 (SUMO-1). *Proc Natl Acad Sci USA* 2000, 97, 14145-14150.

Powell, I. J. Epidemiology and pathophysiology of prostate cancer in African-American men. *J Urol* 2007, 177, 444-449.

Purbey, P. K.; Jayakumar, P. C.; Deepalakshmi, P. D.; Patole, M. S.; Galande, S. GST fusion vector with caspase-6 cleavage site for removal of fusion tag during column purification. *Biotechniques* 2005, 38, 360-364.

Qin, S.; Ferdinand, A. S.; Richie, J. P.; O'Leary, M. P.; Mok, S. C.; Liu, B. C. Chromatofocusing fractionation and two-dimensional difference gel electrophoresis for low abundance serum proteins. *Proteomics* 2005, 5, 3183-3192.

Qu, Y.; Adam, B. L.; Yasui, Y.; Ward, M. D.; Cazares, L. H.; Schellhammer, P. F.; Feng, Z.; Semmes, O. J.; Wright, G. L. Jr. Boosted decision tree analysis of surface-enhanced laser desorption/ionization mass spectral serum profiles discriminates prostate cancer from noncancer patients. *Clin Chem* 2002, 48, 1835-1843.

Quayle, S. N.; Mawji, N. R.; Wang, J.; Sadar, M. D. Decoy molecules of the androgen receptor block the growth of prostate cancer. *Proc Natl Acad Sci USA* 2007, 104, 1331-1336.

Rai, A. J.; Chan, D. W. Cancer proteomics: Serum diagnostics for tumor marker discovery. *Ann N Y Acad Sci* 2004, 1022, 286-294.

Ratnala, V. R.; Swarts, H. G.; Van Oostrum, J.; Leurs, R.; De Groot, H. J.; Bakker, R. A.; De Grip, W. J. Large-scale overproduction, functional purification and ligand affinities of the His-tagged human histamine H1 receptor. *Eur J Biochem* 2004, 271, 2636-2646.

Reid, J.; Kelly, S. M.; Watt, K.; Price, N. C.; Mc Ewan, I. J. Conformational analysis of the androgen receptor amino-terminal domain involved in transactivation. Influence of structure-stabilizing solutes and protein-protein interactions. *J Biol Chem* 2002, 277, 20079-20086.

Reiter, R. E.; Gu, Z.; Watabe, T.; Thomas, G.; Szigeti, K.; Davis, E.; Wahl, M.; Nisitani, S.; Yamashiro, J.; Le Beau, M. M.; Loda, M.; Witte, O. N. Prostate stem cell antigen: A cell surface marker overexpressed in prostate cancer. *Proc Natl Acad Sci* 1998, 95, 1735-1740.

Rogers, C. G.; Yan, G.; Zha, S.; Gonzalgo, M. L.; Isaacs, W. B.; Luo, J.; De Marzo, A. M.; Nelson, W. G.; Pavlovich, C. P. Prostate cancer detection on urinalysis for alpha methylacyl coenzyme a racemase protein. *J Urol* 2004, 172, 1501-1503.

Romanov, V. I.; Durand, D. B.; Petrenko, V. A. Phage display selection of peptides that affect prostate carcinoma cells attachment and invasion. *Prostate* 2001, 47, 239-251.

Ross, P. L.; Huang, Y. N.; Marchese, J. N.; Williamson, B.; Parker, K.; Hattan, S.; Khainovski, N.; Pillai, S.; Dey, S.; Daniels, S.; Purkayastha, S.; Juhasz, P.; Martin, S.; Bartlet-Jones, M.; He, F.; Jacobson, A.; Pappin, D. J. Multiplexed protein quantitation in *Saccharomyces cerevisiae* using amine-reactive isobaric tagging reagents. *Mol Cell Proteomics* 2004, 3, 1154-1169.

Rowland, J. G.; Robson, J. L.; Simon, W. J.; Leung, H. Y.; Slabas, A. R. Evaluation of an *in vitro* model of androgen ablation and identification of the androgen responsive proteome in LNCaP cells. *Proteomics* 2007, 7, 47-63.

Sadar, M. D. Androgen-independent induction of prostate-specific antigen gene expression via cross-talk between the androgen receptor and protein kinase A signal transduction pathways. *J Biol Chem* 1999, 274, 7777-7783.

Sadar, M. D.; Gleave, M. E. Ligand-independent activation of the androgen receptor by the differentiation agent butyrate in human prostate cancer cells. *Cancer Res* 2000, 60, 5825-5831.

Sassenfeld, H. M.; Brewer, S. J. A polypeptide fusion designed for purification of recombinant proteins. *Bio/Technology* 1984, 2, 76-81.

Scott, J. K.; Smith, G. P. Searching for peptide ligands with an epitope library. *Science* 1990, 249, 386-390.

Scheich, C.; Sievert, V.; Bussow, K. An automated method for highthroughput protein purification applied to a comparison of His-tag and GST-tag affinity chromatography. *BMC. Biotechnol* 2003, 3, 12.

Sherwood, E. R.; Berg, L. A.; Mc Ewan, R. N.; Pasciak, R. M.; Kozlowski, J. M.; Lee, C. Twodimensional protein profiles of cultured stromal and epithelial cells from hyperplastic human prostate. *J Cell Biochem* 1989, 40, 201-214.

Schaufele, F.; Carbonell, X.; Guerbadot, M.; Borngraeber, S.; Chapman, M. S.; Ma, A. A.; Miner, J. N.; Diamond, M. I. The structural basis of androgen receptor activation: Intramolecular and intermolecular amino-carboxy interactions. *Proc Natl Acad Sci USA* 2005, 102, 9802-9807.

Schuster, M.; Wasserbauer, E.; Einhauer, A.; Ordner, C.; Jungbauer, A.; Hammerschmidt, F.; Werner, G. Protein expression strategies for identification of novel target proteins. *J Biomol Screen* 2000, 5, 89-97.

Sharma, S. S.; Chong, S.; Harcum, S. W. Simulation of large-scale production of a soluble recombinant protein expressed in *Escherichia coli* using an intein-mediated purification system. *Appl Biochem Biotechnol* 2005, 126, 93-118.

Skerra, A.; Schmidt, T. G. M. Use of the Strep-tag and streptavidin for detection and purification of recombinant proteins. *Methods Enzymol* 2000, 326, 271-304.

Slootstra, J. W.; Kuperus, D.; Pluckthun, A.; Meloen, R. H. Identification of new tag sequences with differential and selective recognition properties for the anti-FLAG monoclonal antibodies M1, M2 and M5. *Mol Divers* 1997, 2, 156-164.

Smyth, D. R.; Mrozkiewicz, M. K.; McGrath, W. J.; Listwan, P.; Kobe, B. Crystal structures of fusion proteins with large-affinity tags. *Protein Sci* 2003, 12, 1313-1322.

Soares, H. D.; Williams, S. A.; Snyder, P. J.; Gao, F.; Stiger, T.; Rohlff, C.; Herath, A.; Sunderland, T.; Putnam, K.; White, W. F. Proteomic approaches in drug discovery and development. *Int Rev Neurobiol* 2004, 61, 97-126.

Söderholm, A. A.; Lehtovuori, P. T.; Nyrönen, T. H. Three-dimensional structure-activity relationships of nonsteroidal ligands in complex with androgen receptor ligand-binding domain. *J Med Chem* 2005, 48, 917-925.

Sreekumar, A.; Laxman, B.; Rhodes, D. R.; Bhagavathula, S.; Harwood, J.; Giacherio, D.; Ghosh, D.; Sanda, M. G.; Rubin, M. A.; Chinnaiyan, A. M. Humoral immune response to alpha-methyl-acyl-CoA racemase and prostate cancer. *J Natl Cancer Inst* 2004, 96, 834-843.

Stacewicz-Sapuntzakis, M.; Borthakur, G.; Burns, J. L.; Bowen, P. E. Correlations of dietary patterns with prostate health. *Mol Nutr Food Res* 2008, 52(1), 114-130.

Stastná, M.; Slais, K. Two-dimensional gel isoelectric focusing. *Electrophoresis* 2005, 26, 3586-3591.

Stephan, C.; Jung, K.; Lein, M.; Sinha, P.; Schnorr, D.; Loening, S. A. Molecular forms of prostate-specific antigen and human kallikrein 2 as promising tools for early diagnosis of prostate cancer. *Cancer Epidemiol Biomarkers Prev* 2000, 9, 1133-1147.

Stephan, C.; Jung, K.; Nakamura, T.; Yousef, G. M.; Kristiansen, G.; Diamandis, E. P. Serum human glandular kallikrein 2 (hK2) for distinguishing stage and grade of prostate cancer. *Int J Urol* 2006, 13, 238-243.

Stephens, D. J.; Banting, G. The use of yeast two-hybrid screens in studies of protein:protein interactions involved in trafficking. *Traffic* 2000, 1, 763-768.

Stewart, I. I.; Thomson, T.; Figeys, D. 18O labeling: A tool for proteomics. *Rapid Commun. Mass Spectrom* 2001, 15, 2456-2465.

Stofko-Hahn, R. E.; Carr, D. W.; Scott, J. D. A single step purification for recombinant proteins. *FEBS Lett* 1992, 302, 274-278.

Tamura, H.; Ishimoto, Y.; Fujikawa, T.; Aoyama, H.; Yoshikawa, H.; Akamatsu, M. Structural basis for androgen receptor agonists and antagonists: Interaction of SPEED 98-listed chemicals and related compounds with the androgen receptor based on an *in vitro* reporter gene assay and 3D-QSAR. *Bioorg Med Chem* 2006, 14, 7160-7174.

Terpe, K. Overview of tag protein fusions: From molecular and biochemical fundamentals to commercial systems. *Appl Microbiol Biotechnol* 2003, 60, 523-533.

Thompson, I. M.; Pauler, D. K.; Goodman, P. J.; Tangen, C. M.; Lucia, M. S.; Parnes, H. L.; Minasian, L. M.; Ford, L. G.; Lippman, S. M.; Crawford, E. D.; Crowley, J. J.; Coltman, C. A. Jr. Prevalence of prostate cancer among men with a prostate-specific antigen level < or =4.0 ng per milliliter. *N Engl J Med* 2004, 350, 2239-2246.

Toby, G. G.; Golemis, E. A. Using the yeast interaction trap and other two-hybrid-based approaches to study protein-protein interactions. *Methods* 2001, 24, 201-217.

Troyer, J. K.; Beckett, M. L.; Wright, G. L. Jr. Detection and characterization of the prostate-specific membrane antigen (PSMA) in tissue extracts and body fluids. *Int J Cancer* 1995, 62, 552-558.

Tyagi, S.; Lal, S. K. Combined transformation and genetic technique verification of protein-protein interactions in the yeast two-hybrid system. *Biochem Biophys Res Commun* 2000, 277, 589-593.

Ueda, T.; Mawji, N. R.; Bruchovsky, N.; Sadar, M. D. Ligand-independent activation of the androgen receptor by interleukin-6 and the role of steroid receptor coactivator-1 in prostate cancer cells. *J Biol Chem* 2002a, 277, 38087-38094.

Ueda, T.; Bruchovsky, N.; Sadar, M. D. Activation of the androgen receptor N-terminal domain by interleukin-6 via MAPK and STAT3 signal transduction pathways. *J Biol Chem* 2002b, 277, 7076-7085.

Uetz, P.; Giot, L.; Cagney, G.; Mansfield, T. A.; Judson, R. S.; Knight, J. R.; Lockshon, D.; Narayan, V.; Srinivasan, M.; Pochart, P.; Qureshi-Emili, A.; Li, Y.; Godwin, B.; Conover, D.; Kalbfleisch, T.; Vijayadamodar, G.; Yang, M.; Johnston, M.; Fields, S.; Rothberg, J. M. A comprehensive analysis of protein-protein interactions in *Saccharomyces cerevisiae*. *Nature* 2000, 403, 623-627.

Unlü, M.; Morgan, M. E.; Minden, J. S. Difference gel electrophoresis: A single gel method for detecting changes in protein extracts. *Electrophoresis* 1997, 18, 2071-2077.

Utleg, A. G.; Yi, E. C.; Xie, T.; Shannon, P.; White, J. T.; Goodlett, D. R.; Hood, L.; Lin, B. Proteomic analysis of human prostasomes. *Prostate* 2003, 56, 150-161.

Vivanco, I.; Sawyers, C. L. The phosphatidylinositol 3-kinase AKT pathway in human cancer. *Nat Rev Cancer* 2002, 2, 489-501.

Wang, G.; Sadar, M. D. Amino-terminus domain of the androgen receptor as a molecular target to prevent the hormonal progression of prostate cancer. *J Cell Biochem* 2006, 98, 36-53.

Wang, M. C.; Papsidero, L. D.; Kuriyama, M.; Valenzuela, L. A.; Murphy, G. P.; Chu, T. M. Prostate antigen: A new potential marker for prostatic cancer. *Prostate* 1981, 2, 89-96.

Wang, Y. K.; Ma, Z.; Quinn, D. F.; Fu, E. W. Inverse 18O labeling mass spectrometry for the rapid identification of marker/target proteins. *Anal Chem* 2001, 73, 3742-3750.

Washburn, M. P.; Wolters, D.; Yates, J. R. 3rd. Large-scale analysis of the yeast proteome by multidimensional protein identification technology. *Nat Biotechnol* 2001, 19, 242-247.

Wellmann, A.; Wollscheid, V.; Lu, H.; Ma, Z. L.; Albers, P.; Schütze, K.; Rohde, V.; Behrens, P.; Dreschers, S.; Ko, Y.; Wernert, N. Analysis of microdissected prostate tissue with ProteinChip arrays-a way to new insights into carcinogenesis and to diagnostic tools. *Int J Mol Med* 2002, 9, 341-347.

Wilson, D. S.; Keefe, A. D.; Szostak, J. W. The use of mRNA display to select high-affinity protein-binding peptides. *Proc Natl Acad Sci USA* 2001, 98, 3750-3755.

Witte, C. P.; Noel, L. D.; Gielbert, J.; Parker, J. E.; Romeis, T. Rapid one step protein purification from plant material using the eight-amino acid StrepII epitope. *Plant Mol Biol* 2004, 55, 135-147.

Wolf, S. S.; Patchev, V. K.; Obendorf, M. A novel variant of the putative demethylase gene, s-JMJD1C, is a coactivator of the AR. *Arch Biochem Biophys* 2007, 460, 56-66.

Wolters, D. A.; Washburn, M. P.; Yates, J. R. 3rd. An automated multidimensional protein identification technology for shotgun proteomics. *Anal Chem* 2001, 73, 5683-5690.

Wong, H. Y.; Burghoorn, J. A.; Van Leeuwen, M.; De Ruiter, P. E.; Schippers, E.; Blok, L. J.; Li, K. W.; Dekker, H. L.; De Jong, L.; Trapman, J.; Grootegoed, J. A.; Brinkmann, A. O. Phosphorylation of androgen receptor isoforms. *Biochem J* 2004, 383, 267-276.

Wright, G. W.; Cazares, L. H.; Leung, S. M.; Nasim, S.; Adam, B. L.; Yip, T. T.; Schellhammer, P. F.; Gong, L.; Vlahou, A. Proteinchip(R) surface enhanced laser desorption/ionization (SELDI) mass spectrometry: A novel protein biochip technology for detection of prostate cancer biomarkers in complex protein mixtures. *Prostate Cancer Prostatic Dis* 1999, 2, 264-276.

Wright, M. E.; Eng, J.; Sherman, J.; Hockenbery, D. M.; Nelson, P. S.; Galitski, T.; Aebersold, R. Identification of androgen-coregulated protein networks from the microsomes of human prostate cancer cells. *Genome Biol* 2003, 5, R4.

Wu, F.; Mo, Y. Y. Ubiquitin-like protein modifications in prostate and breast cancer. *Front Biosci* 2007, 12, 700-711.

Wu, P.; Leinonen, J.; Koivunen, E.; Lankinen, H.; Stenman, U. H. Identification of novel prostate-specific antigen-binding peptides modulating its enzyme activity. *Eur J Biochem* 2000, 267, 6212-6220.

Xia, S. J.; Hao, G. Y.; Tang, X. D. Androgen receptor isoforms in human and rat prostate. *Asian J Androl* 2000, 2, 307-310.

Xiao, D.; Choi, S.; Lee, Y. J.; Singh, S. V. Role of mitogen-activated protein kinases in phenethyl isothiocyanate-induced apoptosis in human prostate cancer cells. *Mol Carcinog* 2005, 43, 130-140.

Xiao, Z.; Jiang, X.; Beckett, M.; Wright, G. L. Jr. Generation of a baculovirus recombinant prostate-specific membrane antigen and its use in the development of a novel protein biochip quantitative immunoassay. *Prot Expression Purif* 2000, 19, 12-21.

Xu, Z.; Bae, W.; Mulchandani, A.; Mehra, R. K.; Chen, W. Heavy metal removal by novel CBD-EC20 sorbents immobilized on cellulose. *Biomacromolecules* 2002, 3, 462-465.

Yan, W.; Chen, S. S. Mass spectrometry-based quantitative proteomic profiling. *Brief Funct Genomic Proteomic* 2005, 4, 27-38.

Yao, X.; Freas, A.; Ramirez, J.; Demirev, P. A.; Fenselau, C. Proteolytic 18O labeling for comparative proteomics: Model studies with two serotypes of adenovirus. *Anal Chem* 2004, 73, 2836-2842.

Yeh, S.; Chang, C. Cloning and characterization of a specific coactivator, ARA70, for the androgen receptor in human prostate cells. *Proc Natl Acad Sci USA* 1996, 93, 5517-5521.

Yeh, S.; Miyamoto, H.; Nishimura, K.; Kang, H.; Ludlow, J.; Hsiao, P.; Wang, C.; Su, C.; Chang, C. Retinoblastoma, a tumor suppressor, is a coactivator for the androgen receptor in human prostate cancer DU145 cells. *Biochem Biophys Res Commun* 1998, 248, 361-367.

Yi, E. C.; Li, X. J.; Cooke, K.; Lee, H.; Raught, B.; Page, A.; Aneliunas, V.; Hieter, P.; Goodlett, D. R.; Aebersold, R. Increased quantitative proteome coverage with (13)C/(12)C-based, acid-cleavable isotope-coded affinity tag reagent and modified data acquisition scheme. *Proteomics* 2005, 5, 380-387.

Zheng, Y.; Xu, Y.; Ye, B.; Lei, J.; Weinstein, M. H.; O'Leary, M. P.; Richie, J. P.; Mok, S. C.; Liu, B. C. Prostate carcinoma tissue proteomics for biomarker discovery. *Cancer* 2003, 98, 2576-2582.

Zhou, H.; Li, X. M.; Meinkoth, J.; Pittman, R. N. Akt regulates cell survival and apoptosis at a post-mitochondrial level. *J Cell Biol* 2000, 151, 483-494.

Zhou, Z. X.; Kemppainen, J. A.; Wilson, E. M. Identification of three proline-directed phosphorylation sites in the human androgen receptor. *Mol Endocrinol* 1995, 9, 605-615.

Zhu, H.; Bilgin, M.; Bangham, R.; Hall, D.; Casamayor, A.; Bertone, P.; Lan, N.; Jansen, R.; Bidlingmaier, S.; Houfek, T.; Mitchell, T.; Miller, P.; Dean, R. A.; Gerstein, M.; Snyder, M. Global analysis of protein activities using proteome chips. *Science* 2001, 293, 2101-2105.

Zhu, H.; Snyder, M. Protein chip technology. *Curr Opin Chem Biol* 2003, 7, 55-63.

Zitzmann, S.; Mier, W.; Schad, A.; Kinscherf, R.; Askoxylakis, V.; Krämer, S.; Altmann, A.; Eisenhut, M.; Haberkorn, U. A new prostate carcinoma binding peptide (DUP-1) for tumor imaging and therapy. *Clin Cancer Res* 2005, 11, 139-146.

Zozulya, S.; Lioubin, M.; Hill, R. J.; Abram, C.; Gishizky, M. L. Mapping signal transduction pathways by phage display. *Nat Biotechnol* 1999, 17, 1193-1198.

CHAPTER 7

RNA INTERFERENCE THERAPEUTICS – PAST, PRESENT, AND FUTURE

JAYAPAL MANIKANDAN, MAHMOOD RASOOL,
MUHAMMAD IMRAN NASEER,
KOTHANDARAMAN NARASIMHAN,
LAILA ABDULLAH DAMIATI,
KALEMEGAM GAUTHAMAN, SAMI BAHLAS, and
PETER NATESAN PUSHPARAJ

CONTENTS

7.1   Introduction 176

7.2   siRNA in Experimental Therapeutics 176

7.3   siRNA Production and Processing Inside the Cell 178

7.4   *Ex Vivo* and *In Vivo* Delivery of siRNAs 179

7.5   *Ex Vivo* and *In Vivo* Efficiency of siRNA 180

7.6   Major Obstacles in siRNA Therapeutics 181

7.7   siRNA in Future Therapeutics 182

7.8   Future Perspectives 183

Ackowledgements 184

Keywords 184

References 185

7.1 INTRODUCTION

RNA interference (RNAi) was first reported by Andrew Fire and Craig Mello in *Caenorhabditis elegans* (Fire *et al.*, 1998). For this discovery, they were bestowed with the Nobel Prize in Physiology or Medicine for the year 2006 (Pushparaj *et al.*, 2008). RNAi exists in plants as a defense strategy against viruses or uncontrolled transposon recruitment (Bernstein *et al.*, 2001). It was observed that dsRNAs can cause specific knockdown of homologous mRNAs (Ahlquist *et al.*, 2002; Schiffelers *et al.*, 2004). Later, it was experimentally documented that small RNAs (21- to 25-nucleotides) were indeed essential for RNAi called as small interfering RNAs (siRNAs) (Bernstein *et al.*, 2001). Moreover, RNAi phenomenon was also recorded in flies (Kennerdell *et al.*, 2000; Ahlquist *et al.*, 2002), and vertebrates (Kennerdell *et al.*, 2000; Li *et al.*, 2000). In this chapter, we discuss the RNAi phenomenon in the past, present, and future in therapeutics.

7.2 siRNA IN EXPERIMENTAL THERAPEUTICS

siRNAs can be exploited in therapeutics to specifically silence genes implicated in transcription and translation of proteins (Pushparaj *et al.*, 2008) that code for enzymes and receptors (Figure 7.1). We and others have extensively reviewed on RNAi and its potential applications in basic research and experimental therapeutics (Pushparaj *et al.*, 2008; Manikandan *et al.*, 2008; Davidson and McCray, 2011). Since siRNAs represent ideal molecules for therapeutic targeting of genes implicated in disease pathways, various Investigational New Drug (IND) submissions had been filed with the United States (US) Food and Drug Administration (FDA) for the approval of clinical trials using siRNAs (Uprichard, 2005).

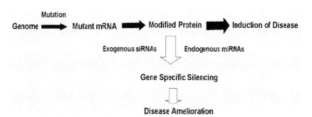

FIGURE 7.1   The basis of RNAi therapeutics. Small interfering RNAs (siRNAs) that include small synthetic dsRNAs and short hairpin RNAs (shRNAs) can be synthesized in laboratory to specifically knockdown the disease-causing genes. They bind and then degrade the mRNA produced by the gene before the mRNA can start producing a harmful protein, which actually causes the illness. Conversely, microRNAs (miRNAs) are synthesized by inherent mechanisms to regulate various vital cellular and molecular processes, within a cell (Pushparaj *et al.*, 2008; Manikandan *et al.*, 2008).

Recent literature reviews on RNAi therapeutics have shown that several siR-NAs are currently tested in various phases of clinical trials for the treatment of cancer, ocular and retinal diseases, antiviral therapy, kidney disorders, lowering of LDL cholesterol, etc. (Pushparaj *et al.*, 2008; Davidson and McCray, 2011). Studies have reported that siRNAs can be exploited in infection, immunity and inflammation, neurobiology, cancer biology, oral biology, medicine, etc. (Pushparaj *et al.*, 2008; Davidson and McCray, 2011).

7.2.1   CANCER

RNAi has been extensively studied in cancer biology and medicine (Kittler and Buchholz, 2003; Lu *et al.*, 2003; Wall and Shi, 2003). Many apoptotic genes such as Bcl-2, p53 and Bcr-Abl (which expresses an oncoprotein in chronic myelogenous leukemia (CML) (Wilda *et al.*, 2002), carcinogenic *k-ras* transcripts, which induces tumor in colon and pancreas (Brummelkamp *et al.*, 2002), can be silenced using RNAi for cancer therapy (Cioca *et al.*, 2003). Combinatorial approach using multiple siRNAs against VEGF-A, VEGFR1 and VEGFR2 has proven to be useful for controlling angiogenesis compared with single siRNAs (Kim *et al.*, 2006). Studies have identified various siRNA targets for the fight against vascular diseases and tumor development (Pushparaj *et al.*, 2008; Manikandan *et al.*, 2008; Davidson and McCray, 2011).

7.2.2   NEUROLOGICAL DISORDERS

Experimental RNAi of neurological diseases/disorders has been recently reviewed (Ramachandran *et al.*, 2014; Deng *et al.*, 2014). Studies have clearly revealed that neurological disease, caused by polyglutamine toxicity, such as Huntington's disease, can be effectively controlled by siRNAs (Xia *et al.*, 2002; wood *et al.*, 2003; Davidson and Paulson, 2004). RNAi targeting of β-secretase (BACE1) decreased β-amyloid peptides Ab140 and Ab142 secretion *in vitro* and reduced cell death. BACE1 is upregulated in Alzheimer's disease (AD) patient's brain and the knockout mice of BACE1 were found to be protected against the development of AD like phenotype (Kao *et al.*, 2003). Furthermore, RNAi targeting of Huntingtin genes in mice reduced the development of Huntington's disease (Pushparaj *et al.*, 2008; Yu *et al.*, 2014).

7.2.3   DOWNSTREAM MODULATORS OF DISEASE

Studies have precisely proved, using hydrodynamic method of siRNA delivery, that liver is the primary organ for siRNA uptake in mice (Zhang *et al.*, 1999;

Chu *et al.*, 2005). RNAi therapeutics is exploited in fas-mediated apoptosis in various hepatic diseases. siRNAs directed against fas diminished the development of fulminant hepatitis and effectively silence fas for 10 days (Song *et al.*, 2003). Furthermore, siRNAs targeting of caspase 8 prevents the development of acute liver failure (ALF) in mice (Contreras *et al.*, 2004) and vouches for the importance of RNAi therapy for liver disorders in the near future (Gitlin *et al.*, 2002; Pushparaj *et al.*, 2008).

7.2.4   VIRAL DISEASES

Studies have shown that siRNAs can be effectively used to target viral genes (McCaffrey *et al.,* 2002; Clayton, 2004). It has been experimentally shown that both polio virus and human immunodeficiency virus (HIV) propagation can be effectively tackled by RNAi strategies and thus viral infections may be reduced by RNAi vectors (Jacque *et al.*, 2002; Lee *et al.*, 2002; Novina *et al.*, 2002; Kapadia *et al.*, 2003). RNAi of fas cell death receptor (Zender *et al.*, 2003) and caspase 8 (Sen *et al.,* 2003) that are involved in apoptotic cascades in ALF due to viral infections have been studied using RNAi therapeutics. Combination of siRNAs targeting the Severe Acute Respiratory Syndrome virus reduced the viral activity and thereby gives a cue for anti-SARS management and possible treatment in the near future (Zheng *et al.*, 2004).

7.2.5   DOMINANT GENE DISORDERS

The familial amyotrophic lateral sclerosis (ALS) or Lou Gehrig's disease can be treated using RNAi strategies in the near future. In ALS, the mutation in Cu/Zn superoxide dismutase (encoded by *SOD1*) leads to the death of motor neurons. RNAi silencing of *SOD1* can reduce the disease progression and death *in vivo* (Raoul *et al.*, 2005; Ralph *et al.*, 2005). Several targeting strategies can be utilized, namely, direct mutation targeting (Gonzalez-Alegre *et al.*, 2003; Miller *et al.*, 2003), indirect mutation targeting (Miller *et al.*, 2003), and targeting aberrant splicing isoforms (Ryther *et al.*, 2004) to specifically silence the mutant SOD1 in ALS.

7.3   siRNA PRODUCTION AND PROCESSING INSIDE THE CELL

The long dsRNAs are cleaved into 21-25 bp siRNAs by an enzyme called Dicer (Hammond *et al.*, 2000; Zamore *et al.*, 2000; Hutvagner and Zamore, 2002; Pushparaj *et al.*, 2008). The 21-25 bp siRNAs are unwound by helicase and incorporated into a multi-subunit RNA-induced silencing complex (RISC), which targets complementary mRNAs for enzymatic degradation (Elbashir *et al.*, 2001) and selectively inhibiting target gene expression (Figure 7.2).

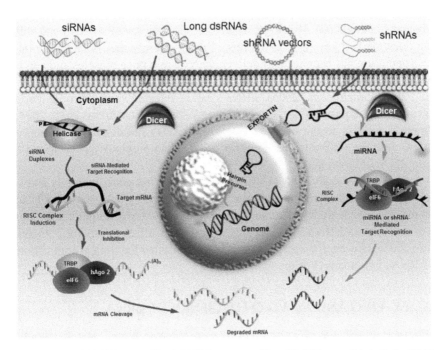

FIGURE 7.2   RNAi Dogma. The dsRNA is processed into 21–25 bp small interfering RNAs (siRNA) by an RNaseIII-like enzyme called Dicer. The siRNAs are unwound by a helicase before entering into a multi-subunit RNA-induced silencing complex (RISC). The target mRNA complementary to the siRNA integrates into RISC and it is cleaved by an endonuclease present within RISC and inhibits the translation of specific mRNA transcripts. However, the synthetic siRNAs (21-25 bp) skip the Dicer step and are directly unwound by the helicase and enter into the RISC for specific degradation of mRNA transcripts. In mammalian cells, both microRNAs (miRNAs) and siRNAs are believed to utilize the same RISC, where they direct the degradation or translational silencing, depending on the complementarity of the target mRNAs (Pushparaj *et al.*, 2008; Manikandan *et al.*, 2008).

7.4   *EX VIVO* AND *IN VIVO* DELIVERY OF siRNAs

We have extensively reviewed the *ex vivo* and *in vivo* siRNA delivery strategies (Pushparaj *et al.*, 2008). Viral vectors such as lentiviral vectors can be effectively used to deliver siRNAs into blood and bone marrow cells (Clayton, 2004). However, the toxicity associated with viral vectors will be a major deterrent for its use in humans (Clayton, 2004; Hannon and Rossi, 2004). The death of a cancer patient who got administered with adenovirus raised the concern about use of viral vectors in RNAi therapeutics in humans (Reid *et al.*, 2002). Cationic liposomes and electroporation have been used to deliver siRNAs and shRNAs *in vitro* and *in vivo* (Calegari *et al.*, 2002; Sorensen *et al.*, 2003; Matsuda and Cepko, 2004). However, tissue specific delivery of siRNAs and shRNAs is certainly a key issue to address in RNAi therapeutics (Clayton, 2004). Electroporation has been used to deliver siRNAs into mice muscles and can be further developed to deliver siRNAs specifically into tissues and organs (Lewis *et al.*, 2002; Golzio *et al.*, 2005). siRNAs can be injected *in vivo* through portal vein with Lipiodol (pale yellow iodized poppy seed oil) (Contreras *et al.*, 2004), intranasal delivery to the lungs (Zhang *et al.*, 2004) and direct application of siRNAs into the rat brain (Iascson *et al.*, 2003) had previously been found to cause RNAi of target genes.

An innovative study by McCaffrey *et al.* (2002) has experimentally shown the *in vivo* delivery of siRNAs in mice liver, using the hydrodynamic transfection method. The majority of studies performed in mice have adopted this method of delivery of dsRNAs into various organs (McCaffrey *et al.*, 2002; Golzio *et al.*, 2005). However, this method cannot be translated for clinical applications of siRNAs in humans (Pushparaj *et al.*, 2008).

7.5   *EX VIVO* AND *IN VIVO* EFFICIENCY OF siRNA

Previous studies have shown that siRNAs are more stable *in vitro* and *in vivo* compared to antisense oligodeoxyribonuceic acids (ODNs) (Bertrand *et al.*, 2002; Pushparaj *et al.*, 2008). However, efficacy differs with various target genes both siRNAs and ODNs (Miyagishi *et al.*, 2003). Since mammalian cells do not have a system, unlike *C. elegans* and plants to amplify and propagate RNAi, the gene knockdown efficiency of siRNAs is transient and the transfected siRNAs are effective up to 7 days and degraded (Pushparaj *et al.*, 2008). It is highly reliant on the rate of cell division (Clayton, 2004). Henceforth, chemical modifications are introduced in siRNA molecules (Figure 7.3) to increase their stability and efficiency in mammalian cells (Braasch *et al.*, 2003). It includes 2'-O-methylation, 2'-fluoropyrimidines (Amarzguioui *et al.*, 2003; Chiu and Rana, 2003; Czauderna *et al.*, 2003), phosphorothioate modifications (replacement of one of the nonbridging oxygen atoms in the phosphodiester bond by sulfur) (Dorsett and Tuschl, 2004). We have recently reviewed the siRNA delivery strategies *in vitro* and *in vivo* (Pushparaj *et al.*, 2008).

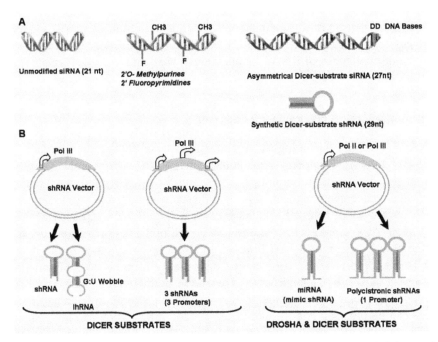

FIGURE 7.3 (A) siRNAs can be chemically modified such as 2'-fluoropyrimidines and 2'-O-methylpurines, for more stability. Asymmetrical Dicer–substrate siRNAs, with two DNA bases (DD) in the blunt end, can also be produced to increase the efficiency. (B) Long hairpin RNAs (lhRNAs) generates multiple Dicer-processed siRNA species. More than one Pol III promoter can be used in one vector to trigger the expression of a variety of shRNAs. Besides, vectors with Pol II or Pol III promoters produce longer precursor RNAs, including miRNA mimics and polycistronic shRNA transcripts, processed by both Dicer and Drosha for exerting RNAi phenomenon (Kim and Rossi, 2007).

7.6   MAJOR OBSTACLES IN siRNA THERAPEUTICS

Delivery of siRNAs to specific target tissue in RNAi therapeutics is always a major challenge as well as an obstacle for effective clinical translation. The activation of interferon (IFN) response, especially when larger dsRNAs are expressed from viral vectors and nonspecific gene silencing are observed in *in vivo* RNAi studies (Pushparaj *et al.*, 2008). The endogenous silencing or inhibition of miRNAs by the exogenous siRNAs is another obstacle observed in RNAi therapeutics. However, by using either small dsRNAs as well as vector-mediated delivery of large dsRNAs can reduce these side effects (Lieberman *et al.*, 2003). The dsRNA activates type 1 IFN response and signal transducer and activator of transcription (STAT) mediated expression of dsRNA dependent protein kinase (PKR) that activates eukaryotic initiation factor 2 (eIF2α) by phosphorylation of its small subunit which in turn inhibits translation and induces apoptosis (Stark *et al.*, 1998). We have previously reviewed in detail the major challenges of RNAi therapeutics *in vitro* and *in vivo* (Pushparaj *et al.*, 2008).

7.7   siRNA IN FUTURE THERAPEUTICS

RNAi is a robust technology to specifically silence the gene expression in living organisms (Pushparaj *et al.*, 2008). Since RNAi is very precise, we can target splice variants, foreign, or mutant genes in experimental therapeutics. Recent advancements in the next generation sequencing (NGS) strategies lead to the deduction of complete human genome will facilitate the precise designing of RNAi vectors against an array of genes implicated in various disease pathologies. Designing RNAi libraries to target specific genes in the mammalian genome provides us a unique opportunity to study loss of function studies *in vitro* and *in vivo*. RNAi is a reversible methodology when compared with gene knockout and transgenic approaches. The expression of a particular gene can be reversed and the attenuated genetic information can be regained without losing background genetic and regulatory machinery. The endogenous gene silencing can be restored once the RNAi vector is removed or silenced. Consequently, the expression of a gene can be attenuated or induced at any given time to characterize the role of a gene or a set of genes in various disease pathologies (Zhou *et al.*, 2002). siRNAs can be delivered into specific targets *in vivo* using protamine-antibody fusion protein linked with specific siRNA and has a huge potential for the development of RNAi therapeutics in the near future (Song *et al.*, 2005). Enormous technological advancements have occurred in the design, synthesis, and purification of siRNAs over the last few years (Pushparaj *et al.*, 2008). High-throughput analysis of thousands of gene targets using RNAi can be carried out at different levels of throughput with or without automation. Recently, RNAi cell microarrays are developed for high-throughput analysis in drug target identification and genome-wide RNAi screening (Wheeler *et al.*, 2005). This method could be used for clinical diagnosis and the development of personalized medicine for various diseases in the near future (Figure 7.4).

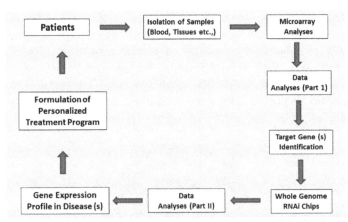

FIGURE 7.4 The potential applications of RNAi in clinical diagnosis and personalized medicine. Genome-wide RNAi screening in mammalian cells could be used for clinical diagnosis of various disease pathologies in the near future. The potential discovery of a gene or a group of genes by microarray experiments can be silenced by high-throughput whole genome RNAi chips for further analyses. Robust computational strategies coupled with high-throughput RNAi in clinics might be considered for designing personalized treatments for patients suffering from various disease pathologies.

7.8 FUTURE PERSPECTIVES

In this chapter, we have critically dissected the therapeutic potential of RNAi in human diseases. Importantly, RNAi has enormous commercial, economic, and scientific prospects in modern era. Besides, RNAi offers researchers a robust, less laborious, and relatively cost-effective tool for investigating biological systems by selectively blocking any gene(s) of interest both *in vitro* and *in vivo*. On the contrary, the study of gene function using gene knockouts, transgenic animal models, and AS-ODNs are costly, laborious, time-consuming and not relatively compatible with high-throughput gene silencing studies. The designing of siRNAs are quiet simple and relatively straight forward approach than other gene silencing molecules such as AS-ODNs. The bioactivity of siRNAs is found to be more than 100 folds higher than AS-ODNs and hence used in high-throughput drug discovery research (Pushparaj *et al.*, 2008; Manikandan *et al.*, 2008).

However, proper siRNA designing strategies using *in silico* methods, optimized transfection and robust expression vectors can further aid in the development of RNAi therapeutics in translational medicine. Nevertheless, RNAi has significantly augmented biomedical research in drug target discovery and functional genomics. Interestingly, the recent discovery of RNA activation (RNAa) phenomenon by small dsRNAs (Li *et al.*, 2006) presents us another dimension to RNAi therapeutics. We have previously reviewed the RNAa phenomenon in biological systems (Pushparaj *et al.*, 2008). The RNAi and RNAa represent two different but complementary concepts of genetic regulation (Figure 7.5) and hence we termed it as RNAi/RNAa as the "Yin and Yang" of RNAome (Pushparaj *et al.*, 2008). In conclusion, RNAi therapeutics in the past and present has shown tremendous potential for translating into future therapy for various human diseases.

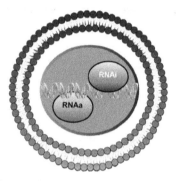

FIGURE 7.5   RNAa/RNAi mechanism observed in cellular and molecular systems exemplifies the "Yin and Yang" phenomenon (adapted and modified from Pushparaj *et al.*, 2008).

ACKOWLEDGEMENTS

This project was supported by the NSTIP strategic technologies program in the Kingdom of Saudi Arabia (KSA) Project Numbers 12-BIO2719-03 and 12-BIO2267-03. The authors also acknowledge the Science and Technology Unit, King Abdulaziz University (KAU) with thanks for technical support. In addition, authors would like to acknowledge the wonderful figures generated by the technical team of Beacon Biosoft (www.beaconbiosoft.com) within a very short time.

KEYWORDS

- **Alzheimer's disease**
- **Apoptotic genes**
- ***Caenorhabditis elegans***
- **Cationic liposomes**
- **Chronic myelogenous leukemia**
- **Dicer**
- **Direct mutation targeting**
- **Electroporation**

- **Endogenous gene silencing**
- **Experimental therapeutics**
- **Fulminant hepatitis**
- **Human immunodeficiency virus**
- **Huntingtin genes**
- **Hydrodynamic transfection**
- **Indirect mutation targeting**
- **Lou Gehrig's disease**
- **Neurological diseases**
- **Next generation sequencing**
- **RNA interference**
- **RNAi therapeutics**
- **RNA-induced silencing complex**
- **siRNA delivery**
- **Small interfering RNAs**
- **Targeting aberrant splicing isoforms**

REFERENCES

Ahlquist, P. RNA-dependent RNA polymerases, viruses, and RNA silencing. *Science* 2002, 296, 1270-1273.

Amarzguioui, M.; Holen, T.; Babaie, E.; Prydz, H. Tolerance for mutations and chemical modifications in a siRNA. *Nucleic Acid Res* 2003, 31, 589-595.

Bernstein, E.; Denli, A. M.; Hannon, G. J. The rest is silence. *RNA* 2001, 7, 1509-1521.

Bertrand, J. R.; Pottier, M.; Vekris, A.; Opolon, P.; Maksimenko, A.; Malvy, C. Comparison of antisense oligonucleotides and siRNAs in cell culture and *in vivo*. *Biochem. Biophys Res Commun* 2002, 296, 1000-1004.

Braasch, D. A.; Jensen, S.; Liu, Y.; Kaur, K.; Arar, K.; White, M. A.; Corey, D. R. RNA interference in mammalian cells by chemically-modified RNA. *Biochemistry* 2003, 42, 7967-7975.

Brummelkamp, T.; Bernards, R.; Agami, R. Stable suppression of tumorigenicity by virus-mediated RNA interference. *Cancer Cell* 2002, 2, 243-247.

Calegari, F.; Haubensak, W.; Yang, D.; Huttner, W. B.; Buchholz, F. Tissue-specific RNA interference in postimplantation mouse embryos with endoribonuclease-prepared short interfering RNA. *Proc Natl Acad Sci USA* 2002, 99, 14236-14240.

Chiu, Y. L.; Rana, T. siRNA function in RNAi: A chemical modification analysis. *RNA* 2003, 9, 1034-1048.

Chiu, Y. L.; Rana, T. M. RNAi in human cells: Basic structural and functional features of small interfering RNA. *Mol Cell* 2002, 10, 549-561.

Chu, Q.; Joseph, M.; Przybylska, M.; Yew, N. S.; Scheule, R. K. Transient siRNA-mediated attenuation of liver expression from an alpha-galactosidase A plasmid reduces subsequent humoral immune responses to the transgene product in mice. *Mol Ther* 2005, 12, 264-273.

Cioca, D.; Aoki, Y.; Kiyosawa, K. RNA interference is a functional pathway with therapeutic potential in human myeloid leukemia cell lines. *Cancer Gene Ther* 2003, 10, 125-133.

Clayton, J. RNA interference: The silent treatment. *Nature* 2004, 431, 599-605.

Contreras, J. L.; Vilatoba, M.; Eckstein, C.; Bilbao, G.; Thompson, J. A.; Eckhoff, D. E. Caspase-8 and caspase-3 small interfering RNA decreases ischemia/reperfusion injury to the liver in mice. *Surgery* 2004, 136, 390-400.

Czauderna, F.; Fechtner, M.; Dames, S.; Aygun, H.; Klippel, A.; Pronk, G. J.; Giese, K.; Kaufmann, J. Structural variations and stabilizing modifications of synthetic siRNAs in mammalian cells. *Nucleic Acid Res* 2003, 31, 2705-2716.

Davidson, B. L.; Paulson, H. L. Molecular medicine for the brain: Silencing of disease genes with RNA interference. *Lancet Neurol* 2004, 3, 145-149.

Davidson, B. L.; McCray, P. B. Current prospects for RNA interference-based therapies. *Nat Rev Genet* 2011, 12, 329-340.

Deng, Y.; Wang, C. C; Choy, K. W.; Du, Q.; Chen, J.; Wang, Q.; Li, L.; Chung, T.K.; Tang, T. Therapeutic potentials of gene silencing by RNA interference: principles, challenges, and new strategies. *Gene* 2014, 38, 217-227.

Dorsett, Y.; Tuschl, T. siRNAs: Applications in functional genomics and potential as therapeutics. *Nat Rev Drug Discov* 2004, 3, 318-329.

Elbashir, S. M.; Lendeckel, W.; Tuschl, T. RNA interference is mediated by 21- and 22-nucleotide RNAs. *Genes Dev* 2001, 15, 188-200.

Fire, A.; Xu, S.; Montgomery, M. K.; Kostas, S. A.; Driver, S. E.; Mello, C. C. Potent and specific genetic interference by double-stranded RNA in *Caenorhabditis elegans*. *Nature* 1998, 391, 806-811.

Gitlin, L.; Karelsky, S.; Andino, R. Short interfering RNA confers intracellular antiviral immunity in human cells. *Nature* 2002, 418, 430-434.

Golzio, M.; Mazzolini, L.; Moller, P.; Rols, M. P.; Teissie, J. Inhibition of gene expression in mice muscle by *in vivo* electrically mediated siRNA delivery. *Gene Ther* 2005, 12, 246-251.

Gonzalez-Alegre, P.; Miller, V.; Davidson, B.; Paulson, H. Toward therapy for DYT1 dystonia: Allele-specific silencing of mutant TorsinA. *Ann Neurol* 2003, 53, 781-787.

Hammond, S.; Bernstein, E.; Beach, D.; Hannon, G. J. An RNA-directed nuclease mediates posttranscriptional gene silencing in Drosophila cells. *Nature* 2000, 404, 293-296.

Hannon, G. J.; Rossi, J. J. Unlocking the potential of the human genome with RNA interference. *Nature* 2004, 431, 371-378.

Hutvagner, G.; Zamore, P. D. RNAi: Nature abhors a double-strand. *Curr Opin Genet Dev* 2002, 12, 225-232.

Iascson, R.; Kull, B.; Salmi, P.; Wahlestedt, C. Lack of efficacy of 'naked' small interfering RNA applied directly to rat brain. *Acta Physiol Scand* 2003, 179, 173-177.

Jacque, J. M.; Triques, K.; Stevenson, M. Modulation of HIV-1 replication by RNA interference. *Nature* 2002, 418, 435-438.

Kao, S. C.; Krichevsky, A.; Kosik, K.; Tsai, L. H. BACE1 suppression by RNA interference in primary cortical neurons. *J Biol Chem* 2003, 279, 1942-1949.

Kapadia, S.; Brideau-Andersen, A.; Chisari, F. Interference of hepatitis C virus RNA replication by short interfering RNAs. *Proc Natl Acad Sci USA* 2003, 100, 2014-2018.

Kennerdell, J. R.; Carthew, R. W. Heritable gene silencing in Drosophila using double-stranded RNA. *Nat Biotechnol* 2000, 18, 896-898.

Kim, B.; Tang, Q.; Biswas, P. S.; Xu, J.; Schiffelers, R. M.; Xie, F. Y.; Ansari, A. M.; Scaria, P. V.; Woodle, M. C.; Lu, P.; Rouse, B. T. Inhibition of ocular angiogenesis by siRNA targeting vascular endothelial growth factor-pathway genes; therapeutic strategy for herpetic stromal keratitis. *Am J Pathol* 2004, 165, 2177-2185.

Kim, D. H.; Rossi, J. J. Strategies for silencing human disease using RNA interference. *Nat Rev Genet*. 2007 8, 173-184.

Kittler, R.; Buchholz, F. RNA interference: Gene silencing in the fast lane. *Semin Cancer Biol* 2003, 13, 259-265.

Lee, N. S.; Dohjima, T.; Bauer, G.; Li, H.; Li, M. J.; Ehsani, A.; Salvaterra, P.; Rossi, J. Expression of small interfering RNAs targeted against HIV-1 rev transcripts in human cells. *Nat Biotechnol* 2002, 20, 500-505.

Lewis, D. L.; Hagstrom, J. E.; Loomis, A. G.; Wolff, J. A.; Herweijer, H. Efficient delivery of siRNA for inhibition of gene expression in postnatal mice. *Nat Genet* 2002, 32, 107-108.

Li, Y. X.; Farrell, M. J.; Liu, R.; Mohanty, N.; Kirby, M. L. Double-stranded RNA injection produces null phenotypes in zebrafish. *Dev Biol* 2000, 217, 394-405.

Li, L. C.; Okino, S. T.; Zhao, H.; Pookot, D.; Place, R. F.; Urakami, S. *et al.* Small dsRNAs induce transcriptional activation in human cells. *Proc Natl Acad Sci USA*, 2006, 103, 17337-17342.

Lieberman, J.; Song, E.; Lee, S. K.; Shankar, P. Interfering with disease: Opportunities and roadblocks to harnessing RNA interference. *Trends Mol Med* 2003, 9, 397-403.

Lu, P. Y.; Xie, F. Y.; Woodle, M. C. siRNA-mediated antitumorigenesis for drug target validation and therapeutics. *Curr Opin Mol Ther* 2003, 5, 225-234.

Manikandan, J.; Aarthi, J. J.; Kumar, S. D.; Pushparaj, P. N. Oncomirs: the potential role of non-coding microRNAs in understanding cancer. *Bioinformation* 2008, 2, 330-334.

Matsuda, T.; Cepko, C. Electroporation and RNA interference in the rodent retina *in vivo* and *in vitro. Proc Natl Acad Sci USA* 2004, 101, 16-22.

McCaffrey, A. P.; Meuse, L.; Pham, T. T.; Conklin, D. S.; Hannon, G. J.; Kay, M. A. RNA interference in adult mice. *Nature* 2002, 418, 38-39.

Miller, V. M.; Xia, H.; Marrs, G. L.; Gouvion, C. M.; Lee, G.; Davidson, B. L.; Paulson, H. L. Allele-specific silencing of dominant disease genes. *Proc Natl Acad Sci USA* 2003, 100, 7195-7200.

Miyagishi, M.; Hayashi, M.; Taira, K. Comparison of the suppressive effects of antisense oligonucleotides and siRNAs directed against the same targets in mammalian cells. *Antisense. Nucleic Acid Drug Dev* 2003, 3, 1-7.

Novina, C. D.; Murray, M. F.; Dykxhoorn, D. M.; Beresford, P. J.; Riess, J.; Lee, S. K.; Collman, R. G.; Lieberman, J.; Shankar, P.; Sharp, P. A. siRNA-directed inhibition of HIV-1 infection. *Nat Med* 2002, 8, 681-686.

Pushparaj, P. N.; Aarthi, J. J.; Manikandan, J.; Kumar, S. D. siRNA, miRNA, and shRNA: *in vivo* applications. *J Dent Res* 2008, 87, 992-1003.

Pushparaj, P. N.; Aarthi, J. J.; Kumar, S. D.; Manikandan, J. RNAi and RNAa - the yin and yang of RNAome. *Bioinformation* 2008, 2, 235-237.

Ralph, G. S.; Radcliffe, P. A.; Day, D. M.; Carthy, J. M.; Leroux, M. A.; Lee, D. C.; Wong, L. F.; Bilsland, L. G.; Greensmith, L.; Kingsman, S. M.; Mitrophanous, K. A.; Mazarakis, N. D.; Azzouz, M. Silencing mutant SOD1 using RNAi protects against neurodegeneration and extends survival in an ALS model. *Nat Med* 2005, 11, 429-433.

Ramachandran P. S.; Keiser M. S.; Davidson B. L. Recent advances in RNA interference therapeutics for CNS diseases. *Neurotherapeutics* 2013, 10, 473-485.

Raoul, C.; Abbas-Terki, T.; Bensadoun, J. C.; Guillot, S.; Haase, G.; Szulc, J.; Henderson, C. E.; Aebischer, P. Lentiviral-mediated silencing of SOD1 through RNA interference retards disease onset and progression in a mouse model of ALS. *Nat Med* 2005, 11, 423-428.

Reid, T.; Warren, R.; Kirn, D. Intravascular adenoviral agents in cancer patients: Lessons from clinical trials. *Cancer Gene Ther* 2002, 9, 979-986.

Ryther, R. C.; Flynt, A. S.; Harris, B. D.; Phillips, J. A.; Patton, J. G. GH1 splicing is regulated by multiple enhancers whose mutation produces a dominant-negative GH isoform that can be degraded by allele-specific siRNA. *Endocrinology* 2004, 145, 2988-2996.

Schiffelers, R. M.; Woodle, M. C.; Scaria, P. Pharmaceutical prospects for RNA interference. *Pharm Res* 2004, 21, 1-7.

Sen, A.; Steele, R.; Ghosh, A. K.; Basu, A.; Ray, R.; Ray, R. B. Inhibition of hepatitis C virus protein expression by RNA interference. *Virus Res* 2003, 96, 27-35.

Song, E.; Lee, S. K.; Wang, J.; Ince, N.; Ouyang, N.; Min, J.; Chen, J.; Shankar, P.; Lieberman, J. RNA interference targeting Fas protects mice from fulminant hepatitis. *Nat Med* 2003, 9, 347-351.

Song, E.; Zhu, P.; Lee, S. K.; Chowdhury, D.; Kussman, S.; Dykxhoorn, D. M.; Feng, Y.; Palliser, D.; Weiner, D. B.; Shankar, P.; Marasco, W. A.; Lieberman, J. Antibody mediated *in vivo* delivery of small interfering RNAs via cell surface receptors. *Nat Biotech* 2005, 23, 709-717.

Sorensen, D. R.; Leirdal, M.; Sioud, M. Gene silencing by systemic delivery of synthetic siRNAs in adult mice. *J Mol Biol* 2003, 327, 761-766.

Stark, G. R.; Kerr, I. M.; Williams, B. R.; Silverman, R. H.; Schreiber, R. D. How cells respond to interferons. *Annu Rev Biochem* 1998, 67, 227-264.

Uprichard, S. L. The therapeutic potential of RNA interference. *FEBS Lett* 2005, 579, 5996-6007.

Wall, N. R.; Shi, Y. Small RNA: Can RNA interference be exploited for therapy? *Lancet* 2003, 362, 1401-1403.

Wheeler, D. B.; Carpenter, A. E.; Sabatini, D. M. Cell microarrays and RNA interference chip away at gene function. *Nat Genet* 2005, 37, 25-30.

Wilda, M.; Fuchs, U.; Wossmann, W.; Borkhardt, A. Killing of leukemic cells with a BCR/ABL fusion gene by RNA interference (RNAi). *Oncogene* 2002, 21, 5716-5724.

Wood, M.; Trulzsch, B.; Abdelgany, A.; Beeson, D. Therapeutic gene silencing in the nervous system. *Hum Mol Gen* 2003, 12, 279-284.

Xia, H.; Mao, Q.; Paulson, H. L.; Davidson, B. L. siRNA mediated gene silencing *in vitro* and *in vivo*. *Nat Biotechnol* 2002, 20, 1006-1010.

Yu, S.; Liang, Y.; Palacino, J.; Difiglia, M.; Lu, B.; Drugging unconventional targets: insights from Huntington's disease. *Trends Pharmacol Sci* 2014, 35, 53-62.

Zamore, P. D.; Tuschl, T.; Sharp, P. A.; Bartel, D. P. RNAi: Double-stranded RNA directs the ATP-dependent cleavage of mRNA at 21 to 23 nucleotide intervals. *Cell* 2000, 101, 25-33.

Zender, L.; Hutker, S.; Liedtke, C.; Tillmann, H. L.; Zender, S.; Mundt, B.; Waltemathe, M.; Gosling, T.; Flemming, P.; Malek, N. P.; Trautwein, C.; Manns, M. P.; Kuhnel, F.; Kubicka, S. Caspase 8 small interfering RNA prevents acute liver failure in mice. *Proc Natl Acad Sci USA* 2003, 100, 7797-7802.

Zhang, G.; Budker, V.; Wolff, J. High levels of foreign gene expression in hepatocytes after tail vein injections of naked plasmid DNA. *Hum Gene Therapy* 1999, 10, 1735-1737.

Zhang, X.; Shan, P.; Jiang, D.; Noble, P. W.; Abraham, N. G.; Kappas, A.; Lee, P. J. Small interfering RNA targeting heme oxygenase-1 enhances ischemia–reperfusion-induced lung apoptosis. *J Biol Chem* 2004, 279, 10677-10684.

Zheng, B. J.; Guan, Y.; Tang, Q.; Du, C.; Xie, F. Y.; He, M. L.; Chan, K. W.; Wong, K. L.; Lader, E.; Woodle, M. C.; Lu, P. Y.; Li, B.; Zhong, N. Prophylactic and therapeutic effects of small interfering RNA targeting SARS-coronavirus. *Antivir Ther* 2004, 9, 365-374.

Zhou, Y.; Ching, Y. P.; Kok, K. H.; Kung, H. F.; Jin, D. Y. Post-transcriptional suppression of gene expression in Xenopus embryos by small interfering RNA. *Nucleic Acids Res* 2002, 30, 1664-1669.

CHAPTER 8

MOLECULAR MECHANISMS OF HEPATITIS C VIRUS ENTRY – IMPACT OF HOST CELL FACTORS FOR INITIATION OF VIRAL INFECTION

MIRJAM B. ZEISEL and THOMAS F. BAUMERT

CONTENTS

8.1   Introduction 190
8.2   Molecular Mechanisms of HCV Binding and Entry into Target Cells 191
8.3   Host Cell Factors Involved in HCV Binding and Attachment 192
8.4   Host Cell Factors Involved in HCV Entry 193
8.5   Conclusion and Future Perspectives.. 197
Acknowledgements... 197
Keywords 197
References... 198

8.1 INTRODUCTION

The hepatitis C virus is a small enveloped positive-strand RNA virus belonging to the genus *Hepacivirus* of the *Flaviviridae* family (Lindenbach *et al.*, 2007). The HCV genome encodes a single precursor polyprotein of about 3,000 amino acids. This protein is cleaved co- and posttranslationally into functional structural and nonstructural proteins by host and viral proteases. The structural proteins assemble into HCV particles of about 55–60 nm in diameter (Kaito *et al.*, 1994; Shimizu *et al.*, 1996; Wakita *et al.*, 2005). HCV is thought to adopt a classical icosahedral scaffold in which the two envelope glycoproteins E1 and E2 are anchored to the host cell-derived double-layer lipid envelope (Penin *et al.*, 2004). E1 and E2 are type I transmembrane glycoproteins forming noncovalent heterodimers. Underneath the envelope lies the nucleocapsid which is probably composed of multiple copies of the core protein in complex with the viral genome (Penin *et al.*, 2004). The nonstructural proteins have various functions involved in viral RNA replication and proteolytic processing: the two viral autoproteases NS2 and the serine protease NS3; the NS4A polypeptide, an essential cofactor for the NS3 protease; the NS4B and NS5A proteins; and finally the NS5B RNA-polymerase (Barth *et al.*, 2006; Moradpour *et al.*, 2007). The nonstructural proteins coordinate viral replication by the formation of a membrane-bound replication complex. Apart from these, an additional protein, termed F (for frameshift)-protein, has been proposed which is encoded by an overlapping reading frame in the core protein coding sequence.

HCV is principally transmitted through blood and blood-derived products (Alter, 1997). It mainly targets the liver thereby inducing hepatitis that may progress to liver cirrhosis and ultimately lead to hepatocellular carcinoma (Alter *et al.*, 1992). 50–80 % of patients develop a chronic HCV infection and 4–20 % of patients with chronic hepatitis C will develop liver cirrhosis within 20 years. In patients with liver cirrhosis the risk to develop HCC is 1–5 % per year (Di Bisceglie *et al.*, 2002; Pawlotsky, 2004). Therapeutic options are still limited and a protective vaccine is not available to date. The current treatment for chronic HCV infection is the combination of pegylated interferon and the oral antiviral drug ribavirin given for 24 or 48 weeks. However, this treatment is characterized by limited efficacy, high cost and substantial side effects (Di Bisceglie *et al.*, 2002; Pawlotsky, 2004). Therefore, more effective and better tolerated therapeutic strategies are needed. The development of novel antiviral strategies depends on a detailed molecular understanding of the viral life cycle. However, studies of the HCV life cycle have long been hampered by the lack of an efficient cell culture system to generate infectious virus *in vitro*. Several model systems have thus been developed for the study of defined aspects of the HCV life cycle such as viral entry, replication, assembly and release (Barth *et al.*, 2006). Recombinant HCV envelope glycoproteins (Pileri *et al.*, 1998), HCV-like particles (HCV-LPs) (Baumert *et al.*, 1998; Wellnitz *et al.*, 2002; Barth *et al.*, 2005) and retroviral HCV pseudotypes (HCVpp) (Bartosch *et al.*, 2003b; Hsu *et al.*, 2003) have been successfully used to analyze virus binding and entry. Most recently, efficient *in vitro* model systems for the production of infectious recombinant virions (HCVcc) have been described (Lindenbach *et al.*, 2005; Wakita *et al.*, 2005; Zhong *et al.*, 2005) that allow to study the complete HCV life cycle. This model system also enables to determine the role of host cell factors in productive HCV infection.

Attachment of the virus to the cell surface followed by viral entry is the first step in a cascade of interactions between the virus and the target cell that is required for successful infection (Marsh *et al.*, 2006). These steps are important determinants of tissue tropism and pathogenesis and thus represent major targets for host neutralizing responses as well as antiviral therapies.

8.2 MOLECULAR MECHANISMS OF HCV BINDING AND ENTRY INTO TARGET CELLS

8.2.1 *VIRAL DETERMINANTS OF HCV BINDING AND ENTRY: ENVELOPE GLYCOPROTEINS E1 AND E2*

HCV envelope glycoproteins E1 and E2 are critical for host cell entry. It has been shown that HCVpp assembled with either E1 or E2 glycoproteins were significantly less infective than HCVpp containing both envelope glycoproteins (Bartosch *et al.*, 2003b) and that HCVcc derived from a E1E2-deleted construct are not infectious (Wakita *et al.*, 2005). In addition, they key role for envelope glycoprotein E2 in virus-host cell interaction was confirmed by studies demonstrating that antibodies targeting both linear and conformational epitopes of envelope glycoprotein E2 are able to inhibit HCV-LP binding, HCVpp entry and HCVcc infection (Baumert *et al.*, 1998; Wellnitz *et al.*, 2002; Bartosch *et al.*, 2003b; Hsu *et al.*, 2003; Barth *et al.*, 2005; Lindenbach *et al.*, 2005; Wakita *et al.*, 2005; Zhong *et al.*, 2005). The exact role of E1 still remains unknown. It has been suggested that E1 may directly interact with cell surface molecules and/ or contribute to proper folding and processing of E2. E1-cell surface interaction may contribute to viral binding and entry as antibodies targeting the N-terminal region of E1 have been shown to inhibit HCV-LP binding (Triyatni *et al.*, 2002; Barth *et al.*, 2005) as well as HCV infection of a B-cell-derived cell line (Keck *et al.*, 2004). In addition to mediating HCV binding to target cells, recent studies suggest that HCV envelope proteins E1 and E2 are probably also involved in viral fusion (Takikawa *et al.*, 2000; Lavillette *et al.*, 2007).

8.2.2   CELLULAR DETERMINANTS OF HCV BINDING AND ENTRY

Using the above mentioned model systems, several cell surface molecules have been identified interacting with HCV during viral binding and entry: CD81 (Pileri *et al.*, 1998), the low-density lipoprotein (LDL) receptor (Agnello *et al.*, 1999), scavenger receptor class B type I (SR-BI) (Scarselli *et al.*, 2002), DC-SIGN (dendritic cell-specific intercellular adhesion molecule 3 grabbing nonintegrin)/L-SIGN (DC-SIGNr, liver and lymph node specific) (Pohlmann *et al.*, 2003; Lozach *et al.*, 2004), highly sulfated heparan sulfate (Barth *et al.*, 2003) and claudin-1 (Evans *et al.*, 2007). Experimental data using HCVpp and HCVcc have confirmed functional roles for heparan sulfate (Koutsoudakis *et al.*, 2006b) as well as CD81 (Lindenbach *et al.*, 2005; Wakita *et al.*, 2005; Koutsoudakis *et al.*, 2006b), claudin-1 (Evans *et al.*, 2007) and SR-BI (Bartosch *et al.*, 2003c; Catanese *et al.*, 2007; Grove *et al.*, 2007; Kapadia *et al.*, 2007; Zeisel *et al.*, 2007b) in HCV binding and entry, respectively.

8.3   HOST CELL FACTORS INVOLVED IN HCV BINDING AND ATTACHMENT

8.3.1   GLYCOSAMINOGLYCANS: HEPARAN SULFATE

Various viruses and other microorganisms may bind to eukaryotic cells by interacting with linear polysaccharide chains attached to cell surface proteins or glycosaminoglycans (GAG). Heparan sulfate is a GAG serving as a binding molecule for several viruses including HCV. Heparin - a structural homolog of highly sulfated heparan sulfate – reduced HCV-LP binding to human hepatoma cells (Barth *et al.*, 2003) and markedly inhibited serum-derived HCV or VSV/ HCV pseudotype binding to target cells (Germi *et al.*, 2002; Basu *et al.*, 2004). In addition, pretreatment of cells with heparinases reduced HCVcc infectivity (Koutsoudakis *et al.*, 2006b) confirming the important role of heparan sulfate in HCV-host cell interaction.

8.3.2   LECTINS: DC-SIGN/L-SIGN

C-type lectins have a dual role and act both as adhesion molecules and as pathogen recognition receptors. The mannose binding C-type lectins DC-SIGN and L-SIGN mediate contact between dendritic cells, T lymphocytes and endothelial cells. DC-SIGN and L-SIGN recognize carbohydrate structures on pathogens (Koppel *et al.*, 2005) and have been shown to bind envelope glycoprotein E2 with high affinity (Gardner *et al.*, 2003; Pohlmann *et al.*, 2003). As neither

DC-SIGN nor L-SIGN are expressed in hepatocytes, these molecules are not directly involved in HCV infection of hepatocytes. However, these lectins are expressed on cells localized close to hepatocytes. DC-SIGN is expressed in Kupffer cells (van Kooyk *et al.*, 2003) and L-SIGN is highly expressed in liver sinusoidal endothelial cells. Coculture model systems have suggested that these lectins may capture circulating HCV and transmit the virus to neighboring hepatocytes to allow infection: L-SIGN and DC-SIGN expressed on human Radji B cells or HeLa cells are able to capture and transmit HCVpp to human hepatoma Huh7 cells (Cormier *et al.*, 2004a; Lozach *et al.*, 2004). This process may thus contribute to the pathogenesis of viral infection by facilitating viral infection of hepatocytes which are not in direct contact with circulating blood.

8.4   HOST CELL FACTORS INVOLVED IN HCV ENTRY

8.4.1   TETRASPANINS: CD81 AND CLAUDIN-1

Tetraspanins are composed of four transmembrane domains, short intracellular domains and two extracellular loops, the small extracellular loop (SEL) and the large extracelullar loop (LEL). These proteins are widely expressed and are involved in regulating numerous cellular processes (Hemler, 2005). CD81 has first been identified as an HCV E2 binding molecule by expression cloning (Pileri *et al.*, 1998). Further studies then demonstrated the important role of CD81 in HCV infection: anti-CD81 antibodies and soluble form of CD81 LEL have been shown to inhibit HCVpp and HCVcc entry into Huh-7 hepatoma cells and human heptocytes (Bartosch *et al.*, 2003b; Hsu *et al.*, 2003; Cormier *et al.*, 2004a; McKeating *et al.*, 2004; Zhang *et al.*, 2004; Lindenbach *et al.*, 2005; Wakita *et al.*, 2005; Zhong *et al.*, 2005) and silencing of CD81 expression in hepatoma cells by small interfering RNAs inhibited HCVpp entry as well as HCVcc infectivity. Moreover, expression of CD81 in hepatoma cell lines that are resistant to HCVpp and HCVcc infection conferred susceptibility to HCV infection (Bartosch *et al.*, 2003c; Cormier *et al.*, 2004a; Zhang *et al.*, 2004; Akazawa *et al.*, 2007) and CD81 expression levels on hepatoma cells correlate with HCV infectivity (Koutsoudakis *et al.*, 2006a; Akazawa *et al.*, 2007). These results provide an explanation for HCV tissue tropism *in vivo* as they suggest that susceptibility to HCV infection may be linked to CD81 density on the cell surface. However, CD81 does not seem to be the determinant for the species specificity of HCV as CD81 from HCV refractory species are able to bind HCV E2 (Meola *et al.*, 2000) and CD81 of various species may confer susceptibility to HCV infection (Flint *et al.*, 2006). Most recently, expression cloning studies identified another member of the tetraspanin family, claudin-1 (CLDN1), as another host cell factor important for HCV infection (Evans *et al.*, 2007). CLDN1 is not a liver-specific molecule but is highly expressed in the liver as well as in other tissues (Furuse *et al.*, 1998). HCV permissiveness correlates with CLDN1 expression. In addition, expression of CLDN1 in nonhepatic 293T cells rendered them susceptible to HCVpp entry and overexpression of this molecule in CD81-deficient HepG2 hepatoma cells increased their HCV permissivity (Evans *et al.*, 2007). Interestingly, murine CLDN1 also supports HCVpp entry, demonstrating that CLDN1, as CD81, is not a determinant for species host range (Evans *et al.*, 2007). Taken together, these results suggest that CLDN1 is not an alternative entry pathway to CD81 (Evans *et al.*, 2007) but rather that several HCV coreceptors may form an HCV receptor complex. As tetraspanins are able to form associations with a wide variety of proteins as well as cholesterol and gangliosides (Hemler, 2005) it is conceivable that several receptors associate in discrete membrane domains to mediate HCV infection.

8.4.2   SCAVENGER RECEPTORS: SR-BI AND SR-BII

SR-BI or CLA-1 (CD36 and LIMPII Analogous-1) is a glycoprotein with a large extracellular loop anchored to the plasma membrane at both the N- and C- termini by transmembrane domains with short extensions into the cytoplasm (Krieger, 2001). SR-BI is a multiligand receptor able to bind native high-density lipoprotein (HDL) and low density lipoproteins (LDL) as well as modified lipoproteins such as oxidized LDL (oxLDL). SR-BI is involved in bidirectional cholesterol transport at the cell membrane. SR-BI is not a liver specific factor but is highly expressed in liver and steroidogenic tissues (Krieger, 2001) as well as human monocyte-derived dendritic cells but not on any other peripheral blood mononuclear cells (Yamada *et al.*, 2005). Class B scavenger receptors may serve as pattern recognition receptors for bacteria as SR-BI and its splicing variant SR-BII have been found to mediate binding and uptake of a broad range of bacteria into nonphagocytic human epithelial cells overexpressing SR-BI and SR-BII (Philips *et al.*, 2005; Vishnyakova *et al.*, 2006). SR-BI was identified as an HCV E2 binding molecule by cross-linking studies (Scarselli *et al.*, 2002). The role of SR-BI in HCV infection of target cells was confirmed by demonstrating that antibodies directed against cell surface expressed SR-BI inhibited binding of recombinant envelope glycoproteins and HCV-like particles to primary hepatocytes (Barth *et al.*, 2005) as well as HCVpp entry (Bartosch *et al.*, 2003c; Lavillette *et al.*, 2005a; Lavillette *et al.*, 2005b). The interaction between SR-BI and HCV seems to be complex and may be modulated by physiological SR-BI ligands, such as HDL or oxLDL: HDL and oxLDL have been shown to enhance and inhibit HCVpp entry, respectively (Bartosch *et al.*, 2005; Voisset *et al.*, 2005; von Hahn *et al.*, 2006), whereas both HDL and LDL inhibited HCV replication in human hepatocytes infected with serum-derived HCV (Molina *et al.*, 2007). Most recently, several studies demonstrated the important role of SR-BI in productive HCV infection using the HCVcc system: overexpression of SR-BI and SR-BII increased HCVcc infectivity (Grove *et al.*, 2007) whereas siRNA-mediated down-regulation of SR-BI in human hepatoma cells reduced HCVcc infection (Zeisel *et al.*, 2007a). Moreover, anti-SR-BI antibodies specifically inhibited HCVcc infection (Catanese *et al.*, 2007; Zeisel al., 2007a). HDL is also able to modulate HCVcc infection: it has been reported that these lipoproteins may enhance (Dreux *et al.*, 2006; Zeisel *et al.*, 2007a) or inhibit (Catanese *et al.*, 2007) HCVcc infection of human hepatoma cells.

8.4.3   LDL RECEPTOR

The LDL receptor (LDLR) transports cholesterol-containing lipoproteins into the cell by endocytosis *via* clathrin-coated pits. LDL and very low-density lipoproteins (VLDL) are the major LDLR ligands. The LDLR was suggested to mediate HCV binding and entry as HCV is able to associate with LDL and VLDL in serum (Thomssen *et al.*, 1992; Andre *et al.*, 2002). The LDLR has been shown to bind and internalize virus-LDL particles (Agnello *et al.*, 1999). Moreover, antibodies directed against LDLR, apoB (contained in LDL) and apoE (contained in VLDL) were able to inhibit HCV endocytosis (Agnello *et al.*, 1999; Wunschmann *et al.*, 2000; Germi *et al.*, 2002; Bartosch *et al.*, 2003b). A recent study suggested that LDLR plays a role in an early step of serum-derived HCV infection of primary human hepatocytes (Molina *et al.*, 2007). However, studies using the HCVpp system where HCV is not associated with lipoproteins suggest that LDLR does not appear to play a role for infection of Huh7 cells with HCVpp (Bartosch *et al.*, 2003b). Further studies using HCVcc and human hepatocytes will allow gaining more insight into the role of LDLR in HCV infection.

8.4.4   MOLECULAR MECHANISMS OF HCV ENTRY

Recent studies using the HCVpp and HCVcc model systems investigated the kinetics of the early steps of interaction between HCV and target cells by determining the ability of different agents interfering with heparan sulfate, CD81, CLDN1, SR-BI, or endosomal acidification to inhibit HCV entry when administered at different intervals during the early phase of infection. These studies suggest that HCV binding and entry is a complex multistep process (Figure 8.1). Heparin inhibited HCVcc infection only when it was present during HCVcc attachment to target cells, suggesting that interaction between heparan sulfate and HCV mainly contributes to viral attachment but not to viral entry into target cells (Koutsoudakis *et al.*, 2006b). In contrast, antibodies directed against CD81 (Cormier *et al.*, 2004b; Koutsoudakis *et al.*, 2006b, Evans *et al.*, 2007) and SR-BI (Zeisel *et al.*, 2007a) markedly inhibited HCVcc infection when added following viral binding and up to 1 h after viral attachment suggesting that CD81 and SR-BI function as HCV entry coreceptors after docking of the virus to attachment molecules. Moreover, as the entry steps mediated by these two receptors occur at a similar time-point (Zeisel *et al.*, 2007a) and as blocking of both receptors inhibited HCVcc infection more potently than blocking each receptor alone (Kapadia *et al.*, 2007; Zeisel *et al.*, 2007a), SR-BI and CD81 most likely act in concert to mediate HCV entry. In addition, kinetic studies using HCVpp entry into Flag-tagged CLDN1 overexpressing 293T cells and anti-Flag antibodies showed that CLDN1 acts at a postbinding step after HCV interaction with CD81 (Evans *et al.*, 2007). Side-by-side studies using HCVcc and antibodies directed against these three cell surface molecules will allow to gain further insights into the kinetics of HCV entry and define the early steps of virus-host cell interaction.

Taken together, these data suggest that initial binding of HCV to target cells may be mediates by GAGs, such as heparan sulfate. The virus is then transferred to a first set of cell surface entry receptors, including SR-BI and CD81, and finally interacts with molecules involved in late steps of viral entry, such as CLDN1, before being endocytosed (Blanchard *et al.*, 2006; Codran *et al.*, 2006; Meertens *et al.*, 2006), transported into endosomes (Meertens *et al.*, 2006) where a pH-dependent fusion process ultimately delivers HCV genome into the cytosol (Blanchard *et al.*, 2006; Tscherne *et al.*, 2006; Lavillette *et al.*, 2007) (Figure 8.1).

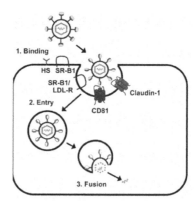

FIGURE 8.1 Model of the hepatitis C virus binding and entry into the host cell. HCV binding and entry is a multistep process:(1) attachment of viral envelope glycoproteins E1 and E2 to host cell surface molecules; (2) entry into the host cell via clathrin-mediated endocytosis; (3) fusion between viral and endosomal membranes leads to the release of the viral genome into the cytoplasm (adapted from Zeisel *et al.*, 2007a).

8.5 CONCLUSION AND FUTURE PERSPECTIVES

Using various HCV model systems, several host cell factors involved in HCV binding and entry have been uncovered. To date, all HCV permissive cell lines that have been identified express CD81, SR-BI and CLDN1 and are of hepatic origin. However, various cell lines of nonhepatic origin which express all HCV receptor candidates are nonpermissive for HCV (Bartosch *et al.*, 2003a; Bartosch *et al.*, 2003b; Bartosch *et al.*, 2003c; Zhang *et al.*, 2004). In addition, none of the HCV receptor candidates has a liver-specific expression profile. This suggests that there are most likely additional liver specific factors mediating HCV infection which are still to be discovered. Interfering with HCV entry holds promise for drug design and discovery as the understanding of molecular mechanisms underlying HCV interaction with the host cell is growing. The complexity of the virus entry process offers several new therapeutic targets.

ACKNOWLEDGEMENTS

This work was supported by grants of the European Union, Brussels, Belgium (QLK-2-CT-2002–01329 and VIRGIL Network of Excellence LSHM-CT-2004–503359); the Institut National de la Santé et de la Recherche Médicale (Inserm), Paris, France; the chair of excellence program of the Agence Nationale de la Recherche (ANR-05-CEXC-008), Paris, France and Agence Nationale de la Recherche sur le SIDA et les Hépatites Virales (ANRS Grant #06221), Paris, France. MBZ is supported by the Inserm Poste Vert program in the frame work of Inserm European Associated Laboratory Freiburg-Strasbourg.

KEYWORDS

- CD81
- Claudin-1
- Heparan sulfate
- Hepatitis C virus
- Hepatocytes
- Host cell factors
- Lectins
- Low-density lipoprotein
- Tetraspanins
- Viral infection

REFERENCES

Agnello, V.; Abel, G.; Elfahal, M.; Knight, G. B.; Zhang, Q. X. Hepatitis C virus and other flaviviridae viruses enter cells via low density lipoprotein receptor. *Proc Natl Acad Sci USA* 1999, 96, 12766-12771.

Akazawa, D.; Date, T.; Morikawa, K.; Murayama, A.; Miyamoto, M.; Kaga, M.; Barth, H.; Baumert, T. F.; Dubuisson, J.; Wakita, T. Cd81 expression is important for heterogeneous Hcv permissiveness of Huh7 cell clones. *J Virol* 2007, 81, 5036-5045.

Alter, M. J. Epidemiology of hepatitis C. *Hepatology* 1997, 26, 62S-65S.

Alter, M. J.; Margolis, H. S.; Krawczynski, K.; Judson, F. N.; Mares, A.; Alexander, W. J.; Hu, P. Y.; Miller, J. K.; Gerber, M. A.; Sampliner, R. E. The natural history of community-acquired hepatitis C in the United States. The Sentinel Counties Chronic non-A, non-B Hepatitis Study Team. *N Engl J Med* 1992, 327, 1899-1905.

Andre, P.; Komurian-Pradel, F.; Deforges, S.; Perret, M.; Berland, J. L.; Sodoyer, M.; Pol, S.; Brechot, C.; Paranhos-Baccala, G.; Lotteau, V. Characterization of low- and very-low-density hepatitis C virus RNA-containing particles. *J Virol* 2002, 76, 6919-6928.

Barth, H.; Cerino, R.; Arcuri, M.; Hoffmann, M.; Schurmann, P.; Adah, M. I.; Gissler, B.; Zhao, X.; Ghisetti, V.; Lavezzo, B.; Blum, H. E.; von Weizsacker, F.; Vitelli, A.; Scarselli, E.; Baumert, T. F. Scavenger receptor class B type I and hepatitis C virus infection of primary tupaia hepatocytes. *J Virol* 2005, 79, 5774-5785.

Barth, H.; Liang, T. J.; Baumert, T. F. Hepatitis C virus entry: Molecular biology and clinical implications. *Hepatology* 2006, 44, 527-535.

Barth, H.; Schäfer, C.; Adah, M. I.; Zhang, F.; Linhardt, R. J.; Toyoda, H.; Kinoshita-Toyoda, A.; Toida, T.; Van Kuppevelt, T. H.; Depla, E.; Weizsäcker, F. V.; Blum, H. E.; Baumert, T. F. Cellular binding of hepatitis C virus envelope glycoprotein E2 requires cell surface heparan sulfate. *J Biol Chem* 2003, 278, 41003-41012.

Bartosch, B.; Bukh, J.; Meunier, J. C.; Granier, C.; Engle, R. E.; Blackwelder, W. C.; Emerson, S. U.; Cosset, F. L.; Purcell, R. H. *In vitro* assay for neutralizing antibody to hepatitis C virus: Evidence for broadly conserved neutralization epitopes. *Proc Natl Acad Sci USA* 2003a, 100, 14199-14204.

Bartosch, B.; Dubuisson, J.; Cosset, F. L. Infectious hepatitis C virus pseudo-particles containing functional E1-E2 envelope protein complexes. *J Exp Med* 2003b, 197, 633-642.

Bartosch, B.; Verney, G.; Dreux, M.; Donot, P.; Morice, Y.; Penin, F.; Pawlotsky, J. M.; Lavillette, D.; Cosset, F. L. An interplay between hypervariable region 1 of the hepatitis C virus E2 glycoprotein, the scavenger receptor BI, and high-density lipoprotein promotes both enhancement of infection and protection against neutralizing antibodies. *J Virol* 2005, 79, 8217-8229.

Bartosch, B.; Vitelli, A.; Granier, C.; Goujon, C.; Dubuisson, J.; Pascale, S.; Scarselli, E.; Cortese, R.; Nicosia, A.; Cosset, F. L. Cell entry of hepatitis C virus requires a set of co-receptors that include the CD81 tetraspanin and the SR-B1 scavenger receptor. *J Biol Chem* 2003c, 278, 41624-41630.

Basu, A.; Beyene, A.; Meyer, K.; Ray, R. The hypervariable region 1 of the E2 glycoprotein of hepatitis C virus binds to glycosaminoglycans, but this binding does not lead to infection in a pseudotype system. *J Virol* 2004, 78, 4478-4486.

Baumert, T. F.; Ito, S.; Wong, D. T.; Liang, T. J. Hepatitis C virus structural proteins assemble into virus like particles in insect cells. *J Virol* 1998, 72, 3827-3836.

Blanchard, E.; Belouzard, S.; Goueslain, L.; Wakita, T.; Dubuisson, J.; Wychowski, C.; Rouille, Y. Hepatitis C virus entry depends on clathrin-mediated endocytosis. *J Virol* 2006, 80, 6964-6972.

Catanese, M. T.; Graziani, R.; von Hahn, T.; Moreau, M.; Huby, T.; Paonessa, G.; Santini, C.; Luzzago, A.; Rice, C. M.; Cortese, R.; Vitelli, A.; Nicosia, A. High-avidity monoclonal antibodies against the human scavenger class B type I receptor efficiently block hepatitis C virus infection in the presence of high-density lipoprotein. *J Virol* 2007, 81, 8063-8071.

Codran, A.; Royer, C.; Jaeck, D.; Bastien-Valle, M.; Baumert, T. F.; Kieny, M. P.; Pereira, C. A.; Martin, J. P. Entry of hepatitis C virus pseudotypes into primary human hepatocytes by clathrin-dependent endocytosis. *J Gen Virol* 2006, 87, 2583-2593.

Cormier, E. G.; Durso, R. J.; Tsamis, F.; Boussemart, L.; Manix, C.; Olson, W. C.; Gardner, J. P.; Dragic, T. L-SIGN (CD209L) and DC-SIGN (CD209) mediate transinfection of liver cells by hepatitis C virus. *Proc Natl Acad Sci USA* 2004a, 101, 14067-14072.

Cormier, E. G.; Tsamis, F.; Kajumo, F.; Durso, R. J.; Gardner, J. P.; Dragic, T. CD81 is an entry coreceptor for hepatitis C virus. *Proc Natl Acad Sci USA* 2004b, 101, 7270-7274.

Di Bisceglie, A. M.; Hoofnagle, J. H. Optimal therapy of hepatitis C. *Hepatology* 2002, 36, S121-127.

Dreux, M.; Pietschmann, T.; Granier, C.; Voisset, C.; Ricard-Blum, S.; Mangeot, P. E.; Keck, Z.; Foung, S.; Vu-Dac, N.; Dubuisson, J.; Bartenschlager, R.; Lavillette, D.; Cosset, F. L. High density lipoprotein inhibits hepatitis C virus-neutralizing antibodies by stimulating cell entry via activation of the scavenger receptor BI. *J Biol Chem* 2006, 281, 18285-18295.

Evans, M. J.; von Hahn, T.; Tscherne, D. M.; Syder, A. J.; Panis, M.; Wolk, B.; Hatziioannou, T.; McKeating, J. A.; Bieniasz, P. D.; Rice, C. M. Claudin-1 is a hepatitis C virus co-receptor required for a late step in entry. *Nature* 2007, 446(7137), 801-805.

Flint, M.; von Hahn, T.; Zhang, J.; Farquhar, M.; Jones, C. T.; Balfe, P.; Rice, C. M.; McKeating, J. A. Diverse CD81 proteins support hepatitis C virus infection. *J Virol* 2006, 80, 11331-11342.

Furuse, M.; Fujita, K.; Hiiragi, T.; Fujimoto, K.; Tsukita, S. Claudin-1 and -2: Novel integral membrane proteins localizing at tight junctions with no sequence similarity to occludin. *J Cell Biol* 1998, 141, 1539-1550.

Gardner, J. P.; Durso, R. J.; Arrigale, R. R.; Donovan, G. P.; Maddon, P. J.; Dragic, T.; Olson, W. C. L-SIGN (CD 209L) is a liver-specific capture receptor for hepatitis C virus. *Proc Natl Acad Sci USA* 2003, 100, 4498-4503.

Germi, R.; Crance, J. M.; Garin, D.; Guimet, J.; Lortat-Jacob, H.; Ruigrok, R. W.; Zarski, J. P.; Drouet, E. Cellular glycosaminoglycans and low density lipoprotein receptor are involved in hepatitis C virus adsorption. *J Med Virol* 2002, 68, 206-215.

Grove, J.; Huby, T.; Stamataki, Z.; Vanwolleghem, T.; Meuleman, P.; Farquhar, M.; Schwarz, A.; Moreau, M.; Owen, J. S.; Leroux-Roels, G.; Balfe, P.; McKeating, J. A. Scavenger receptor BI and BII expression levels modulate Hepatitis C virus infectivity. *J Virol* 2007, 81, 3162-3169.

Hemler, M. E. Tetraspanin functions and associated microdomains. *Nat Rev Mol Cell Biol* 2005, 6, 801-811.

Hsu, M.; Zhang, J.; Flint, M.; Logvinoff, C.; Cheng-Mayer, C.; Rice, C. M.; McKeating, J. A. Hepatitis C virus glycoproteins mediate pH-dependent cell entry of pseudotyped retroviral particles. *Proc Natl Acad Sci USA* 2003, 100, 7271-7276.

Kaito, M.; Watanabe, S.; Tsukiyama-Kohara, K.; Yamaguchi, K.; Kobayashi, Y.; Konishi, M.; Yokoi, M.; Ishida, S.; Suzuki, S.; Kohara, M. Hepatitis C virus particle detected by immunoelectron microscopic study. *J Gen Virol* 1994, 75, 1755-1760.

Kapadia, S. B.; Barth, H.; Baumert, T.; McKeating, J. A.; Chisari, F. V. Initiation of Hepatitis C Virus infection is dependent on cholesterol and cooperativity between CD81 and scavenger receptor B type I. *J Virol* 2007, 81, 374-383.

Keck, Z. Y.; Sung, V. M.; Perkins, S.; Rowe, J.; Paul, S.; Liang, T. J.; Lai, M. M.; Foung, S. K. Human monoclonal antibody to hepatitis C virus E1 glycoprotein that blocks virus attachment and viral infectivity. *J Virol* 2004, 78, 7257-7263.

Koppel, E. A.; van Gisbergen, K. P.; Geijtenbeek, T. B.; van Kooyk, Y. Distinct functions of DC-SIGN and its homologues L-SIGN (DC-SIGNR) and mSIGNR1 in pathogen recognition and immune regulation. *Cell Microbiol* 2005, 7, 157-165.

Koutsoudakis, G.; Herrmann, E.; Kallis, S.; Bartenschlager, R.; Pietschmann, T. The level of CD81 cell surface expression is a key determinant for productive entry of Hepatitis C Virus into host cells. *J Virol* 2006a, 81, 588-598.

Koutsoudakis, G.; Kaul, A.; Steinmann, E.; Kallis, S.; Lohmann, V.; Pietschmann, T.; Bartenschlager, R. Characterization of the early steps of hepatitis C virus infection by using luciferase reporter viruses. *J Virol* 2006b, 80, 5308-5320.

Krieger, M. Scavenger receptor class B type I is a multiligand HDL receptor that influences diverse physiologic systems. *J Clin Invest* 2001, 108, 793-797.

Lavillette, D.; Morice, Y.; Germanidis, G.; Donot, P.; Soulier, A.; Pagkalos, E.; Sakellariou, G.; Intrator, L.; Bartosch, B.; Pawlotsky, J. M.; Cosset, F. L. Human serum facilitates hepatitis C virus infection, and neutralizing responses inversely correlate with viral replication kinetics at the acute phase of hepatitis C virus infection. *J Virol* 2005a, 79, 6023-6034.

Lavillette, D.; Pecheur, E. I.; Donot, P.; Fresquet, J.; Molle, J.; Corbau, R.; Dreux, M.; Penin, F.; Cosset, F. L. Characterization of fusion determinants points to the involvement of three discrete regions of both E1 and E2 glycoproteins in the membrane fusion process of hepatitis C virus. *J Virol* 2007, 81, 8752-8765.

Lavillette, D.; Tarr, A. W.; Voisset, C.; Donot, P.; Bartosch, B.; Bain, C.; Patel, A. H.; Dubuisson, J.; Ball, J. K.; Cosset, F. L. Characterization of host-range and cell entry properties of the major genotypes and subtypes of hepatitis C virus. *Hepatology* 2005b, 41, 265-274.

Lindenbach, B. D.; Evans, M. J.; Syder, A. J.; Wolk, B.; Tellinghuisen, T. L.; Liu, C. C.; Maruyama, T.; Hynes, R. O.; Burton, D. R.; McKeating, J. A.; Rice, C. M. Complete replication of hepatitis C virus in cell culture. *Science* 2005, 309, 623-626.

Lindenbach, B. D.; Thiel, H. J.; Rice, C. M. Flaviviridae: The viruses and their replication. In *Fields Virology*, Knipe D.M.; Howley P.M.; Eds., Lippincott-Raven: Philadelphia, 2007; pp. 1101-1152.

Lozach, P. Y.; Amara, A.; Bartosch, B.; Virelizier, J. L.; Arenzana-Seisdedos, F.; Cosset, F. L.; Altmeyer, R. C-type lectins L-SIGN and DC-SIGN capture and transmit infectious hepatitis C virus pseudotype particles. *J Biol Chem* 2004, 279, 32035-32045.

Marsh, M.; Helenius, A. Virus entry: Open sesame. *Cell* 2006, 124, 729-740.

McKeating, J. A.; Zhang, L. Q.; Logvinoff, C.; Flint, M.; Zhang, J.; Yu, J.; Butera, D.; Ho, D. D.; Dustin, L. B.; Rice, C. M.; Balfe, P. Diverse hepatitis C virus glycoproteins mediate viral infection in a CD81-dependent manner. *J Virol* 2004, 78, 8496-8505.

Meertens, L.; Bertaux, C.; Dragic, T. Hepatitis C virus entry requires a critical postinternalization step and delivery to early endosomes via clathrin-coated vesicles. *J Virol* 2006, 80, 11571-11578.

Meola, A.; Sbardellati, A.; Bruni Ercole, B.; Cerretani, M.; Pezzanera, M.; Ceccacci, A.; Vitelli, A.; Levy, S.; Nicosia, A.; Traboni, C.; McKeating, J.; Scarselli, E. Binding of hepatitis C virus E2 glycoprotein to CD81 does not correlate with species permissiveness to infection. *J Virol* 2000, 74, 5933-5938.

Molina, S.; Castet, V.; Fournier-Wirth, C.; Pichard-Garcia, L.; Avner, R.; Harats, D.; Roitelman, J.; Barbaras, R.; Graber, P.; Ghersa, P.; Smolarsky, M.; Funaro, A.; Malavasi, F.; Larrey, D.; Coste,

J.; Fabre, J. M.; Sa-Cunha, A.; Maurel, P. The low-density lipoprotein receptor plays a role in the infection of primary human hepatocytes by hepatitis C virus. *J Hepatol* 2007, 46, 411-419.

Moradpour, D.; Penin, F.; Rice, C. M. Replication of hepatitis C virus. *Nature Reviews* 2007, 5, 453-463.

Pawlotsky, J. M. Treating hepatitis C in "difficult-to-treat" patients. *N Engl J Med* 2004, 351, 422-423.

Penin, F.; Dubuisson, J.; Rey, F. A.; Moradpour, D.; Pawlotsky, J. M. Structural biology of hepatitis C virus. *Hepatology* 2004, 39, 5-19.

Philips, J. A.; Rubin, E. J.; Perrimon, N. Drosophila RNAi screen reveals CD36 family member required for mycobacterial infection. *Science* 2005, 309, 1251-1253.

Pileri, P.; Uematsu, Y.; Campagnoli, S.; Galli, G.; Falugi, F.; Petracca, R.; Weiner, A. J.; Houghton, M.; Rosa, D.; Grandi, G.; Abrignani, S. Binding of hepatitis C virus to CD81. *Science* 1998, 282, 938-941.

Pohlmann, S.; Zhang, J.; Baribaud, F.; Chen, Z.; Leslie, G. J.; Lin, G.; Granelli-Piperno, A.; Doms, R. W.; Rice, C. M.; McKeating, J. A. Hepatitis C Virus glycoproteins interact with DC-SIGN and DC-SIGNR. *J Virol* 2003, 77, 4070-4080.

Scarselli, E.; Ansuini, H.; Cerino, R.; Roccasecca, R. M.; Acali, S.; Filocamo, G.; Traboni, C.; Nicosia, A.; Cortese, R.; Vitelli, A. The human scavenger receptor class B type I is a novel candidate receptor for the hepatitis C virus. *EMBO J* 2002, 21, 5017-5025.

Shimizu, Y. K.; Feinstone, S. M.; Kohara, M.; Purcell, R.; Yoshikura, H. Hepatitis C virus: Detection of intracellular virus particles by electron microscopy. *Hepatology* 1996, 23, 205-209.

Takikawa, S.; Ishii, K.; Aizaki, H.; Suzuki, T.; Asakura, H.; Matsuura, Y.; Miyamura, T. Cell fusion activity of hepatitis C virus envelope proteins. *J Virol* 2000, 74, 5066-5074.

Thomssen, R.; Bonk, S.; Propfe, C.; Heermann, K. -H.; Koechel, H. G.; Uy, A. Association of hepatitis C virus in human sera with beta-lipoprotein. *Med Microbiol Immunol* 1992, 181, 293-300.

Triyatni, M.; Saunier, B.; Maruvada, P.; Davis, A. R.; Ulianich, L.; Heller, T.; Patel, A.; Kohn, L. D.; Liang, T. J. Interaction of hepatitis C virus-like particles and cells: A model system for studying viral binding and entry. *J Virol* 2002, 76, 9335-9344.

Tscherne, D. M.; Jones, C. T.; Evans, M. J.; Lindenbach, B. D.; McKeating, J. A.; Rice, C. M. Time- and temperature-dependent activation of hepatitis C virus for low-pH-triggered entry. *J Virol* 2006, 80, 1734-1741.

van Kooyk, Y.; Geijtenbeek, T. B. DC-SIGN: Escape mechanism for pathogens. *Nat Rev Immunol* 2003, 3, 697-709.

Vishnyakova, T. G.; Kurlander, R.; Bocharov, A. V.; Baranova, I. N.; Chen, Z.; Abu-Asab, M. S.; Tsokos, M.; Malide, D.; Basso, F.; Remaley, A.; Csako, G.; Eggerman, T. L.; Patterson, A. P. CLA-1 and its splicing variant CLA-2 mediate bacterial adhesion and cytosolic bacterial invasion in mammalian cells. *Proc Natl Acad Sci USA* 2006, 103, 16888-16893.

Voisset, C.; Callens, N.; Blanchard, E.; Op De Beeck, A.; Dubuisson, J.; Vu-Dac, N. High density lipoproteins facilitate hepatitis C virus entry through the scavenger receptor class B type I. *J Biol Chem* 2005, 280, 7793-7799.

von Hahn, T.; Lindenbach, B. D.; Boullier, A.; Quehenberger, O.; Paulson, M.; Rice, C. M.; McKeating, J. A. Oxidized low-density lipoprotein inhibits hepatitis C virus cell entry in human hepatoma cells. *Hepatology* 2006, 43, 932-942.

Wakita, T.; Pietschmann, T.; Kato, T.; Date, T.; Miyamoto, M.; Zhao, Z.; Murthy, K.; Habermann, A.; Krausslich, H. G.; Mizokami, M.; Bartenschlager, R.; Liang, T. J. Production of infectious hepatitis C virus in tissue culture from a cloned viral genome. *Nat Med* 2005, 11, 791-796.

Wellnitz, S.; Klumpp, B.; Barth, H.; Ito, S.; Depla, E.; Dubuisson, J.; Blum, H. E.; Baumert, T. F. Binding of hepatitis C virus-like particles derived from infectious clone H77C to defined human cell lines. *J Virol* 2002, 76, 1181-1193.

Wunschmann, S.; Medh, J. D.; Klinzmann, D.; Schmidt, W. N.; Stapleton, J. T. Characterization of hepatitis C virus (HCV) and HCV E2 interactions with CD81 and the low-density lipoprotein receptor. *J Virol* 2000, 74, 10055-10062.

Yamada, E.; Montoya, M.; Schuettler, C. G.; Hickling, T. P.; Tarr, A. W.; Vitelli, A.; Dubuisson, J.; Patel, A. H.; Ball, J. K.; Borrow, P. Analysis of the binding of hepatitis C virus genotype 1a and 1b E2 glycoproteins to peripheral blood mononuclear cell subsets. *J Gen Virol* 2005, 86, 2507-2512.

Zeisel, M. B.; Fafi-Kremer, S.; Fofana, I.; Barth, H.; Stoll-Keller, F.; Doffoel, M.; Baumert, T. F. Neutralizing antibodies in hepatitis C virus infection. *World J Gastroenterol* 2007a, 13, 4824-4830.

Zeisel, M. B.; Koutsoudakis, G.; Schnober, E. K.; Haberstroh, A.; Blum, H. E.; Cosset, F.-L.; Wakita, T.; Jaeck, D.; Doffoel, M.; Royer, C.; Soulier, E.; Schvoerer, E.; Schuster, C.; Stoll-Keller, F.; Bartenschlager, R.; Pietschmann, T.; Barth, H.; Baumert, T. F. Scavenger receptor BI is a key host factor for Hepatitis C virus infection required for an entry step closely linked to CD81. *Hepatology* 2007b, 46(6), 1722-1731.

Zhang, J.; Randall, G.; Higginbottom, A.; Monk, P.; Rice, C. M.; McKeating, J. A. CD81 is required for hepatitis C virus glycoprotein-mediated viral infection. *J Virol* 2004, 78, 1448-1455.

Zhong, J.; Gastaminza, P.; Cheng, G.; Kapadia, S.; Kato, T.; Burton, D. R.; Wieland, S. F.; Uprichard, S. L.; Wakita, T.; Chisari, F. V. 2005. Robust hepatitis C virus infection *in vitro*. *Proc Natl Acad Sci USA* 2004, 102, 9294-9299.

CHAPTER 9

MOLECULAR PHYLOGENETICS: A TOOL FOR ELUCIDATION OF EVOLUTIONARY PROCESSES FROM BIOLOGICAL DATA

MARTINA TALIANOVA

CONTENTS

9.1   Introduction 204
9.2   Evolutionary Models Facilitate Estimation of Evolutionary Change 207
9.3   Molecular Phylogenetics 215
Acknowledgements... 242
Keywords 242
References... 243

9.1 INTRODUCTION

Sequencing projects provide scientists with large amounts of molecular data, mostly in the form of nucleotide or amino acid sequences. The increasing availability of entire and partial genome sequences allows researchers to compare sequences of specific parts of genes, complete genes, groups of genes, and entire genomes with regard to DNA sequence homologies and genome organization. These data contain information about the structure, function, biological, and biochemical properties of the molecules, and they can also help us to reveal mechanisms affecting the evolutionary processes acting on the sequences. The information stored in biological sequences can be extracted using various techniques, and the comparison of DNA sequences of genes and genomes can provide us with insights into the possible evolutionary relationships of genetic information among living organisms. Phylogenetic analysis can reveal genetic variability both within and between species, and help us to assess evolutionary relationships between more distantly related species (Arber, 2000).

Molecular phylogenetics belongs to a steadily evolving area of analytical techniques with a significant use among various biological domains. It has been successfully applied not only in the molecular systematics, but also in comparative genomics (e.g., analysis of gene homology, gene family diversification (Spaethe and Briscoe, 2004; Remington *et al.*, 2004; Yoo *et al.*, 2006), prediction of relationships between structure and function of proteins (Miki *et al.*, 2005; Theobald and Wuttke, 2005; Cai and Zhang, 2006), pathogen diversity mapping (Zozio *et al.*, 2005; Kauser *et al.*, 2005; Devi *et al.*, 2006), epidemiology (Klein *et al.*, 2006; Zhang *et al.*, 2006), biodiversity studies (Sušnik *et al.*, 2006; Martínez-Solano *et al.*, 2006) and many others.

9.1.1 EVOLUTION AND VARIATION IN BIOLOGICAL SEQUENCES, AND MOLECULAR CLOCK

DNA sequences are complex sources of genetic variation. The majority of mechanisms that contribute to genetic variation can be grouped into three qualitatively different natural strategies (Arber, 2000). The first category is local changes in DNA sequences, which cause nucleotide substitutions – deletions and insertions of one or a few base pairs, eventually a local jumbling of few base pairs. The second, is rearrangements of DNA segments within genome, such as duplications, deletions, or transpositions of mobile genetic elements (DNA inversions, deletions formation, etc.). The third, is the acquisition of DNA by horizontal gene transfer, which allows the recipient organism to share the evolutionary inventions made by other organisms. The site patterns in the alignment of nucleic acid (amino acid) sequences carry information about the evolutionary history of the corresponding gene or species, which can be used to construct the phylogenetic tree of speciation or trace the divergence of different organismal lineages (although, the history of the genes does not always correspond to the history of the corresponding species). Sequences that are identical at most positions imply that the species are closely related, and therefore that they are closer to each other in the phylogenetic tree. More diverse sequences imply greater evolutionary distinction (Steel, 2005).

In 1965, Zuckerkandl and Pauling (1965) proposed a theory of a *molecular clock*, which assumes an approximately constant rate of molecular evolution over time for all proteins in all lineages. They observed that the rate of evolution for proteins (e.g., hemoglobin) were relatively constant among different mammalian species. The molecular clock hypothesis had large impact on the field of molecular biology. First, if proteins evolved at constant rate, they could be used to reconstruct phylogenetic relationships among different organisms and to estimate dates of their divergence and specification. The surprising observation of rate constancy was explained by the effective neutrality of most changes to genes (Kimura, 1983). The neutral theory proposed that changes in the amino acid sequence have little influence on the fitness of an organism, and so the rates of change are not affected by natural selection. Thus, the rate of molecular evolution should equate to the neutral mutation rate, independent of factors such as environmental changes and population sizes. In the last two decades, DNA sequences have become more available and have revealed that the molecular clock does not hold for most genes or species groups, except when sequences are from closely related species (Yang, 2002). This opened up a question about the dating of divergence time, which is the most common use of the clock. The dates computed using molecular sequences were in controversy with the fossil records. A part of the discrepancy probably arises from the incompleteness of fossil data, and another part seems to arise from inaccuracies in molecular date estimation, which is very sensitive to violation of the molecular clock. From a mathematical point of view, the molecular clock is considered to be a stochastic process. Molecular changes accumulate randomly according to the Poisson distribution, so that random fluctuations are expected, although the underlying rate is constant over time. Further, different genes or proteins (or even different regions of the same gene) may have very different evolutionary rates—their clocks tick at different rates. Under the theory of neutrality, this can be explained by the fact that different genes are under different selective constraints. The number of neutral mutations will be fewer among genes with very strong selective constraints, and so these genes will evolve at lower evolutionary rates. The last assumption that holds for molecular clock, is that it is not expected to be universal; instead it should only be applied to a group of species (e.g., clock holds for a particular gene within mammals) (Yang, 2002).

9.1.2 DATA FOR MOLECULAR PHYLOGENETICS

Molecular phylogenetics uses features either in the form of biomolecular sequences or various genome-level data (Boore, 2006). Different sources of data each have their strengths and weaknesses that may vary according to the studied taxa or according to the questions one might wish to investigate. To reconstruct phylogenetic relationships and to estimate the divergence times among taxa according to sequence data, both DNA and amino acid sequences are used. Usually, DNA sequences are used to infer relatively recent events, whereas amino acid sequences, because they are more conserved, are then used for more ancient events since DNA sequences are usually too divergent to make accurate estimates, (Hedges, 2002). Molecular phylogenetics based on genome-level characters utilizes features such as gene-order data, position of mobile genetic elements, genome rearrangements, structural data of protein subunits, and secondary structure of rRNAs and tRNAs. These kinds of data have the potential to resolve many fundamental, controversial relationships for which no other data are useful (Boore, 2006).

Methods used in phylogenetic reconstruction require an input in the form of a set of characters (i.e., the character matrix) (Figure 9.1A), or in the form of a distance matrix (Figure 9.1B). Sequence characters are obtained from aligned DNA or protein sequences where each column in the alignment represents a character. Thus, characters obtained from the sequence data have either four states (nucleotide sequences) or twenty-one states (amino-acid sequences) (Lenzini and Marianelli, 1997). For the methods based on distance matrix data, the input is represented by a symmetric matrix M_{nxn} indicating distances between pairs of organisms, with entry M_{ij} representing the distance between organism i and organism j. Distance matrix M is computed usually by transforming the data in the character form into distance values.

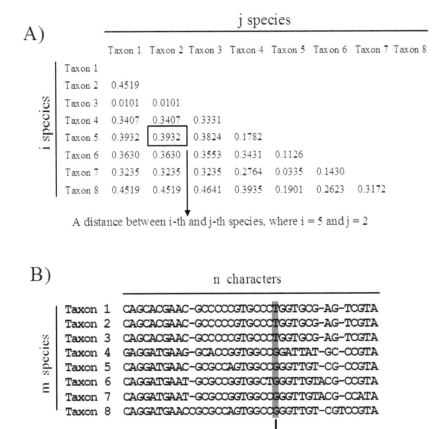

FIGURE 9.1   (A) Distance matrix represents distances between i-th and j-th species. The matrix is symmetric and lower triangular (i.e., the elements above the main diagonal are the same as elements below the main diagonal and thus are usually not displayed). The values on the main diagonal are equal to zero (the distance of the species from itself is zero) and thus are usually not displayed. (B) Alignment of DNA sequences represents a character matrix of the size m x n. Each column (i.e., each site in the alignment) corresponds to one character.

9.2   EVOLUTIONARY MODELS FACILITATE ESTIMATION OF EVOLUTIONARY CHANGE

Mathematical models have been accommodated in various biological areas (e.g., ecology, population genetics, physiology) to estimate behavior of the biological processes. Some processes are very simple and thus require simple mathematical expressions; others are complex and require complex models. A model incorporates parameters, representing factors taking part in an underlying process, and mathematical expressions that put the parameters into mutual relationships to describe and formalize the mechanism.

In molecular phylogenetics, evolutionary models play a key role. Models actually describe, how individual nucleotides (amino acids) evolve and can be used to estimate the amount and character of evolutionary change at a given nucleotide site through time. If we assume that the modeled process has generated the analyzed data, we are not only able to reconstruct evolutionary relationships between the sequences, but we are also able to resolve properties of an underlying evolutionary mechanism and to test various evolutionary hypotheses.

Although, there exist a variety of methods for inferring evolutionary relationships, none of them is assumption-free. The assumptions are either implicit (e.g., the evolution of lineages is tree-like; mutations are independent and identically distributed – i.i.d. assumption; mutations can revert to a previous state – reversibility; the evolution is parsimonious, etc.) (Kelchner and Thomas, 2006), or the assumptions are mathematically explicit – they are formulated as "parameters of the model". Since the numerical values of the parameters are usually not known *a priori*, they need to be estimated from the model. Examples of such parameters are the lengths of branches in a tree, the frequency of each base (amino acid) in the sequence (averaged over all sequence sites and over the tree), the rate of substitution of each nucleotide (i.e., relative tendencies of bases to be substituted for one another), or parameters estimating heterogeneity of substitution rates (i.e., among-site-variation), etc. Each estimable parameter represents a specific feature of sequence evolution and has a logical biological foundation for its use. Evolutionary models combine implicit assumptions with explicit parameters; although usually in context of phylogeny estimation, it is the explicit parametrical component that is referred to as the "model". Most of the methods for reconstructing phylogenetic relationships (e.g., the distance-based methods, maximum-likelihood, Bayesian inference methods) require explicit evolutionary models that, when applied to nucleotide or amino-acid sequence data facilitate the estimation of the parameters of the corresponding evolutionary process, as well as the reconstruction of evolutionary relationships.

9.2.1   SURVEY OF COMMONLY USED MODELS IN MOLECULAR PHYLOGENETICS

Since nucleotide substitutions are the primary unit of sequence differences, the basic class of models are so called *substitution models*. The majority of the substitution models used in phylogenetic estimation are based on the first order Markov stochastic process (Lió and Goldman, 1998). These models assume that the mechanism of evolution is a stochastic (random) Markov process that acts either on four states (changes between nucleotides A, T, C, G) (Figure 9.2) or 20 states (changes in amino acids). Markov processes have several important properties: (1) a lack of memory - substitution events are not influenced by the previous substitution at the same site, (2) homogeneity – the pattern of substitution remains the same among all lineages, in different parts of the tree, (3) stationarity – the nucleotide frequences have remained more or less the same during the evolution, (4) the assumption of independent and identical distribution of characters along the sequence (i.i.d.) - each nucleotide changes irrespectively and independently to other nucleotides in the sequence. Another assumption often used to simplify numerical calculations is reversibility - mutations can revert to a previous state.

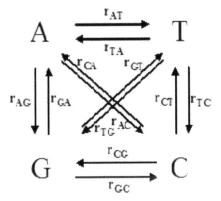

FIGURE 9.2 Markov process in nucleotide sequences causes changes at its four states (individual nucleotides A, T, G, C). r_{ij} represents the rate of substitution from nucleotide i to nucleotide j.

Consider a stochastic model of DNA or amino-acid sequence evolution. We assume an independence of evolution at different sequence sites, which allows us to work with probabilities $P_{ij}(t)$ that the base i will have changed to base j after a time t (i and j take the values 1, ..., 4 to represent the nucleotides A, T, C, G for DNA sequences and 1, ..., 20 for amino acid sequences) (Lió and Goldman, 1998). In this model, a matrix of instantaneous substitution rates $Q = \{q_{ij}\}$ is given, where each element q_{ij} represents the rate of substitution from nucleotide i to nucleotide j. Time (or distance or branch length) is then measured by the expected number of nucleotide substitutions per site. Given the rate matrix Q, the matrix of transition probabilities over time t can be calculated as $P(t) = \{p_{ij}(t)\} = e^{Qt}$, where $p_{ij}(t)$ is the probability that nucleotide i changes into nucleotide j after time t (Liò and Goldman, 1998). A standard numerical algorithm can be used to calculate the eigenvalues and eigenvectors of Q to calculate $P(t)$ (Yang, 1994a).

The most commonly used substitution models in molecular phylogenetics belong to the class of General Time Reversible models (GTR) (time reversibility means, that the substitution from a base i to j is treated equally as substitution from j to i). Examples of important parameters used to distinguish among various models are base frequencies, the frequency of transitions versus transversions, or the rate of evolution. The simplest, the Jukes and Cantor model (JC) (Jukes and Cantor, 1969) assumes equal base frequencies and no difference between transitions and transversions. The HKY model (Hasegawa et al., 1985) considers that the four nucleotides occur at different frequencies and distinguishes between transition and transversion frequencies. Other commonly used models belonging to the GTR class are the Kimura (K80) model (Kimura, 1980) known as the K2P model or the Felsenstein model (F81) (Felsenstein, 1981b) (Figure 9.3 displays matrices of instantaneous substitution rates Q for the simplest models).

JC model:

$$Q = \begin{array}{c} \\ A \\ T \\ C \\ G \end{array}\begin{array}{c} \begin{array}{cccc} A & T & C & G \end{array} \\ \left[\begin{array}{cccc} \cdot & \alpha & \alpha & \alpha \\ \alpha & \cdot & \alpha & \alpha \\ \alpha & \alpha & \cdot & \alpha \\ \alpha & \alpha & \alpha & \cdot \end{array}\right] \end{array}$$

F81 model:

$$Q = \begin{array}{c} \\ A \\ T \\ C \\ G \end{array}\begin{array}{c} \begin{array}{cccc} A & T & C & G \end{array} \\ \left[\begin{array}{cccc} \cdot & \alpha\pi_T & \alpha\pi_C & \alpha\pi_G \\ \alpha\pi_A & \cdot & \alpha\pi_C & \alpha\pi_G \\ \alpha\pi_A & \alpha\pi_T & \cdot & \alpha\pi_G \\ \alpha\pi_A & \alpha\pi_T & \alpha\pi_C & \cdot \end{array}\right] \end{array}$$

K2P model:

$$Q = \begin{array}{c} \\ A \\ T \\ C \\ G \end{array}\begin{array}{c} \begin{array}{cccc} A & T & C & G \end{array} \\ \left[\begin{array}{cccc} \cdot & \beta & \beta & \alpha \\ \beta & \cdot & \alpha & \beta \\ \beta & \alpha & \cdot & \beta \\ \alpha & \beta & \beta & \cdot \end{array}\right] \end{array}$$

HKY model:

$$Q = \begin{array}{c} \\ A \\ T \\ C \\ G \end{array}\begin{array}{c} \begin{array}{cccc} A & T & C & G \end{array} \\ \left[\begin{array}{cccc} \cdot & \beta\pi_T & \beta\pi_C & \alpha\pi_G \\ \beta\pi_A & \cdot & \alpha\pi_C & \beta\pi_G \\ \beta\pi_A & \alpha\pi_T & \cdot & \beta\pi_G \\ \alpha\pi_A & \beta\pi_T & \beta\pi_C & \cdot \end{array}\right] \end{array}$$

FIGURE 9.3 Matrices of instantaneous substitution rates for the simplest substitution models: the JC model with parameter α representing equal rates of evolution for each nucleotide. The K2P model distinguishes between the rate of transitions per site (α) and transversions (β). The F81 model takes into account different frequencies of individual bases, adding 3 free parameters π_i to JC model, $i = $ A, T, C, G. The HKY model implements different transition/transversion rates into the F81 model.

The previously described models assume that the columns of characters in the sequence alignment are independent and identically distributed, which is very often not appropriate. It is known that these assumptions are often violated and that the number of substitutions at a site is strongly dependent on its position along the DNA sequence (Li and Graur, 1991). One obvious reason for this is that there are different functional constraints on each site. There exist several different approaches to account for the site-dependent variation in the number of substitutions; both discrete and continuous models have been created to describe this variation. Discrete models usually assume two or more rate classes. In the discrete Hidden Markov model (Felsenstein and Churchil, 1996), the rates of evolution at different sites are assumed to be drawn from a finite set of possible rates. This method allows for substitution rates to differ between sites in the molecular sequence and for correlations to exist between the rates of neighboring sites. Gamma-distributed rate models (Uzzel and Corbin, 1971; Wakeley, 1994; Yang, 1994a), which are suitable for analysis of observed patterns of rate variation in coding regions of both nuclear DNA and mtDNA (Yang, 1994a) are also very popular. Further, "covarion models" are used to deal with sites that vary together (Fitch and Markowitz, 1970; Lockhart et al., 1996; Tuffley and Steel, 1998). The idea behind this approach is that at any given time some sites are invariable owing to functional or structural constraints. However, as mutations are fixed elsewhere in the sequence, these constraints may change, and sites that were previously invariable may become variable and *vice versa*. There have been also developed specific models useful for ribosomal RNA analysis (Tillier and Collins, 1995; Smith et al., 2004) in which loop regions are treated differently than stem regions (the stem regions are treated as hydrogen-bonded pairs).

The computation of evolutionary distances from amino acid sequences is similar to the computation of distances from nucleic acid sequences, and various models accounting for amino acid substitutions have been proposed (e.g., Poisson correction model (Zuckerkandl and Pauling, 1965; Dickerson, 1971); codon mutation model (Goldman and Yang, 1994; Yang and Nielsen, 2000); model of protein evolution (Dayhoff, 1972; Dayhoff et al., 1978) and many others). Once a model is specified with its parameters and the data have been collected, it is necessary to evaluate the goodness-of-fit of the model, that is, how well the model fits the observed pattern of data.

9.2.2 MODEL SELECTION AND EVALUATION

A basic dilemma when starting with phylogenetic analyses is to decide which evolutionary model to choose—how complex the model should be for a given problem. A model with too few parameters may be too unrealistic given certain data and thus can give improper results. Nevertheless, increasing the number of parameters of the model to improve its fit to this data does not necessarily improve the phylogeny estimation. The reason is that parameter-rich models must estimate more parameters from the same amount of data. The more parameters that must be estimated from the finite dataset, the higher is the variance associated with each parameter, and the lower the precision of parameter estimates can be (Kelchner and Thomas, 2006). This situation is called "overfitting" of the data by the model.

The advantage of using models with explicit assumptions about the substitution process is that it enables us to compare alternative models of evolution in a statistical context. Choosing an appropriate substitution model is a crucial step. The importance of model-fit depends on the phylogenetic technique used and also on questions to be investigated (Kelchner and Thomas, 2006) (Figure 9.4). For example, estimating only tree topology (which is also one of the parameters) is relatively robust to the violation of model assumptions. In contrast, estimating parameters such as substitution rates, branch lengths, genetic distance (Yang, 1994b; Buckley *et al.*, 2001) or estimating the support of the tree topology using bootstrap techniques (Felsenstein, 1985; Bos and Posada, 2005) can be more drastically affected by the violation of model assumptions. The outcome of one model compared to another often differs such that the choice of the model can often be more important than the method of phylogenetic reconstruction (Bos and Posada, 2005).

FIGURE 9.4 Importance of goodness of fit of the model (according to Kelchner and Thomas, 2006).

Goodness-of-fit of the model to the data is assessed by finding parameter values of a model that best fit the data. This procedure is called parameter estimation. The commonly used methods for parameter estimation and model evaluation are based on statistical approaches. We distinguish between two classes of statistical testing of substitution models (Posada and Crandall, 2001).

The first category of tests is designed to compare two (or more) different models. We can consider them using two different statistical approaches: the likelihood framework or the Bayesian framework. Under each of these methods, the model is first chosen based on an initial fixed-tree topology that is computed using a fast, approximate tree-building approach (e.g., neighbor-joining or parsimony (Huelsenbeck and Crandall, 1997; Frati *et al.*, 1997). Then, the search over the tree space is performed using the chosen model to find the best estimate of the tree (branch lengths and topology) and the best estimate for model parameters (Abdo *et al.*, 2005). In a likelihood framework, the fit of the model to the data is measured by using the likelihood function (Edwards, 1972; Felsenstein, 1981a). The likelihood function is proportional to the probability of observing data (D) given the model (M), P(D|M). Parameter values are sought that maximize the likelihood function and thus make the data most probable. The most popular tests in this framework are various versions of likelihood ratio tests (LRTs) (Felsenstein, 1988; Yang *et al.*, 1994; Frati *et al.*, 1997; Huelsenbeck and Crandall, 1997), where the test statistics of LRT is following:

$$\delta = 2 \ (\ln L1 - \ln L0)$$

where $\ln L1$ (resp. $\ln L0$) is the maximized log-likelihood score of the more complex (resp. more simple) model. In LRTs, given one dataset, the maximized log-likelihood estimates of two nested models are compared (the alternative model l1 (i.e., the more complex model) is compared to the null model l0 (i.e., the simpler model)). In this case, δ usually follows asymptotically $\chi 2$ distribution under the null hypothesis with n degrees of freedom, where n is the difference in the number of free parameters between the null and alternative model. The p-value is then used to determine the significance. The models are nested in the case, when they are all special cases of the most general model in the candidate class (e.g., models belonging to the class of General Time Reversible models). For example, it is not possible to compare any of GTR models with covarion models.

Akaike information criterion (AIC) (Akaike, 1974) also uses the likelihood function to provide a measure of fit between the model and the data. In addition, AIC includes a penalty for overparametrization (i.e., it penalizes models with too many parameters). The AIC for a particular model i is calculated as follows:

$$AICi = -2 \ \ln Li + 2ki,$$

where $\ln Li$ is the maximum log-likelihood of the of the data under i-th model and ki is the number of parameters in model i. An advantage of AIC is that the AIC value is computed for an individual model, and thus AIC values can be used to compare nonnested models. The model with the lowest AIC value is the one that is preferred. AIC can be modified to enable correction for small sample size (small is defined as n/ki ≤ 40; where n indicates number of sequence sites (Sullivan and Joyce, 2005).

In a Bayesian framework, testing using Bayes factors (Kass and Raftery, 1995) and Bayesian information criterion (BIC) (Schwarz, 1974) is commonly used. Bayes factors (BF) quantify the relative support of two competing models given the observed data. BF is the Bayesian analog of the LRT that can be applied to nested, as well as to nonnested models. In BF, model likelihoods P(D|M) are calculated by integrating (not maximizing as in maximum-likelihood framework) over all possible parameter values (Posada and Buckley, 2004). BFs compare the support provided by the data for two competing models i and j:

$$\text{Bij} = \frac{P(D|M_i)}{P(D|M_j)},$$

Evidence for Mi is considered very strong if Bij >150; Bij <1 supports Mj over Mi (Raftery, 1996). Bayesian information criterion provides and approximation to the natural logarithm of the Bayes factor, especially when the sample size is large and competing models are nested (Kass and Wasserman, 1994). The BIC is suitable to test nested and nonnested models. BIC for a particular model i is calculated as follows:

$$\text{BICi} = -2\text{lnLi} + \text{kilnn};$$

where lnLi is the maximum log-likelihood of the data under i-th model, ki is the number of free parameters in model i, and n is the sample size (i.e., the sequence length). The smaller the BIC, the better the model fits to the data. Although, BIC is similar to AIC, it differs in penalizing sample size as well as the number of estimable parameters (Kelchner and Thomas, 2006). BIC is more likely to select less complex models than the models selected by Bayes factors (Posada and Buckley, 2004).

A relatively new approach is model testing using decision theory (Minin et al., 2003). This approach focuses on phylogenetic performance, the ability of the model to estimate branch lengths that determine its quality, and not the overall fit of the model to the data.

The second way to assess models is to test the overall adequacy of a particular model. Parametric bootstrapping using Monte Carlo simulation (Goldman, 1993; Whelan et al., 2001) is often used to generate the null distribution for the δ statistics from LRT (in cases when χ2 distribution cannot be used). In this procedure, the null distribution of test statistic is calculated by simulating many datasets. The maximum-likelihood values of the model parameters for the simulations are estimated from the original data under the assumption that null hypothesis are correct. For each simulated dataset, the likelihood ratio is calculated by maximizing the likelihood under the null and alternative hypotheses. The proportion of replicates in which the likelihood ratio calculated using the original data is exceeded by the ratio calculated using the simulated data represents the significance level of the test.

9.3 MOLECULAR PHYLOGENETICS

All living organisms are related by the common ancestry creating a "tree of life". Evolutionary biologists have put enormous effort into resolving the pattern of relationships in the tree of life. Phylogenies, the evolutionary histories of groups of organisms, are one of the most widely used tools throughout the life sciences. They are also used as objects of research in systematics, evolutionary biology, epidemiology, etc. The evolutionary process not only determines relationships among species, but also allows prediction of structural, physiological, and biochemical properties (Chambers *et al.*, 2000) and thus many biological processes can be effectively modeled and summarized in this fashion. Most of the methods used in phylogenetic inference construct phylogenetic trees.

The reconstruction of evolutionary relationships in molecular phylogenetics is a hierarchical process (Steel, 2005) (Figure 9.5). First, analyzed sequences are aligned to determine site-by-site homologies and to detect DNA or amino acid differences. Correct alignment of the sequences is a critical component of phylogenetic analysis, but can be difficult when (1) there are a large number of sequences to compare, (2) the sequences are too long, or (3) the sequences have are too divergent. Second, a mathematical model describing the evolution of the sequences is established. The model can be built empirically, using properties calculated through comparisons of observed sequences, or parametrically, using chemical or biological properties of DNA and amino acids. Evolutionary models enable the estimation of the evolutionary distance between two homologous sequences. Evolutionary distance is measured by the expected number of nucleotide substitutions per site that have occurred in the evolutionary pathway between two compared sequences and their most recent common ancestor (i.e., evolutionary distance = evolutionary rate x evolutionary time (Pybus, 2006)). Distances may be represented as branch lengths in a phylogenetic tree stretching from the extant sequences at the tips of the tree to the hypothetical ancestral sequences, which are generally not known, at the internal nodes.

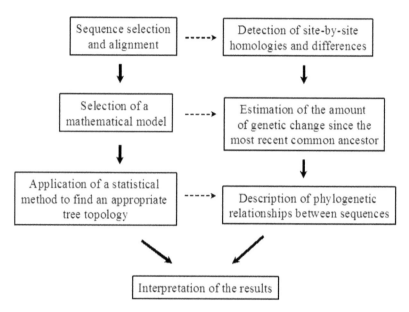

FIGURE 9.5 A basic hierarchy of phylogenetic reconstruction.

Third, a suitable statistical method is applied to find the tree topology and branch lengths that best describe the phylogenetic relationships of the sequences. The last step consists of the evaluation of the accuracy of the tree and the interpretation of the results.

9.3.1 PHYLOGENIES AND PHYLOGENETIC TREES

Phylogenetic relationships resemble the structure of a tree which is an acyclic connected graph used to model the actual evolutionary history of a group of sequences or organisms. A phylogenetic tree (phylogeny) (Figure 9.6A) represents the actual pattern of historical relationships that we try to estimate: it is a model of the evolution of a set of taxa (species, genes, strains). The extant taxa being compared are placed on the terminal nodes (leaves or tips) of the tree (i.e., leaf-labeled tree). These taxa are often called Operation Taxonomic Units (OTUs). Internal nodes of the tree correspond to the speciation events from the most recent common ancestor, and they represent the hypothetical ancestor of the pair of taxa connected by this node. Phylogenetic trees are usually bifurcating (i.e., the nodes are of the degree two). This means that at the time of speciation or gene duplication the sequence splits into two descendant sequences. However, sometimes multifurcating tree with nodes of the degree more than two may appear, for example when interior branches show no nucleotide substitution, and thus the evolutionary distance represented by the length of the branches is equal to zero (Nei and Kumar, 2000). Nodes are joined by branches (edges). The branch length often represents the amount of evolution that took place between two nodes in the tree, which is usually calculated by comparing the differences between sequences, or elapsed evolutionary time, which is measured in years. Evolutionary trees in which branches are denoted by the length values are often called weighted trees. According to the meaning of the branch lengths, it is possible to distinguish between various kinds of phylogenetic trees (Figure 9.6).

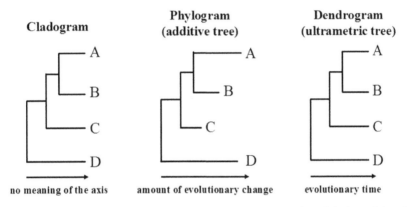

FIGURE 9.6   The types of phylogenetic trees regarding the meaning of the branch lengths.

Further, the trees can be either rooted (Figure 9.7B) or unrooted (Figure 9.7C) with a specific topology representing the branching pattern of a tree (Page and Holmes, 1998). A rooted phylogenetic tree is a directed tree with a unique node (i.e., the root) representing the most recent common ancestor from which all the sequences at the tips of the tree descended. Hence, placing the root on the tree enables us to estimate the direction of the evolutionary process that corresponds to the evolutionary time. This means that given a pair of nodes connected by a branch, the node closer to the root is an ascendant of the node further away from the root. For example, trees constructed using the molecular clock assumption are always rooted. An unrooted tree lacks this directionality; it only displays the relationships between the taxa. Furthermore, an unrooted tree does not provide any information about whether neighboring taxa are really close relatives. There are several options for how to place a root on a tree (Swofford and Begle, 1993). For example, by enforcing molecular clock, midpoint rooting, in which the root of an unrooted tree is placed at the midpoint of the longest distance between two taxa in a tree) (Farris, 1972), Lundberg rooting in which the root is placed at the point where the outgroup/hypothetical ancestor would attach most parsimoniously, or rooting the tree using an outgroup, a taxon, that is clearly less related to the rest of the group than are any two members in the group to each other to root the tree. The outgroup chooses the root so that the ingroup, the group of taxa for which we want to construct phylogenetic relationships, is monophyletic (the group is monophyletic, if it consists of an inferred common ancestor and all its descendants). However, several problems arise with outgroup rooting, for example, in choosing how divergent the outgroup should be, or whether to use one or more outgroups (Felsenstein, 2004).

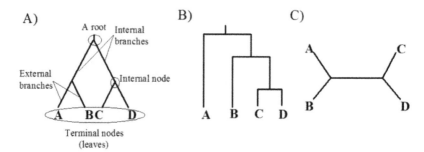

FIGURE 9.7 (A) Description of individual parts of a phylogenetic tree. (B) A rooted tree (rooting enforced by molecular clock) (i.e., an equal amount of character change is observable from the root to the tips of the tree), (C) an unrooted tree.

9.3.2 TREE-BUILDING METHODS IN MOLECULAR PHYLOGENETICS

The choice of an appropriate tree-building method is one of the most complex issues during the process of phylogenetic construction. A useful way of how to classify these methods is according to (1) how the methods handle the data, and (2) what approach is taken to construct the optimal phylogenetic tree (Page and Holmes, 1998).

The first group of method-classification may be separated by distance-based methods and character-based methods. Distance-based methods compute pairwise distances according to some measure typically by transforming the sequence data to distance values; then, the actual data in the form of sequence alignment (character matrix) are neglected and the fixed distances are used in the construction of the tree. Trees derived using character-based methods are optimized according to the distribution of actual data patterns in relation to a specified character in the character matrix.

The second classification scheme distinguishes methods according to how they build a tree. Clustering methods produce a tree by following series of steps (i.e., clustering algorithm). These methods are computationally very fast, but often produce only a single tree and thus do not enable the comparison of competing tree topologies. Criterion-based methods use an optimality criterion (e.g., the least number of changes in the tree) for comparing alternative trees to one another and deciding which one fits better. Criteria-based methods decide among each possible tree by assigning a score to each one. It is thus possible to rank resulting phylogenies in order of preference. This conveys to the user an immediate knowledge about the strength of support for that tree.

9.3.2.1 COMPARISON CRITERIA FOR TREE-BUILDING METHODS

Given the range of tree-building methods available, it is important to be able to decide which ones are better than others. Steel and Penny (2000) treat the method of inferring evolutionary trees as being composed of three largely independent parts: (1) the choice of optimality criterion, (2) the search strategy over the space of trees, and 3) assumptions about the model of evolution. Penny *et al.* (1992) identified five properties that a tree-building method should have:

The efficiency (also called power) of a method is a measure of how close the parameter estimates are to the true value as the amount of the data increases (i.e., the variance of the estimate is small). Efficiency may be measured in terms of the number of characters required to find the correct solution at a given sample size (the number of nucleotides needed to find the correct tree with a probability of >0.99 ranges from about 200 to more than 109, depending on the method used (Hillis, 1995)). The consistency of the phylogenetic method given evolutionary model refers to, whether the method will converge on the true tree with the addition of more and more data in the dataset. However, a more practical question is how the method will perform given limited data under more realistic conditions (Hillis *et al.*, 1994). The robustness (support) is the sensitivity of a method to violation of its underlying parameters or to perturbation in the data. These three criteria are the most relevant when assessing accuracy of the tree-building method. Further important criteria are computational speed, and discriminating ability (falsifiability) of the method, whether the data are able, in principle, to reject the model.

The performance of phylogenetic methods can be tested using either known phylogenies (Hillis *et al.*, 1992) or by computer simulations. The availability of known phylogenies provides the most compelling evidence of whether the used method is successful, though the knowledge of the actual evolutionary relationships of organisms or sequences is rather rare. An alternative way to test phylogenetic methods is by computer simulation. In this case, computer programs, supplied with a tree topology and evolutionary model, simulate sequences according to this information. Tree-building methods then use the simulated data, and their performance is assessed according to their ability to recover the original tree topology. Simulation is a good tool for comparing alternative methods because the model for generating the data is known and the differences in performance can thus be readily interpreted (Hillis, 1995). It is worth noting that each method has its limitations: the ideal method that satisfies all criteria is not likely to exist, and the performance of the methods is still a challenge in the field of phylogenetics.

9.3.2.2 DISTANCE BASED METHODS

Distance methods require knowledge of evolutionary distance (i.e., the number of changes that have occurred along the branches between two sequences) between all pairs of taxa (usually in the form of a matrix) as an input. These methods are based on the assumption that evolutionary distances between taxa represent their evolutionary relationships. The distance (branch length) d_i can be considered as a function of evolutionary time t_i elapsed along the branch i with a rate of nucleotide substitution r_i, such that $d_i = r_i t_i$. The value of $r_i t_i$ represents the expected number of changes along the branch, including those changes that are visible as well as those that are unobservable to us (multiple changes). The phylogenetic tree is then constructed according to the relationship between the distance values, and the original data is no longer used. It is also worth noting that distances are entities that must satisfy certain mathematical and statistical requirements (Page and Holmes, 1998). In general, distance methods can be distinguished according to (1) whether they seek a tree that best accounts for the observed distances, or (2) whether they seek the tree with the minimum sum of branch lengths (Felsenstein, 2004).

The least-squares methods (Cavalli-Sforza and Edwards, 1967; Fitch and Margoliash, 1967) minimize the discrepancy between observed and expected distances using the least-squares criterion. The minimum evolution method (Rzhetsky and Nei, 1992) uses two criteria at once: the branch lengths are determined using the least-squares criterion, and the tree topology is determined using the minimum evolution criterion (i.e., the total branch length of the tree should be minimal). These methods search for a tree with the best value according to the criterion on which they are based.

Another group of distance methods is based on a clustering procedure. It is a basic technique that constructs a tree by successively deciding which pair of terminal nodes should be neighbors, and so reducing the size of the input at each step when the nearest neighbors are clustered together and a new node is created. The differences are in how the distances are estimated, how the neighborhood decision is made, and how the distance matrix is then modified. The Unweighted pair-group method using arithmetic averages – UPGMA (Sokal and Michener, 1958) is based on the assumption that sequences evolve according to a molecular clock (i.e., in all lineages the evolutionary rate is the same). When the data do not evolve according to the molecular clock assumption, UPGMA will fail to recover the true tree. The most commonly distance-based method used nowadays in molecular phylogenetics is the Neighbour-joining method (NJ) (Saitou and Nei, 1987).

9.3.2.3   NEIGHBOR-JOINING METHOD

NJ is a very popular method of phylogeny reconstruction, and it is probably the most widely used distance-based algorithm in practice (Mailund *et al.*, 2006). NJ takes a distance matrix based on pair-wise alignment scores as an input, and for the tree reconstruction it uses a precise clustering algorithm rather than any optimality criterium. It starts with a star-like tree (Figure 9.8A) with only one internal node connecting all sequences in analysis (without previous clustering). Then a pair of the nearest neighbors is chosen under a principle of minimum evolution (i.e., it reconstructs a phylogenetic tree from evolutionary distances so that the sum of the branch lengths in the total tree is minimized) (Figure 9.8B). This process continues until a bifurcating tree results. The method provides both the topology and branch lengths of the phylogenetic tree.

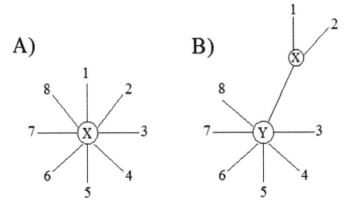

FIGURE 9.8   Construction of a phylogenetic tree using the neighbor-joining method. (A) A starlike tree with no hierarchical structure; (B) a tree in which taxon 1 and 2 are clustered.

With a running time of O(n3) on n taxa (Studier and Kepler, 1988), it is relatively fast and thus suitable for large datasets. Empirical work shows it to be reasonably accurate, at least for cases in which evolutionary rates among lineages do not vary rapidly. NJ is a model-based method (i.e., it requires explicit evolutionary model to compute evolutionary distances). The distances can then be corrected according to assumptions about the nature of evolutionary changes (e.g., correction for multiple substitutions, such as A→T→A, which are usually not observable). In distance methods that assume that sequences evolve according to molecular clock, the correction for multiple substitutions is not necessary (Rzhetsky and Sitnikova, 1996). Distance methods without such an assumption (neighbor joining method) use a specific evolutionary model to correct for unobservable changes. On the other hand, the goodness-of-fit of the distances is substantially dependend upon the choice of the model (nearly correct model will produce nearly correct distances). The neighbor-joining method does not require a uniform molecular rate, and thus is less sensitive to different branch lengths (in contrast to UPGMA), and it can be used to quickly generate a good tree topology that can be used as an initial tree for the optimization with other methods, such as the maximum-likelihood method. An objection to the distance-based methods in general is that converting the data from alignment to the distances results in the loss of the information, although studies show that this loss is remarkably small (Felsenstein, 2004).

9.3.2.4 FURTHER APPROACHES IN DISTANCE-BASED METHOD

Many new distance-based approaches have been proposed. For example BIONJ (Gasquel, 1997) or Weighbor (Bruno et al., 2000), which are improved, though approximate versions of Neighbor-joining method. Further, methods based on quartets of species (i.e., methods that use all possible combinations of four-species subtrees to infer the overall tree topology) are the Short Quartet Method (Erdös et al., 1997), the Split decomposition method (Bandelt and Dress, 1992) or Disk-covering methods (Huson et al., 1998). Although these methods seem to be promising, there has not yet been enough experimental performance analysis on real data sets for the methods to be well understood.

9.3.2.5 CHARACTER-BASED METHODS

In contrast to distance-based methods, character-based (discrete) methods operate directly on the aligned sequences, or on functions derived from the sequences, rather than on pairwise distances. Thus, they endeavor to avoid the loss of information that occurs when sequences are converted into distances.

9.3.2.6   MAXIMUM PARSIMONY

Maximum parsimony (MP) methods (Edwards and Cavalli-Sforza, 1963; Camin and Sokal, 1965; Fitch, 1977) are very popular methods of tree inference, mainly because they are easy to understand. The justification of MP is based on the principle known as Occam's razor: when explaining the observations, the simpler evolutionary hypothesis is preferred to the more complicated one. In MP, we are trying to maximize the similarity arising from the common ancestry of the sequences. If any character does not fit the phylogenetic relationship, it indicates that similarity arose due to the homoplasy and not due to the origin from a common ancestor. MP performs a site-by-site analysis and identifies a tree (or a tree) that involves the smallest number of changes necessary to explain the differences among the data under investigation. The optimality score (often referred as the tree length) is the minimum number of changes for a tree. The best hypothesis is a tree requiring the fewest changes (i.e., the most parsimonious one). Parsimony based approaches usually involve two separate components: (1) a procedure to find the minimum number of changes needed to explain the data for a given tree topology, and (2) a search through the space of trees to find the optimal topology.

9.3.2.7   FINDING THE MINIMUM NUMBER OF CHANGES

Instead of directly building a tree, MP assigns a "cost" (i.e., the score) to each tree. The aim is to find a tree that minimizes the cost. The characters at the internal nodes are also determined such that they give the minimum cost function value for a given tree topology. In traditional parsimony (Fitch, 1971), the cost function is 1 if two characters are different and 0 if they are the same. MP searches all possible tree topologies for the optimal (minimal) tree; however, the number of unrooted trees that have to be analyzed rapidly increases with the number of OTUs (having 10 OTUs, 34 459 425 unrooted trees have to be searched). For each tree topology, MP computes a minimum number of nucleotide changes that are required to explain the observed site pattern (Figure 9.9). However, not all sites in the sequence alignment are used for analysis. MP works only with informative sites, sites that enable us to distinguish between different tree topologies to choose the most parsimonious between them, while uninformative (invariant) sites are eliminated from the analysis (e.g., site with GGGG is invariant because there is no change, as well as CACC is invariant, because it always requires only one change for any tree topology). Invariant sites are uninformative only in terms of maximum parsimony, in other methods such as distance methods or maximum-likelihood they are informative. The number of changes is then summed over all informative sites in the alignment to give a parsimony score for each tree topology. The topology with the smallest total number of changes is taken as the estimate of the phylogeny – the most parsimonious tree.

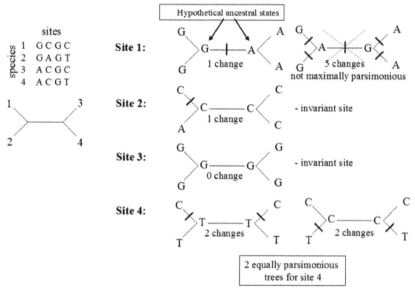

FIGURE 9.9   Reconstruction of the most parsimonious trees for given sites 1–4 for a given tree topology.

It is important to note that branch lengths are not considered in parsimony analysis. Multiple equally parsimonious trees may each arrive at the final state in a different way. Thus, the number of steps along individual branches (or the length of each branch) is not determined, and only the total number of steps for a tree is considered. Traditional parsimony is an unweighted MP method. It assumes that nucleotide or amino acid substitutions occur with equal (or nearly equal) probability, though in reality, certain substitutions occur more often than the other (e.g., transitions vs. transversions). Therefore, it is useful to put different weights to different types of substitutions when the minimum number of substitutions for a given topology is computed (Nei and Kumar, 2000) or to differently weight changes that occur at various codon positions. Weighted parsimony also enables us to give weights to sites which evolve at different rates (Farris, 1969; Swofford *et al.*, 1996).

Although MP methods are often claimed as being "assumption-free", this is not true. MP does not require a selection of an explicit model, but it does make implicit assumptions about the nature of evolutionary change, assuming independence and specific model of character transformation that is often called a step-matrix or weighting scheme. A commonly used weighting scheme is to assign equal weights to each character change. This was originally viewed as strength of the MP methods; however, it does not enable the investigation of the biological implications of different evolutionary models. MP methods may give more reliable trees than other methods when the extent of sequence divergence is low (Miyamoto and Cracraft, 1991), when there are no backward or parallel substitutions, when the rate of nucleotide substitution low and the number of nucleotides examined is large (Sourdis and Nei, 1988). Furthermore, the parsimony analysis is very useful for some types of molecular data (e.g., insertion sequences, insertions/deletions, gene order or short interspersed nuclear elements—SINEs (Steel and Penny, 2000).

However, in reality, nucleotide sequences are often subject to backward and parallel substitutions, and the number of nucleotides is rather small. In this case, MP methods tend to give incorrect topologies. In some cases, MP may suffer from inconsistencies (i.e., an increasing amount of data does not tend to give the true topology) giving incorrect topologies when the rate of nucleotide substitution varies extensively with evolutionary lineage (Felsenstein, 1978). This can also happen when the rate is constant within all lineages (Hendy and Penny, 1989; Zharkikh and Li, 1993; Takezaki and Nei, 1994; Kim, 1996). This problem is known as long-branch attraction (Hendy and Penny, 1989; Bergsten, 2005) (in contrast to the short-branch attraction (Nei, 1996). Long-branch (respectively short-branch) attraction occurs, when rapidly (respectively slowly) evolving lineages are inferred to be closely related, regardless of their true evolutionary relationships. The combinations of conditions within which these problems occur are called "Felsenstein zone" (Huelsenbeck, 1997). However, long-branch attraction is not only an issue of maximum parsimony; other methods (e.g., maximum-likelihood) also suffer under this problem. It can be effectively solved by using methods that incorporate differential rates of substitution.

9.3.3 VARIOUS APPROACHES IN PARSIMONY METHODS

There exist a variety of techniques based on maximum parsimony that are applicable on various biological data. In molecular phylogenetics, the traditional maximum parsimony approach is usually applied on multistate characters, such as nucleotide sequences with four states, with the assumption that one state can change to any other. This concept is called Wagner parsimony. Although, there are many methods that can also work for binary states. Perhaps the simplest parsimony method is Camin-Sokal parsimony (Camin and Sokal, 1965), where there are only changes from the state 0 to the state 1 without any reversals. This model is suitable when exploring the evolution of small deletions in DNA, where the reversals do not occur. Another MP approach is called Dollo parsimony (Le Quesne, 1974). It allows characters to be lost, but never gained again in the same form. This method can be applied for the analysis of gain and loss of restriction sites in DNA (Felsenstein, 2004). Polymorphism parsimony (Farris, 1978; Felsenstein, 1979) assumes that parallel changes of state are not independent. Although commonly used when analyzing morphological features, it is easily applicable in explaining changes in karyotype due chromosome inversions. All mentioned variants of parsimony also enable the reconstruction of the ancestral state for each character. This is usually done by applying the Sankoff algorithm (Sankoff, 1975; Sankoff and Rousseau, 1975), which is similar to the Fitch algorithm (Fitch, 1971) used in traditional parsimony to compute the minimum number of changes for each tree topology.

9.3.4 MAXIMUM LIKELIHOOD

The maximum-likelihood (ML) method for phylogenetic inference (Cavalli-Sforza and Edwards, 1967; Neyman, 1971; Felsenstein, 1981a) is a well-established statistical method of parameter estimation in phylogenetics that is very efficient in extracting information from the data. Maximum likelihood methods are very powerful tools of statistical inference for several reasons: (1) they use all of the available data, (2) they can incorporate various evolutionary models, (3) they permit estimation of parameters and their associated standard errors, and (4) they allow the comparison of different evolutionary hypotheses (i.e., evolutionary models or phylogenetic trees) through their maximum-likelihood statistics (for examples of evolutionary hypotheses refer Figure 9.10).

Is a DNA substitution model adequate to explain the data?	H_0: Assume a particular model of DNA substitution H_1: Assume a multinomial distribution for the frequencies of site patterns.
Is the DNA substitution process identical among genomics regions?	H_0: Assume that the substitution parameters are the same among genomic regions H_1: Allow substitution parameters to vary among genomic regions
Are DNA substitution rates constant among sites?	H_0: Assume equal rates among sites H_1: Allow among-site rate heterogeneity
Are phylogenies estimated from different data congruent?	H_0: Assume that all data partitions have the same phylogeny H_1: Allow different phylogenies to underlie different partitions
Is a specified taxonomic group monophyletic?	H_0: Assume that a group is monophyletic H_1: Relax the constraint of monophyly
Are phylogenies for hosts and parasites consistent with a common history?	H_0: Assume an identical phylogeny for associated hosts and parasites H_1: Allow different phylogenies for hosts and parasites

FIGURE 9.10 Examples of hypotheses about evolutionary processes that can be tested using likelihood framework (according to Huelsenbeck and Rannala, 1997).

In phylogenetic inference, the likelihood is an entity proportional to the probability of observing the data (e.g., a set of aligned nucleotide sequences or rather the number of site patterns that occur in a set of aligned sequences) under a given phylogenetic tree and a specified model of evolution: $L = P(Data|tree, model)$. The principles of maximum-likelihood suggest that the explanation that makes the observed outcome the most likely (i.e., the most probable) is the one to be preferred. In other words, if we consider that a phylogenetic tree represents the pathway of sequence evolution, we are looking for a tree (an evolutionary pathway) that gave rise to our data with the highest probability. ML requires an explicit evolutionary model for the data to specify the probability of observing different site patterns. A phylogenetic model consists of (1) a tree of relationships between sequences, (2) branch lengths of the tree representing expected number of changes per site, and (3) a model of the substitution process. The probability of observing the data under the assumed model changes depending on the parameter values of the model. ML chooses the parameter values that maximize the probability of observing the data. Examples of the parameters are the branch lengths (interspeciation times and rates of mutation representing how much evolution took place along each branch), parameters associated with the substitution matrix (e.g., transition/transversion bias), or parameters that describe how rates vary across sites (Steel and Penny, 2000).

9.3.5 FINDING A TREE TOPOLOGY WITH MAXIMUM LIKELIHOOD

To sketch the procedure of maximum-likelihood computation for a phylogenetic tree, let us consider following example (inspired by Nei and Kumar, 2000). Let us assume a tree with four taxa (M, N, O, P) and a dataset of aligned DNA sequences corresponding to each taxon with a length of n nucleotides (the alignment is without any insertions or deletions). For the k-th site (k = 1,..,n) in the alignment, we observe a nucleotide i (i = A, T, G, C) for each taxon. The nucleotides at the nodes x, y, and z are not known *a priori* (Figure 9.11A).

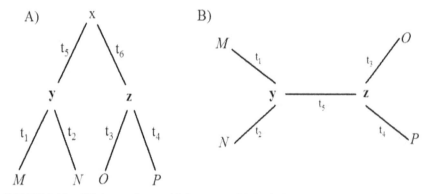

FIGURE 9.11 (A) A rooted tree with four taxon M, N, O, P. (B) A corresponding unrooted tree.

Let us consider k-th nucleotide site and let Pij(t) be the probability that nucleotide i becomes nucleotide j within the time t at a given site (i and j indicate any of A, T, G, C) (evolutionary time is measured in term of branch length, i.e., the expected number of substitution per branch di = riti). The branch lengths are considered parameters, and the estimate is computed by maximizing the likelihood function for a given set of observed nucleotides. The likelihood function for a k-th site is given by

$$lk = pxPxy(t5)PyMi(t1)PyNi(t2)Pxz(t6)PzOi(t3)PzPi(t4)$$

where px is the prior probability that node x has a nucleotide i. px is usually calculated from the substitution model as the equilibrium frequency of nucleotide i in the entire set of sequences.

Pij(t) is computed according to the specific model of substitution. To simplify the computation, we assume reversibility of the substitution process (this enables us to eliminate the root from the tree in Figure 9.11A and to work with the tree in Figure 9.11B, where t5+t6 = t5). We assume that evolutionary change starts from the node y, and we can then write the equation for lk as follows:

$$lk = pyPyMi(t1)PyNi(t2)Pyz(t5)Pxz(t6)PzOi(t3)PzPi(t4)$$

Since in practice we do not know character states at nodes y and z, the likelihood will be the sum of the above quantity over all possible nucleotides at the nodes y and z:

$$L_k = \sum_y P_y \left[P_{yM_i}\left(t_1\right) P_{yN_i}\left(t_2\right) \right] \left[\sum_z P_{yz}\left(t_5\right) P_{zO_i}\left(t_3\right) P_{zP_i}\left(t_4\right) \right]$$

The above calculations consider only one nucleotide site. The likelihood for the entire sequence is thus the sum of Lk's for each site, and the log-likelihood of the entire tree then becomes

$$\ln L = \sum_{k=1}^{n} \ln L_k$$

We can change the parameter ti (i.e., branch lengths) to obtain the maximal value of lnL (having maximum lnL, in turn we obtained the maximum-likelihood estimates of branch lengths for the tree topology). To obtain likelihood values for all four possible taxon tree topologies, we have to do the above calculations individually for all of them. For each tree, the combination of branch lengths and other parameters that maximizes the likelihood of the tree (by maximizing the likelihood function) is found. This is done numerically, usually by the Newton-Raphson method. The topology with the highest likelihood value is then considered to be the maximum-likelihood estimate of the phylogeny. It is obvious that visiting all possible trees and calculating the likelihood for each of them is computationally a very expensive procedure. There are two tasks when searching in a space of trees with branch lengths (Felsenstein, 2004). The first is to find the optimum branch lengths for each given tree topology. This step can be considerably simplified using various algorithms (Felsenstein, 1981a; Adachi and Hasegawa, 1996). To speed up the maximum-likelihood estimation, it is also possible to determine the initial tree topology using some non-ML method (the best seems to be the Neighbor-joining method that is fast and suitable even for large datasets (Strimmer and von Haeseler, 1996) and then to optimize branch lengths. The second task involves searching the space of tree topologies for the one with branch lengths that gives it the highest likelihood.

The development of the methods based on maximum-likelihood goes hand in hand with progress in computational techniques as well as with the refinement of evolutionary models to approach biological realism. What makes ML very attractive is the ease with which hypotheses can be formulated and tested (Huelsenbeck and Crandall, 1997). It is also very powerful in extracting the information from the data. ML can handle complex models that can incorporate various biologically important processes. The maximum-likelihood estimates are asymptotically efficient, meaning that the variance of a maximum-likelihood estimate is equal to the variance of any unbiased estimate as the sample size increases (Huelsenbeck and Crandall, 1997). However, maximum-likelihood estimates are often biased (Huelsenbeck and Crandall, 1997). A technical difficulty concerns computing the likelihood itself; it is computationally very time consuming, and there may exist more than one maximum-likelihood value for a given tree (i.e., the existence of local optima) making it difficult to guarantee that the likelihood value for that tree is maximal. Further, ML examines all possible topologies in searching for the ML-tree. Since the number of all possible topologies rapidly increases with the number of sequences, it makes ML extremely time consuming for analyses of larger datasets. Competing methods of analysis, such as maximum parsimony, involve much less computation, but they do not make use of all the data. Consequently, every advance in computing speed makes maximum-likelihood more and more attractive.

9.3.5.1   SEARCH FOR AN OPTIMAL TREE

The procedure of phylogeny reconstruction is in a computational context a combinatorial optimization problem (Weber et al., 2006). Since commonly used methods (MP, ML, Bayesian inference) explore each possible tree, the combinatorial aspect involves a search among all such trees (i.e., a search through a tree space). Combinatorial optimization problems are known to be NP-hard (i.e., there exists no efficient algorithm to solve the problem in polynomial time). The optimization is represented by a special function (maximum-likelihood function, maximum parsimony criterion) that provides an approximate measure of how close the reconstructed phylogenetic tree is to the real phylogenetic relationships (Nei, 1996; Swofford et al., 1996). This means that the preferred tree is determined by assigning an optimality score to all possible topologies according to a certain procedure and choosing the topology that shows the highest or the lowest optimal score.

In the case of maximum parsimony, the minimum number of evolutionary changes is computed for each topology, and the one with the smallest value is chosen as the preferred (i.e., most parsimonious) tree. In maximum-likelihood methods, in contrast, for each topology the log-likelihood value (lnL) represents the optimality score, and the tree showing the highest lnL is chosen as the maximum-likelihood estimate of a phylogenetic tree. There are various algorithms, which search the tree space and attempt to find the optimal tree based on the pre-specified criteria and objective function (Fitch, 1971; Swofford and Maddison, 1987). These methods include an exhaustive search (i.e., testing every possible tree topology), which is applicable only when the number of taxa is small (less than 10). Therefore, methods such as MP and ML rely on heuristics (i.e., algorithms that are empirically shown to work, but not always, or that give nearly the right answer) that obtain optimal trees in reasonable computing time. Different algorithms of the heuristic search for MP trees are now available (Maddison and Maddison, 1992; Swofford and Begle, 1993). The most popular algorithm is the stepwise addition of taxa algorithm and branch swapping procedures (e.g., nearest-neighbor interchanges (NNI), subtree pruning regrafting (SPR) or tree bisection-reconnection (TBR) (Swofford and Begle, 1993). Other successful algorithms for solving hard optimization problems are genetic algorithms (Holland, 1975) or simulated annealing (Metropolis *et al.*, 1953).

Finding the ML tree is computationally very intensive, because the tree likelihood depends not only on the tree topology but also on numerical parameters (e.g., branch lengths, substitution rates, etc.). Therefore, it is necessary to employ heuristic methods. Many of them are similar to the heuristics used for obtaining MP trees, though the efficiencies of these algorithms in obtaining the correct topology are not always the same (Takahashi and Nei, 2000). Stochastic approaches for optimizing each tree have been described (based on Markov chain Monte Carlo methods in Bayesian inference), but iterative "hill-climbing" methods are usually considered faster and sufficient for numerous combinatorial optimization problems (Aarts and Lenstra, 1997). One of the most important algorithmic progresses to get optimized evolutionary models and trees using ML is represented by PHYML package (Guindon and Gascuel, 2003). With computational speed equal to distance- or parsimony-based methods and with the high topological accuracy, it is very suitable also for large datasets and it also greatly facilitates the building of multiple trees in bootstrap analysis.

9.3.5.2   CONSENSUS TREES

It may happen that it is not possible to determine the final topology unambiguously (e.g., we have several optimal topologies) or it may be that we have two or more different tree topologies for the same group of taxa constructed using different data. We may be interested in (1) how to obtain one overall estimate of the tree topology, or (2) whether it is possible to assess the extent of difference between the trees. One way to solve this problem is to create a composite tree that represents all the trees. Such a composite tree is called a consensus tree. There exist several ways how to construct a consensus tree (Felsenstein, 2004). The most commonly used are the strict consensus method (Rohlf, 1982) and majority-rule consensus method (Margush and McMorris, 1981). The strict consensus method constructs a tree that contains all groups that occur on all trees (Figure 9.12A). Majority-rule consensus method creates a tree that consists of those groups that are present in a majority of the trees (Figure 9.12B). It is possible to set up the parameter ℓ that represents a percentage of time that a group is represented in the trees whose consensus is being created (e.g., if $\ell = 80$, the majority-rule tree will consist of groups that occur more than 80 % of the time). It is important to note that consensus tree does not take into account the strength of the evidence supporting the clades in each original trees. Consensus trees should therefore not be interpreted as the primary hypothesis of a phylogeny as are the original trees, but rather as a summary of such hypotheses (Felsenstein, 2004). To be able to measure differences between trees, several distance measures have been proposed, such as the symmetric difference (Bourque, 1978), the quartet distance (Estabrook et al., 1985) or the nearest-neighbor interchange distance (Waterman and Smith, 1978).

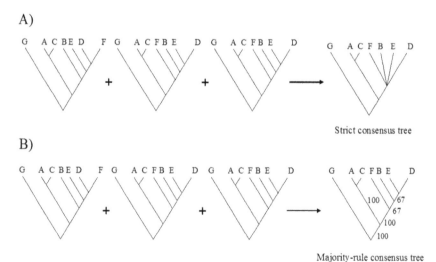

FIGURE 9.12   Reconstruction of (A) strict consensus tree and (B) majority-consensus tree (according to Felsenstein, 2004).

9.3.5.3   BAYESIAN INFERRENCE

Bayesian analysis is closely related to maximum-likelihood methods. Although, it is not a recent invention, it has only been recently recognized in the area of phylogenetic analysis, and has been applied to numerous interesting problems, such as the estimation of ancestral states on a phylogeny (Yang *et al.*, 1995), the search for positively seleted sites (Nielsen and Yang, 1998), inferences about the history of cospeciation (Huelsenbeck *et al.*, 2000a), or divergence time estimation (Thorne *et al.*, 1998). As with maximum-likelihood methods, Bayesian estimation of phylogenies (Larget and Simon, 1999; Huelsenbeck *et al.*, 2001a) requires data and the same stochastic substitution models consisting of tree topology, branch lengths, and parameters (Figure 9.13). Bayesian analysis is based on the entity called the posterior probability of a tree, P[Tree i|Data], which is a probability of the tree conditional on the observed data. The posterior probability represents a combination of the prior probability (i.e., the knowledge about the evolutionary process that we have before the data have been sampled) (P[Tree i]), and a likelihood for each tree (P[Data|Tree i]) (Figure 9.10) using Bayes's formula

$$P[Tree \mid Data] = \frac{P[Data \mid Tree] \times P[Tree]}{P[Data]}$$

Parameters of interest are viewed as a posterior probability distribution (i.e., a range of probability values which a parameter can obtain) rather than a point ("best") estimate (i.e., the best guess about the value of the parameter), as in MP or ML. The tree with the highest posterior probability represents the best estimate of the phylogeny. For example, imagine that we are tossing a coin. Our prior is that the coin is fair, and so we expect that after 1000 tosses we would obtain 50 % of heads and 50 % of tails. However, after tossing the coin 1000 times, we observed 210 heads and 790 tails, so the posterior probability of tossing head or tail is not 0.5 and we can conclude that the prior assumption (i.e., the coin is fair) was not right.

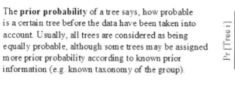

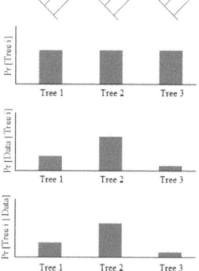

The **prior probability** of a tree says, how probable is a certain tree before the data have been taken into account. Usually, all trees are considered as being equally probable, although some trees may be assigned more prior probability according to known prior information (e.g. known taxonomy of the group).

The **likelihood** is proportional to the probability of the data given a tree. This entity requires a model with specific assumptions regarding the evolutionary process generating the data.

The **posterior probability** of a tree says, how probable is a certain tree given the data. It is obtained by combining prior probability and likelihood for each tree using Bayes' formula.

FIGURE 9.13   The cornerstones of Bayesian phylogenetics (according to Huelsenbeck *et al.*, 2001).

9.3.5.4   *COMPUTATION OF POSTERIOR PROBABILITIES*

Although, the posterior probability is easy to formulate, in practice it involves a summation over all trees and, for each tree, integration over all possible combinations of branch lengths and substitution model parameter values, which can be infeasible even for fairly small trees (Huelsenbeck *et al.*, 2001):

$$P[Tree \mid Data] = \frac{P[Data \mid Tree] \times P[Tree]}{\int P[Data \mid Tree] \times P[tree]}$$

The most popular method, which helps to solve these problems is the Markov Chain Monte Carlo (MCMC) technique also called Metropolis-Hastings algorithm (Metropolis *et al.*, 1953; Hastings, 1970). With the development of powerful computers, MCMC has enabled a revolution in Bayesian phylogenetic inference, as well as a way to study many problems in evolutionary biology and genetics. The basic principle is to construct a Markov chain with the state space created by the parameters we wish to estimate (i.e., parameters in the used evolutionary model, such as tree topology, branch lengths, substitution rates, etc.) and a stationary distribution that is the joint posterior probability distribution of the parameters. The Markov chain represents a random walk through the parameter space so that any probability distribution can be approximated by periodically sampling the values. Each step in a Markov chain consists of random modification of the tree topology, branch lengths, and other parameters in the evolutionary model. If the posterior probability computed for a proposed step is higher than the probability of the current tree and parameter values, the proposed step is accepted, and the new tree becomes the subject for further perturbations. The principle is that for a properly constructed Markov chain, the proportion of the time that any tree is visited is an approximation of the posterior probability of that tree (Tierney, 1994). For example, if we are asking, what is the probability that a certain clade in the tree is true, we can run a Markov chain and let it sample 100000 trees. If the clade of interest is present in 85000 trees, then the probability (given the data) that this clade is true is approximately 0.85. It is immediately obvious that Bayesian analysis provides a natural way to incorporate uncertainty into a phylogeny – the probability that a tree is correct is the posterior probability of the tree conditioned on the data and the model that are assumed to be correct. Posterior probabilities can be approximated using MCMC even under complex substitution models in much shorter time than in maximum-likelihood bootstrapping (Huelsenbeck and Rannala, 2004), which is the closest analog to a Bayesian analysis of a phylogeny (Larget and Simon, 1999).

Bayesian inference of phylogeny brings new perspective to numerous problems in evolutionary biology and phylogenetics. It suggests a natural way to accommodate uncertainty into phylogenies and provides an intuitive measure of support for trees and a practical way to estimate large phylogenies using a statistical approach (Huelsenbeck et al., 2000b). The posterior probability is, in contrast to bootstrap values, the true probability of the support for the tree. The most exciting aspect of MCMC Bayesian inference is its computational efficiency. The method allows the incorporation of complex models of DNA substitution, and other aspects of evolution. This increases the realism of the models, improving the accuracy of the methods. Bayesian analysis searches for the best set of trees (unlike ML, which seeks for the single most likely tree) represented by the consensus tree (Larget and Simon, 1999). Further, Bayesian methods enable the analysis of large datasets, since instead of searching for the optimal tree, it samples trees according to their posterior probabilities.

Although Bayesian analysis using MCMC is an elegant method for solving many problems, it is relatively new, and there are a number of unresolved questions that have not yet been solved (Huelsenbeck et al., 2002). A practical but serious problem is the convergence of the Markov chain – the determination of how long to run the chain to produce a reasonable sample from the posterior distribution. Another inconvenient problem is the discrepancy between Bayesian posterior probabilities and nonparametric bootstrap test values (Huelsenbeck and

Rannala, 2004). A third problem is that Bayesian analysis requires the specification of any prior knowledge and thus all parameters in the evolutionary model (topology, branch lengths, substitution parameters) must be associated with prior probability distributions. There exist three classes of priors on trees (Felsenstein, 2004): (1) priors from a birth-death process, (2) flat priors, and (3) priors using an arbitrary distribution for branch lengths (e.g., exponential distribution). What prior to use is a serious issue of Bayesian inference.

9.3.5.5   PHYLOGENETIC NETWORKS

The previously described methods for phylogenetic inference produce trees. Phylogenetic trees often oversimplify the mechanisms of evolution, since they do not account for events such as horizontal gene transfer in bacteria, hybrid speciation, or recombination events creating specific links among organisms and thus contributing to the reticulate evolution. Phylogenetic networks enable us to model evolutionary processes of organisms where such reticulations (i.e., nontree events) ocurred. A network is represented by directed acyclic graph, where due to reticulations edges connecting nodes from different branches of a tree arise. Despite the importance of phylogenetic networks, for a long time the reconstruction of reticulate evolution has been troublesome. In recent years, a lot of effort has been taken to address the construction of networks and their evaluation (Maddison, 1997; Linder *et al.*, 2004; Makarenkov *et al.*, 2004; Nakhleh *et al.*, 2005; Huson and Bryant, 2006; Jin *et al.*, 2006).

9.3.5.6   ASSESSMENT OF THE ACCURACY OF THE PHYLOGENETIC TREE

With the increasing emphasis in biology on the reconstruction of phylogenetic trees, questions have arisen as to how confident the phylogenetic tree should be and how support for phylogenetic trees should be measured. To answer this question, several methods for assessing accuracy and precision of the tree-building methods have been proposed (Felsenstein, 2004). In general, there are two types of statistical tests of a branching pattern of phylogenetic trees (Sitnikova, 1996). The first class involves testing estimates of branch length, represented as deviations of estimated branch lengths from the true (expected) branch lengths. These tests include for example, the interior-branch test (Nei *et al.*, 1985; Li, 1989; Rzhetsky and Nei, 1992), and its bootstrapped version (Dopazo, 1994). These tests are applicable only to trees produced by distance methods. Another option is to use tests based on likelihood ratio test (LRT) (Felsenstein, 1988; An-isimova and Gascuel, 2006), which compare an alternative hypothesis of a positive branch length to the nested null hypothesis with the branch of interest constrained to a zero-length. The second class of tests examines how the resulting topology differs from the true or expected topology. The most commonly used test of the reliability of an inferred tree is a nonparametric bootstrap test (Felsenstein, 1985). In the bootstrap test (Figure 9.14), the reliability of an inferred tree is evaluated by using a bootstrap resampling technique (Efron, 1982). This technique provides assessments of "confidence" for each clade of an observed tree based on the proportion of bootstrap trees showing that same clade. In this method, numerous (usually 500-2000) "pseudoreplicate" data sets are generated from the original data matrix of the same size as the original data matrix by resampling characters (columns) with replacement from the original dataset. Subsequently, applying any of the treebuilding techniques on each pseudodataset as well as on original dataset, trees are built and the topologies are screened. The proportion (percentage) of bootstrap replicates in which a given clade (i.e., the branching pattern) appears is called a bootstrap value. Bootstrap values indicate the statistical support for the given clade and have been proposed as a confidence interval for the accuracy of the clade. Results are usually displayed in the form of consensus tree (usually a majority-rule consensus tree), where each clade is evaluated by a bootstrap value. Bootstrap results are commonly interpreted as a measure of the probability that a phylogenetic estimate represents the true phylogeny. In general, bootstrap values of 95 %, 99 % or higher are considered as a significant support for a particular branching order. Although bootstrapping is commonly used, it should be interpreted cautiously. It only provides an indication of the degree of support given by a particular technique for a particular clade. In cases in which the given method is positively misleading, the bootstrap values will support a clade, even if the probability that the clade is real may be quite low (Hillis and Bull, 1993). Under many conditions, bootstrapping can be used as a highly conservative measure of accuracy, but the magnitude of bias will differ from branch to branch and study to study. Therefore the values cannot be directly compared among studies (Hillis and Bull, 1993).

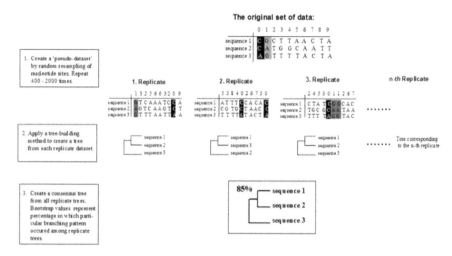

FIGURE 9.14   A scheme of nonparametric bootstrap test (according to Baldauf, 2003).

Another method commonly used in assessing the sampling error in phylogenetic trees construction is the Jack-knife nonparametric method (Efron, 1982) similar to nonparametric bootstrapping, or Bayesian inference (Huelsenbeck *et al.*, 2001a; Huelsenbeck *et al.*, 2002). In Bayesian phylogenetics, parameters such as the tree topology, branch lengths, or substitution parameters, are modeled as posterior probability distributions. The posterior probability of a tree can be interpreted as the probability that the tree is correct or that a clade is true (Huelsenbeck *et al.*, 2001a). Since the posterior probability is conditional on the information content asserted in the prior view, on the data and the model, it cannot be considered as a universal measure of the probability of the truth (Simmons *et al.*, 2004).

9.3.6   HOMOLOGY VERSUS HOMOPLASY

The majority of the disciplines using phylogenetic reconstruction are focused in phylogenetic relationships between species. However, most phylogenetic studies use methodologies for estimation of gene trees assuming that gene trees will be congruent with species trees. The reliable inference of phylogenetic relationships is only possible using homologous characters, e.g., characters whose similarity arose after divergence from their common ancestor. There exist several types of homology (Fitch, 2000). For example, orthology is a type of homology arising from the sequence divergence followed after speciation. Orthology gives rise to sequences whose true phylogeny corresponds to the true phylogeny of the corresponding organisms (only orthologous genes have this feature). Paralogy means that sequence divergence followed after gene duplication. Paralogous genes may descend and diverge while existing side by side in the same lineage. Mixing orthologous and paralogous genes leads to a tree that does not correspond to the species tree. Xenology refers to the presence of horizontal gene transfer, thus the descent is not from parent to offspring. Such events have a great impact on the tree topology, especially if they are not recognized. Another relationship between characters is analogy. Analogous characters show similarity that arose due to convergent and parallel processes (i.e., they do not share the common ancestor) (Figure 9.15). The processes of gene duplication, horizontal gene transfer, convergent, and parallel evolution are sources of homoplasy, which refers to the relationship of any two identical character states that arose independently given a specific tree. Homoplasy is known to create incongruent relationships in a phylogenetic tree tieing together similar but unrelated taxa. Homoplasy can provide misleading evidence of the phylogenetic relationship if mistakenly interpreted as homology. That is why it is very important to distinguish reliable evidence of homology from confusing evidence of homoplasy and much attention should be paid to proper sampling of the data for the analysis.

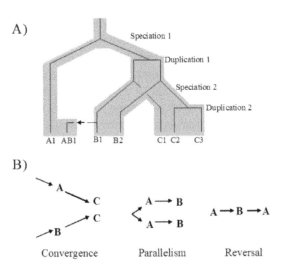

FIGURE 9.15 Various types of evolutionary relationship between characters (according to Fitch, 2000). (A) The hypothetical evolution of a gene (black lines) in a species tree (gray background) from the common ancestor, evolving into three species (A,B,C) (according to Fitch, 2000). There are two speciation and two duplication events. The genes C2 and C3 are paralogous to each other. Both are orthologous to B2, while paralogous to B1 and orthologous to A1. AB1 gene originated due to horizontal gene transfer from species B to species A, and is thus xenologous to all other genes. (B) Other sources of homoplasy.

9.3.6.1   *GENE TREES VERSUS SPECIES TREES*

In many biological fields, there is a focus is on reconstructing phylogenetic trees of species so as to understand the evolutionary relationships between them. With the accumulation of sequence data, with the development of powerful statistical methods and more realistic evolutionary models, scientists are getting closer and closer to resolving and understanding the tree of life. Although, the vast majority of methodologies construct gene trees assuming that gene trees simply reflect the relationships between species, it has become clear that this is often not true due to complex relationships between genes caused by processes such as horizontal gene transfer, interspecific gene flow, or by gene duplication or loss (Page and Charleston, 1997). Thus it is questionable whether sequences or gene trees are better to use as direct estimator of the species tree. Several approaches have been suggested to infer species trees from the gene trees. The combined-data approach, also known as the total evidence approach (Kluge, 1989; Kluge and Wolf, 1993; Nixon and Carpenter, 1996), is based on the assumption that nucleotides can be used as direct estimators of species trees, which should be estimated using the whole sequence of the genome. In this approach, sequences from all available genes are concatenated into a single alignment (supermatrix) (this could also include other characters, such as morphology). However, since it assumes that nucleotides are independent estimators of the species trees, the goodness of species tree estimate may be influenced by the length of the sequences (the longer the sequences, the better the estimate of the species tree). Thus, the species tree estimation is more likely to be determined by the genes with long sequences rather than by the short genes and it can be biased if the gene trees for the long genes have wrong topologies (Slowinski and Page, 1999). Another method for inferring species trees is the gene parsimony method (Maddison, 1997; Page and Charleston, 1997). The justification of this method is that having both a correct gene tree and a species tree that do not correspond to each other, one can reconstruct the circumstances under which both trees could be true. The algorithm searches for a combination of species tree and gene tree that minimizes the total number of substitutions (e.g., gene duplications and gene losses). This approach enables one to determine the number of substitution events that took place in the tree and where in evolutionary time they occurred. This is especially useful for analysis of large gene families. Duplication events also provide an alternative way to locate the root of the tree (Felsenstein, 2004). This is useful in cases when a suitable outgroup is missing. Supertree methods (Bininda-Edmonds, 2005) are able to combine diverse sources of information, using phylogenetic trees as the primary data as opposed to using character data, such as characters in sequence alignment. These methods construct a comprehensive phylogeny by combining source trees that only partially overlap in their taxon sets (in contrast to consensus trees that combine only trees labeled with the same taxons). However, this approach does not rely on any biological justification for explaining incongruence between genes or input trees (Edwards *et al.*, 2007). A very recent methodology uses a Bayesian hierarchical model to estimate the phylogenetic relationships of a group of species using multiple estimated gene tree distributions (e.g., those originating in a Bayesian analysis of DNA sequences) (Liu and Pearl, 2007). It enables us to employ substitution models as in traditional phylogenetic methods, together with coalescent theory to explain genealogical signals from species trees to gene trees and from gene trees to sequence data. In this way, a complete stochastic model is formed to simultaneously estimate gene trees, species trees, ancestral population sizes, and species divergence times.

9.3.7 *PHYLOGENIES FOR LARGE DATASETS*

Phylogenetic inference has become very important in the study of molecular evolution and genomics. The ever-increasing amount of orthologous genes from various species and sequences from multigene families put extreme importance on the development of phylogenetic methods that can handle huge amounts of data. With the increase in the data, the number of possible tree topologies increases astronomically, and the probability of finding the optimal tree rapidly diminishes. As a result, the reconstruction of phylogenetic trees for very large datasets is computationally very difficult.

Various approaches have been proposed for any of the tree-building methods to analyze large datasets. The neighbor-joining method with its substantial computational speed can be effectively used to generate initial phylogenies of large datasets (Tamura *et al.*, 2004). Considering maximum parsimony and maximum-likelihood methods, reconstructing optimal phylogenies is a very challenging task. Current methods usually rely on heuristics, such as hill-climbing, simulated annealing, genetic algorithms, or divide-and-conquer algorithms, which allow the analysis of large datasets in a reasonable time. For maximum parsimony, heuristics implemented in the program TNT (Tree analysis using New Technology) (Goloboff *et al.*, 2000) perform very well. Another solution for MP and ML trees is using Disk Covering Methods (Huson *et al.*, 1998). For maximum-likelihood based analyses, PHYML software based on the nearest-neighbor-interchange heuristic is very effective for reconstructing large phylogenies in reasonable time. Bayesian methods are also considered very suitable to deal with large datasets (e.g., in the implementation in MrBayes (Huelsenbeck and Ronquist, 2001b)). A substantial increase in the speed of the analysis can be also achieved by using clusters of computers using parallel tools and methods

(Keane *et al.*, 2005). This divides the problem into smaller subproblems, where each subproblem is allocated between parallel platforms.

ACKNOWLEDGEMENTS

This work was supported by the Grant Agency of the Czech Republic (grant numbers 204/05/P505 and 521/06/0056).

KEYWORDS

- Molecular phylogenetics
- Akaike information criterion
- Bayesian analysis
- Biomolecular sequences
- Comparative genomics
- Covarion models
- Evolutionary change
- Evolutionary models
- General time reversible models
- Hidden Markov model
- Homology
- Homoplasy
- Markov process
- Maximum likelihood
- Maximum parsimony
- Molecular clock
- Molecular systematics
- Neighbor-joining method
- Phylogenetic inference
- Poisson distribution
- Substitution models
- Xenology

REFERENCES

Aarts, E.; Lenstra, J. K. *Local search in combinatorial optimization*, Wiley: Chichester, UK, 1997.

Abdo, Z.; Minin, V. N.; Joyce, P.; Sullivan, J. Accounting for uncertainty in the tree topology has little effect on the decision-theoretic approach to model selection in phylogeny estimation. *Mol Biol Evol* 2005, 22, 691-703.

Adachi, J.; Hasegawa, M. MOLPHY version 2.3: Programs for molecular phylogenetics based on maximum likelihood. *Comput Sci Monogr* 1996, 28, 1-150.

Akaike, H. A new look at the statistical model identification. *IEEE Trans Automat Contr AC* 1974, 19, 716-723.

Anisimova, A.; Gascuel, O. Approximate likelihood-ratio test for branches: A fast, accurate, and powerful alternative. *Syst Biol* 2006, 55, 539-552.

Arber, W. Genetic variation: Molecular mechanisms and impact on microbial evolution. *FEMS Microbiol Rev* 2000, 24, 1-7.

Baldauf, S. L. Phylogeny for the faint of heart: A tutorial. *Trends Genet* 2003, 19, 345-351.

Bandelt, H. J.; Dress A. W. Split decomposition: A new and useful approach to phylogenetic analysis of distance data. *Mol Phylogenet Evol* 1992, 1, 242-252.

Bergsten, J. A review of long-branch attraction. *Cladistics* 2005, 2, 163-193.

Bininda-Emonds, O. R. P. Supertree construction in the genomic age. *Methods Enzymol* 2005, 395, 745-757.

Boore, J. L. The use of genome-level characters for phylogenetic reconstruction. *Trends Ecol Evol* 2006, 21, 439-446.

Bos, D. H.; Posada, D. Using models of nucleotide evolution to build phylogenetic trees. *Dev Comp Immunol* 2005, 29, 211-227.

Bourque, M. Arbres de Steiner et reseaux dont certains somments sont à localisation variable. *PhD Dissertation*, Université de Montréal : Montréal, Quebec, 1978.

Bruno, W. J.; Socci, N. D.; Halpern, A. L. Weighted Neighbor Joining: A likelihood-based approach to distance-based phylogeny reconstruction. *Mol Biol Evol* 2000, 17, 189-197.

Buckley, T. R.; Simon, C.; Chambers, G. K. Exploring among-site rate variation models in a maximum likelihood framework using empirical data: The effects of model assumptions on estimates of topology, edge lengths, and bootstrap support. *Syst Biol* 2001, 50, 67-86.

Cai, X.; Zhang, Y. Molecular evolution of the ankyrin gene family. *Mol Biol Evol* 2006, 23, 550-558.

Camin, J.; Sokal, R. A method for deducting branching sequences in phylogeny. *Evolution* 1965, 19, 311-326.

Chambers, J. K.; Macdonald, L. E.; Sarau, H. M.; Ames, R. S.; Freeman, K.; Foley, J. J.; Zhu, Y.; McLaughlin, M. M.; Murdock, P.; McMillan, L.; Trill, J.; Swift, A.; Aiyar, N.; Taylor, P.; Vawter, L.; Naheed, S.; Szekeres, P.; Hervieu, G.; Scott, C.; Watson, J. M.; Murphy, A.; Duzic, E.; Klein, C.; Bergsma, D. J.; Wilson, S.; Livi, P. A G protein-coupled receptor for UDP-glucose. *J Biol Chem* 2000, 15, 10767-10771.

Cavalli-Sforza, L. L.; Edwards, A. W. F. Phylogenetic analysis: Models and estimation procedures. *Am J Hum Genet* 1967, 19, 233-257.

Dayhoff, M. O. *Atlas of protein sequence and structure, Vol. 5, Supplement 3*, National Biomedical Research Foundation, Washington, DC., 1972.

Dayhoff, M. O.; Schwartz, R. M.; Orcutt, B. C. A model of evolutionary change in proteins. In *Atlas of protein sequences and structure*, Dayhoff, M. O., ed.; National Biomedical Research Foundation: Silver Spring, MD, Washington, DC, 1978; pp. 345-352.

Devi. M. S.; Ahmed, I.; Khan, A. A.; Rahman, S. A.; Alvi, A.; Sechi, L. A.; Ahmed, N. Genomes of *Helicobacter pylori* from native Peruvians suggest admixture of ancestral and modern lineages and reveal a western type cag-pathogenicity islands. *BMC Genomics* 2006, 7, 191.

Dickerson, R. E. The structure of cytochrome c and the rates of molecular evolution. *J Mol Evol* 1971, 1, 26-45.

Dopazo, J. Estimating errors and confidence intervals for branch lengths in phylogenetic trees by a bootstrap, approach. *J Mol Evol* 1994, 38, 300-304.

Edwards, A.W. F. *Likelihood*, Cambridge University Press: Cambridge, 1972.

Edwards, A.W. F.; Cavalli-Sforza, L. L. The reconstruction of evolution. *Heredity* 1963, 18, 553.

Edwards, S. V.; Liu, L.; Pearl, D. K. High-resolution species trees without concatenation. *Proc Natl Acad Sci USA* 2007, 104, 5936-5941.

Efron, B. *The jackknife, the bootstrap and other resampling plans*. Society for Industrial and Applied Mathematics: Philadelphia, 1982.

Erdös, P.L.; Steel, M.; Szekeley, L.; Warnow, T. Inferring big trees from short sequences. *Proceedings of International Congress on Automata Languages and Programming*, 1997.

Estabrook, G. F.; McMorris, F. R.; Meacham, C. A. Comparison of undirected phylogenetic trees based on subtrees of 4 evolutionary units. *Syst Zool* 1985, 34, 193-200.

Farris, J. S. A successive approximations approach to character weighting. *Syst Zool* 1969, 18, 374-385.

Farris, J. S. Estimating phylogenetic trees from distance distance matrices. *Am Nat* 1972, 106, 645-668.

Farris, J. S. Inferring phylogenetic trees from chromosome inversion data. *Syst Zool* 1978, 27, 275-284.

Felsenstein, J. Cases in which parsimony or compatibility methods will be positively misleading. *Syst Zool* 1978, 27, 401-410.

Felsenstein, J. Alternative methods of phylogenetic inference and their interrelationship. *Syst Zool* 1979, 28, 49-62.

Felsenstein, J. Evolutionary trees from DNA sequences: A maximum likelihood approach. *J Mol Evol* 1981a, 17, 368-376.

Felsenstein, J. A. likelihood approach to character weighting and what it tells us about parsimony and compatibility. *Biol J Linnaean Soc* 1981b, 16, 183-196.

Felsenstein, J. Confidence limits on phylogenies: An approach using the bootstrap. *Evolution* 1985, 39, 783-791.

Felsenstein, J. Phylogenies from molecular sequences: Inference and reliability. *Annu Rev Genet* 1988, 22, 521-565.

Felsenstein, J.; Churchill G. A. A hidden Markov model approach to variation among sites in rate of evolution. *Mol Biol Evol* 1996, 13, 93-104.

Felsenstein, J. *Inferring phylogenies*, Sinauer Associates: Sunderland, Massachusets, 2004.

Fitch, W. M. Toward defining the course of evolution: Minimum change for a specific tree topology. *Syst Zool* 1971, 20, 406-416.

Fitch, W. M. On the problem of discovering the most parsimonious tree. *Am Nat* 1977, 111, 223-257.

Fitch, W. M. Homology: A personal view on some of the problems. *Trends Genet* 2000, 16, 227-231.

Fitch, W. M.; Margoliash, E. Construction of phylogenetic trees. *Science* 1967, 155, 279-284.

Fitch, W. M.; Markowitz, E. An improved method for determining codon variability in a gene and its application to the rate of fixation of mutations in evolution. *Biochem Genet* 1970, 4, 579-593.

Frati, F.; Simon, C.; Sullivan, J.; Swofford, D. L. Gene evolution and phylogeny of the mitochondrial cytochrome oxidase gene in *Collembola*. *J Mol Evol* 1997, 44, 145-158.

Gasquel, P. BIONJ: An improved version of the NJ algorithm based on a simple model of sequence data. *Mol Biol Evol* 1997, 14, 685-695.

Goldman, N. Statistical tests of models of DNA substitution. *J Mol Evol* 1993, 36, 182-198.

Goldman, N.; Yang, Z. A codon-based model of nucleotide substitution for protein-coding DNA sequences. *Mol Biol Evol* 1994, 11, 725-736.

Goloboff, P. A.; Farris, S.; Nixon, K. TNT (Tree analysis using New Technology) (BETA) ver. Xxx, Tucumán, Argentina, 2000.

Goodman, M.; Czelusniak, J.; Moore, G. W.; Romero-Herrera, A. E.; Matsuda, G. Fitting the gene lineage into its species lineage: A parsimony strategy illustrated by cladograms constructed from globin sequences. *Syst Zool* 1979, 28, 132-168.

Guigó, R.; Muchnik, I.; Smith, T. Reconstruction of ancient molecular phylogeny. *Mol Phy Evol* 1996, 6, 189-213.

Guindon, S.; Gasquel, O. A simple, fast, and accurate algorithm to estimate large phylogenies by maximum likelihood. *Syst Biol* 2003, 52, 696-704.

Hasegawa, M.; Kishino, H.; Yano, T. Dating of the human-ape splitting by a molecular clock of mitochondrial DNA. *J Mol Evol* 1985, 22, 160-174.

Hastings, W. K. Monte Carlo sampling methods using Markov chains and their applications. *Biometrika* 1970, 57, 97-109.

Hedges, S. B. Early life (book review). *Q Rev Biol* 2002, 77, 445.

Hendy, M. D.; Penny, D. A framework for the quantitative study of evolutionary trees. *Syst Zool* 1989, 38, 297-309.

Hillis, D. M. Approaches for assessing phylogenetic accuracy. *Syst Biol* 1995, 44, 3-16.

Hillis, D. M.; Bull, J. J.; White, M. E.; Badgett, M. R.; Molineux, I. J. Experimental phylogenetics: Generation of a known phylogeny. *Science* 1992, 255, 589-592.

Hillis, D. M.; Bull, J. J. An empirical test of bootstrapping as a method for assessing confidence in phylogenetic analysis. *Syst Biol* 1993, 42, 182-192.

Hillis, D. M.; Huelsenbeck, J. P. To tree the truth: Biological and numerical simulation of phylogeny. *Soc Gen Physiol Ser* 1994, 49, 55-67.

Hillis, D. M.; Huelsenbeck, J. P.; Swofford, D. L. Hobgoblin of phylogenetic information. *Nature* 1994, 369, 363-364.

Holland, J. H. *Adaptation in natural and artificial systems*, University of Michigan Press: Ann Arbor, 1975.

Huelsenbeck, J. P. Is the Felsenstein zone a fly trap? *Syst Biol* 1997, 46, 69-74.

Huelsenbeck, J. P.; Crandall, K. A. Phylogeny estimation and hypothesis testing using maximum likelihood. *Annu Rev Ecol Syst* 1997, 28, 437-466.

Huelsenbeck, J. P.; Rannala, B. Frequentist properties of Bayesian posterior probabilities of phylogenetic trees under simple and complex substitution models. *Syst Biol* 2004, 53, 904-913.

Huelsenbeck, J. P.; Rannala, B.; Larget, B. A Bayesian framework for the analysis of cospeciation. *Evolution* 2000a, 54, 353-364.

Huelsenbeck, J. P.; Rannala, B.; Masly, J. P. Accomodating phylogenetic uncertainty in evolutionary studies. *Science* 2000b, 288, 2349-2350.

Huelsenbeck, J. P.; Ronquist, F.; Nielsen, R.; Bollback, J. P. Bayesian inference of phylogeny and its impact on evolutionary biology. *Science* 2001a, 294, 2310-2314.

Huelsenbeck, J. P.; Ronquist, F. MrBayes: Bayesian inference of phylogenetic trees. *Bioinformatics* 2001b, 17, 754-755.

Huelsenbeck, J. P.; Larget, B.; Miller, R. E.; Ronquist, F. Potential applications and pitfalls of Bayesian inference of phylogeny. *Syst Biol* 2002, 51, 673-688.

Huson, D. H.; Nettles, S.; Parida, L.; Warnow, T.; Yooseph, S. The disk-covering method for tree reconstruction. *Proceedings of "Algorithms and Experiments" (ALEX)*, Trento, Italy, 1998; pp. 9-11.

Huson, D. H.; Bryant, D. Application of phylogenetic networks in evolutionary studies. *Mol Biol Evol* 2006, 23, 254-267.

Jin, G.; Nakhleh, L.; Snir, S.; Tuller, T. Maximum likelihood of phylogenetic networks. *Bioinformatics* 2006, 22, 2604-2611.

Jukes, T. H.; Cantor, C. R. Evolution of protein molecules. In *Mammalian protein metabolism*, Munro, H. N., ed.; Academic Press: New York, 1969; pp. 21-132.

Kass, R. E.; Wasserman, L. *A reference Bayesian test for nested hypotheses and its relationship to the Schwartz criterion.* Department of Statistics, Carnegie Mellon University, Pittsburgh, Pennsylvania, p. 16.

Kass, R. E.; Raftery, A. E. Bayes factors and model uncertainty. *J Am Stat Assoc* 1995, 90, 773-795.

Kauser, F.; Hussain, M. A.; Ahmed, I.; Srinivas, S.; Devi, S. M.; Majeed, A. A.; Rao, K. R.; Khan, A. A.; Sechi, L. A.; Ahmed, N. Comparative genomics of *Helicobacter pylori* isolates recovered from ulcer disease patients in England. *BMC Microbiol* 2005, 5, 32.

Keane, T. M.; Naughton, T. J.; Tavers, S. A.; McInerney, J. O.; McCormack, G. P. DPRml: Distributed phylogeny reconstruction by maximum likelihood. *Bioinformatics* 2005, 21, 969-974.

Kelchner, S. A.; Thomas M. A. Model use in phylogenetics: Nine key questions. *Trends Ecol Evol* 2006, 22, 87-94.

Kim, J. General inconsistency conditions for maximum parsimony: Effects of branch lengths and increasing numbers of taxa. *Syst Biol* 1996, 45, 363-374.

Kimura, M. A. simple method for estimating evolutionary rate of base substitutions through comparative studies of nucleotide sequences. *J Mol Evol* 1980, 16, 111-120.

Kimura, M. *The neutral theory of molecular evolution.* Cambridge University Press: Cambridge, 1983.

Klein, J.; Parlak, U.; Ozyoruk, F.; Christensen, L. S. The molecular epidemiology of Foot-and-Mouth disease virus serotypes A and O from 1998 to 2004 in Turkey. *BMC Vet Res* 2006, 2, 35.

Kluge, A. G. A concern for evidence and phylogenetic hypothesis of relationships among Epicrates (Boidae, Serpentes.). *Syst Zool* 1989, 38, 7-25.

Kluge, A. G.; Wolf A. J. Cladistics: What's in a word. *Cladistics* 1993, 9, 183-199.

Larget, B.; Simon, D. L. Markov Chain Monte Carlo algorithms for the Bayesian analysis of phylogenetic trees. *Mol Biol Evol* 1999, 16, 750-759.

Lenzini, G.; Marianelli, S. Algorithms for phylogeny reconstruction in a new mathematical model. *Calcolo* 1997, 34, 1-24.

Le Quesne, W. J. The uniquely evolved character concept and its cladistic application. *Syst Zool* 1974, 23, 513-517.

Li, W. H. A statistical test of phylogenies estimated from sequence data. *Mol Biol Evol* 1989, 6, 424-435.

Li, W. H.; Graur D. *Fundamentals of molecular evolution*. Sinauer Associates, Sunderland, Massachusets. 1991.

Linder, C. R.; Moret, B. M. E.; Nakhleh, L.; Warnow, T. Network (reticulate) evolution: Biology, models, and algorithms. In: *Proceedings of the PSB04*, The Big Island: Hawaii, 2004.

Liu, L.; Pearl, D. K. Species trees from gene trees: Reconstructing Bayesian posterior distribution of a species phylogeny using estimated gene tree distributions. *Syst Biol* 2007, 56, 504-514.

Lio, P.; Goldman, N. Models of molecular evolution and phylogeny. *Genome Res* 1998, 8, 1223-1244.

Lockhart, P. J.; Larkum, A. W. D.; Steel, M. A.; Waddell, P. J.; Penny, D. Evolution of chlorophyll and bacteriochlorophyll: The problem of invariant sites in sequence analysis. *Proc Natl Acad Sci USA* 1996, 93, 1930-1934.

Maddison, W. P.; Maddison, D. R. *MacClade: Analysis of phylogeny and character evolution*, Sinauer Associates: Sunderland, Massachusetts, 1992.

Maddison, W.P. Gene trees in species trees. *Syst Biol* 1997, 46, 523-536.

Mailund, T.; Brodal, G. S.; Fagerberg, R.; Pedersen, C. N. S.; Philips, D. Recrafting the neighbor-joining method. *BMC Bioinformatics* 2006, 7, 29.

Makarenkov, V.; Legendre, P.; Desdevises, Y. Modelling phylogenetic relationships using reticulated networks. *Zool Sci* 2004, 33, 89-96.

Margush, T.; McMorris, F. R. Consensus *n*-trees. *Bull Math Biol* 1981, 43, 239-244.

Martínez-Solano, I.; Teixeira, J.; Buckley, D.; García-París, M. Mitochondrial DNA phylogeography of *Lissotriton boscai* (Caudata, Salamandridae): Evidence for old, multiple refudia in an Iberian endemic. *Mol Ecol* 2006, 15, 3375-3388.

Metropolis, N.; Rosenbluth, A. W.; Rosenbluth, M. N.; Teller, A. H.; Teller, E. Equation of state calculations by fast computing machines. *J Chem Phys* 1953, 21, 1087-1092.

Miki, H.; Okada, Y.; Hirokawa, N. Analysis of the kinesin superfamily: Insights into structure and function. *Trends Cell Biol* 2005, 15, 467-476.

Minin, V.; Abdo, Z.; Joyce, P.; Sullivan, J. Performance-based selection of likelihood models for phylogeny estimation. *Syst Biol* 2003, 52, 674-683.

Mirkin, B.; Muchnik, I.; Smitt, T. A biologically meaningful model for comparing molecular phylogenies. *J Comput Biol* 1995, 2, 493-507.

Miyamoto, M. M. Cracraft, J. Phylogenetic inference, DNA sequence analysis, and the future of molecular systematics. In *Phylogenetic analysis of DNA sequences*, Miyamoto, M.; Cracraft, J. eds., Oxford University Press: New York, 1991; pp. 3-17.

Nakhleh, L.; Jin, G.; Zhao, F.; Mellor-Crummeny, J. Reconstructing phylogenetic networks using maximum parsimony. *Proc IEEE Comput Syst Bioinform Conf (CSB2005)*, 2005, pp. 93-102.

Nei, M. Phylogenetic analysis in molecular evolutionary genetics. *Annu Rev Genet* 1996, 30, 371-403.

Nei, M.; Stephens, J. C.; Saitou N. Methods for computing the standard errors of branching points in an evolutionary tree and their application to molecular data from humans and apes. *Mol Biol Evol* 1985, 2, 66-85.

Nei, M.; Kumar, S. *Molecular Evolution and Phylogenetics*, Oxford University Press: Oxford, 2000.

Neyman, J. In: *Molecular studies of evolution: A source of novel statistical problems*, Gupta, S. S.; Yackel, J., eds., New York, 1971, pp. 1-27.

Nielsen, R.; Yang, Z. Likelihood models for detecting positively selected amino acid sites and applications to the HIV-1 envelope gene. *Genetics* 1998, 148, 929-936.

Nixon, K. C.; Carpenter, J. M. On simultaneous analysis. *Cladistics* 1996, 12, 221-241.

Page, R. D. M. Maps between trees and cladistic analysis of historical associations among genes, organisms, and areas. *Syst Biol* 1994, 43, 58-77.

Page, R. D. M.; Charleston, M. A. From gene to organismal phylogeny: Reconciled trees and the gene tree/species tree problem. *Mol Phyl Evol* 1997, 7, 231-240.

Page, R .D. M.; Holmes E. C. *Molecular evolution: A phylogenetic approach*, Blackwell Science: Oxford, 1998.

Penny, D.; Hendy, M. D.; Steel, M. A. Progress with methods for constructing evolutionary trees. *Trends Ecol Evol* 1992, 7, 73-79.

Posada, D.; Crandall, K. A. Selecting models of nucleotide substitution: An application to human immunodeficiency virus 1 (HIV-1). *Mol Biol Evol* 2001, 18, 897-906.

Posada, D.; Buckley, T. R. Model selection and model averaging in phylogenetics: Advantages of Akaike information criterion and Bayesian approaches over likelihood ratio tests. *Syst Biol* 2004, 53, 793-808.

Pybus, O. G. Model selection and the molecular clock. *PLOS Biol* 2006, 4, e151.

Raftery, A. E. Hypotheses testing and model selection. In *Markov Chain Monte Carlo in practice*, Gilks, W. R.; Richardson, S.; Spiegelhalter, D. J., eds.; Chapman and Hall: London, 1996, pp. 163-187.

Remington, D. L.; Vision, T. J.; Guilfoyle, T. J.; Reed, J. W. Contrasting modes of diversification in the Aux/IAA and ARF gene families. *Plant Physiol* 2004, 135, 1738-1752.

Rohlf, F. J. Consensus indices for comparing classifications. *Math Biosci* 1982, 59, 131-144.

Rzhetsky, A.; Nei, M. A simple method for estimating and testing minimum-evolution trees. *Mol Biol Evol* 1992, 9, 945-986.

Rzhetsky, A.; Sitnikova, T. When is it safe to use an oversimpligied substitution model in tree-making? *Mol Biol Evol* 1996, 13, 1255-1265.

Saitou, N.; Nei, M. The Neighbor-Joining method: A new method for reconstructing phylogenetic trees. *Mol Biol Evol* 1987, 4, 406-425.

Sankoff, D. D. Minimal mutation trees of sequences. *SIAM J Appl Math* 1975, 28, 35-42.

Sankoff, D. D.; Rousseau P. Locating the vertices of a Steiner tree in arbitrary space. *Mathematical Programming* 1975, 9, 240-246.

Schwarz, G. Estimating the dimension of a model. *Ann Stat* 1974, 6, 461-464.

Simmons, M. P.; Pickett, K. M.; Miya, M. How meaningful are Bayesian support values? *Mol Biol Evol* 2004, 21, 188-199.

Sitnikova, T. Bootstrap method of interior-branch test for phylogenetic trees. *Mol Biol Evol* 1996, 13, 605-611.

Sokal, R. R.; Michener C.D. A statistical method for evaluating systematic relationships. *Univ Kan Sci Bull* 1958, 38, 1409-1438.

Sourdis, J.; Nei, M. Relative efficiencies of the maximum parsimony and distance-matrix methods in obtaining the correct phylogenetic tree. *J Mol Evol* 1988, 5, 298-311.

Smith, A. D.; Lui, T. W. H.; Tillier, E. R. M. Empirical models for substitution in ribosomal RNA. *Mol Biol Evol* 2004, 21, 419-427.

Spaethe, J.; Briscoe, D. A. Early duplication and functional diversification of the opsin gene family in insects. *Mol Biol Evol* 2004, 21, 1583-1594.

Steel, M. Should phylogenetic models be trying to fit an elephant? *Trends Genet* 2005, 21, 307-309.

Steel, M.; Penny, D. Parsimony, Likelihood, and the role of models in molecular phylogenetics. *Mol Biol Evol* 2000, 17, 839-850.

Strimmer, K. S.; von Haeseler, A. Quartet-puzzling: A quartet maximum-likelihood method for reconstructing tree topologies. *Mol Biol Evol* 1996, 13, 964-969.

Studier, J. A.; Keppler, K. J. A note on the Neighbor-Joining method of Saitou and Nei. *Mol Biol Evol* 1988, 5, 729-731.

Sullivan, J.; Joyce, P. Model selection in phylogenetics. *Annu Rev Ecol Evol Syst* 2005, 36, 445-466.

Swofford, D. L.; Begle, D. P. *PAUP: Phylogenetic analysis using parsimony. Version 3.1.1. User's manual*, Laboratory of Molecular Systematics: Smithsonian Institution, Washington, DC, 1993.

Swofford, D. L.; Maddison, W. P. Reconstructing ancestral character states under Wagner parsimony. *Math Biosci* 1987, 87, 199-299.

Swofford, D. L.; Olsen, G. J.; Waddel, P. J.; Hillis, D. M. *Phylogenetic inference*. In *Molecular systematics*, 2nd ed.; Hillis, D. M.; Moritz, C.; Mable, B. K., eds., Sinauer Associates: Sunderland, Massachusetts, 1996, pp. 407-514.

Sušnik, S.; Snoj, A.; Wilson, I. F.; Mrdak, D.; Weiss, S. Historical demography of brown trout (*Salmo trutta*) in the Adriatic drainage including the putative *S. letnica* endemic to Lake Ohrid. *Mol Phylogenet Evol* 2006, 44, 63-76.

Takahashi, K.; Nei, M. Efficiencies of fast algorithms of phylogenetic inference under the criteria of maximum parsimony, minimum evolution and maximum likelihood when a large number of sequences are used. *Mol Biol Evol* 2000, 17, 1251-1258.

Takezaki, N.; Nei, M. Inconsistency of the maximum parsimony method when the rate of nucleotide substitution is constant. *J Mol Evol* 1994, 39, 210-218.

Tamura, K.; Nei, M.; Kumar, S. Prospects for inferring very large phylogenies by using the neighbor-joining method. *Proc Natl Acad Sci USA* 2004, 101, 11030-11035.

Theobald, D. L.; Wuttke, D. S. Divergent evolution within protein superfolds inferred from profile-based phylogenetics. *J Mol Biol* 2005, 354, 722-737.

Tierney, L. Markov chains for exploring posterior distributions (with discussion). *Ann Stat* 1994, 22, 1701-1762.

Tillier, E. R. M.; Collins, R. A. Neighbor-joining and maximum likelihood with RNA sequences: Addressing the interdependence of sites. *Mol Biol Evol* 1995, 12, 7-15.

Thorne, J.; Kishino, H.; Painter, J. S. Estimating the rate of evolution of the rate of molecular evolution. *Mol Biol Evol* 1998, 15, 1647-1657.

Tuffley, C.; Steel, M. A. Modelling the covarion hypothesis of nucleotide substitution. *Math Biosci* 1998, 147, 63-91.

Uzzel, T.; Corbin, K. Fitting discrete probability distributions to evolutionary events. *Science* 1971, 172, 1089-1096.

Wakeley, J. Substitution rate variation among sites and the estimation of transition bias. *Mol Biol Evol* 1994, 11, 436-442.

Waterman, M. S.; Smith, T. F. On the similarity of dendrograms. *J Theor Biol* 1978, 73, 789-800.

Weber, G.; Ohno-Machado, L.; Shieber, S. Representation in stochastic search for phylogenetic tree reconstruction. *J Biomed Inform* 2006, 39, 43-50.

Whelan, S.; Lio, P.; Goldman, N. Molecular phylogenetics state of the art methods for looking into past. *Trends Genet* 2001, 17, 262-272.

Yang, Z. Estimating the pattern of nucleotide substitution. *J Mol Evol* 1994a, 39, 105-111.

Yang, Z. Statistical properties of the maximum likelihood method of phylogenetic estimation and comparison with distance matrix methods. *Syst Biol* 1994b, 43, 329-342.

Yang, Z. Molecular clock. In *Oxford Encyclopedia of Evolution*, Pagel, M., Ed.; Oxford University Press: Oxford, 2002, pp. 747-750.

Yang, Z.; Goldman, N.; Friday, A. Maximum likelihood trees from DNA sequences: A peculiar statistical estimation problem. *Syst Biol* 1994, 44, 384-399.

Yang, Z.; Kumar, S.; Nei, M. A new method of inference of ancestral nucleotide and amino acid sequences. *Genetics* 1995, 141, 1641-1650.

Yang, Z.; Nielsen, R. Estimating synonymous and nonsynonymous substitution rates under realistic evolutionary models. *Mol Biol Evol* 2000, 15, 32-43.

Yoo, M. J.; Albert, V. A.; Soltis, P. S.; Soltis, D. E. Phylogenetic diversification of glycogen synthase kinase 3/SHAGGY-like kinase genes in plants. *BMC Plant Biol* 2006, 6, 3

Zharkikh, A.; Li, W. H. Inconsistency of the maximum-parsimony method: The case of five taxa with a molecular clock. *Syst Biol* 1993, 42, 113-125.

Zhang, C. Y.; Wei, J. F.; He, S. H. Adaptive evolution of the spike gene of SARS coronavirus: Changes in positively selected sites in different epidemic groups. *BMC Microbiol* 2006, 6, 88.

Zozio, T.; Allix, C.; Gunal, S.; Saribas, Z.; Alp, A.; Durmaz, R.; Fauville-Dufaux, M.; Rastogi, N.; Sola, C. Genotyping of *Mycobacterium tuberculosis* clinical isolates in two cities of Turkey: Description of a new family of genotypes that is phylogeographically specific for Asia Minor. *BMC Microbiol* 2005, 5, 44.

Zuckerkandl, E.; Pauling, L. Evolutionary divergence and convergence in proteins. In *Evolving genes and proteins*, Bryson, V.; Vogel, H., eds., Academic Press: New York, 1965, pp. 97-166.

CHAPTER 10

NEW INSIGHTS INTO THE ROLE OF PROTEIN PRENYLATION AND PROCESSING IN PLANT DEVELOPMENT, PHYTOHORMONE SIGNALING, AND SECONDARY METABOLISM

DRING N. CROWELL and DEVARAJAN THANGADURAI

CONTENTS

10.1   Introduction... 252

10.2   Cellular Mechanisms for Prenylation and Post-Prenylation
Processing of Plant Proteins 254

10.3   Identification and Functional Characterization of Prenylated
Plant Proteins 259

10.4   ROPs... 262

10.5   RABs... 263

10.6   ANJ1 263

10.7   V-SNARES 263

10.8   AIG1 264

10.9   APETALA1 265

10.10 NAP1 267

10.11 MUBs.. 267

10.12 AUX2-11 268

10.13 Other Biological Roles for Protein Prenylation in Plants................. 269

Keywords 270

References.. 271

10.1   INTRODUCTION

Protein prenylation is the posttranslational, covalent attachment of an isoprenoid lipid to a protein. Usually, prenylated proteins are modified by thioether linkage of a carboxyl-terminal cysteine residue to a fifteen carbon farnesyl or a twenty carbon geranylgeranyl group (Clarke, 1992; Zhang and Casey, 1996; Randall and Crowell, 1999; Rodríguez-Concepción et al., 1999; Crowell, 2000; Galichet and Gruissem, 2003). Three distinct protein prenyltransferases, with different substrate preferences, have been described in plants (e.g., Arabidopsis, tomato, tobacco, pea), animals, and yeast. Protein farnesyltransferase (PFT) is a heterodimeric enzyme consisting of α- and β- subunits. This enzyme catalyzes the transfer of a farnesyl moiety from farnesyl diphosphate (FPP) to the cysteine residue of a carboxyl-terminal CaaX motif, where "C" is cysteine, "a" is an aliphatic amino acid, and "X" is usually methionine, glutamine, alanine, cysteine or serine (Reid et al., 2004). PFT is a zinc metalloenzyme and requires magnesium for activity (Clarke, 1992; Zhang and Casey, 1996; Crowell, 2000). Protein geranylgeranyltransferase type I (PGGTI) is also a heterodimeric enzyme consisting of an α - subunit identical to that of PFT and a distinct β-subunit. This enzyme catalyzes the transfer of a geranylgeranyl moiety from geranylgeranyl diphosphate (GGPP) to the cysteine residue of a carboxyl-terminal CaaX motif, where " X" is leucine or isoleucine (Reid et al., 2004). Like PFT, PGGTI is a zinc metalloenzyme, but does not require magnesium for activity (Clarke, 1992; Zhang and Casey, 1996; Crowell, 2000). Protein geranylgeranyltransferase type II (PGGTII), or Rab geranylgeranyltransferase, consists of distinct α - and β- subunits and catalyzes the transfer of a geranylgeranyl moiety from GGPP to one or two cysteine residues at the carboxyl terminus of a Rab GTPase that is associated with the Rab Escort Protein (REP). Rab GTPases typically have a CCXX, XCXC, XCCX, or XXCC motif at the carboxyl terminus, although some possess a carboxyl-terminal CaaX motif (Clarke, 1992; Zhang and Casey, 1996; Crowell, 2000; Leung et al., 2006; Leung et al., 2007).

Proteins modified in the cytoplasmic compartment by either PFT or PGGTI undergo further processing in the endoplasmic reticulum (Clarke, 1992; Zhang and Casey, 1996; Crowell, 2000). First, the three amino acids at the carboxyl terminus are proteolytically removed by one of two specific zinc metalloproteases (CaaX proteases) and, second, the prenylcysteine residue at the new carboxyl terminus is methylated by a specific prenylcysteine methyltransferase (PCMT; Plate 10.I) (Clarke et al., 1988; Gutierrez et al., 1989; Hancock et al., 1989; Ong et al., 1989; Kawata et al., 1990; Yamane et al., 1990; Fujiyama et al., 1991; Hrycyna et al., 1991; Clarke, 1992; Sapperstein et al., 1994; Fukada, 1995; Boyartchuk et al., 1997; Tam et al., 1998; Crowell, 2000). Some, but not all, Rab proteins are also modified by proteolysis and/or carboxyl methylation (Wei *et al.*, 1992; Leung *et al.*, 2006; Leung *et al.*, 2007). In *Saccharomyces cerevisiae*, the protein products of the *STE24* and *RCE1* genes catalyze the proteolysis of prenylated CaaX sequences, and the protein product of the *STE14* gene catalyzes the methylation of the resulting carboxyl-terminal prenylcysteine residue (Hrycyna and Clarke, 1990; Hrycyna *et al.*, 1991; Hrycyna and Clarke, 1992; Ashby *et al.*, 1993; Sapperstein *et al.*, 1994; Boyartchuk *et al.*, 1997; Ashby, 1998; Romano *et al.*, 1998; Schmidt *et al.*, 1998; Tam *et al.*, 1998; Desrosiers *et al.*, 1999; Trueblood *et al.*, 2000). In some cases, prenylated proteins are also palmitoylated at upstream cysteine residues via thioester linkage (Hancock *et al.*, 1989).

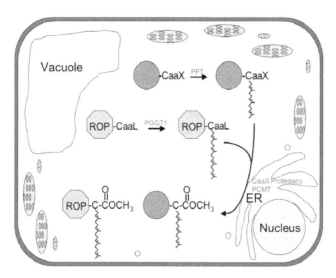

PLATE 10.I   Prenylation of proteins with carboxyl terminal CaaX motifs in plant cells. A substrate of PFT is represented by a blue circle. A substrate of PGGT I (ROP) is represented by a green octagon. PFT, PGGT I, CaaX proteases, and PCMT are indicated in red lettering.

Prenylation is necessary for protein-membrane interactions and, in some cases, protein-protein interactions (Clarke, 1992; Zhang and Casey, 1996; Crowell, 2000). Examples of protein-protein interactions that are prenylation-dependent include the interactions of 1) heterotrimeric G-protein α - subunits with $\beta\gamma$-subunits (Dietrich *et al.*, 2003), 2) yeast Ras2 with adenylate cyclase (Kuroda *et al.*, 1993), 3) RAB6 with RAB GDI (GDP-dissociation inhibitor) (Beranger *et al.*, 1994), and 4) human K-ras4(B) with hSOS1 (a guanine nucleotide exchange factor, or GEF) (Porfiri *et al.*, 1994). However, prenylation is often necessary but insufficient for protein-membrane or protein-protein interactions. Carboxyl terminal proteolysis, methylation, and either an upstream polybasic domain or fatty-acylated cysteine residue, are often required for the functions and interactions of prenylated proteins with membranes and/or other proteins (Hancock *et al.*, 1989; Hancock *et al.*, 1991; Sapperstein *et al.*, 1994; Parish *et al.*, 1995; Parish *et al.*, 1996; Bergo *et al.*, 2000; Rodríguez-Concepción *et al.*, 2000; Chen *et al.*, 2000; Bergo *et al.*, 2001; Bergo *et al.*, 2002; Bergo *et al.*, 2004; Bergo *et al.*, 2004). Indeed, in some cases, carboxyl methylation and S-acylation have been shown to regulate the localization and function of prenylated proteins, which is not surprising given the reversibility of these modifications. For example, prenylcysteine methylation is necessary for efficient plasma membrane localization of a prenylated calmodulin-like protein from petunia (CaM53) (Rodríguez-Concepción *et al.*, 2000). Prenylcysteine methylation was also recently shown to be necessary for the membrane association of Arabidopsis ROP proteins and for negative regulation of ABA signaling (Huizinga D.H., Omosegbon O., Crowell D.N., unpublished data). In another example, activation of a geranylgeranylated ROP GTPase from Arabidopsis (AtROP6) in the presence of GTPγS was shown to correlate with fatty acylation and association of the protein with the detergent-resistant membrane fraction (lipid raft fraction) (Sorek *et al.*, 2007). This is an interesting result in view of the fact that most prenylated proteins are excluded from lipid rafts (Melkonian *et al.*, 1999). But proteolysis and carboxyl methylation are not, in all cases, required for proper localization and function of prenylated proteins. In a recent report, postprenylation processing was shown to be necessary for the localization of farnesylated Ras, but not geranylgeranylated Rho GTPases in mammalian cells (Michaelson *et al.*, 2005).

10.2 CELLULAR MECHANISMS FOR PRENYLATION AND POST-PRENYLATION PROCESSING OF PLANT PROTEINS

10.2.1 *PROTEIN PRENYLTRANSFERASES*

Protein prenylation in plants was first demonstrated in suspension-cultured tobacco cells (Randall *et al.*, 1993). In this study, *in vivo* protein prenylation was detected by [^{14}C]mevalonate-labeling of tobacco cultures and *in vitro* protein prenylation confirmed the presence of PFT and PGGTI activities in cell free extracts of cultured tobacco cells (Randall *et al.*, 1993). PFT, PGGTI, and PGGTII activities and/or subunits were subsequently identified and characterized from pea (*Pisum sativum*) (Yang *et al.*, 1993; Qian *et al.*, 1996; Zhou *et al.*, 1997), tobacco (Zhou *et al.*, 1996), tomato (*Lycopersicon esculentum*) (Loraine *et al.*, 1996; Schmitt *et al.*, 1996; Yalovsky *et al.*, 1997), and Arabidopsis (Cutler *et al.*, 1996; Pei *et al.*, 1998; Running *et al.*, 2004; Johnson *et al.*, 2005). In

Arabidopsis, the *ERA1* (*ENHANCED RESPONSE TO ABA 1*, At5g40280) gene codes for the β-subunit of PFT (Cutler *et al.*, 1996; Pei *et al.*, 1998). Knock-out mutations in this gene cause an enhanced response to abscisic acid (ABA) in seed germination and stomatal closure assays, suggesting that at least one farnesylated protein functions as a negative regulator of ABA signaling (Cutler *et al.*, 1996; Pei *et al.*, 1998). The *era1* phenotype is associated with increased anion channel activation following ABA treatment of guard cells and is sufficient to cause reduced water loss during drought stress (Pei *et al.*, 1998). Indeed, transgenic canola plants transformed with an *ERA1* antisense construct under the control of the drought-inducible rd29A promoter were drought tolerant with respect to seed yield, but without effects on seed composition (Wang *et al.*, 2005). Subsequently, et a1 mutants were found to be allelic to *wiggum* mutants, which exhibit enlarged meristems and supernumerary floral organs (Running *et al.*, 1998; Bonetta *et al.*, 2000; Yalovsky *et al.*, 2000; Ziegelhoffer *et al.*, 2000). Thus, protein farnesylation also has a role in meristem development (i.e., negative regulation of meristem size). Similarly, loss-of-function mutations in the Arabidopsis *PLP* (*PLURIPETALA*, At3g59380) gene, which encodes the shared α-subunit of PFT and PGGT I, cause dramatically larger meristems with disruptions in the pattern of cell layers and flowers with significantly greater numbers of floral organs, especially petals (Running *et al.*, 2004). Thus, protein prenylation is essential for normal control of cellular proliferation within shoot and floral meristems. In contrast, knock-out mutations in the *GGB* (*GERANYLGERANYL-TRANSFERASE BETA*, At2g39550) gene, which encodes the β-subunit of PGGTI, did not cause any developmental phenotypes (Johnson *et al.*, 2005). However, *ggb* plants exhibited an enhanced response to ABA in stomatal closure assays, which correlated with a reduced rate of water loss from excised leaves. Interestingly, *ggb* plants did not exhibit an enhanced response to ABA in seed germination assays (Johnson *et al.*, 2005). Knock-out mutations in the *GGB* gene also caused an enhanced response to auxin-induced lateral root formation, but auxin inhibition of primary root growth was unaffected. Thus, PGGTI negatively regulates ABA signaling in stomata, but not seeds, and negatively regulates auxin-induced lateral root initiation, but does not affect auxin inhibition of primary root growth (Johnson *et al.*, 2005). These results establish roles for geranylgeranylated proteins in tissue-specific ABA and auxin responses. *ERA1*, *PLP*, and *GGB* are all single-copy genes in Arabidopsis and functional redundancy has been demonstrated for PFT and PGGTI (Trueblood *et al.*, 1993). For instance, the floral phenotype of *plp* plants is much more severe than that of *era1* plants, suggesting that PGGTI compensates for loss of PFT in *era1* plants (Bonetta *et al.*, 2000; Yalovsky *et al.*, 2000; Ziegelhoffer *et al.*, 2000; Running *et al.*, 2004). In addition, overexpression of the *GGB* gene partially suppressed the et a1 phenotype (Johnson *et al.*, 2005).

The Arabidopsis genome contains two PGGTII α-subunit (At4g24490, At5g41820) and two PGGTII β-subunit genes (At3g12070, At5g12210). Homozygous mutants with T-DNA insertions in either of the two PGGTII α-subunit genes lacked an obvious developmental phenotype, suggesting that the two genes are functionally redundant (D.N. Crowell, unpublished observations). These mutants have been crossed and the F1 plants, which are expected to produce double homozygotes at a frequency of 0.0625 (1/16) in the F2 generation, are currently being analyzed. Given the important role of Rab proteins in the secretory pathway, PGGTII double homozygotes are expected to be embryolethal. The Arabidopsis genome also contains one Rab Escort Protein gene (At3g06540). The Arabidopsis REP gene is expressed primarily in leaves and flowers and its protein product has been shown to interact with Rab3A *in vitro* (Wojtas *et al.*, 2007).

Recently, protein polyisoprenylation was reported in Arabidopsis (Gutkowska *et al.*, 2004). [^{3}H]mevalonate-labeling and fractionation of Arabidopsis seedlings revealed labeled proteins in nuclear, chloroplast, mitochondrial, ER, and cytoplasm/light vesicle fractions, although cross-contamination could account for the appearance of labeled proteins in some of these fractions. Nevertheless, methyl iodide treatment released primarily farnesol and geranylgeraniol from these fractions, consistent with the cysteinyl thioether linkage typical of prenylated proteins. Treatment with KOH, on the other hand, released polyprenols, including Pren-9, Pren-10, Dolichol-14, -15, -16, and -17, suggesting ester linkage of long chain polyisoprenoids to proteins in all fractions except the nuclear fraction. The functional significance of, and prenyltransferases responsible for, protein polyisoprenylation remain unknown (Gutkowska *et al.*, 2004). Polyisoprenylated proteins in plant mitochondria and chloroplasts have been reported but the identities and functions of these proteins remain elusive (Shipton *et al.*, 1995; Parmryd *et al.*, 1997; Parmryd *et al.*, 1999).

10.2.2 PROCESSING OF PRENYLATED PROTEINS IN PLANTS

As described above, most prenylated proteins undergo further processing, which includes proteolysis and carboxyl methylation. In Arabidopsis, orthologs of the *S. cerevisiae STE24* and *RCE1* genes have been identified (Bracha *et al.*, 2002; Cadiñanos *et al.*, 2003). These genes, called *AtSTE24* (At4g01320) and *AtFACE-2* (At2g36305), respectively, encode CaaX proteases that cleave the "aaX" portion of the CaaX motif after prenylation of the cysteine residue. Two prenylcysteine methyltransferase (PCMT) genes corresponding to the *S. cerevisiae STE14* gene have also been identified (Crowell *et al.*, 1998; Crowell and Kennedy, 2001; Narasimha Chary *et al.*, 2002). These genes, called *AtSTE14A*

(At5g23320) and *AtSTE14B* (At5g08335), encode prenylcysteine methyltrans-ferases that catalyze the methylation of biologically relevant prenylcysteine substrates (i.e., farnesylcysteine and geranylgeranylcysteine, but not geranyl-cysteine) using S-adenosyl-L-methionine as a methyl donor. However, the two enzymes exhibit dramatically different kinetic properties. Specifically, the *At-STE14B* gene encodes an enzyme with lower apparent K_ms for prenylcysteine substrates and higher specific activities than the *AtSTE14A*-encoded enzyme (Narasimha Chary *et al.*, 2002). Furthermore, *AtSTE14A* and *AtSTE14B* exhibit different patterns of expression in the tissues and organs of intact Arabidopsis plants, as judged by semiquantitative RT-PCR, and promoter-GUS constructs reveal vastly different activities associated with the *AtSTE14A* and *AtSTE14B* promoters (i.e., *AtSTE14B* promoter activity is much greater and more widely distributed in Arabidopsis tissues and organs) (Narasimha Chary *et al.*, 2002).

Recently, a specific prenylcysteine methylesterase (PCME) activity was de-scribed in Arabidopsis membrane fractions and the corresponding gene identi-fied (Deem *et al.*, 2006). This gene encodes a protein with significant relatedness to sterol esterases, insect juvenile hormone esterases, and carboxyl ester lipases, and possesses a predicted trans-membrane domain and carboxyl esterase type-B serine active site. Moreover, recombinant PCME expressed in *S. cerevisiae* catalyzed the specific hydrolysis of biologically relevant prenylcysteine methyl ester substrates. Because the reversible methylation and demethylation of pre-nylated proteins could potentially regulate the membrane association and func-tion of prenylated negative regulators of ABA signaling, we analyzed transgenic Arabidopsis plants harboring a *PCMT* overexpression construct and other plants with T-DNA insertions in the *PCME* gene. In both cases, the plants exhibited significant ABA insensitivity, suggesting that 1) *PCMT* is a negative regulator of ABA signaling, 2) *PCME* is a positive regulator of ABA signaling, and 3) prenylated negative regulators of ABA signaling are incompletely methylated *in planta* (i.e., if they were fully methylated, overexpression of *PCMT* would not have resulted in an ABA insensitive phenotype) (Huizinga D.H., Omosegbon O., Crowell D.N., unpublished data). Furthermore, the *PCME* gene was shown to be ABA-inducible. These data establish a positive feedback loop whereby ABA induction of the *PCME* gene causes the demethylation and inactivation of prenylated negative regulators of ABA signaling, resulting in ABA hypersensi-tivity (Huizinga D.H., Omosegbon O., Crowell D.N., unpublished data). This makes biological sense given the long-term role of ABA in seed dormancy and drought responsiveness. Moreover, these results are consistent with the observa-tion that ABA induces the expression of ABA biosynthetic genes (Xiong *et al.*, 2002; Barrero *et al.*, 2006). Thus, ABA is under positive feedback control at the level of ABA synthesis and perception (Figure 10.1).

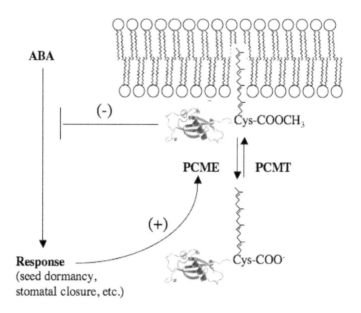

FIGURE 10.1 Model for ABA regulation of ABA responsiveness in *Arabidopsis* via induction of PCME expression and demethylation (i.e., inactivation) of prenylated negative regulators of ABA signaling. ABA promotes seed dormancy and stomatal closure, but also increases the expression of the *PCME* gene, which is a positive regulator of ABA signaling.

In a recent report, Arabidopsis plants were shown to contain a specific farnesylcysteine lyase (FCLY) activity that oxidizes farnesylcysteine to farnesal and cysteine and the corresponding gene was identified (Crowell *et al.*, 2007). Farnesal is subsequently reduced to farnesol by an NAD(P)H-dependent reductase/NAD(P)-dependent dehydrogenase. FCLY activity is necessary for the oxidation of farnesylcysteine released upon degradation of prenylated proteins. Indeed, mutants with reduced FCLY activity exhibited an enhanced response to ABA, presumably via accumulation of farnesylcysteine, which is a competitive inhibitor of PCMT (Crowell *et al.*, 2007). Thus, like *PCMT*, the *FCLY* gene is a negative regulator of ABA signaling. Unlike mammalian prenylcysteine lyases, which oxidize farnesylcysteine and geranylgeranylcysteine with comparable efficiencies (Zhang *et al.*, 1997; Tschantz *et al.*, 1999; Tschantz *et al.*, 2001; Beigneux *et al.*, 2002; Digits *et al.*, 2002), Arabidopsis FCLY exhibits specificity for farnesylcysteine, suggesting that geranylgeranylcysteine is metabolized by a different mechanism.

10.3   IDENTIFICATION AND FUNCTIONAL CHARACTERIZATION OF PRENYLATED PLANT PROTEINS

The first systematic screen for prenylated plant proteins was performed in the mid-1990s, years before the sequencing of the Arabidopsis genome and other plant genomes made *in silico* searches for proteins with carboxyl-terminal CaaX motifs possible (Biermann *et al.*, 1994). This early screening effort took advantage of the fact that 1) few, if any, *E. coli* proteins are spuriously prenylated *in vitro* using purified PFT or PGGTI or plant cell extracts containing PFT and PGGTI, and 2) protein prenylation only requires a carboxyl-terminal CaaX sequence without any requirements for secondary structure. Thus, "prenyl screening" was accomplished by introducing a plant (Arabidopsis, tobacco, or soybean) cDNA library into an *E. coli* expression vector, plating the cDNA expression library on LB-agar plates, making lifts of the colonies onto nylon filters and placing the filters, colony-side up, on fresh LB-agar plates containing IPTG (to induce colony growth and recombinant protein expression). Colonies were then lysed *in situ* by placing the filters, colony-side up, on a sheet of Whatman #1 paper saturated with 150 mM NaCl, 100 mM Tris-HCl, pH 8.0, 5 mM MgCl$_2$, 2 µg/ml DNaseI, and 50 µg/ml lysozyme, and then transferring them to a second sheet of Whatman paper saturated with 150 mM NaCl, 0.1 % SDS, 0.1 M NaOH. The filters were then neutralized on a third sheet of Whatman paper saturated with 150 mM NaCl, 100 Tris-HCl, pH 6.5, air dried, and blocked by incubation in the presence of Tris-buffered saline containing 20 mg/ml bovine serum albumin and 0.05 % Tween 20. Prenylation reactions were performed by placing the filters in a reaction mixture containing [^{3}H]FPP or [^{3}H]GGPP and a source of protein prenyltransferase activity in 50 mM Hepes, pH 7.5, 20 mM MgCl$_2$, 5 mM dithiothreitol (5 mL/filter). Unincorporated label was removed by successive washes in 50 % ethanol and 95 % ethanol, and cDNA-encoded, *in vitro* prenylated, proteins were identified using Amplify fluorographic reagent and Kodak XAR-5 film. This procedure resulted in the identification of both known and novel prenylated proteins (Biermann *et al.*, 1994; Biermann *et al.*, 1996; Dykema *et al.*, 1999). Among the known proteins identified were members of the ROP family of small GTPases (in Arabidopsis, 8 out of 11 ROP proteins are geranylgeranylated), members of the RAB family of small GTPases, and the prenylated plant molecular chaperone ANJ1 (an ortholog of the protein encoded by the *E. coli dnaJ* gene). These findings implicated protein prenylation in signal transduction and cytoskeletal organization (i.e., ROP proteins), vesicle transport (i.e., RAB proteins), and the plant response to heat shock (i.e., ANJ1) (Biermann *et al.*, 1996; Dykema *et al.*, 1999). The previously unknown prenylated proteins that were identified included a prenylated v-SNARE protein related to the Ykt6p protein of *S. cerevisiae*, a prenylated guanylate-binding protein called AIG1 implicated in the response of Arabidopsis plants containing the *RPS2* resistance gene to bacterial pathogens carrying the *avrRpt2* avirulence gene, and a family of prenylated transition metal-binding proteins (i.e., ATFP3) (Dykema *et al.*, 1999). These findings suggest roles for protein prenylation in vesicle docking and fusion (i.e., Ykt6p), plant disease responses (i.e., AIG1), and regulation of heavy metal homeostasis (i.e., ATFP3).

With the sequencing of various plant genomes, many gene products with carboxyl-terminal CaaX sequences have been identified as known or putative prenylated proteins (Rodríguez-Concepción *et al.*, 1999; Crowell, 2000; Galichet and Gruissem, 2003; Maurer-Stroh *et al.*, 2007). These include ROP proteins, RAB proteins, Membrane-anchored Ub-fold (MUB) proteins, peroxisome biogenesis factor-19 (PEX19), G-protein γ-subunits (in Arabidopsis encoded by the *AGG1* and *AGG2* genes), nucleosome assembly protein-1 (NAP-1;1), APETALA1, S-locus cysteine-rich proteins, glutathione transporter proteins, calmodulin-like proteins, and AUX2-11 (Table 10.1). These predictions suggest that protein prenylation affects many cellular processes in addition to those described above, including protein ubiquitination, peroxisome biogenesis, nucleosome assembly, transcriptional regulation of floral meristem development, self-incompatibility, glutathione transport, calcium signaling, and auxin signaling. It is important to point out, however, that these proteins may not all be prenylated, depending on subcellular location and accessibility of the carboxyl terminus to protein prenyltransferases. What follows is a description of what is currently known about some of these proteins.

TABLE 10.1   *In silico* screening for clusters of putative prenylated plant proteins (Maurer-Stroh *et al.*, 2007).

Protein	Farnesylated (F) and/or Geranyl-geranylated (GG)	Species	Function
ROP GTPases	GG	Arabidopsis – ROP1-8, rice, cotton, etc.	Actin organization, polarized growth, hormone signaling, etc.
RAB GT-Pases	GG	Arabidopsis, tomato, etc.	Vesicle formation, transport, and docking to target membranes
CCH Copper Chaperone-related	F, GG	Arabidopsis – ATFP3, Cdl19	Heavy metal homeostasis

TABLE 10.1 *(Continued)*

Protein	Farnesylated (F) and/or Geranyl-geranylated (GG)	Species	Function
Ub-fold proteins	F, GG	Arabidopsis, rice, etc.	Unknown
Synapto-brevin-like SNARE protein (Ykt6-like)	GG	Arabidopsis – ATGP1, tobacco – NTGP1	Vesicle docking and fusion to target membranes
DnaJ-like heat shock chaperones	F	Arabidopsis – AtJ2, AtJ3; Atriplex, tobacco	Heat shock molecular chaperones
Pex19-like	F	Arabidopsis	Peroxisome biogenesis
G protein γ subunit	GG	Arabidopsis – AGG1, AGG2	ABA, auxin signaling, and plant defense
Nucleosome assembly protein, NAP1-like	F	Arabidopsis	Cell division and expansion
AIG1-like; plant resistance to bacteria	F, GG	Arabidopsis	*RPS2* – mediated plant defense against bacteria carrying *avrRpt2*
Floral homeotic MADS box protein APETALA1	F	Arabidopsis	Floral meristem identity, sepal, and petal identity (A function)
S-locus cysteine-rich protein	F	*Brassica oleracea* – SCR3, SCR16	Self-incompatibility
Glutathione, peptide transporter, Isp4-like protein	F, GG	Arabidopsis	Glutathione, peptide transport
Calmodulin-related	GG (can also be farnesylated *in vitro*)	Petunia – CaM53; rice – CaM61	Calcium signaling
AUX2-11 (IAA4), AUX/IAA protein	GG	Arabidopsis	Negative regulation of auxin signaling

10.4   ROPS

In Arabidopsis, there are eleven *ROP* genes (Yang, 2002; Vernoud *et al.*, 2003).
Eight of these encode proteins, called Type I ROPs (ROP1-8), that are predicted
to be geranylgeranylated (i.e., they bear a CaaL motif at the carboxyl terminus),
whereas the other three encode proteins, called Type II ROPs (ROP9-11), that
are S-acylated at two or three carboxyl-terminal cysteines (Lavy *et al.*, 2002;
Yang, 2002; Vernoud *et al.*, 2003; Lavy and Yalovsky, 2006). The prenylated,
Type I ROP proteins have been shown to be involved in many cellular processes,
including actin organization (ROP1-6: Fu *et al.*, 2001; Lemichez *et al.*, 2001; Fu
et al., 2002; Yang, 2002; Gu *et al.*, 2003; Gu *et al.*, 2005), pollen tube tip growth
(ROP1: Lin *et al.*, 1996; Li *et al.*, 1999; Fu *et al.*, 2001; Yang, 2002; Gu *et
al.*, 2003; Gu *et al.*, 2005), negative regulation of ABA signaling in guard cells
(ROP6: Lemichez *et al.*, 2001; Yang, 2002) and seeds (ROP2: Li *et al.*, 2001;
Yang, 2002), auxin-induced lateral root initiation (ROP2: Li *et al.*, 2001; Yang,
2002) and root hair formation (ROP2, ROP4, ROP6, NtRac1: Molendijk *et al.*,
2001; Jones *et al.*, 2002; Tao *et al.*, 2002; Yang, 2002), auxin-induced gene expression (ROP1, ROP3, ROP6, ROP11, NtRac1: Tao *et al.*, 2002), directional
cell expansion during organogenesis (ROP2: Fu *et al.*, 2002; Yang, 2002), embryo development (ROP2: Li *et al.*, 2001), seedling development (ROP2: Li *et
al.*, 2001), shoot growth and apical dominance (ROP2: Li *et al.*, 2001), brassino-
lide-mediated hypocotyl elongation (ROP2: Li *et al.*, 2001; Yang, 2002), H_2O_2
production (OsRac1, Cotton Rac13: Potikha *et al.*, 1999; Ono *et al.*, 2001; Yang,
2002; Molendijk *et al.*, 2004; Morel *et al.*, 2004), secondary cell wall formation (Cotton Rac13: Potikha *et al.*, 1999), and defense responses (OsRac1: Ono
et al., 2001; Suharsono *et al.*, 2002). ROPs are themselves regulated by guanine nucleotide exchange factors (GEFs), GTPase-activating proteins (GAPs),
and GDP dissociation inhibitor proteins (GDIs) (Yang, 2002; Molendijk *et al.*,
2004; Raftopoulou and Hall, 2004). In addition, a ROP GTPase was found to
be associated with active CLAVATA1 (CLV1) complexes, suggesting that CLA-
VATA3 activation of CLV1 leads to activation of ROP (Trotochaud *et al.*, 1999).
Thus, ROP GTPase(s) appear to have a role in meristem development. Tobacco
NtRac1 is activated by auxin (Tao *et al.*, 2002). Thus, auxin signaling appears
to operate, at least in part, via activation of specific ROPs, which positively
regulate auxin-induced lateral root initiation, root hair formation, and gene expression (Tao *et al.*, 2002; Tao *et al.*, 2005). In contrast, Arabidopsis ROP6 is
inactivated by ABA (Lemichez *et al.*, 2001). Thus, ABA signaling in guard cells
operates, at least in part, via inactivation of ROP6, a negative regulator of ABA
signaling (Lemichez *et al.*, 2001).

10.5   RABS

RAB proteins are members of the small GTPase family that participate in the formation, transport, and docking of vesicles to target membranes in the secretory pathway (Jürgens and Geldner, 2002; Rutherford and Moore, 2002; Deneka *et al.*, 2003; Vernoud *et al.*, 2003; Molendijk *et al.*, 2004). RABs have been shown to be necessary for docking of ER vesicles to cis-Golgi, post-Golgi vesicles to endosomes and vacuoles, retrograde Golgi vesicles to ER, and secretory vesicles to the plasma membrane (i.e., polarized secretion) (Rutherford and Moore, 2002; Jürgens and Geldner, 2002; Deneka *et al.*, 2003; Vernoud *et al.*, 2003; Molendijk *et al.*, 2004). Mutations in, and localization of, RAB proteins have implicated this large class of GTPases in numerous cellular functions, including targeted secretion of cell wall components (Preuss *et al.*, 2004; Preuss *et al.*, 2006; Samaj *et al.*, 2006), polar growth of root hairs and pollen tubes (Preuss *et al.*, 2004; Preuss *et al.*, 2006; Samaj *et al.*, 2006), sterol endocytosis (Grebe *et al.*, 2003), targeting of cargo proteins to the vacuole (Sohn *et al.*, 2003; Bolte *et al.*, 2004), and ethylene signaling (Moshkov *et al.*, 2003). RABs are also involved in plant development and adaptation to environmental stresses (Bolte *et al.*, 2000; Lu *et al.*, 2001). Prenylation and, in some cases, carboxyl methylation, are required for RAB protein localization and function (Ueda *et al.*, 2001; Vernoud *et al.*, 2003).

10.6   ANJ1

ANJ1 is a farnesylated molecular chaperone with similarity to the DnaJ protein of *E. coli* (Zhu *et al.*, 1993; Dykema *et al.*, 1999). This 47 kDa protein, is heat-shock inducible and thought to be involved in Hsp70-mediated protein folding and import into organelles at elevated temperatures (Zhu *et al.*, 1993). Orthologs of ANJ1 from a number of plant species have been shown to be prenylated (Dykema *et al.*, 1999). The Arabidopsis genome contains two genes (At5g22060, At3g44110) that are predicted to encode farnesylated DnaJ-like proteins (AtJ2 and AtJ3, respectively).

10.7   V-SNARES

One of the proteins identified in the "prenyl screen" described above was an Arabidopsis geranylgeranylated protein called ATGP1 (*Arabidopsis thaliana* geranylgeranylated protein 1) (Biermann *et al.*, 1994; Dykema *et al.*, 1999). ATGP1 exhibited signifiant relatedness to the *S. cerevisiae* v-SNARE protein Ykt6p (McNew *et al.*, 1997; Zhang and Hong, 2001). Ykt6p is a member of the docking complex that mediates fusion of ER vesicles to the Golgi apparatus, although Ykt6p has also been implicated in vacuolar vesicle fusion (Søgaard *et al.*, 1994; Ungermann *et al.*, 1999; Kweon *et al.*, 2003). Whereas the two Arabidopsis Ykt6-like proteins are geranylgeranylated (At5g58060, At5g58180), Ykt6p is farnesylated (McNew *et al.*, 1997). Nevertheless, these proteins are thought to be membrane-anchored via the isoprenoid tail because they lack a predicted trans-membrane domain like other v-SNARE proteins. v-SNARE and RAB proteins function in concert to mediate specific docking and fusion of vesicles to target membranes (Søgaard *et al.*, 1994; Clague, 1999).

10.8 AIG1

Another protein identified by "prenyl screening" was a tobacco protein called NTGP4 (*Nicotiana tabacum* geranylgeranylated protein 4). NTGP4 exhibited relatedness to the Arabidopsis AIG1 protein, which possesses a consensus CaaX motif at its carboxyl terminus (Biermann *et al.*, 1994; Dykema *et al.*, 1999). This protein has been shown to be involved in the response of Arabidopsis plants containing the *RPS2* resistance gene to bacterial pathogens carrying the *avrRpt2* avirulence gene (Reuber and Ausubel, 1996). AIG1 contains a nucleotide binding domain with sequence relatedness to that of mammalian interferon-induced guanylate binding proteins (Bourne *et al.*, 1991; Cheng *et al.*, 1991). Interestingly, the Arabidopsis genome contains two clusters of AIG1-like genes. The cluster on chromosome 1 includes the At1g33890, At1g33900, At1g33910, At1g33930, At1g33950, At1g33960 (AIG1), and At1g33970 genes, which bear the carboxyl-terminal motifs CSIL, CNML, CNIL, CNIL, CNIL, CSIL, and CINL, respectively. The cluster on chromosome 4 includes the At4g09930, At4g09940, and At4g09950 genes, which bear the carboxyl-terminal motifs CIIL, CIIM, and CTVL, respectively. Thus, nine of the ten AIG1-like proteins are predicted to be geranylgeranylated, and one is predicted to be farnesylated. These findings implicate prenylated proteins in the plant response to pathogen attack and suggest multiple, perhaps redundant, roles for AIG-1-like proteins in plant defense (Reuber and Ausubel, 1996).

10.8.1 HEAVY METAL BINDING PROTEINS

ATFP3 (*Arabidopsis thaliana* farnesylated protein 3) was one of several related proteins identified by prenyl screening that was predicted to be farnesylated (Biermann *et al.*, 1994; Dykema *et al.*, 1999). These proteins bear at least one CXXC motif upstream of the CaaX sequence and were empirically shown to bind heavy metals, including Cu^{2+} and Zn^{2+}. ATFP3 is identical to Cdl19, which was originally identified as the product of a Cd^{2+}-responsive gene, but was later shown to be responsive to Hg^{2+}, Fe^{2+}, and Cu^{2+} as well (Suzuki *et al*., 2002). The protein product of this gene bound directly to Cd^{2+} and conferred Cd^{2+} tolerance when expressed in recombinant *S. cerevisiae* cells. A GFP-Cdl19 fusion protein expressed in tobacco BY-2 cells was plasma membrane localized, whereas a mutant GFP-Cdl19M with the cysteine of the CaaX motif changed to glycine was not. In addition, overexpression of Cdl19 in transgenic Arabidopsis plants conferred Cd^{2+} tolerance, suggesting that Cdl19 is involved in heavy metal homeostasis or detoxification (Suzuki *et al*., 2002). Three Arabidopsis genes are predicted to encode prenylated heavy metal-binding proteins; two farnesylated proteins (At5g63530, At5g24580) and one geranylgeranylated protein (At5g50740).

10.8.2   CALMODULIN-LIKE PROTEINS

Calmodulin-like proteins from petunia (CaM53) and rice (CaM61) have been shown to be prenylated, suggesting a role for protein prenylation in calcium signal transduction (Rodríguez-Concepción *et al*., 1999; Xiao *et al*., 1999). The petunia protein, called CaM53, is functional *in vitro* and in recombinant *S. cerevisiae* cells, and ectopic expression of the corresponding gene in *Nicotiana benthamiana* caused significant growth inhibition, stem thickening, leaf curling, chlorosis, and necrosis. However, a mutant CaM53 with the cysteine of the CTIL motif changed to serine did not cause stunting and necrosis, suggesting that these effects are dependent on CaM53 prenylation (Rodríguez-Concepción *et al*., 1999). CaM53 was efficiently geranylgeranylated and farnesylated *in vitro* by recombinant plant PGGTI, demonstrating isoprenoid cross-specificity for PGGTI in the presence of this protein substrate. In addition, the polybasic domain upstream of the CaaX motif of CaM53 increased the efficiency of CaM53 prenylation. GFP fused to CaM53 was plasma membrane localized, whereas GFP fused to the mutant STIL form was nuclear (Caldelari *et al*., 2001), and plasma membrane localization was shown to be dependent on both prenylation and carboxyl methylation (Rodríguez-Concepción *et al*., 1999; Rodríguez-Concepción *et al*., 2000).

10.9   APETALA1

The Arabidopsis shoot apical meristem (SAM), after the transition to reproductive development, is called an inflorescence meristem (IM). The Arabidopsis IM is indeterminate and produces secondary inflorescence meristems and floral meristems (FM) on its flanks by cell division activity within the peripheral zone

(Doerner, 1999). A floral meristem, which is determinate and develops into a single flower composed of four sepals, four petals, six stamens, and two carpels, forms on the flank of the Arabidopsis IM at sites of high *LFY* (*LEAFY*) and *AP1* (*APETALA1*) gene expression (Boss *et al.*, 2004; Putterill *et al.*, 2004). *LFY* and *AP1* both have floral meristem identity function, but *LFY* is expressed earlier in flower development than *AP1* (Hempel *et al.*, 1997; Boss *et al.*, 2004; Putterill *et al.*, 2004). However, *LFY* activates the *AP1* gene and *vice versa*, forming a positive feedback loop within the FM that maintains floral meristem identity (Liljegren *et al.*, 1999). The *TFL1* (*TERMINAL FLOWER 1*) gene represses *LFY* and *AP1* expression in the IM to preserve the indeterminate nature of the IM, whereas *LFY* and *AP1* repress *TFL1* gene expression in the floral meristem to maintain determinacy and promote the development of a flower (Liljegren *et al.*, 1999). Consistent with its floral meristem identity function, *AP1* is expressed throughout the floral meristem in stage 1 and 2 flowers (Gustafson-Brown *et al.*, 1994; Simon *et al.*, 1996). However, in stage 3 flowers, *AP1* expression becomes restricted to the first and second whorls of the developing flower, where it assumes a role in specification of sepals and petals, respectively (Gustafson-Brown *et al.*, 1994). The flowers of *ap1* mutants lack petals and show a partial conversion to inflorescence meristem fate, with bracts formed instead of sepals and secondary flowers occasionally developing in the axils of the bracts (Bowman *et al.*, 1993). AP1 is one of many MADS domain transcription factors governing flower development. The AP1 carboxyl-terminal domain is not only interesting because of its transactivation and protein interaction functions, but also because it is prenylated (Rodríguez-Concepción *et al.*, 1999; Yalovsky *et al.*, 2000). AP1 possesses a CFAA motif at the carboxyl terminus and it, or a carboxyl-terminal fragment of it, is prenylated by PFT both *in vitro* and *in vivo* (Yalovsky *et al.*, 2000). Moreover, prenylation of AP1 may be necessary for meristem identity function because overproduction of a mutant ap1 protein that cannot be prenylated (ap1mS, where the CaaX cysteine is changed to serine) in wild type Arabidopsis, or overproduction of wild type AP1 in a PFT mutant of Arabidopsis (*era1*), did not result in a terminal flower phenotype (Yalovsky *et al.*, 2000). Conversely, overproduction of wild type AP1 in wild type Arabidopsis caused the inflorescence meristem to assume floral meristem identity, resulting in a terminal flower phenotype. While these results suggest that AP1 is farnesylated and that farnesylation may be necessary for the floral meristem identity function of AP1, other data suggest that prenylation does not affect AP1 function. For example, Krizek and Meyerowitz (1996) demonstrated that the carboxyl terminus of AP1 can be replaced by the carboxyl terminus of the AG (AGAMOUS) protein, an unprenylated MADS domain transcription factor involved in stamen and carpel development, and the resulting protein was capable of rescuing an *ap1* mutant and causing a terminal flower overexpression phenotype. In addition, Ng and Yanofsky (2001) demonstrated that a VP16 transactivation domain fused to the carboxyl terminus of AP1, which disrupts the carboxyl-terminal location of the CaaX motif, result in a functional protein. These results indicate that prenylation is not required for AP1 function. Furthermore, if prenylation were required for the floral organ identity function of AP1, *plp* mutants of Arabidopsis would be expected to exhibit an apetala phenotype (no petals). However, as described above, *plp* mutants exhibit the opposite phenotype (extra petals) (Running *et al.*, 2004). Together, these results suggest that prenylation may attenuate the floral organ identity function of AP1, may be important only in the context of an AP1 carboxyl-terminal domain, and may differentially affect the floral meristem identity and floral organ identity functions of AP1. At present, it is not known how prenylation affects the biochemical properties of AP1, such as DNA binding, interactions with other MADS domain transcription factors, subnuclear localization, cell-to-cell movement, and activation of downstream target genes.

10.10   NAP1

Protein prenylation was shown to be associated with cell division in the mid-1990s (Morehead *et al.*, 1995; Qian *et al.*, 1996), but the prenylated protein(s) responsible for this association were not known. However, in a recent report, Arabidopsis nucleosome assembly protein 1 (AtNAP1;1) was shown to be a farnesylated protein that regulates cell proliferation and expansion during leaf development (Galichet and Gruissem, 2006). Farnesylated, but not unfarnesylated, AtNAP1;1 rescued a *nap1* mutant of *S. cerevisiae* but, surprisingly, AtNAP1;1 localization was not dependent on farnesylation (i.e., a mutant At-NAP1;1 with the cysteine of the CaaX motif changed to serine did not exhibit altered localization in plant cells). More importantly, ectopic overexpression of AtNAP-1;1 disrupted leaf development in a farnesylation-dependent manner (Galichet and Gruissem, 2006). Three Arabidopsis genes are predicted to encode farnesylated paralogs of NAP1 – AtNAP1;1, AtNAP1;2, and AtNAP1;3 (At4g26110, At2g19480, and At5g56950, respectively).

10.11   MUBS

A unique class of ubiquitin-fold proteins with carboxyl-terminal CaaX sequences was recently reported (Downes *et al.*, 2006). These Membrane-anchored Ub-fold proteins (MUBs) were found in Arabidopsis, rice, and a variety of animal and fungal species. In Arabidopsis, six MUBs were described. Five of these are either farnesylated (AtMUB1 and AtMUB4) or geranylgeranylated (AtMUB3,

AtMUB5, and AtMUB6), while the sixth (AtMUB2) is palmitoylated. Ubiqui-tin-fold proteins are posttranslational modifiers involved in a variety of cellular processes, including regulated protein degradation (Ub and SUMO, which com-petes with Ub; Melchior, 2000), activation of the Cullin subunit of SCF (Skip-Cullin-F-box) complexes (RUB1; Petroski and Deshaies, 2005), and autophagy (ATG-8 and ATG-12; Ohsumi, 2001). Given that MUBs have a carboxyl-ter-minal prenylcysteine, it is unlikely that they are posttranslational protein modi-fiers. However, it is possible that MUBs serve as docking sites on the plasma membrane for other proteins (Downes *et al.*, 2006).

10.12   AUX2-11

AtAUX2-11, also called IAA4, is a member of the AUX/IAA family of transcrip-tional repressors (Wyatt *et al.*, 1993). AUX/IAA proteins repress ARF-mediated transcription of auxin-responsive genes. In response to auxin, which binds the F-box subunit (TIR1) of the SCF^{TIR1} complex (Dharmasiri *et al.*, 2005), AUX/IAA proteins are ubiquitinated and degraded, allowing for transcription of auxin-regulated genes, including AUX/IAA genes (Dharmasiri and Estelle, 2002). Thus, auxin-induced ubiquitination and degradation of AUX/IAA proteins leads to the activation of auxin-regulated genes and reestablishment of an AUX/IAA pool. This negative feedback loop ensures that auxin responses are transient. AUX2-11 is unique among the AUX/IAA proteins because it has a carboxyl-terminal CGGL motif, which has been shown to be a substrate for PGGTI (Caldelari *et al.*, 2001). Interestingly, *ggb* mutants lacking PGGTI activity exhibit an enhanced response to auxin-induced lateral root initiation, possibly via inactivation of AUX2-11 and increased transcription of auxin-regulated genes (Johnson *et al.*, 2005).

10.12.1   HETEROTRIMERIC G PROTEIN γ-SUBUNITS

The Arabidopsis genome contains only one heterotrimeric G-protein α-subunit gene (*GPA1*, At2g26300), one G-protein β-subunit gene (*AGB1*, At4g34460), and two G-protein γ-subunit genes (*AGG1*, At3g63420 and *AGG2*, At3g22942). AGG1 and AGG2 are both geranylgeranylated, and AGG2 is also S-acylated (Zeng *et al.*, 2007). Consequently, both are associated with the plasma mem-brane, but AGG1 is also associated with internal membranes (Zeng *et al.*, 2007). AGG1 and AGG2 can be prenylated by both PFT and PGGTI, illustrating the functional redundancy between PFT and PGGTI in Arabidopsis, and do not re-quire a functional Gα subunit for plasma membrane localization (Zeng *et al.*, 2007). AGG1-deficient mutants exhibited decreased resistance to necrotrophic pathogens, reduced expression of the plant defensin PDF1.2, and hyposensitivity to methyl jasmonate (Trusov *et al.*, 2007). However, both AGG1- and AGG2-deficient mutants exhibited an enhanced response to auxin-induced lateral root formation, suggesting that both function to repress auxin-mediated initiation of lateral roots. As described above, *ggb* mutants lacking PGGTI exhibit an enhanced response to auxin-induced lateral root formation (Johnson *et al.*, 2005; Trusov *et al.*, 2007), a phenotype that may, in part, be due to inactivation of AGG1 and AGG2 (as well as AUX2-11). Two G-protein-coupled receptors (GCR1 and GCR2) have been described and both appear to affect ABA signaling (Pandey and Assmann, 2004; Liu *et al.*, 2007). Whereas GCR1 is a negative regulator of GPA1-mediated ABA signaling in guard cells, GCR1 acts in concert with GPA1 and AGB1 to negatively regulate ABA signaling during seed germination (Pandey and Assmann, 2004). Interestingly, *ggb* mutants lacking PGGTI exhibit an enhanced response to ABA in guard cells, but not seeds (Johnson *et al.*, 2005). This finding suggests that geranylgeranylation, and plasma membrane localization, of AGG1 and/or AGG2 may be necessary for negative regulation of GPA1-mediated ABA responses in guard cells, but not in seeds where GCR1, GPA1, and AGB1 function together to negatively regulate ABA signaling. GCR2 has been definitively shown to be an ABA receptor but, unlike GCR1, GCR2 positively regulates ABA signaling (Liu *et al.*, 2007). Given these findings, it seems likely that *AGG1* and/or *AGG2* mutants exhibit an enhanced response to ABA in guard cells, but this has not been reported.

10.13   OTHER BIOLOGICAL ROLES FOR PROTEIN PRENYLATION IN PLANTS

The results discussed above implicate protein prenylation in seed germination, seedling development, leaf development, flower development, phytohormone (ABA, Auxin, and Brassinolide) signaling, calcium signaling, actin organization, vesicle docking and fusion, H_2O_2 production, pathogen defense, secondary cell wall formation, response to heat shock, and heavy metal homeostasis and detoxification.

A role for protein prenylation in secondary metabolism was also recently reported. In *Catharanthus roseus*, which produces monoterpenoid indole alkaloids (ajmalicine, serpentine, vincristine, and vinblastine), PFT and PGGTI were found to be necessary for induction of genes in the 2-*C*-methyl-erythritol 4-phosphate pathway (MEP pathway, otherwise known as the plastidial isoprenoid biosynthetic pathway) following auxin depletion of *C. roseus* suspension cultures (Courdavault *et al.*, 2005). This conclusion was based on the observation that RNAi-mediated decreases in PFT β-subunit or PGGTI β-subunit gene expression caused a decrease in MEP pathway gene expression and monoterpenoid indole alkaloid production. In addition to supplying the isoprenoid portion of the monoterpenoid indole alkaloids in *C. roseus*, the MEP pathway is responsible for the production of *trans*-zeatin, monoterpenes (C10), diterpenes (C20), tetraterpenes (C40), phytol, carotenoids, plastoquinone-9, gibberellin, abscisic acid, and possibly the GGPP for protein geranylgeranylation (Lichtenthaler, 2000; Gerber *et al.*, 2007). The mevalonate (MVA) pathway, on the other hand, is located in the cytoplasmic compartment and is responsible for the biosynthesis of sesquiterpenes (C15), triterpenes (C30), phytosterols, and steroids. The MVA pathway also provides the FPP for protein farnesylation (Crowell, 2000).

KEYWORDS

- ABA signaling
- ABA synthesis
- Actin organization
- Arabidopsis genome
- Calmodulin-like proteins
- Carboxyl methylation
- Demethylation
- Escort protein gene
- Farnesylcysteine
- Functional redundancy
- Geranylgeranylcysteine
- Heterodimeric enzyme
- Mammalian cells
- Phytohormone signaling
- Phytosterols
- Plant development
- Plant proteins
- Postprenylation processing
- Prenyl screening
- Prenylcysteine methylation
- Protein prenylation
- Protein-membrane interactions
- Protein-protein interactions
- Proteolysis
- Secondary metabolism
- Vesicle docking

REFERENCES

Ashby, M. N. CaaX converting enzymes. *Curr Opin Lipidol* 1998, 9, 99-102.

Ashby, M. N.; Errada, P. R.; Boyartchuk, V. L.; Rine, J. Isolation and DNA sequence of the STE14 gene encoding farnesyl cysteine: Carboxyl methyltransferase. *Yeast* 1993, 9, 907-913.

Barrero, J. M.; Rodríguez, P. L.; Quesada, V.; Piqueras, P.; Ponce, M. R.; Micol, J. L. Both abscisic acid (ABA)-dependent and ABA-independent pathways govern the induction of NCED3, AAO3 and ABA1 in response to salt stress. *Plant Cell and Environ* 2006, 29, 2000-2008.

Beigneux, A.; Withycombe, S. K.; Digits, J. A.; Tschantz, W .R.; Weinbaum, C .A.; Griffey, S .M.; Bergo, M.; Casey P. J.; Young S. G. Prenylcysteine lyase deficiency in mice results in the accumulation of farnesylcysteine and geranylgeranylcysteine in brain and liver. *J Biol Chem* 2002, 277, 38358-38363.

Beranger, F.; Cadwallader, K.; Porfiri, E.; Powers, S.; Evans, T.; De Gunzburg, J.; Hancock, J. F. 1994. Determination of structural requirements for the interaction of Rab6 with RabGDI and Rab geranylgeranyltransferase. *J Biol Chem* 2002, 269, 13637-13643.

Bergo, M. O.; Ambroziak, P.; Gregory, C.; George, A.; Otto, J. C.; Kim, E.; Nagase, H.; Casey, P. J.; Balmain, A.; Young, S. G. Absence of the CAAX endoprotease Rce1: Effects on cell growth and transformation. *Mol Cell Biol* 2002, 22, 171-181.

Bergo, M. O.; Gavino, B. J.; Hong, C.; Beigneux, A. P.; Mcmahon, M.; Casey, P. J.; Young, S. G. Inactivation of Icmt inhibits transformation by oncogenic K-Ras and B-Raf. *J Clin Invest* 2004, 113, 539-550.

Bergo, M. O.; Leung, G. K.; Ambroziak, P.; Otto, J. C.; Casey, P. J.; Gomes, A. Q.; Seabra, M. C.; Young, S. G. Isoprenylcysteine carboxyl methyltransferase deficiency in mice. *J Biol Chem* 2001, 276, 5841-5845.

Bergo, M. O.; Leung, G. K.; Ambroziak, P.; Otto, J. C.; Casey, P. J.; Young, S. G. Targeted inactivation of the isoprenylcysteine carboxyl methyltransferase gene causes mislocalization of K-Ras in mammalian cells. *J Biol Chem* 2000, 275, 17605-17610.

Bergo, M. O.; Lieu, H. D.; Gavino, B. J.; Ambroziak, P.; Otto, J. C.; Casey, P. J.; Walker, Q. M.; Young, S. G. On the physiological importance of endoproteolysis of CAAX proteins: Heart-specific RCE1 knockout mice develop a lethal cardiomyopathy. *J Biol Chem* 2004, 279, 4729-4736.

Biermann, B.; Randall, S. K.; Crowell, D. N. Identification and isoprenylation of plant GTP-binding proteins. *Plant Mol Biol* 1996, 31, 1021-1028.

Biermann, B. J.; Morehead, T. A.; Tate, S. E.; Price, J. R.; Randall, S. K.; Crowell, D.N. 1994. Novel isoprenylated proteins identified by an expression library screen. *J Biol Chem* 1996, 269, 25251-25254.

Bolte, S.; Brown. S.; Satiat-Jeunemaitre, B. The N-myristoylated Rab GTPase m-RABmc is involved in post-Golgi trafficking events to the lytic vacuole in plant cells. *J Cell Sci* 2004, 117, 943-954.

Bolte, S.; Schiene, K.; Dietz, K. J. Characterization of a small GTP-binding protein of the rab5 family in *Mesembryanthemum crystallinum* with increased level of expression during early salt stress. *Plant Mol Biol* 2000, 42, 923-936.

Bonetta, D.; Bayliss, P.; Sun, S.; Sage, T.; Mccourt, P. Farnesylation is involved in meristem organization in Arabidopsis. *Planta* 2000, 211, 182-190.

Boss, P. K.; Bastow, R. M.; Mylne, J. S.; Dean, C. Multiple pathways in the decision to flower: Enabling, promoting, and resetting. *Plant Cell* 2004, 16, S18-S31.

Bourne, H. R.; Sanders, D. A.; Mccormick, F. The GTPase superfamily: Conserved structure and molecular mechanism. *Nature* 1991, 349, 117-127.

Bowman, J. L.; Alvarez, J.; Weigel, D.; Meyerowitz, E.; Smyth, D. R. Control of flower development in *Arabidopsis thaliana* by APETALA1 and interacting genes. *Development* 1993, 119, 721-743.

Boyartchuk, V. L.; Ashby, M. N.; Rine, J. Modulation of Ras and a-factor function by carboxyl-terminal proteolysis. *Science* 1997, 275, 1796-1800.

Bracha, K.; Lavy, M.; Yalovsky, S. The Arabidopsis AtSTE24 is a CAAX protease with broad substrate specificity. *J Biol Chem* 2002, 277:29856-29864.

Cadiñanos, J.; Varela, I.; Mandel, D. A.; Schmidt, W. K.; Díaz-Perales, A.; López-Otín, C.; Freije, J. M. AtFACE-2, a functional prenylated protein protease from *Arabidopsis thaliana* related to mammalian Ras-converting enzymes. *J Biol Chem* 2003, 278, 42091-42097.

Caldelari, D.; Sternberg, H.; Rodríguez-Concepción, M.; Gruissem, W.; Yalovsky, S. Efficient prenylation by a plant geranylgeranyltransferase-I requires a functional CaaL box motif and a proximal polybasic domain. *Plant Physiol* 2001, 126, 1416-1429.

Chen, Z.; Otto, J. C.; Bergo, M. O.; Young, S. G.; Casey, P. J. The C-terminal polylysine region and methylation of K-Ras are critical for the interaction between K-Ras and microtubules. *J Biol Chem* 2000, 275, 41251-41257.

Cheng, Y. S.; Patterson, C. E.; Staeheli, P. Interferon-induced guanylate-binding proteins lack an N(T)KXD consensus motif and bind GMP in addition to GDP and GTP. *Mol Cell Biol* 1991, 11, 4717-4725.

Clague, M. J. Membrane transport: Take your fusion partners. *Curr Biol* 1999, 9, R258-R260.

Clarke, S. Protein isoprenylation and methylation at carboxyl-terminal cysteine residues. *Annu Rev Bioche* 1992, 61, 355-386.

Clarke, S.; Vogel, J. P.; Deschenes, R. J.; Stock, J. Posttranslational modification of the Ha-ras oncogene protein: Evidence for a third class of protein carboxyl methyltransferases. *Proc Natl Acad Sci USA* 1988, 85, 4643-4647.

Courdavault, V.; Burlat, V.; St-Pierre, B.; Giglioli-Guivarc'h, N. Characterisation of CaaX-prenyltransferases in *Catharanthus roseus*: Relationships with the expression of genes involved in the early stages of monoterpenoid biosynthetic pathway. *Plant Sci* 2005, 168, 1097-1107.

Courdavault, V.; Thiersault, M.; Courtois, M.; Gantet, P.; Oudin, A.; Doireau, P.; St-Pierre, B.; Giglioli-Guivarch, N. CaaX-prenyltransferases are essential for expression of genes involved in the early stages of monoterpenoid biosynthetic pathway in *Catharanthus roseus* cells. *Plant Mol Biol* 2005, 57, 855-870.

Crowell, D. N. Functional implications of protein isoprenylation in plants. *Prog Lipid Res* 2000, 39, 393-408.

Crowell, D. N.; Huizinga, D. H.; Deem, A. K.; Trobaugh, C.; Denton, R.; Sen, S. E. *Arabidopsis thaliana* plants possess a specific farnesylcysteine lyase that is involved in detoxification and recycling of farnesylcysteine. *Plant J* 2007, 50, 839-847.

Crowell, D. N.; Kennedy M. Identification and functional expression in yeast of a prenylcysteine alpha-carboxyl methyltransferase gene from *Arabidopsis thaliana*. *Plant Mol Biol* 2001, 45, 469-476.

Crowell, D. N.; Randall, S. K.; Sen, S. E. Prenycysteine alpha-carboxyl methyltransferase in suspension-cultured tobacco cells. *Plant Physiol* 1998, 118, 115-123.

Cutler, S.; Ghassemian, M.; Bonetta, D.; Cooney, S.; Mccourt, P. A protein farnesyl transferase involved in abscisic acid signal transduction in Arabidopsis. *Science* 1996, 273, 1239-1241.

Deem, A. K.; Bultema, R. L.; Crowell, D. N. Prenylcysteine methylesterase in *Arabidopsis thaliana*. *Gene* 2006, 380, 159-166.

Deneka, M.; Neeft, M.; Van Der Sluijs, P. Regulation of membrane transport by Rab GTPases. *Crit Rev Biochem Mol Biol* 2003, 38, 121-142.

Desrosiers, R. R.; Nuguyen, Q. T.; Beliveau, R. The carboxyl methyltransferase modifying G proteins is a metalloenzyme. *Biochem Biophys Res Commun* 1999, 261, 790-797.

Dharmasiri, N.; Dharmasiri, S.; Estelle, M. The F-box protein TIR1 is an auxin receptor. *Nature* 2005, 435, 441-445.

Dharmasiri, S.; Estelle, M. The role of regulated protein degradation in auxin response. *Plant Mol Biol* 2002, 49, 401-409.

Dietrich, A.; Scheer, A.; Illenberger, D.; Kloog, Y.; Henis, Y. I.; Gierschik, P. Studies on G-protein alpha.betagamma heterotrimer formation reveal a putative S-prenyl-binding site in the alpha subunit. *Biochem J* 2003, 376, 449-456.

Digits, J. A.; Pyun, H. J.; Coates, R. M.; Casey, P. J. Stereospecificity and kinetic mechanism of human prenylcysteine lyase, an unusual thioether oxidase. *J Biol Chem* 2002, 277, 41086-41093.

Doerner, P. Shoot meristems: Intercellular signals keep the balance. *Curr Biol* 1999, 9, R377-R380.

Downes, B. P.; Saracco, S. A.; Lee, S. S.; Crowell, D. N.; Vierstra, R. D. MUBs, a family of ubiquitin-fold proteins that are plasma membrane-anchored by prenylation. *J Biol Chem* 2006, 281, 27145-27157.

Dykema, P. E.; Sipes, P. R.; Marie, A.; Biermann, B. J.; Crowell, D. N.; Randall, S. K. A new class of proteins capable of binding transition metals. *Plant Mol Biol* 1999, 41, 139-150.

Fu, Y.; Li, H.; Yang, Z. The ROP2 GTPase controls the formation of cortical fine F-actin and the early phase of directional cell expansion during Arabidopsis organogenesis. *Plant Cell* 2002, 14, 777-794.

Fu, Y.; Wu, G.; Yang, Z. Rop GTPase-dependent dynamics of tip-localized F-actin controls tip growth in pollen tubes. *J Cell Biol* 2001, 152, 1019-1032.

Fujiyama, A.; Tsunasawa, S.; Tamanoi, F.; Sakiyama, F. S-farnesylation and methyl esterification of C-terminal domain of yeast RAS2 protein prior to fatty acid acylation. *J Biol Chem* 1991, 266, 17926-17931.

Fukada, Y. Prenylation and carboxylmethylation of G-protein gamma subunit. *Methods Enzymol* 1995, 250, 91-105.

Galichet, A.; Gruissem, W. Protein farnesylation in plants-conserved mechanisms but different targets. *Curr Opin Plant Biol* 2003, 6, 530-535.

Galichet, A.; Gruissem, W. Developmentally controlled farnesylation modulates AtNAP1;1 function in cell proliferation and cell expansion during Arabidopsis leaf development. *Plant Physiol* 2006, 142, 1412-1426.

Gerber, E.; Hemmerlin, A.; Hartmann, M.; Heintz, D.; Crowell, D. N.; Rohmer, M.; Bach, T. J. Evidence for the plastidial origin of the isoprenyl residue used for the geranylgeranylation of a chimeric green fluorescent protein in BY-2 cells: A test system for inhibitors. *Proc. 8th Eur. Symp. Plant Isoprenoids* 2007; p. 12.

Grebe, M.; Xu, J.; Mobius, W.; Ueda, T.; Nakano, A.; Geuze, H. J.; Rook, M. B.; Scheres, B. Arabidopsis sterol endocytosis involves actin-mediated trafficking via ARA6-positive early endosomes. *Curr Biol* 2003, 13, 1378-1387.

Gu, Y.; Fu, Y.; Dowd, P.; Li, S.; Vernoud, V.; Gilroy, S.; Yang, Z. A Rho family GTPase controls actin dynamics and tip growth via two counteracting downstream pathways in pollen tubes. *J Cell Biol* 2005, 169, 127-138.

Gu, Y.; Vernoud, V.; Fu, Y.; Yang, Z. ROP GTPase regulation of pollen tube growth through the dynamics of tip-localized F-actin. *J Exp Bot* 2003, 54, 93-101.

Gustafson-Brown, C.; Savidge, B.; Yanofsky, M. F. Regulation of the Arabidopsis floral homeotic gene APETALA1. *Cell* 1994, 76, 131-143.

Gutierrez, L.; Magee, A. I.; Marshall, C. J.; Hancock, J. F. Post-translational processing of p21ras is two-step and involves carboxyl-methylation and carboxy-terminal proteolysis. *EMBO J* 1989, 8, 1093-1098.

Gutkowska, M.; Biekowski, T.; Hung, V. S.; Wanke, M.; Hertel, J.; Danikiewicz, W.; Swiezewska, E. Proteins are polyisoprenylated in *Arabidopsis thaliana*. *Biochem Biophys Res Commun* 2004, 322, 998-1004.

Hancock, J. F.; Cadwallader, K.; Marshall, C. J. Methylation and proteolysis are essential for efficient membrane binding of prenylated p21K-ras(B). *EMBO J* 1991, 10, 641-646.

Hancock, J. F.; Cadwallader, K.; Paterson, H.; Marshall, C. J. A CAAX or a CAAL motif and a second signal are sufficient for plasma membrane targeting of ras proteins. *EMBO J* 1991, 10, 4033-4039.

Hancock, J. F.; Magee, A. I.; Childs, J. E.; Marshall, C. J. All ras proteins are polyisoprenylated but only some are palmitoylated. *Cell* 1989, 57, 1167-1177.

Hempel, F. D.; Weigel, D.; Mandel, M. A.; Ditta, G.; Zambryski, P. C.; Feldman, L. J.; Yanofsky, M. F. Floral determination and expression of floral regulatory genes in Arabidopsis. *Development* 1997, 124, 3845-3853.

Hrycyna, C. A.; Clarke, S. Farnesyl cysteine C-terminal methyltransferase activity is dependent upon the STE14 gene product in *Saccharomyces cerevisiae*. *Mol Cell Biol* 1990, 10, 5071-5076.

Hrycyna, C. A.; Clarke, S. Maturation of isoprenylated proteins in *Saccharomyces cerevisiae*. Multiple activities catalyze the cleavage of the three carboxyl-terminal amino acids from farnesylated substrates *in vitro*. *J Biol Chem* 1992, 267, 10457-10464.

Hrycyna, C. A.; Sapperstein, S. K.; Clarke, S.; Michaelis, S. The *Saccharomyces cerevisiae* STE14 gene encodes a methyltransferase that mediates C-terminal methylation of a-factor and RAS proteins. *EMBO J* 1991, 10, 1699-1709.

Johnson, C. D.; Chary, S. N.; Chernoff, E. A.; Zeng, Q.; Running, M. P.; Crowell, D. N. Protein geranylgeranyltransferase I is involved in specific aspects of abscisic acid and auxin signaling in Arabidopsis. *Plant Physiol* 2005, 139, 722-733.

Jones, M. A.; Shen, J. J.; Fu, Y.; Li, H.; Yang, Z.; Grierson, C. S. The Arabidopsis Rop2 GTPase is a positive regulator of both root hair initiation and tip growth. *Plant Cell* 2002, 14, 763-776.

Jürgens, G.; Geldner, N. Protein secretion in plants: From the trans-Golgi network to the outer space. *Traffic* 2002, 3, 605-613.

Kawata, M.; Farnsworth, C. C.; Yoshida, Y.; Gelb, M. H.; Glomset, J. A.; Takai, Y. Post translationally processed structure of the human platelet protein smg p21B: Evidence for geranylgeranylation and carboxyl methylation of the C-terminal cysteine. *Proc Natl Acad Sci USA* 1990, 87, 8960-8964.

Krizek, B. A.; Meyerowitz, E. M. Mapping the protein regions responsible for the functional specificities of the Arabidopsis MADS domain organ-identity proteins. *Proc Natl Acad Sci USA* 1996, 93, 4063-4070.

Kuroda, Y.; Suzuki, N.; Kataoka, T. The effect of posttranslational modifications on the interaction of Ras2 with adenylyl cyclase. *Science* 1993, 259, 683-686.

Kweon, Y.; Rothe, A.; Conibear, E.; Stevens, T. H. Ykt6p is a multifunctional yeast R-SNARE that is required for multiple membrane transport pathways to the vacuole. *Mol Cell Biol* 2003, 14, 1868-1881.

Lavy, M.; Bracha-Drori, K.; Sternberg, H.; Yalovsky, S. A cell-specific, prenylation-independent mechanism regulates targeting of type II RACs. *Plant Cell* 2002, 14, 2431-2450.

Lavy, M.; Yalovsky, S. Association of Arabidopsis type-II ROPs with the plasma membrane requires a conserved C-terminal sequence motif and a proximal polybasic domain. *Plant J* 2006, 46, 934-947.

Lemichez, E.; Wu, Y.; Sanchez, J. P.; Mettouchi, A.; Mathur, J.; Chua, N. H. Inactivation of AtRac1 by abscisic acid is essential for stomatal closure. *Genes Dev* 2001, 15, 1808-1816.

Leung, K. F.; Baron, R.; Ali, B. R.; Magee, A. I.; Seabra, M. C. Rab GTPases containing a CAAX motif are processed post-geranylgeranylation by proteolysis and methylation. *J Biol Chem* 2007, 282, 1487-1497.

Leung, K. F.; Baron, R.; Seabra, M. C. Thematic review series: Lipid posttranslational modifications. geranylgeranylation of Rab GTPases. *J Lipid Res* 2006, 47, 467-475.

Li, H.; Lin, Y.; Heath, R. M.; Zhu, M. X.; Yang, Z. Control of pollen tube tip growth by a Rop GTPase-dependent pathway that leads to tip-localized calcium influx. *Plant Cell* 1999, 11, 1731-1742.

Li, H.; Shen, J. J.; Zheng, Z. L.; Lin, Y.; Yang, Z. The Rop GTPase switch controls multiple developmental processes in Arabidopsis. *Plant Physiol* 2001, 126, 670-684.

Lichtenthaler, H. K. Non-mevalonate isoprenoid biosynthesis: Enzymes, genes and inhibitors. *Biochem Soc Trans* 2000, 28, 785-789.

Liljegren, S. J.; Gustafson-Brown, C.; Pinyopich, A.; Ditta, G. S.; Yanofsky, M. F. Interactions among APETALA1, LEAFY, and TERMINAL FLOWER1 specify meristem fate. *Plant Cell* 1999, 1, 11007-1018.

Lin, Y.; Wang, Y.; Zhu, J. K, Yang, Z. Localization of a Rho GTPase implies a role in tip growth and movement of the generative cell in pollen tubes. *Plant Cell* 1996, 8, 293-303.

Liu, X.; Yue, Y.; Li, B.; Nie, Y.; Li, W.; Wu, W. H.; Ma, L. A G protein-coupled receptor is a plasma membrane receptor for the plant hormone abscisic acid. *Science* 2007, 315, 1712-1716.

Loraine, A. E.; Yalovsky, S.; Fabry, S.; Gruissem, W. Tomato Rab1A homologs as molecular tools for studying Rab geranylgeranyl transferase in plant cells. *Plant Physiol* 1996, 110, 1337-1347.

Lu, C.; Zainal, Z.; Tiucker, G. A.; Lycett, G. W. Developmental abnormalities and reduced fruit softening in tomato plants expressing an antisense Rab11 GTPase gene. *Plant Cell* 2001, 13, 1819-1833.

Maurer-Stroh, S.; Koranda, M.; Benetka, W.; Schneider, G.; Sirota, F. L.; Eisenhaber, F. Towards complete sets of farnesylated and geranylgeranylated proteins. *Comp Biol* 2007, 3, 634-648.

Mcnew, J. A.; Søgaard, M.; Lampen, N. M.; Machida, S.; Ye, R. R.; Lacomis, L.; Tempst, P.; Rothman, J. E.; Sollner, T. H. Ykt6, a prenylated SNARE essential for endoplasmic reticulum-Golgi transport. *J Biol Chem* 1997, 272, 17776-17783.

Melchior, F. SUMO-nonclassical ubiquitin. *Annu Rev Cell Dev Biol* 2000, 16, 591-626.

Melkonian, K. A.; Ostermeyer, A. G.; Chen, J. Z.; Roth, M. G.; Brown, D. A Role of lipid modifications in targeting proteins to detergent-resistant membrane rafts. Many raft proteins are acylated, while few are prenylated. *J Biol Chem* 1999, 274, 3910-3917.

Michaelson, D.; Ali, W.; Chiu, V. K.; Bergo, M.; Silletti, J.; Wright, L.; Young, S. G.; Philips, M. Postprenylation CAAX processing is required for proper localization of Ras but not Rho GTPases. *Mol Biol Cell* 2005, 16, 1606-1616.

Molendijk, A. J.; Bischoff, F.; Rajendrakumar, C. S.; Friml, J.; Braun, M.; Gilroy, S.; Palme, K. 2001. *Arabidopsis thaliana* Rop GTPases are localized to tips of root hairs and control polar growth. *EMBO J* 2005, 20, 2779-2788.

Molendijk, A. J.; Ruperti, B.; Palme, K. Small GTPases in vesicle trafficking. *Curr Opin Plant Biol* 2004, 7, 694.

Morehead, T. A.; Biermann, B. J.; Crowell, D. N.; Randall, S. K. Changes in protein isoprenylation during the growth of suspension-cultured tobacco cells. *Plant Physiol,* 1995, 109, 277-284.

Morel, J.; Fromentin, J.; Blein, J. P.; Simon-Plas, F.; Elmayan, T. Rac regulation of NtrbohD, the oxidase responsible for the oxidative burst in elicited tobacco cell. *Plant J* 2004, 37, 282-293.

Moshkov, I. E.; Mur, L. A.; Novikova, G. V.; Smith, A. R.; Hall, M. A. Ethylene regulates monomeric GTP-binding protein gene expression and activity in Arabidopsis. *Plant Physiol* 2003, 131, 1705-1717.

Moshkov, I. E.; Novikova, G. V.; Mur, L. A.; Smith, A. R.; Hall, M. A. Ethylene rapidly up-regulates the activities of both monomeric GTP-binding proteins and protein kinase(s) in epicotyls of pea. *Plant Physiol* 2003, 131, 1718-1726.

Narasimha Chary, S.; Bultema, R. L.; Packard, C. E.; Crowell, D. N. Prenylcysteine alpha-carboxyl methyltransferase expression and function in *Arabidopsis thaliana. Plant J* 2002, 32, 735-747.

Ng, M.; Yanofsky, M. F. Activation of the Arabidopsis B class homeotic genes by APETALA1. *Plant Cell* 2001, 13, 739-753.

Ohsumi, Y. Molecular dissection of autophagy: Two ubiquitin-like systems. *Nature Rev Mol Cell Biol* 2001, 2, 211-216.

Ong, O. C.; Ota, I. M.; Clarke, S.; Fung, B. K. The membrane binding domain of rod cGMP phosphodiesterase is posttranslationally modified by methyl esterification at a C-terminal cysteine. *Proc Natl Acad Sci USA* 1989, 86, 9238-9242.

Ono, E.; Wong, H. L.; Kawasaki, T.; Kodama, O.; Shimamoto, K. Essential role of the small GTPase Rac in disease resistance of rice. *Proc Natl Acad Sci USA* 2001, 98, 759-764.

Pandey, S.; Assmann, S. M. The Arabidopsis putative G protein-coupled receptor GCR1 interacts with the G protein alpha subunit GPA1 and regulates abscisic acid signaling. *Plant Cell* 2004, 16, 1616-1632.

Parish, C. A.; Smrcka, A. V.; Rando, R. R. Functional significance of beta gamma-subunit carboxymethylation for the activation of phospholipase C and phosphoinositide 3-kinase. *Biochemistry* 1995, 34, 7722-7727.

Parish, C. A.; Smrcka, A. V.; Rando, R. R. The role of G protein methylation in the function of a geranylgeranylated beta gamma isoform. *Biochemistry* 1996, 35, 7499-7505.

Parmryd, I.; Andersson, B.; Dallner, G. Protein prenylation in spinach chloroplasts. *Proc Natl Acad Sci USA* 1999, 96, 10074-10079.

Parmryd, I.; Shipton, C. A.; Swiezewska, E.; Dallner, G.; Andersson, B. Chloroplastic prenylated proteins. *FEBS Lett* 1997, 414, 527-531.

Pei, Z. M.; Ghassemian, M.; Kwak, C. M.; Mccourt, P.; Schroeder, J. I. Role of farnesyltransferase in ABA regulation of guard cell anion channels and plant water loss. *Science* 1998, 282, 287-290.

Petroski, M. D.; Deshaies, R. J. Function and regulation of cullin-RING ubiquitin ligases. *Nature Rev Mol Cell Biol* 2005, 6, 9-20.

Porfiri, E.; Evans, T.; Chardin, P.; Hancock, J. F. Prenylation of Ras proteins is required for efficient hSOS1-promoted guanine nucleotide exchange. *J Biol Chem* 1994, 269, 22672-22677.

Potikha, T. S.; Collins, C. C.; Johnson, D. I.; Delmer, D. P.; Levine, A. The involvement of hydrogen peroxide in the differentiation of secondary walls in cotton fibers. *Plant Physiol* 1999, 119, 849-858.

Preuss, M. L.; Schmitz, A. J.; Thole, J. M.; Bonner, H. K. S.; Otegui, M. S.; Nielsen, E. A role for the RabA4b effector protein PI-4Kbeta1 in polarized expansion of root hair cells in *Arabidopsis thaliana. J Cell Biol* 2006, 172, 991-998.

Preuss, M. L.; Serna, J.; Falbel, T. G.; Bednarek, S. Y.; Nielsen, E. The Arabidopsis Rab GTPase RabA4b localizes to the tips of growing root hair cells. *Plant Cell* 2004, 16, 1589-1603.

Putterill, J.; Laurie, R.; Macknight, R. It's time to flower: The genetic control of flowering time. *Bioessays* 2004, 26, 363-373.

Qian, D.; Zhou, D.; Ju, R.; Cramer, C. L.; Yang, Z. Protein farnesyltransferase in plants: Molecular characterization and involvement in cell cycle control. *Plant Cell* 1996, 8, 2381-2394.

Raftopoulou, M.; Hall, A. Cell migration: Rho GTPases lead the way. *Dev Biol* 2004, 265, 23-32.

Randall, S. K.; Crowell, D. N. Protein isoprenylation in plants. *Crit Rev Biochem Mol Biol* 1999, 34, 325-338.

Randall, S. K.; Marshall, M. S.; Crowell, D. N. Protein isoprenylation in suspension-cultured tobacco cells. *Plant Cell* 1993, 5, 433-442.

Reid, T. S.; Terry, K. L.; Casey, P. J.; Beese, L. S. Crystallographic analysis of CaaX prenyltransferases complexed with substrates defines rules of protein substrate selectivity. *J Mol Biol* 2004, 343, 417-433.

Reuber, T. L.; Ausubel, F. M. Isolation of Arabidopsis genes that differentiate between resistance responses mediated by the RPS2 and RPM1 disease resistance genes. *Plant Cell* 1996, 8, 241-249.

Rodríguez-Concepción, M.; Toledo-Ortiz, G.; Yalovsky, S.; Caldelari, D.; Gruissem, W. Carboxylmethylation of prenylated calmodulin CaM53 is required for efficient plasma membrane targeting of the protein. *Plant J* 2000, 24, 775-784.

Rodríguez-Concepción, M.; Yalovsky, S.; Gruissem, W. Protein prenylation in plants: Old friends and new targets. *Plant Mol Biol* 1999, 39, 865-870.

Rodríguez-Concepción, M.; Yalovsky, S.; Zik, M.; Fromm, H.; Gruissem, W. The prenylation status of a novel plant calmodulin directs plasma membrane or nuclear localization of the protein. *EMBO J* 1999, 18, 1996-2007.

Romano, J. D.; Schmidt, W.K.; Michaelis, S. The *Saccharomyces cerevisiae* prenylcysteine carboxyl methyltransferase Ste14p is in the endoplasmic reticulum membrane. *Mol Biol Cell* 1998, 9, 2231-2247.

Running, M. P.; Fletcher, J. C.; Meyerowitz, E. M. The WIGGUM gene is required for proper regulation of floral meristem size in Arabidopsis. *Development* 1998, 125, 2545-2553.

Running, M. P.; Lavy, M.; Sternberg, H.; Galichet, A.; Gruissem, W.; Hake, S.; Ori, N.; Yalovsky, S. Enlarged meristems and delayed growth in plp mutants result from lack of CaaX prenyltransferases. *Proc Natl Acad Sci USA* 2004, 101, 7815-7820.

Rutherford, S.; Moore, I. The Arabidopsis Rab GTPase family: Another enigma variation. *Curr Opin Plant Biol* 2002, 5, 518-528.

Samaj, J.; Muller, J.; Beck, M.; Bohm, N.; Menzel, D. Vesicular trafficking, cytoskeleton and signalling in root hairs and pollen tubes. *Trends Plant Sci* 2006, 11, 594-600.

Sapperstein, S.; Berkower, C.; Michaelis, S. Nucleotide sequence of the yeast STE14 gene, which encodes farnesylcysteine carboxyl methyltransferase, and demonstration of its essential role in a-factor export. *Mol Cell Biol* 1994, 14, 1438-1449.

Schmidt, W. K.; Tam, A.; Fujimura-Kamada, K.; Michaelis, S. Endoplasmic reticulum membrane localization of Rce1p and Ste24p, yeast proteases involved in carboxyl-terminal CAAX protein processing and amino-terminal a-factor cleavage. *Proc Natl Acad Sci USA* 1998, 95, 11175-11180.

Schmitt, D.; Callan, K.; Gruissem, W. Molecular and biochemical characterization of tomato farnesyl-protein transferase. *Plant Physiol* 1996, 112, 767-777.

Shipton, C. A.; Parmryd, I.; Swiezewska, E.; Andersson, B.; Dallner, G. Isoprenylation of plant proteins *in vivo*. Isoprenylated proteins are abundant in the mitochondria and nuclei of spinach. *J Biol Chem* 270, 566-572.

Simon, R.; Igeno, M. I.; Coupland, G. 1996. Activation of floral meristem identity genes in Arabidopsis. *Nature* 1995, 384, 59-62.

Søgaard, M.; Tani, K.; Ye, R. R.; Geromanos, S.; Tempst, P.; Kirchhausen, T.; Rothman, J. E.; Sollner, T. A rab protein is required for the assembly of SNARE complexes in the docking of transport vesicles. *Cell* 1994, 78, 937-948.

Sohn, E. J.; Kim, E. S.; Zhao, M.; Kim, S. J.; Kim, H.; Kim, Y. W.; Lee, Y. J.; Hillmer, S.; Sohn, U.; Jiang, L.; Hwang, I. Rha1, an Arabidopsis Rab5 homolog, plays a critical role in the vacuolar trafficking of soluble cargo proteins. *Plant Cell* 2003, 15, 1057-1070.

Sorek, N.; Poraty, L.; Sternberg, H.; Bar, E.; Lewinsohn, E.; Yalovsky, S. Activation status-coupled transient S acylation determines membrane partitioning of a plant Rho-related GTPase. *Mol Cell Biol* 2007, 27, 2144-2154.

Suharsono, U.; Fujisawa, Y.; Kawasaki, T.; Iwasaki, Y.; Satoh, H.; Shimamoto, K. The heterotrimeric G protein alpha subunit acts upstream of the small GTPase Rac in disease resistance of rice. *Proc Natl Acad Sci USA* 2002, 99, 13307-13312.

Suzuki, N. Y.; Koizumi, N.; Sano, H. Functional characterization of a heavy metal binding protein Cdl19 from Arabidopsis. *Plant J* 2002, 32, 165-173.

Tam, A.; Nouvet, F. J.; Fujimura-Kamada, K.; Slunt, H.; Sisodia, S. S.; Michaelis, S. Dual roles for Ste24p in yeast a-factor maturation: NH2-terminal proteolysis and COOH-terminal CAAX processing. *J Cell Biol* 1998, 142, 635-649.

Tao, L. Z.; Cheung, A. Y.; Nibau, C.; Wu, H. M. RAC GTPases in tobacco and Arabidopsis mediate auxin-induced formation of proteolytically active nuclear protein bodies that contain AUX/IAA proteins. *Plant Cell* 2005, 17, 2369-2383.

Tao, L. Z.; Cheung, A. Y.; Wu, H. M. Plant Rac-like GTPases are activated by auxin and mediate auxin-responsive gene expression. *Plant Cell* 2002, 14, 2745-2760.

Trotochaud, A. E.; Hao, T.; Wu, G.; Yang, Z.; Clark, S. E. The CLAVATA1 receptor-like kinase requires CLAVATA3 for the assembly into a signalling complex that includes KAPP and a Rho-related protein. *Plant Cell* 1999, 11, 393-405.

Trueblood, C. E.; Boyartchuk, V. L.; Picologlou, E. A.; Rozema, D.; Poulter, C. D.; Rine, J. The CaaX proteases, Afc1p and Rce1p, have overlapping but distinct substrate specificities. *Mol Cell Biol* 2000, 20, 4381-4392.

Trueblood, C. E.; Ohya, Y.; Rine, J. Genetic evidence for *in vivo* cross-specificity of the CaaX-box protein prenyltransferases farnesyltransferase and geranylgeranyltransferase-I in *Saccharomyces cerevisiae*. *Mol Cell Biol* 1993, 13, 4260-4275.

Trusov, Y.; Rookes, J. E.; Tilbrook, K.; Chakravorty, D.; Mason, M. G.; Anderson, D.; Chen, J. G.; Jones, A. M.; Botella, J. R. Heterotrimeric G protein gamma subunits provide functional selectivity in G beta-gamma dimer signaling in Arabidopsis. *Plant Cell* 2007, 19, 1235-1250.

Tschantz, W. R.; Digits, J. A.; Pyun, H. J.; Coates, R. M.; Casey, P. J. Lysosomal prenylcysteine lyase is a FAD-dependent thioether oxidase. *J Biol Chem* 276, 2321-2324.

Tschantz, W. R.; Zhang, L.; Casey, P. J. Cloning, expression, and cellular localization of a human prenylcysteine lyase. *J Biol Chem* 1999, 274, 35802-35808.

Ueda, T.; Yamaguchi, M.; Uchimiya, H.; Nakano, A. Ara6, a plant-unique novel type Rab GTPase, functions in the endocytic pathway of *Arabidopsis thaliana*. *EMBO J* 2001, 20, 4730-4741.

Ungermann, C.; Von Mollard, G. F.; Jensen, O. N.; Margolis, N.; Stevens, T. H.; Wickner, W. Three v-SNAREs and two t-SNAREs, present in a pentameric cis-SNARE complex on isolated vacuoles, are essential for homotypic fusion. *J Cell Biol* 1999, 145, 1435-1442.

Vernoud, V.; Horton, A. C.; Yang, Z.; Nielson, E. Analysis of the small GTPase gene superfamily of Arabidopsis. *Plant Physiol* 2003, 131, 1191-1208.

Wang, Y.; Ying, J.; Kuzma, M.; Chalifoux, M.; Sample, A.; Mcarthur, C.; Uchacz, T.; Sarvas, C.; Wan, J.; Dennis, D. T.; Mccourt, P.; Huang, Y. Molecular tailoring of farnesylation for plant drought tolerance and yield protection. *Plant J* 2005, 43, 413-424.

Wei, C.; Lutz, R.; Sinensky, M.; Macara, I. G. p23rab2, a ras-like GTPase with a -GGGCC C-terminus, is isoprenylated but not detectably carboxymethylated in NIH3T3 cells. *Oncogene* 1992, 7, 467-473.

Wojtas, M.; Swiezewski, S.; Sarnowski, T. J.; Plochocka, D.; Chelstowska, A.; Tolmachova, T.; Swiezewska, E. 2007. Cloning and characterization of Rab Escort Protein (REP) from *Arabidopsis thaliana*. *Cell Biol Int* 31, 246-251.

Wyatt, R. E.; Ainley, W. M.; Nagao, R. T.; Conner, T. W.; Key, J. L. Expression of the Arabidopsis AtAux2-11 auxin-responsive gene in transgenic plants. *Plant Mol Biol* 1993, 22, 731-749.

Xiao, C.; Xin, H.; Dong, A.; Sun, C.; Cao, K. A novel calmodulin-like protein gene in rice which has an unusual prolonged C-terminal sequence carrying a putative prenylation site. *DNA Res* 1999, 6, 179-181.

Xiong, L.; Lee, H.; Ishitani, M.; Zhu, J. K. Regulation of osmotic stress-responsive gene expression by the LOS6/ABA1 locus in Arabidopsis. *J Biol Chem* 2002, 277, 8588-8596.

Yalovsky, S.; Kulukian, A.; Rodríguez-Concepción, M.; Young, C. A.; Gruissem, W. Functional requirement of plant farnesyltransferase during development in Arabidopsis. *Plant Cell* 2000, 12, 1267-1278.

Yalovsky, S.; Rodríguez-Concepción, M.; Bracha, K.; Toledo-Ortiz, G.; Gruissem, W. Prenylation of the floral transcription factor APETALA1 modulates its function. *Plant Cell* 2000, 12, 1257-1266.

Yalovsky, S.; Trueblood, C. E.; Callan, K. L.; Narita, J. O.; Jenkins, S. M.; Rine, J.; Gruissem, W. Plant farnesyltransferase can restore yeast Ras signaling and mating. *Mol Cell Biol* 17, 1986-1994.

Yamane, H. K.; Farnsworth, C. C.; Xie, H. Y.; Howald, W.; Fung, B. K.; Clarke, S.; Gelb, M. H.; Glomset, J. A. 1990. Brain G protein gamma subunits contain an all-trans-geranylgeranylcysteine methyl ester at their carboxyl termini. *Proc Natl Acad Sci USA* 1997, 87, 5868-5872.

Yang, Z. Small GTPases: Versatile signaling switches in plants. *Plant Cell* 2002, 14:S375-388.

Yang, Z.; Cramer, C. L.; Watson, J. C. Protein farnesyltransferase in plants. Molecular cloning and expression of a homolog of the beta subunit from the garden pea. *Plant Physiol* 1993, 101, 667-674.

Zeng, Q.; Wang, X. J.; Running, M. P. Dual lipid modification of Arabidopsis G gamma-subunits is required for efficient plasma membrane targeting. *Plant Physiol* 2007, 143, 1119-1131.

Zhang, F. L.; Casey, P. J. Protein prenylation: Molecular mechanisms and functional consequences. *Annu Rev Biochem* 1996, 65, 241-269.

Zhang, L.; Tschantz, W. R.; Casey, P. J. Isolation and characterization of a prenylcysteine lyase from bovine brain. *J Biol Chem* 1997, 272, 23354-23359.

Zhang, T.; Hong, W. Ykt6 forms a SNARE complex with syntaxin 5, GS28, and Bet1and participates in a late stage in endoplasmic reticulum-Golgi transport. *J Biol Chem* 276, 27480-27487.

Zhou, D.; Qian, D.; Cramer, C. L.; Yang, Z. Developmental and environmental regulation of tissue- and cell-specific expression for a pea protein farnesyltransferase gene in transgenic plants. *Plant J* 1997, 12, 921-930.

Zhou, D.; Yang, Z.; Cramer, C. L. A cDNA clone encoding the beta subunit of protein farnesyltransferase from *Nicotiana glutinosa* (Accession No. U73203). *Plant Physiol* 1996, 112, 1398-1399.

Zhu, J. K, Bressan, R. A.; Hasegawa, P. M. Isoprenylation of the plant molecular chaperone ANJ1 facilitates membrane association and function at high temperature. *Proc Natl Acad Sci USA* 1993, 90, 8557-8561.

Zhu, J. K, Shi, J.; Bressan, R. A.; Hasegawa, P. M. Expression of an Atriplex nummularia gene encoding a protein homologous to the bacterial molecular chaperone DNA. *J Plant Cell* 1993, 5, 341-349.

Ziegelhoffer, E. C.; Medrano, L. J.; Meyerowitz, E. M. Cloning of the Arabidopsis WIGGUM gene identifies a role for farnesylation in meristem development. *Proc Natl Acad Sci USA* 2000, 97, 7633-7638.

CHAPTER 11

MANGOMICS: INFORMATION SYSTEMS SUPPORTING ADVANCED MANGO BREEDING

DAVID J. INNES, NATALIE L. DILLON, HEATHER E. SMYTH,
MIRKO KARAN, TIMOTHY A. HOLTON, IAN S.E. BALLY, and
RALF G. DIETZGEN

CONTENTS

11.1 Introduction 282
11.2 Information Systems 283
11.3 Nucleotide Sequence Data 284
11.4 Molecular Marker Discovery 289
11.5 Biochemical Pathways 293
11.6 Phenotypes 296
11.7 Flavour and Aroma Chemistry 298
11.8 Tools and Applications 300
11.9 Conclusion and Future Perspectives 303
Keywords 303
References 304

11.1   INTRODUCTION

Mango (*Mangifera indica* L.) is an important tropical fruit crop. It is thought to be an allotetraploid species (2n = 40 chromosomes) with an estimated genome size of 439 mega base pairs (Mukherjee, 1950; Arumuganathan and Earle, 1991). Total worldwide mango production exceeded 38 million tons in 2010. Of the ~100 mango-producing countries, India was the largest single producer providing 42 % of worldwide production; over 16 million tons in 2010. Overall, 75 % of mangoes produced annually come from India, China, Thailand, Pakistan, Mexico, and Indonesia.

The Australian mango industry is predominantly based on the variety Kensington Pride, which despite its popularity with consumers has a number of agronomically undesirable characteristics that impact fruit quality and productivity. Issues include excessive tree vigor, irregular fruit production, short shelf life, inconsistent external coloring, susceptibility to sap burn and some fungal diseases. Fortunately, other important varieties maintained in the Australian National Mango Genbank have desirable attributes that when used as breeding parents can contribute to mango tree and fruit quality improvements (Kulkarni *et al.*, 2002). Traditional approaches to develop mangoes with increased quality and nutrition, pest and disease resistance and improved tree architecture may take years to lead to the release of a new variety. Like in other horticultural fruit crops such as kiwifruit and apple, use of genetic and genomic data to support breeding efforts can achieve mango improvement more efficiently through marker-assisted selection.

The horticultural properties (Bally, 2006; Bally *et al.*, 2009; Bally, 2011) and biotechnological advances (Krishna and Singh, 2007; Dietzgen *et al.*, 2012) in mango breeding and production have been well reviewed in recent years. These reviews focus on the advances in tissue culture technologies as well as some applications of established marker systems such as microsatellites, random amplification of polymorphic DNA (RAPD), restriction fragment (RFLP) and amplified fragment (AFLP) length polymorphisms. There is, however, little scientific literature on the application of "omics" technologies in mango apart from an analysis of the mango leaf proteome (Renuse *et al.*, 2012). A query into sequence resources housed at the National Centre for Biotechnology Information (NCBI) revealed that as of 2012 there were only 622 nucleotide and 1,665 expressed sequence tag (EST) sequences publically available. By way of comparison, there is in excess of 55,000 ESTs publically available for kiwifruit (*Actinidia deliciosa*) and many other similar comparisons can be drawn.

The Queensland Mango Genomics Initiative is a multidisciplinary project involving researchers with expertise in mango breeding, variety selection, physiology, molecular biology, biochemistry, genetics, bioinformatics, and food technology (Figure 11.1). These disciplines generate distinct data sets, each of which provides a great capacity for expediting mango improvement (Dietzgen *et al.*, 2009). It is the linking and interactions between these data sets that has the potential to deliver powerful molecular and bioinformatic tools to the mango industry.

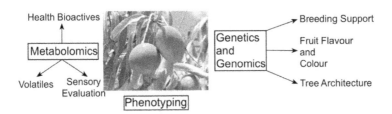

FIGURE 11.1 Overview of the Queensland Mango Genomics Initiative and its approaches.

11.2 INFORMATION SYSTEMS

The Mango Genomics Information System (MGIS) was built using industry standard, open source technologies. The key criterion upon which the design strategies were selected was that all researchers involved in the Mango Genomics Initiative should have access to a complete suite of datasets and tools irrespective of geographical location or research agency. Web-based databases and tools naturally facilitate this level of information sharing and networking. All version control management occurs at a single site, the web host, so there is no need to push data files and interface updates to all users. These updates happen seamlessly and are available as deployed to the web server.

The web site was built using PHP 5.3 connected to a MySQL version 5.5 database and is served via an Apache 2 web server. Most routine database queries are built as stored procedures while complex dynamic SQL statements are built in PHP and the MySQL databases are accessed through the mysqli module. Client-side scripting with JavaScript helps give the web site a native desktop application look and feel as well as reducing the number of web page refreshes. Graphs are not prerendered images but rather they are drawn live using PHPlot. Security is moderated by encrypted user-password authentications and all site activity is logged. Prepared statements are used to circumvent SQL injection attacks. While there are security concerns when placing data in the public domain, databases on the web also promote future collaborations. User-access controls built into the system allows external parties to view genes of interest, for example, though full access to sequence information can be restricted until formal partnerships are entered into. The following sections contain information about the datasets accessed through MGIS and provide some examples of how these data and tools are applied to mango research and development of improved varieties.

11.3   NUCLEOTIDE SEQUENCE DATA

DNA sequencing technologies and platforms are advancing at a great rate. Read lengths are getting longer, accuracy is improving and cost per base is diminishing. Throughout the Mango Genomics Initiative, we have used a combination of long read capillary Sanger sequencing and second generation platforms such as Roche 454 and Illumina GAII to partially profile the mango genome and proteome (Metzker, 2010). By examining these sequences we may be able to identify sequence and expression variations between varieties that can be exploited in breeding programs.

11.3.1   EXPRESSED SEQUENCE TAGS

Expressed sequence tags are typically single pass sequences of cloned cDNAs derived from transcribed genes (transcripts) present in tissue samples. The sequences identified in this manner are subject to tissue specificity, temporal variation and response to a wide range of biotic and abiotic stresses. The limited mango gene sequences available to date suggested that selecting a range of tissues from mature, fruit bearing trees would be a suitable choice for generating initial reference EST sets. The cDNA libraries were prepared from RNA extracted from Kensington Pride variety red leaf, fruit skin and flesh, flower, and root tissues. A library was similarly prepared from RNA extracted from the red leaves from Irwin variety trees. The libraries were constructed in pSPORT 1 plasmid vector using the Invitrogen SuperScript Plasmid System for cDNA Synthesis and Cloning. This system resulted in directionally cloned cDNAs, which facilitated the production of 5' EST sequences. As these cDNAs were prepared using oligo-dT, the use of a directional cloning strategy minimized sequencing through the difficult-to-sequence polyA tails. Automated bacterial colony picking of cDNA containing clones and single pass capillary sequencing was done at the Australian Genome Research Facility (AGRF; Brisbane, Australia).

Prior to analyzing and annotating the sequence data, the raw chromatograms were processed to remove cloning vector sequences and low quality base calls. The pSPORT 1 vector sequence was added to a CrossMatch database and vector clipping and quality score trimming of the raw chromatograms was performed within the Staden package (Staden *et al.*, 2000) at the Queensland Facility for Advanced Bioinformatics (QFAB; Brisbane, Australia). During the clipping process, polyA sequences were retained. The experiment file (exp) output from the Staden package, a tagged text file, was parsed using a custom PERL script to produce multifasta sequence and quality files containing only the vector trimmed and quality clipped sequence reads. In addition to source file and output file names, this PERL script also takes an optional sequence length parameter. Invoking this parameter allowed filtering of smaller EST sequences. This analysis was done for each of the EST libraries as well as for a set of reads from all libraries combined. The nomenclature used for each EST meant that it could easily be traced back to the source variety and tissue type.

The parsed and length-filtered sequence sets were assembled into contiguous sequences using cap3 (Huang and Madan, 1999). For each library, the contigs and unassembled singletons were merged to produce a unigene set. In order to easily track the source of each contigs, the sequences were renamed to append the mango variety and tissue type as a prefix to each contig name. The resulting sequence and assembly information was parsed into the database. These sequence sets served as source data for gene discovery using comparative genomics as well as the elucidation of putative single nucleotide polymorphism (SNP) and simple sequence repeat (SSR) markers applicable to mapping and marker-assisted breeding.

11.3.2   GENOMIC SEQUENCES

A complete genome sequence for mango is not currently available. The study of the genome offers some advantages over expressed sequences. With the complete complement of genes present in the genome, the tissue specificity, temporal expression patterns and gene expression patterns in response to external stresses that confound transcriptome analysis are overcome. The challenges with whole genome analysis of orphan species, such as mango, is that *de novo* assemblies can be difficult and a range of sequencing strategies are required to produce a high quality draft sequence. These strategies include shotgun sequencing of paired-end libraries with large (Illumina mate-pair) and small (Illumina pair-end) fragment sizes, transcriptome sequencing and, where possible, additional resources such as bacterial artificial chromosome (BAC) clones and physical mapping data.

At the time we commenced the analysis of the mango genome in 2005, it was not feasible to generate a complete genome sequence as the costs outweighed the benefits. There are a range of genomics approaches that can be applied to the study of species for which complete genome sequences are not available. Low-pass "shotgun" sequencing has been applied to the identification of gene sequences in plant species (Wicker *et al.*, 2008) and the ever-decreasing costs has made genotyping by sequencing a viable approach to identifying sequence variations between varieties.

An alternative set of approaches is an adaptation of established marker technologies such as restriction fragment length and amplified fragment length polymorphisms. Unlike the shotgun approach that randomly size-fragments the genome and sequences a pool of these random fragments, these methods combine next-generation sequencing with fragmentation of the genome about specific genomic landmarks, namely restriction endonuclease recognition sequences. A range of new complexity reduction methods exist including reduced representation libraries (RRLs; Maughan *et al.*, 2009), complexity reduction of polymorphic sequences (CRoPS; van Orsouw *et al.*, 2007) and restriction-site-associated DNA sequencing (RAD-seq; Willing *et al.*, 2011). This is also the strategy used by Diversity Array Technologies for their DArT markers (Jaccoud *et al.*, 2001). Davey *et al.* (2011) reviews genotyping approaches using next-generation sequencing.

The use of restriction-site based fragmentation methods requires no prior genome sequence knowledge and so is well suited to nonmodel species such as mango. The major advantage that sequencing restriction fragments offers over sequencing pools of random fragments is that a high proportion of fragments will be in common between varieties. This promotes the rapid discovery of SNP and SSR markers between varieties that can be applied to breeding programs and mapping populations. The platforms that fill the high output sequencing space vary in approach and output. Generally, SOLiD (Life Technologies) and Illumina (Solexa) produce greater sequence output than Roche 454-pyrosequencing, at the expense of read length. Conversely, 454-pyrosequencing produces fewer numbers of reads, and lower output overall, but generates much longer single reads.

One of the primary outcomes of the Mango Genomics Initiative was to generate molecular markers that could be applied to mapping and marker-assisted selection. These applications require the conversion of putative SNP and SSR markers into either PCR or higher throughput screening techniques such as Sequenom MassArray, Illumina GoldenGate or capillary fragment analysis. The design of these assays requires a degree of sequence flanking the polymorphic sites so the increased read length of the 454-pyrosequencing chemistry was chosen over the shorter, higher-output platforms. The length of fragments and size of the fragment pool can be tailored by the choice of restriction endonucleases. Classically, AFLP protocols use a combination of restriction enzymes such that one has a 6-base recognition sequence and a second enzyme with a more frequent 4-base recognition sequence. In the post-genomics era, it is possible to model the pool of reduced complexity fragments that would be produced by *in silico* digests of model species' genome sequences with different combinations of enzymes (Figure 11.2). Not only will this give insights into the size of the fragment pool that will be available for sequencing, but can be extended using tools such as the Basic Local Alignment Search Tool (BLAST; Altschul *et al.*, 1997) to identify the number of fragments that will be in genic regions. Finally, depending on the size of the fragment pool and the desired depth of coverage, samples can be multiplexed by incorporating a specific DNA sequence barcode during the preparation of each RRL.

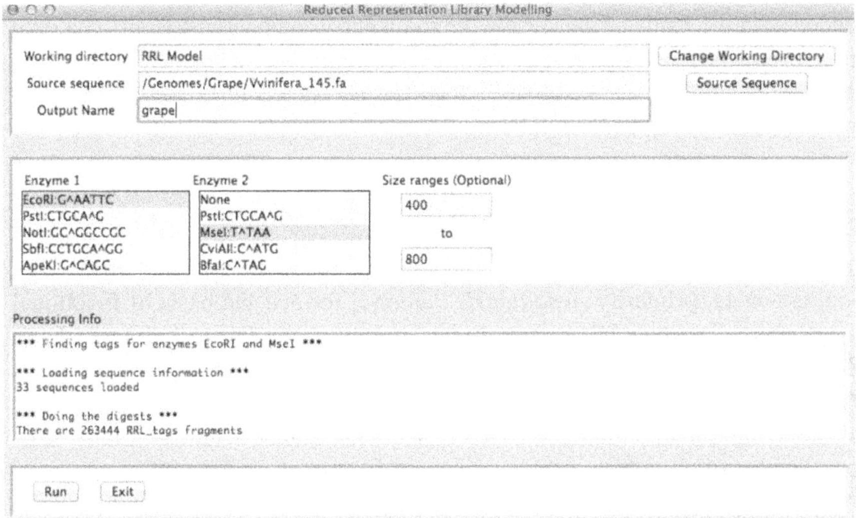

FIGURE 11.2   Tool for *in silico* predictions of reduced representation library fragments.

An overview of the preparation of RRLs for 454-pyrosequencing is provided by Maughan *et al.* (2009) and the libraries prepared from mango DNA samples for Kensington Pride and Irwin species essentially followed this method. Sequencing was done at the AGRF (Brisbane, Australia) using the Titanium chemistry on a Roche GS-FLX pyrosequencer. Even though the process of generating RRLs requires PCR amplification, these types of libraries have commonalities to both shotgun and amplicon approaches when it comes to sequencing as the number of individual amplicons is typically measured in the thousands.

Integrated RAD-seq tools, such as Rainbow (Chong *et al.*, 2012), were not available when these libraries were being sequenced so the raw sequence reads were preprocessed using PERL scripts in order to remove reads that failed to meet a minimum length requirement of typically 100 bases. As the templates of these sequences were produced using restriction endonucleases, the terminal bases of each fragment were also verified against the expected bases for the restriction enzyme combination. The cleaned sets of reads were clustered using cap3. For this clustering, the parameters were much more stringent than those used for assembling the EST unigenes. This is due to the nature of the fragmentation such that no long contiguous sequences will be formed as each fragment represents a discrete segment of the genome. Compare this with random genome fragmentation where there is an expectation that individual short reads will be extended into longer contiguous sequences. As with the EST sequences, contigs, and unassembled single sequences were combined to produce a unique sequence set comprising genomic survey and partial gene sequences each appended with the name of the variety the sequence was generated from. The final unique sequence set, along with assembly information was placed in the database and passed into the SSR and SNP molecular marker discovery pipelines in order to identify sequence variations.

11.3.3 SEQUENCE ANNOTATION

Comparative analysis against the annotated sequences of well-characterized plant genomes, primarily *Arabidopsis thaliana*, formed the basis of functional assignment of mango DNA sequences. Tools such as BLAST, InterProScan (Zdobnov and Apweiler, 2001) and the gene ontology tools blast2go (Conesa *et al.*, 2005) and Mercator (Thimm *et al.*, 2004; Crowhurst *et al.*, 2008) were used to assign putative function based on sequence and structural similarity. The compiled set of unique sequences generated from each EST and RRL were annotated using blastx (Altschul *et al.*, 1997) against *A. thaliana* peptides from the *Arabidopsis* Information Resource (TAIR8; Swarbreck *et al.*, 2008) and a reference set of plant proteins extracted from sequence databases housed at NCBI. These protein datasets were chosen in order to produce a comprehensive and reliable set of annotations. The use of the *A. thaliana* peptides provides several other advantages. Firstly, the *Arabidopsis* genome is the most thoroughly investigated in the plant kingdom and is now in its tenth revision (Lamesch *et al.*, 2012). Using high quality species-centric databases such as TAIR, also allows leveraging of information from other resources such as AraCyc (Mueller *et al.*, 2003) and AraPath (Lai *et al.*, 2012) for analyzing pathways as well as the breadth of comparative knowledge and tools provided in databases accessible through the Plant Metabolic Network (www.plantcyc.org) and Phytozome (www.phytozome.net; Goodstein *et al.*, 2012). Complementing the *Arabidopsis* resource with a high quality reference set of protein sequences from representa-

tives across the plant kingdom serves not only to reinforce the annotations assigned to sequences based on matches to *Arabidopsis* but also to assign potential
function to gene sequences with no match to peptides in the TAIR dataset. Functional annotations were related to sequences and contig assemblies in the database. Powerful sequence search facilities provide a powerful way of collating
sequences based on sequence name, mango variety, and description keywords.

11.4   MOLECULAR MARKER DISCOVERY

Molecular markers are sequence variations which, when queried, provide information about the sample being assessed. This information may contribute to
the knowledge of relatedness between trees in a breeding program or at a global
level in diversity studies. Sequence variations linked to phenotypes can be used
to make selections and track the pedigree of that selection through subsequent
rounds of genetic crosses. There is a large range of molecular marker technologies that have been applied to mango. In 2007, Vasanthaiah *et al.* provided an
overview of the application of molecular markers as applied to genetic analysis in mangoes. This review covered the use of non-DNA, isozyme typing, as
well as a suite of DNA marker methods which included random amplification
of polymorphic DNA and simple sequence repeat markers in genetic diversity
applications. The increased availability of high information output sequencing
technologies has shifted the focus from examining random banding patterns,
such as those produced in RAPDs, to interrogating specific nucleotides within
the genome and typing panels of such variations in order to produce a genetic
profile associated with either traits or genetic interrelatedness.

11.4.1   SIMPLE SEQUENCE REPEAT DISCOVERY

Simple sequence repeats, short tandem repeats (STRs) or microsatellites are
repeating units of two (dinucleotide repeat) to six (hexanucleotide repeat) base
pair DNA sequences. These codominantly inherited sequences, believed to arise
as a result of slipped-strand mispairing during replication (Levinson and Gutman, 1987), have been the basis of human forensic and paternity testing since
replacing RFLPs in the mid-1990s. In agriculture, SSR markers have been applied to mapping, parentage assignment and verification, as well as inferring
phylogenetic relationships between species and varieties. Microsatellite enrichment libraries have been prepared from mango genomic DNA and used to examine the genetic relationship between 28 mango genotypes from diverse geographic regions (Viruel *et al.*, 2005). SSR variations within expressed sequences
(EST-SSRs) can have a pronounced effect on the structure of the protein product

and can serve as potential molecular markers for traits influenced by these gene products.

Computational tools, such as SPUTNIK (Abajian, 1994), facilitate the identification of simple sequence repeat motifs within DNA sequences. These original tools have subsequently been extended to automatically design PCR primers in the sequences flanking these motifs. Tools, such as SSRprimer (Jewell *et al.*, 2006), use the functionality of SPUTNIK to identify SSR sequences and feed the output into primer3 (Rozen and Skaletsky, 2000), an industry standard PCR primer design tool, to generate ready to use assays.

The mango EST unigene and RRL unique sequences were scanned for putative SSR motifs using an in-house Python script and the primer3 core (Table 11.1). Runs of homopolymers were ignored and the minimum length of dinucleotide repeats was set at six units and four repeat units for tri-, tetra-, penta- and hexanucleotide repeats. EST-SSR containing sequences were processed through the primer3 stage three times in order to generate suitable primers for PCR amplicons ranging from 150 to 250 base pairs, 250 to 350 base pairs and 350 to 450 base pairs. This allows for more complex and higher density multiplexed SSR panels to be assembled. In total, 7 % of the mango EST sequences contained SSRs while only 1 % of the RRL sequences contained repeat motifs. The number of SSR containing sequences where PCR primers could be designed differed considerably between the datasets with 31 % of the RRL-SSRs resulting in suitable assays compared to 72 % for the longer EST-SSR sequences (Innes *et al.*, 2009). Putative SSRs, along with assay information, are indicated in the Mango Genomics database and primer characteristics such as sequence and Tm are displayed in a dedicated panel (Figure 11.3).

TABLE 11.1   Summary of putative SSRs identified within EST and RRL sequences*.

Putative SSRs	Di	Tri	Tetra	Penta	Hexa	Total
KP red leaf	84	347	12	3	8	454
KP fruit skin	60	210	19	1	8	298
KP flower	51	245	9	9	12	326
KP root	39	355	22	2	20	438
Irwin red leaf	62	210	8	4	2	286
EST total	296	1367	70	19	50	1802
Putative SSRs	Di	Tri	Tetra	Penta	Hexa	Total
RRL (*Eco*RI/*Mse*I)	186	349	72	18	13	638
RRL (*Pst*I/*Cvi*AII)	309	476	73	12	9	879
Genomic Total	495	825	145	30	22	1517

TABLE 11.1   *(Continued)*

Putative SSRs	Di	Tri	Tetra	Penta	Hexa	Total
Putative SSR Assays						
KP red leaf	42	262	10	1	6	321
KP fruit skin	34	152	9	1	7	203
KP flower	16	206	1	2	10	235
KP root	30	279	16	2	18	345
Irwin red leaf	33	167	5	1	2	208
EST total	155	1066	41	7	43	1312
RRL (*Eco*RI/*Mse*I)	32	104	10	3	3	152
RRL (*Pst*I/*Cvi*AII)	108	200	17	3	3	331
Genomic Total	140	304	27	6	6	483

*SSRs were mined from EST sequence libraries from Kensington Pride (KP) red leaf, fruit skin, flower, and root tissue. EST sequences were also generated from red leaf tissue collected from the Irwin variety. Genomic SSRs were identified from KP genomics DNA digested with two enzyme combinations: *Eco*RI/*Mse*I and *Pst*I/*Cvi*AII (Innes *et al.*, 2009).

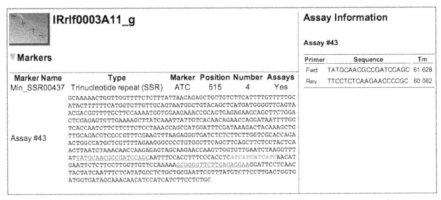

FIGURE 11.3   EST sequence containing a trinucleotide repeat and associated PCR primers; the SSR motif is marked in italics whilst the location of the forward and reverse primer sequences are underlined.

11.4.2   SINGLE NUCLEOTIDE POLYMORPHISM DISCOVERY

Single nucleotide polymorphisms are the most common type of sequence variation. It is estimated that SNPs occur at approximately every 700 base pairs across the *Bos taurus* genome, every 285 base pairs across the *Bos indicus* genome (Bovine HapMap Consortium *et al.*, 2009) and at the time of writing there was over 5 million rice SNPs listed in databases at NCBI. SNPs are mostly biallelic and generally do not have the power of a single microsatellite locus but are far more abundant and evenly spaced throughout the genome making them excellent candidates as molecular markers. For example, flesh color in papaya has been linked to SNPs in the chromoplast-specific lycopene β-cyclase (Blas *et al.*, 2010; Devitt *et al.*, 2010). From a technology perspective, high-throughput SNP typing platforms make the analysis of large number of SNP variants significantly simpler than assessing SSR markers by capillary electrophoresis of fluorophore-labeled PCR products.

AutoSNPdb (Barker *et al.*, 2003) and QualitySNP (Tang *et al.*, 2006) are just two open source examples of software solutions for mining SNP data from contig assembly data. Commercial software packages, such as Geneious (Biomatters, New Zealand) and CLC Genomics Workbench (CLC Bio, Denmark), provide an integrated approach to sequence assembly, annotation, and variant discovery using a combination of custom and industry standard algorithms and approaches. SNPs within the Sanger sequenced mango ESTs were identified using QualitySNP directly from the ace files produced during the cap3 assembly processes.

Identifying high quality SNPs from pyrosequenced RRLs can require some additional processing. Pyrosequencing converts the serial addition of nucleotides (flows) complementary to the template sequence into a series of base calls by measuring light emission. These emissions result from the activities of luciferase. The luciferase substrate, ATP, is generated by sulfurylase and the pyrophosphate released during the incorporation of nucleotides (Ronaghi, 2001). Prior to novel sequence being generated, a four base key comprising one of each base is used to reference the light intensities that correlate to the incorporation of a single nucleotide. When multiple nucleotides of the same type are added in a single flow, the intensity of light emissions should be proportional to the number of bases added. In some cases, ambiguous intensities can result in an either over- or underestimation of the number of bases being added. This pyrosequencing noise can confound SNP identification and downstream SNP assay design so it is important to remove these artifacts prior to generating SNP lists. Current methodologies for "denoising" pyrosequencing data come from metagenomics research using microbial 16S ribosomal subunit gene sequences for organism identification and inferring phylogeny (Quince *et al.*, 2009; Caporaso *et al.*, 2010; Schloss *et al.*, 2011). These projects rely on accurate sequence alignments to predict the relationships between organisms and for the identification of potentially new species. Potential sequence aberrations resulting from the insertion of spurious bases were removed using tools developed locally. In the absence of a reference sequence, gaps in each contig consensus were interrogated against each individual reads in the contig assembly. These gaps were subsequently removed where the sequence context suggested they were the result of pyrosequencing noise. This was determined by examining the nucleotides flanking the gap for the presence of short homopolymeric runs. Once all the gaps in the consensus sequence have been examined, a clean assembly was produced and passed into the SNP discovery pipeline. An overview of the sequence analysis and identification pipeline is provided in Figure 11.4. SNP information was parsed out of the marker discovery pipeline into the Mango Genomics database. On viewing the sequence assemblies, the read alignments can be reduced to haplotypes displaying only variations in a consecutive block (Figure 11.5).

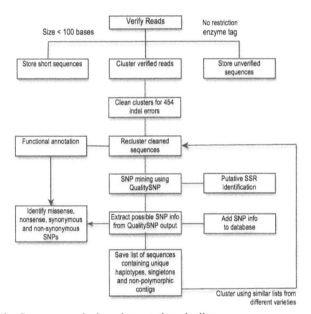

FIGURE 11.4   Sequence analysis and annotation pipeline.

11.5   BIOCHEMICAL PATHWAYS

Many mango fruit traits, including flesh color and flavor, can be linked to discrete sets of biochemical compounds. During the ripening process, fruit soften, change color and exhibit other altered sensory characteristics. Many of these changes are mediated by gene products in well-characterized pathways and these pathways are conserved between plant species. Using sequence annotations based on *A. thaliana*, together with information available in AraCyc, the MGIS allows extensive searching of biochemical pathways based on a pathway name, enzyme commission number, or the name of either an enzyme of a compound. The system supports an extensive list of compound synonyms in order to alleviate the need to provide the exact compound name as it appears in the AraCyc database. The database was used to search for sequences of genes involved in the development of fruit color. A search for the anthocyanin and carotenoid biosynthesis pathways identified a number of sequences for enzymes involved in these pathways within the Mango Genomics database. These results are summarized in Table 11.2 (Innes *et al.*, 2009) and the results of these searches have been applied to microarray and target enrichment experiments to aid in further understanding the roles these play in color development.

TABLE 11.2 Candidate gene sequences for anthocyanin and carotenoid biosynthesis identified within expressed sequence tag (EST) and reduced representation (RRL) libraries (Innes *et al.*, 2009).

Arabidopsis Gene	Gene Function/Activity	EST	RRL
Candidate Anthocyanin Genes			
TT4	chalcone synthase	√	√
TT5	chalcone isomerase	√	√
F3H	flavanone 3-hydroxylase	√	
TT3	dihydroflavonol 4-reductase	√	√
TT8	flavonoid 3-hydroxylase	√	√
TT18	anthocyanidin synthase	√	√
UGT73B2	flavonoid 3-glucosyltransferase	√	√
AACT1	anthocyanin 5-aromatic acyltransferase	√	√
ANL2	anthocyaninless	√	√
PAP2	Myb transcription factor		√
ATAN11	WD-repeat family protein		√
Candidate Carotenoid Genes			
PSY	Phytoene synthase	√	√
PDS	Phytoene desaturase	√	√
ZDS	Zeta-carotene desaturase		√
CRTISO	Carotenoid isomerase		√
LCY-B	Lycopene beta-cyclase	√	√

TABLE 11.2   *(Continued)*

Arabidopsis Gene	Gene Function/Activity	EST	RRL
LCY-E	Lycopene epsilon-cyclase	√	√
CHY1	Beta-carotene hydroxylase		√
LUT5	Carotene beta-hydroxylase		√
ZEP	Zeaxanthin epoxidase	√	√
NCED	9-cis-epoxycarotenoid dioxygenase	√	√

In addition to desirable tropical fruit traits, mangoes are a potential source of bioactive compounds with a range of health promoting activities (Wilkinson *et al.*, 2008; Wilkinson *et al.*, 2011; Taing *et al.*, 2012). Flavonoids, such a quercetin, are found in a wide variety of fruits and nuts, including mango. These compounds exhibit a range of bioactivities including acting as antihypertensives, anti-inflammatories, and vasodilators as well as preventing the oxidation of low-density lipoproteins. By contrast, mangiferin, also found in mango, is found in a smaller number of plant species. Separation sciences such as liquid chromatography and mass spectrometry are allowing researchers to identify chemical compounds present within a range of mango fruit skin and pulp fractions (Wilkinson *et al.*, 2011). These studies examine subsets of the mango metabolome, sets of small molecules produced by the network of chemical reactions occurring within the mango cells.

The facility to search for chemical reactions, enzymes, and compounds makes sequence and molecular marker data associated with these compounds easily available. Figureure 11.5 shows an example search for the compound quercetin. The initial search results display the name of the compound, the name of the enzyme that facilitates the reaction, and the *Arabidopsis* gene identifier (Figure 11.5a). A mango icon next to the *Arabidopsis* gene id indicates that a homologous mango sequence is present in the database. Clicking on either the enzyme name or *Arabidopsis* gene id will open a new browser tab displaying the gene's PlantCyc entry housed at the Plant Metabolic Network. Selecting the mango icon opens a summary of the sequences stored in the Mango Genomics database that have the matching *Arabidopsis* gene id annotation (Figure 11.5b). In addition to the sequence name and length, this screen also displays the sequence depth for assembled contigs and if any sequence variations or putative molecular markers are associated with each sequence. From here, selected sequences can also be retrieved in fasta format for further analysis in third party software tools. Clicking on a sequence name opens a tab containing specific information pertaining to the sequence (Figure 11.5c). This data includes a fasta version of the sequence, the complete blast annotation profile, an assembly viewer for sequence contigs and a summary of any sequence variations. A checkbox presented under the contig viewer allows the assembled sequences to be reduced to haplotypes which, when used in conjunction with the variations list, make identifying and selecting putative SNPs very simple.

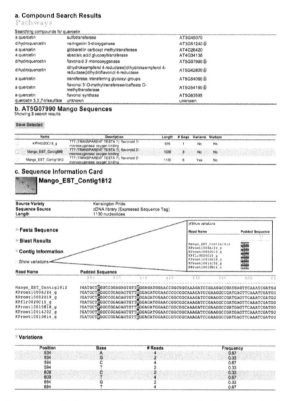

FIGURE 11.5 Search results for quercetin. a: search results for quercetin; b: a screen capture showing the result of clicking on the mango icon next to AT5G07990; c: a sample sequence information card. The inset shows the output when the "show variations" checkbox is ticked.

11.6 PHENOTYPES

Facilities such as The Plant Accelerator (South Australia, Australia) and The High Resolution Plant Phenomics Centre (Canberra, Australia) provide technology platforms for high-throughput plant phenotying capable of handling hundreds of plants per day. These facilities, which combine highly configurable growth environments with multiple imaging, image analysis and sensing technologies, are especially well suited to phenotyping crops such as wheat and barley as well as *Arabidopsis* (Furbank and Tester, 2011). Horticultural species, such as mango, provide challenges in phenotyping, which are not easily handled by current high-throughput methodologies. Characteristics such as large tree architecture and long generation times mean that phenotypic data is largely manually collected and curated. Repeated measurements are made for phenotypes with significance and application to consumer traits (fruit weight, skin color, blush color and intensity, pulp color and flavor), production traits (tree bloom intensity) and tree architecture characteristics such as trunk diameter and branch angles.

Detailed phenotypic measurements made across multiple production seasons are essential for identifying putative QTLs and developing molecular markers through association mapping. While managing the phenotypic data of a small number of parental lines used in a breeding program may be feasible, the complexity of assessing the performance of hundreds of progeny for multiple phenotypes can be greatly simplified using information management and data presentation strategies. In addition to being able to use familiar search strategies to identify accessions that meet any combination of phenotypic criteria, The Mango Genomics database uses a heat map approach for displaying phenotypic variations for a large number of accessions simultaneously (Figure 11.6). This presentation of phenotypic data provides a single location where breeders and researchers can quickly assess phenotypes and make selections for further breeding or larger scale production.

a. Fruit Weight

Fruit Weight ⬥ View

Legend

90.6	191.5	292.5	393.4	494.3	595.3	696.2	797.1	898.1	999.0
52	50	55	1274	63	1266	9013	64	4009	1247
1268	9007	9005	1204	1269	9009	1232	1208	1240	9004
1235	1239	1261	1264	1241	1201	56	1257	1242	9011
1227	1207	1218	66	53	9001	1229	1212	5021	1236
1253	1255	1246	1211	1250	8005	9006	1252	1228	1200
9000	58	65	1222	1230	1213	1254	9002	1251	1202
1233	1224	54	1217	9015	1215	1221	1259	1234	9012
8003	1226	1263	1243	1223	60	62	1214	8004	1238
1270	9008	5020	59	1216	1210	1199	51	49	1271
1249	57	1273	1248	9003	1198	1209	1203	1237	9010
1272	1262	1260	1220	1219	1265	1244	1258	1205	1267
1231									

b. Percent Blush

Percent Blush ⬥ View

Legend

0.0	8.4	16.8	25.1	33.5	41.9	50.3	58.6	67.0	75.4
4009	1226	1211	9015	1212	1267	1253	1198	1224	63
9003	1231	1254	1229	1205	1274	1236	1214	1209	1217
1216	1202	9011	62	1230	1265	1227	1234	55	5021
1207	1271	1252	1268	1215	1228	9013	1221	1233	1273
1199	1242	1248	1237	1263	1243	5020	1223	1232	1235
1266	1208	1200	9009	58	1238	1210	1220	1222	1255
8004	9005	53	1218	9012	1264	1219	1260	1244	51
1270	1250	1259	9004	1241	1204	1251	1203	56	8003
1262	1261	49	9008	9001	54	9006	1272	9002	60
8005	65	1240	59	1239	57	1249	1258	9010	1213
64	1269	66	1246	9007	50	9000	1247	1257	1201
52									

FIGURE 11.6 Heat Map approach to viewing phenotypic information for two different phenotypes.

11.7   FLAVOUR AND AROMA CHEMISTRY

Mangoes exude a large range of volatile compounds that contribute to the highly desirable mango aroma and flavor. Approximately 70 % to 90 % of these compounds are either monoterpenes or sesquiterpenes. Variability in volatile constituents has been demonstrated between cultivars (MacLeod and Pieris, 1984) and has been shown to change during development and maturation (Pandit *et al.*, 2009). These volatiles are typically analyzed by gas chromatography (GC) or GC in combination with mass spectrometry (GC-MS) on extracted fruit pulp, skin, or other tissues. Some researchers profile mango flavor volatiles using trained panel tasters (Suwonsichon *et al.*, 2012) while others use chemosensors such as the zNose (Li *et al.*, 2009). The sensory data collected as part of the Mango Genomics Initiative come from GC-MS traces derived from mango pulp extracts prepared at the eating-ripe stage. These measurements were replicated for extracts from 20 parental lines and hundreds of hybrids. Over 8,000 raw data points per GC trace were collected and stored in the Mango Genomics database (Dietzgen *et al.*, 2009). Storing the raw data rather than generating prerendered trace images allows for a range of enhanced functionality, most notably the ability to zoom in on particular regions of the trace as well as performing data manipulations in real time. Currently, any number of GC traces for individual samples can be viewed. This is useful for looking at the sensory chemical profile of a hybrid progeny as well as its parents. If only two samples are selected, the trace information provided changes slightly. In addition to displaying the traces for the individual samples, a third trace resulting from the subtraction of one trace from the other gives a concise view of what compounds are differentially present and the magnitude of the difference. For example, peaks above the x-axis in this difference graph are in excess in the first sample selected whilst those below the x-axis are more prevalent in the second selected sample (Figure 11.7). The mass spectrometry data, which is used to identify the compounds under each peak, is used to create a HTML image map which overlays the traces. These compounds are related to pathway data, gene sequences and variations stored in the database allowing researchers to select a peak on a trace and have access to all other information related to that peak. By using these chemical fingerprints and their characteristic differences allows researchers to quickly find genes and variations within genes or their expression, that may be involved in the different sensory experiences that drive consumer preferences.

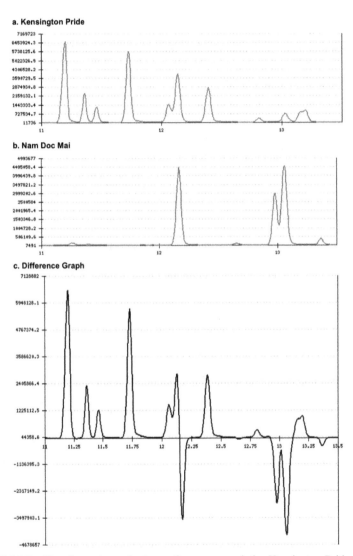

FIGURE 11.7 Gas chromatography traces for mango varieties Kensington Pride and Nam Doc Mai. a: Kensington Pride GC trace displaying peaks with retention times between 11 and 13.5 minutes; b: Nam Doc Mai GC trace displaying peaks with retention times between 11 and 13.5 minutes; c: a graph of the differences between Kensington Pride and Nam Doc Mai samples. Peaks above the x-axis are more prevalent in the Kensington Pride variety whilst those below the x-axis are more specific to Nam Doc Mai.

11.8   TOOLS AND APPLICATIONS

The Mango Genomics information is a large repository of sequence, genotype, phenotype, and consumer chemistry data, which has been subsequently used in a range of applications to enhance production of new mango varieties. Sequence data has been mined in order to construct microarrays for examining changes in gene expression in leaf, fruit skin, fruit flesh and flower tissues. Using largely EST sequence data, multiple probes per transcript were designed and micro-array analysis undertaken to profile gene expression changes during the fruit development and ripening processes.

Solution and solid phase oligonucleotide hybridization methodologies such as Roche NimbleGen and Agilent SureSelect when used in conjunction with next-generation sequencing platforms allow deep resequencing of a large number of candidate genes putatively involved in production and consumer traits. Sequence data housed in the Mango Genomics database was incorporated in a custom pipeline for the design of oligonucleotide probes applicable for SureSelect hybrid selection. This pipeline took curated candidate gene sequences and produced a validated set of tiled oligonucleotide probe sequences. The probe set generated was suitable for submitting to the SureSelect eArray design tool. The processing pipeline ensured all source sequences were in the same orientation and all probe sequences were compared to the entire mango genomics dataset to ensure any nonspecific and inter-probe interactions were eliminated. Finally, probes comprising low complexity sequences were identified using blast and removed from the probe set (Figure 11.8). The probe sequences produced from this pipeline were used to produce biotinylated cRNA baits that have been used to resequence 280 candidate genes for color, flavor, and tree architecture from over a dozen mango varieties. An example assembly of a section of captured sequence shown in Figure 11.9 demonstrates how additional sequence, in this case as small intron, is obtained from capturing fragmented mango DNA using these SureSelect baits.

a. Tiled Oligos Across the Full Sequence Length

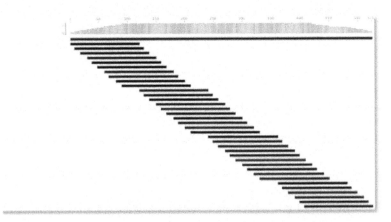

b. Tiled Oligos for Sequence with Excluded Region

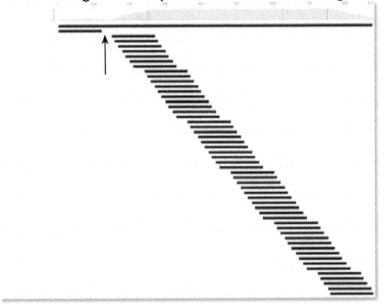

FIGURE 11.8 Design of tiled sequence enrichment oligonucleotides for sequencing mango color and flavor candidate genes. The short horizontal lines depict regions covered by tiled oligonucleotides whilst the long bar represents the source gene sequence. a: a tiled array for a gene sequence without inter-probe complementarity or low complexity sequences; b: a tiled array for a gene sequence with oligonucleotide complementarity. The arrow marks a gap in the source sequence where no probes were designed to avoid undesirable or unpredictable probe interactions.

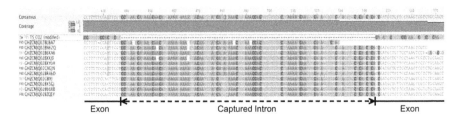

FIGURE 11.9   Assembly of enriched sequence reads shows the capturing of additional sequence. Exonic sequences are marked by the unshaded bases in the assembly diagram while shaded bases are used to illustrate the small intron. As the probes were designed against cDNA sequences, the intron represents additional sequence information obtained by the hybridization of probes against fragmented genomic DNA.

The process of generating new commercial mango varieties involves the production of hybrid populations following the selection of suitable parental lines. The generation of these hybrids lines is a labor-intensive process requiring hand pollination of each recipient plant with pollen from the donor. This is essential to eliminate interspecific hybrids and self-pollination in varieties that are not self-incompatible. An alternative approach is to produce hybrids through open pollination and screen the progeny seedlings using molecular markers specific to each parent in order to select progeny that are the result of the desired cross. We have used SSR markers to screen open pollinated seedlings and have shown that these low cost markers can be readily applied to the identification of hybrid progeny (unpublished data; Figure 11.10).

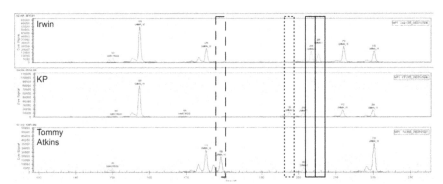

FIGURE 11.10   SSR markers used to distinguish parental lines in the mango breeding program. a: Irwin, solid lined boxes indicate alleles that discriminate Irwin; b: Kensington Pride (KP), dotted line boxes highlight an allele for discriminating KP from the other parental lines; c: Tommy Atkins, the dashed rectangle shows an example of an allele specific for Tommy Atkins.

11.9 CONCLUSION AND FUTURE PERSPECTIVES

The interactive analysis of the linked diverse dataset in the MGIS provides a powerful bioinformatics search tool for improved understanding of mango variety relationships, genetic diversity, and biological and chemical traits. This in turn will be a valuable tool to assist breeders in the selection of progeny with desired characteristics using molecular markers for specific traits already at the seedlings stage, thereby considerably improving breeding efficiency. As with any computer-based tool, user acceptance is critical to the success of the application. The science of human-computer interaction and current dogma for user interface design has been integral to the development of the MGIS. Behind the scenes, the database houses in excess of 375,000 genomic and 25,000 EST sequences and annotations, 42,000 phenotypic measurements, and over 6.4 million GC data points. It has the capacity for further growth and through the inputs of researchers with diverse backgrounds and demands, it is our aim to produce a series of tools and frameworks that meet the needs of a dynamic breeding program which itself aims to provide improved varieties to the global mango market.

KEYWORDS

- Bioinformatics
- Expressed sequence tags
- Genomics
- *Mangifera indica* L.
- Mango genomic information system
- Molecular markers
- Plant phenomics
- Quantitative trait loci
- Shotgun
- Short tandem repeats
- Simple sequence repeats
- Single nucleotide polymorphisms
- Transcriptome

REFERENCES

Abajian, C.; Sputnik, 1994, http://www.abajian.com/sputnik/.

Altschul, S. F.; Madden, T. L.; Schäffer, A. A.; Zhang, J.; Zhang, Z.; Miller, W.; Lipman, D. J. Gapped BLAST and PSI-BLAST: a new generation of protein database search programs. *Nucleic Acids Res* 1997, 25, 3389-3402.

Arumuganathan, K.; Earle, E. D. Nuclear DNA content of some important plant species. *Plant Mol Biol Rep* 1991, 9, 208-218.

Bally, I. S. E. *Mangifera indica* (mango). In *Species Profiles for Pacific Island Agroforestry*, Elevitch, C. R., Ed.; Permanent Agriculture Resources (PAR): Hōlualoa, Hawaii, USA.

Bally, I. S. E. Advances in research and development of mango industry. *Rev Bras Frutic* 2011, 33, 57-63.

Bally, I. S. E.; Lu, P.; Johnson, P. R. *Breeding Plantation Tree Crops: Tropical Species*, Ed.; Jain, S. M.; Priyadarshan, P. M., Springer: New York, 2009.

Barker, G.; Batley, J.; Sullivan, H.; Edwards, K. J.; Edwards, D. Redundancy based detection of sequence polymorphisms in expressed sequence tag data using autoSNP. *Bioinformatics* 2003, 19, 421-422.

Blas, A. L.; Ming, R.; Liu, Z.; Veatch, O. J.; Paull, R. E.; Moore, P. H.; Yu, Q. Cloning of the papaya chromoplast-specific lycopene beta-cyclase, CpCYC-b, controlling fruit flesh color reveals conserved microsynteny and a recombination hot spot. *Plant Physiol* 2010, 152, 2013-2022.

Bovine HapMap Consortium.; Gibbs, R. A.; Taylorm, J. F.; Van Tassell, C. P.; Barendse, W.; Eversole, K. A.; Gill, C. A.; Green, R. D.; Hamernik, D. L.; Kappes, S. M.; Lien, S.; Matukumalli, L. K.; McEwan, J. C.; Nazareth, L. V.; Schnabel, R. D.; Weinstock, G. M.; Wheeler, D. A.; Ajmone-Marsan, P.; Boettcher, P. J.; Caetano, A. R.; Garcia, J. F.; Hanotte, O.; Mariani, P.; Skow, L. C.; Sonstegard, T. S.; Williams, J. L.; Diallo, B.; Hailemariam, L.; Martinez, M. L.; Morris, C. A.; Silva, L. O.; Spelman, R. J.; Mulatu, W.; Zhao, K.; Abbey, C. A.; Agaba, M.; Araujo, F. R.; Bunch, R. J.; Burton, J.; Gorni, C.; Olivier, H.; Harrison, B. E.; Luff, B.; Machado, M. A.; Mwakaya, J.; Plastow, G.; Sim, W.; Smith, T.; Thomas, M. B.; Valentini, A.; Williams, P.; Womack, J.; Woolliams, J. A.; Liu, Y.; Qin, X.; Worley, K. C.; Gao, C.; Jiang, H.; Moore, S. S.; Ren, Y.; Song, X. Z.; Bustamante, C. D.; Hernandez, R. D.; Muzny, D. M.; Patil, S.; San Lucas, A.; Fu, Q.; Kent, M. P.; Vega, R.; Matukumalli, A.; McWilliam, S.; Sclep, G.; Bryc, K.; Choi, J.; Gao, H.; Grefenstette, J. J.; Murdoch, B.; Stella, A.; Villa-Angulo, R.; Wright, M.; Aerts, J.; Jann, O.; Negrini, R.; Goddard, M. E.; Hayes, B. J.; Bradley, D. G.; Barbosa da Silva, M.; Lau, L. P.; Liu, G. E.; Lynn, D. J.; Panzitta, F.; Dodds, K. G. Genome-wide survey of SNP variation uncovers the genetic structure of cattle breeds. *Science* 2009, 324, 528-532.

Caporaso, J. G.; Kuczynski, J.; Stombaugh, J.; Bittinger, K.; Bushman, F. D.; Costello, E. K.; Fierer, N.; Peña, A. G.; Goodrich, J. K.; Gordon, J. I.; Huttley, G. A.; Kelley, S. T.; Knights, D.; Koenig, J. E.; Ley, R. E.; Lozupone, C. A.; McDonald, D.; Muegge, B. D.; Pirrung, M.; Reeder, J.; Sevinsky, J. R.; Turnbaugh, P. J.; Walters, W. A.; Widmann, J.; Yatsunenko, T. QIIME allows analysis of high-throughput community sequencing data. *Nat Methods* 2010, 7, 335-336.

Chong, Z.; Ruan, J.; Wu, C. I. Rainbow: an integrated tool for efficient clustering and assembling RAD-seq reads. *Bioinformatics* 2012, 28, 2732-2737.

Conesa, A.; Götz, S.; García-Gómez, J. M.; Terol, J.; Talón, M.; Robles, M. Blast2GO: a universal tool for annotation, visualization and analysis in functional genomics research. *Bioinformatics* 2005, 15(21), 3674-3676.

Crowhurst, R. N.; Gleave, A. P.; MacRae, E. A.; Ampomah-Dwamena, C.; Atkinson, R. G.; Beuning, L. L.; Bulley, S. M.; Chagne, D.; Marsh, K. B.; Matich, A. J.; Montefiori, M.; Newcomb, R. D.; Schaffer, R. J.; Usadel, B.; Allan, A. C.; Boldingh, H. L.; Bowen, J. H.; Davy, M. W.; Eckloff,

R.; Ferguson, A. R.; Fraser, L. G.; Gera, E.; Hellens, R. P.; Janssen, B. J.; Klages, K.; Lo, K. R.; MacDiarmid, R. M.; Nain, B.; McNeilage, M. A.; Rassam, M.; Richardson, A. C.; Rikkerink, E. H. A.; Ross, G. S.; Schröder, R.; Snowden, K. C.; Souleyre, E. J. F.; Templeton, M. T.; Walton, E. F.; Wang, D.; Wang, M. Y.; Wang, Y. Y.; Wood, M.; Wu, R.; Yauk, Y.; Laing, W. A. Analysis of expressed sequence tags from Actinidia: applications of a cross species EST database for gene discovery in the areas of flavor, health, color and ripening. *Bmc Genomics* 2008, 9, 351.

Davey, J. W.; Hohenlohe, P. A.; Etter, P. D.; Boone, J. Q.; Catchen, J. M.; Blaxter, M. L. Genome-wide genetic marker discovery and genotyping using next-generation sequencing. *Nat Rev Genet* 2011, 12, 499-510.

Devitt, L. C.; Fanning, K.; Dietzgen, R. G.; Holton, T. A. Isolation and functional characterization of a lycopene beta-cyclase gene that controls fruit colour of papaya (*Carica papaya* L.). *J Exp Bot* 2010, 61, 33-39.

Dietzgen, R.; Bally, I. S. E.; Devitt, L. C.; Dillon, N. L.; Fanning, K.; Gidley, M.; Holton, T. A.; Innes, D.; Karan, M.; Sheik-Jabbari, J.; Smyth, H. Mango genetics underpin efficient breeding for variety improvement, Dunmall, T., Ed.; The Seventh Australian Mango Conference, AMIA, Cairns, Queensland, Australia, 2009, pp. 10-12.

Furbank, R. T.; Tester, M. Phenomics–technologies to relieve the phenotyping bottleneck. *Trends Plant Sci* 2011, 16, 635-644.

Goodstein, D. M.; Shu, S.; Howson, R.; Neupane, R.; Hayes, R. D.; Fazo, J.; Mitros, T.; Dirks, W.; Hellsten, U.; Putnam, N.; Rokhsar, D. S. Phytozome: a comparative platform for green plant genomics. *Nucleic Acids Res* 2012, 40, D1178-D1186.

Huang, X.; Madan, A. CAP3: A DNA sequence assembly program. *Genome Res* 1999, 9, 868-877.

Innes, D. J.; Devitt, L. C.; Karan, M.; Dillon, N. L.; Sheikh-Jabbari, J.; Bally, I. S. E.; Dietzgen, R. G.; Holton, T. A. Mango Genomics, *14th Australasian Plant Breeding Conference*, Cairns, Australia, 2009.

Jaccoud, D.; Peng, K.; Feinstein, D.; Kilian, A. Diversity Arrays: a solid state technology for sequence information independent genotyping. *Nucleic Acids Res* 2001, 15, 29(4), e25.

Jewell, E.; Robinson, A.; Savage, D.; Erwin, T.; Love, C. G.; Lim, G. A. C.; Li, X.; Batley, J.; Spangenberg, G. C.; Edwards, D. SSRPrimer and SSR Taxonomy Tree: Biome SSR discovery. *Nucleic Acids Res* 2006, 34, W656-9.

Krishna, H.; Singh, S. K. Biotechnological advances in mango (*Mangifera indica* L.) and their future implication in crop improvement: a review. *Biotechnol Adv* 2007, 25, 223-243.

Kulkarni, V. J.; Bally, I. S. E.; Brettell, R. I. S.; Johnson, P. R.; Hamilton, D. The Australian National Mango Breeding Program-In search of improved cultivars for the new millennium. *Acta Horticulturae* 2002, 575, 287-294.

Lai, L.; Liberzon, A.; Hennessey, J.; Jiang, G.; Qi, J.; Mesirov, J. P.; Ge, S. X. AraPath: a knowledgebase for pathway analysis in Arabidopsis. *Bioinformatics*, 2012, 1, 28(17), 2291-2292.

Lamesch, P.; Berardini, T. Z.; Li, D.; Swarbreck, D.; Wilks, C.; Sasidharan, R.; Muller, R.; Dreher, K.; Alexander, D. L.; Garcia-Hernandez, M.; Karthikeyan, A. S.; Lee, C. H.; Nelson, W. D.; Ploetz, L.; Singh, S.; Wensel, A.; Huala, E. The Arabidopsis Information Resource (TAIR): improved gene annotation and new tools. *Nucleic Acids Res* 2012, 40, D1202-1210.

Levinson, G.; Gutman, G. A. Slipped-strand mispairing: a major mechanism for DNA sequence evolution. *Molecular Biology and Evolution* 1987, 4, 203-221.

Li, Z.; Wang, N.; Vijaya Raghavan, G. S.; Vigneault, C. Ripeness and rot evaluation of "Tommy Atkins" mango fruit through volatiles detection. *Journal of Food Engineering* 2009, 91, 319-324.

MacLeod, A. J.; Pieris, N. M. Comparison of the volatile components of some mango cultivars. *Phytochemistry* 1984, 23, 361-366.

Maughan, P. J.; Udall, J A.; Yourstone, S. M. SNP Discovery via Genome Reduction, Barcoding, and 454-Pyrosequencing in Amaranth. *The Plant Genome* 2009, 2, 260-270.

Metzker, M. L. Sequencing technologies - the next generation. *Nat Rev Genet* 2010, 11, 31-46.

Mueller, L. A.; Zhang, P.; Rhee, S. Y. AraCyc: a biochemical pathway database for Arabidopsis. *Plant Physiol* 2003, 132, 453-460.

Mukherjee, S. K. Mango: its allopolyploid nature. *Nature* 1950, 166, 196-197.

Pandit, S. S.; Kulkarni, R. S.; Chidley, H. G.; Giri, A. P.; Pujari, K. H.; Köllner, T. G.; Degenhardt, J.; Gershenzon, J.; Gupta, V. S. Changes in volatile composition during fruit development and ripening of "Alphonso" mango. *J Sci Food Agric* 2009, 89, 2071-2081.

Quince, C.; Lanzén, A.; Curtis, T. P.; Davenport, R. J.; Hall, N.; Head, I. M.; Read, L. F.; Sloan, W. T. Accurate determination of microbial diversity from 454 pyrosequencing data. *Nat Methods* 2009, 6, 639-641.

Renuse, S.; Harsha, H. C.; Kumar, P.; Acharya, P. K.; Sharma, J.; Goel, R.; Kumar, G. S. S.; Raju, R.; Prasad, T. S. K.; Slotta, T.; Pandey, A. Proteomic analysis of an unsequenced plant - *Mangifera indica*. *J Proteomics* 2012, 75, 5793-5796.

Ronaghi, M. Pyrosequencing sheds light on DNA sequencing. *Genome Res* 2001,11, 3-11.

Rozen, S.; Skaletsky, H. Primer3 on the WWW for general users and for biologist programmers. *Methods Mol Biol* 2000, 132, 365-386.

Schloss, P. D.; Gevers, D.; Westcott, S. L. Reducing the effects of PCR amplification and sequencing artifacts on 16S rRNA-based studies. *PLoS One* 2011, 6, e27310.

Staden, R.; Beal, K. F.; Bonfield, J. K. The Staden package, 1998. *Methods Mol Biol* 2000, 132, 115-130.

Suwonsichon, S.; Chambers, E.; Kongpensook, V.; Oupadissakoon, C. Sensory lexicon for mango as affected by cultivars and stages of ripeness. *J Sens Stud* 2012, 27, 148-160.

Swarbreck, D.; Wilks, C.; Lamesch, P.; Berardini, T. Z.; Garcia-Hernandez, M.; Foerster, H.; Li, D.; Meyer, T.; Muller, R.; Ploetz, L.; Radenbaugh, A.; Singh, S.; Swing, V.; Tissier, C.; Zhang, P.; Huala, E. The Arabidopsis Information Resource (TAIR): gene structure and function annotation. *Nucleic Acids Res* 2008, 36, D1009-1014.

Taing, M. W.; Pierson, J. T.; Hoang, V. L. T.; Shaw, P. N.; Dietzgen, R. G.; Gidley, M.; Roberts-Thomson, S. J.; Monteith, G. R. Mango fruit peel and flesh extracts affect adipogenesis in 3T3-L1 cells. *Food Funct* 2012, 3(8), 828-836.

Tang, J.; Vosman, B.; Voorrips, R. E.; van der Linden, C. G.; Leunissen, J. A. M. QualitySNP: a pipeline for detecting single nucleotide polymorphisms and insertions/deletions in EST data from diploid and polyploid species. *BMC Bioinformatics* 2006, 7, 438.

Thimm, O.; Bläsing, O.; Gibon, Y.; Nagel, A.; Meyer, S.; Krüger, P.; Selbig, J.; Müller, L. A.; Rhee, S. Y.; Stitt, M. Mapman: a user-driven tool to display genomics data sets onto diagrams of metabolic pathways and other biological processes. *The Plant Journal* 2004, 37, 914-939.

van Orsouw, N. J.; Hogers, R. C. J.; Janssen, A.; Yalcin, F.; Snoeijers, S.; Verstege, E.; Schneiders, H.; van der Poel, H.; van Oeveren, J.; Verstegen, H.; van Eijk, M. J. T. Complexity reduction of polymorphic sequences (CRoPS): a novel approach for large-scale polymorphism discovery in complex genomes. *PLoS One* 2007, 2, e1172.

Viruel, M. A.; Escribano, P.; Barbieri, M.; Ferri, M. Fingerprinting, embryo type and geographic differentiation in mango (*Mangifera indica* L., Anacardiaceae) with microsatellites. *Molecular Breeding* 2005, 15, 383-393.

Wicker, T.; Narechania, A.; Sabot, F.; Stein, J.; Vu, G. T. H.; Graner, A.; Ware, D.; Stein, N. Low-pass shotgun sequencing of the barley genome facilitates rapid identification of genes, conserved non-coding sequences and novel repeats. *BMC Genomics* 2008, 9, 518.

Wilkinson, A. S.; Flanagan, B. M.; Pierson, J. T.; Hewavitharana, A. K.; Dietzgen, R. G.; Shaw, P. N.; Roberts-Thomson, S. J.; Monteith, G. R.; Gidley, M. J. Bioactivity of mango flesh and peel extracts on peroxisome proliferator-activated receptor γ [PPARγ] activation and MCF-7 cell proliferation: fraction and fruit variability. *J Food Sci* 2011, 76, H11-8.

Wilkinson, A. S.; Monteith, G. R.; Shaw, P. N.; Lin, C. N.; Gidley, M. J.; Roberts-Thomson, S. J. Effects of the mango components mangiferin and quercetin and the putative mangiferin metabolite norathyriol on the transactivation of peroxisome proliferator-activated receptor isoforms. *J Agric Food Chem* 2008, 56, 3037-3042.

Willing, E. M.; Hoffmann, M.; Klein, J. D.; Weigel, D.; Dreyer, C. Paired-end RAD-seq for *de novo* assembly and marker design without available reference. *Bioinformatics* 2011, 27, 2187-2193.

Zdobnov, E. M.; Apweiler, R. InterProScan-an integration platform for the signature-recognition methods in InterPro. *Bioinformatics* 2001, 17(9), 847-848.

CHAPTER 12

ENVIRONMENTAL GENOMICS: THE IMPACT OF TRANSGENIC CROPS ON SOIL QUALITY, MICROBIAL DIVERSITY, AND PLANT-ASSOCIATED COMMUNITIES

JEYABALAN SANGEETHA, DEVARAJAN THANGADURAI,
MUNISWAMY DAVID, ROOPA T. SOMANATH,
ABHISHEK C. MUNDARAGI, and DIGAMBARAPPA P. BIRADAR

CONTENTS

12.1 Introduction 310
12.2 International Scenario on Transgenic Crops 311
12.3 Indian Scenario on Transgenic Crops.................................... 312
12.4 Impact of Transgenic Crops on Soil Quality 314
12.5 Importance of Soil Microbial Diversity 314
12.6 Impact of Transgenic Crops on Soil Microbial Diversity 315
12.7 Molecular Tools for the Analysis of Soil Microbial Diversity 316
12.8 Impact of Transgenic Cotton: A Case Study 322
12.9 Conclusion and Future Perspectives....................................... 323
Keywords 323
References... 324

12.1 INTRODUCTION

Current developments in biotechnology with respect to newer tools and novel technologies in food, agriculture, and environmental sector have changed our understanding of potential benefits and harmful effects of modern approaches in crop productivity, food consumption and environmental preservation. Several classes of modern approaches have been developed, in order to raise food production to meet the demands of ever expanding global population. Among these, development of newer drops with disease resistance against bacterial, viral, and fungal pathogens, crops with drought, temperature, and cold tolerance, better nutritional and fruit quality and of having increased shelf life, all had greater attention among the scientific community and general public due to their potentially visible impact on the society (Butschi *et al.*, 2009). These viable technologies have been increasingly used to develop even better energy and feed crops in the steady change in climate scenario (Butschi *et al.*, 2009). The increase in usage of chemical fertilizers and synthetic pesticides in the past few decades have shown much alterations in the environmental quality and soil fertility in particular (Single *et al.*, 2003; Girvan *et al.*, 2004; Saeki and Toyota 2004; Marschner and Baumann, 2003).

With the advent of molecular marker assisted and transgenic technologies, several disease resistant and drought tolerant crops have been developed so for and are planted across the globe hoping that the crop productivity can be greatly improved to feed the global population. There are two main reasons for the development of transgenic crops. Firstly, insects and pests such as fruit borer, tomato, and diamond black moth in cauliflower, fruit, and shoot borer in brinjal are uncontrollable and create challenge for farmers with the available conventional methods ultimately it reduces the yield of fruits and vegetables. Secondly pesticide residues on fruits and vegetables are exceeds the permissible limit which leads to health hazards to consumers and farmers. Though there is an enormous potential in the usage of these molecular approaches, also of the greater environmental, cultural, ethical, and concerns on the negative impact of the so called transgenic foods. These transgenic goods appears to have deleterious effects on the soil quality, serious effect on nontarget organisms and possible role in altering the soil geochemical cycles and thereby affecting the soil microbiota. In this chapter, we summarize the known impact of transgenic crops on rhizosphere microbial diversity and its impact on the soil microbial community. Molecular methods used for the assessment of microorganisms are also discussed with the future perspectives of transgenic crops on soil environment.

12.2   INTERNATIONAL SCENARIO ON TRANSGENIC CROPS

Great progress has been achieved in agricultural biotechnology after four decades of first GMO produced (1973). 15 years after commercialization of transgenic crops, there are still only two genetic traits that are available and only four crops that represent more than 99 % acreage are soybean, maize, cotton, and canola. Based on 2009 reports, the leading country in using transgenic crop is United States of America with 57.7 million hectares, followed by Argentina (19.1 million hectares), Brazil (15 million hectares), Canada (7 million hectares), India (6.2 million hectares) and China (3.8 million hectares). In Europe the cultivation of transgenic crop is very less and in limit. In 2007 they used only 110,000 hectares for the cultivation of transgenic crops (Butschi et al., 2009).

Unique opportunity of engineering plants with desired genetic traits has been achieved based on the rapid developments in the areas of recombinant DNA technology and plant tissue culture. In 1996, first transgenic plants such as cotton, corn, and soybean were engineered for insect resistance. After discovery of genetic traits, in three decades about 52.6 million hectares have been planted with herbicide and insect tolerance such as cotton, corn, soybean, papaya, potato, and canola. Industrialized countries are occupying 74 % of the transgenic area sown than developing countries which has 26 %. Most of the transgenic area is covered in USA (68 %), Argentina (22 %), Canada (6 %) and China (3 %) based on the country wise list of the distribution of the transgenic area (http://www.tifac.org). In 1996, Bt cotton was first commercialized in USA and followed by Australia in 1996, Argentina and China in 1997, Mexico and South Africa in 1998, Colombia and India in 2002 (Mohan and Manjunath, 2002). At first time, Bt plant commercially planted in 1996 in Mexico and United States. During 2000, Mexico reached 26,300 hectares of Bt cotton area which is one third of the country's cotton growing area. Monsanto and Delta & Pineland Co., have marketed NuCOTN 33B and NuCOTN 35B Bt cotton varieties in six countries including USA and Mexico from 1996 (Traxler and Godoy-Avila, 2004).

At present, recombinant technology covers around 30–40 % of the cotton area in USA and China. Recent studies demonstrate that USA, South Africa, Mexico, and Chinese Bt adopters realize significant pesticide and cost savings (Pray et al., 2002; Traxler and Godoy-Avila, 2004). However, still hesitation among the society related to the technology's impact and sustainability under different agroecological and socioeconomic conditions are going on (Qaim et al., 2003).

For the first time in 1996, USA started to cultivate Bt maize for commercial purpose and these plants are resistant toward lepidopteran. In 2003, introduced genetically modified maize having resistance toward beetle grubs. By the end of 2009 to 2010, Bt maize has been cultivated in 63 % acres for domestic use.

The different Bt maize growers are from Illinois, Minnesota, and Wisconsin. Now 80 % maize produced in US is genetically modified. Genetically modified maize crops started to grow from 1997 in Europe especially Spain involves in large production, say 79,269 hectares compared to Portugal and Germany. Overall, Europe produce 173 million tones ensilage maize, 56 million tons of grain maize and 10 million tons are imported to Argentina. Genetically modified technology also adopted in India for different crops, in that maize is also one. Monsanto involved in research engagements with India for corn breeding and testing. Thousands of small farmers started to use Bt white maize in South Africa. Bt Alfalfa for commercial purpose was done in USA planting 20,000 hectares in the year 2005. Bt Sugar beet was first time grown in USA and Canada. Papaya resistance toward PRSV (papaya ring spot virus) was planted in 2000 hectares in Hawaii and at China 6275 hectares. GM soybeans were produced worldwide among them USA produces more than Brazil, Argentina, Chain, India, and Paraguay. Europe imports 40 million tons of soybeans from Brazil and USA. From 1996 Bt rapeseed was grown at 5.1 million hectares in Europe whereas it is very less in US and in Australia. All around the world, GM rapeseed used is having resistance toward weed. Bt rice also started to produce in different countries around the world, particularly in US at 2006 by USDA (US Department of Agriculture) which produced transgenic rice variety called LL601. Bt rice in India is in active pipeline, it may take 3 years to complete the process. USA for the first time approved potato producing Bt toxin, which is safe to the environment in 1995. Bt tomato variety 5345 which is resistance toward pest was first done by US in 1998. Bt brinjal in India was produced to make pest resistance that was developed by Maharashtra Hybrid Seed Company (MAHYCO) and this was planted in India, Bangladesh, and in Philippines.

12.3   INDIAN SCENARIO ON TRANSGENIC CROPS

India is among the few developing countries which have instituted biosafety regulations and incorporated them in national laws as far back in 1989. The Department of Biotechnology (DBT) is the nodal agency under the Ministry of Science and Technology, Government of India which deals with all aspects of transgenic research, while the Ministry of Environment and Forest deals with the commercial release of the transgenics. Under the Indian Council of Agricultural Research (ICAR), the National Bureau of Plant Genetic Resources (NB-PGR) plays a pivotal role in the import and quarantine processing of transgenic planting material in the country (Mangal *et al.*, 2003).

India has the largest planted area in the world and it was the first to develop transgenic cotton hybrids. 92.8 % of its areas are used to grow crops. India is one of the largest cotton growing countries in the world, entered the club of GMO countries during 2002, with 8.7 million hectares which is equivalent to 25 % of the world cotton growing area. In 2002, India received approval to release of the three transgenic Bt-cotton hybrids. With this agreement Bt cotton growing countries increased from 6 in 1996 to 16 in 2012. However, transgenic planting material such as variety CISA 614 and HD123 for research purposes was imported during 2004–2007 in 32 locations of India and recorded mean seed cotton yield of 2204 Kg/Ha and 1834 Kg/Ha respectively (Mangal et al., 2003). The variety CISA 164 recorded highest seed cotton yield of 3792 Kg/Ha in agronomy trials (Gotmare, 2010).

Central Institute of Cotton Research (CICR), India has developed CISA 310 of *Gossypium arboretum* for cultivation due to its superiority in both seed cotton and lint yield; fiber quality and less insect damage in North zone and notified vide Gazette of India No. 171, 2010 for cultivation. This variety has been tested in Punjab, Haryana, and Rajasthan during 2000–2004 and has recorded overall mean seed cotton yield of 21.7 Q/ha as against 19.7 Q/ha RG8. It has constantly out yielded the zonal check and the maximum increase was upto 40.1 % in irrigated conditions of north zone with mean increase of 10.1 %. The variety CISA 310 recorded a mean ginning outturn of 36.5 % and was higher than zonal check and qualifying varieties. Less damage by diseases and pests was recorded in this variety (Gotmare, 2010).

The Indian Council of Agricultural Research (ICAR) has developed a new transgenic variety of cotton, which is under trial at different locations. The new cotton variety will be known as "Bikaneri Narma". The seed which Indian farmers use is actually a hybrid variety. They have to be dependent on seed companies for transgenic cotton. The new variety will be available to them in 2 years' time. Government of India approving commercial cultivation of Bt-cotton, it can now be expected that in the future other transgenic crops will also be approved for commercialization, because it is clear that developing countries, including India need genetically modified crops to maintain sustainable agricultural production. The Government of India policies (Vision 2010 of DBT, Vision 2020 of ICAR and other documents of other ministries) and scientific climate in India are strongly supportive of the applications of biotechnological tools, including transgenic for achieving sustainable improvement in agricultural productivity.

In India 6.3 million small farmers started to grow Bt crops. Some researchers are involved in Bt crop research in developing countries especially in India. Their main concentration in research development was taken in Bt cotton and Bt brinjal. In 2002 Bt cotton was approved for commercial cultivation and named as Bollgard. From the last 6 years, the growth is increasing and production was taken more than 75 % in different places in India. The main work carried out in India for the development of Bt cotton are Dow Agro Sciences, JK, Agrigenetics Ltd, Mahyco, Metahelix, and CICR. Bt genes involved in production of cotton plants are *cry1Ac, cry1F, cry1EC, cry2A, CP4EPSPS,* and *cry1C.* Bt brinjal research process was carried out in Tamil Nadu Agricultural University, Maharashtra Hybrid Seeds Company and University of Agricultural Sciences, Bangalore having resistance toward insect pest by using *cry1Ac* gene. Bt rice research process was done in Maharashtra Hybrid Seeds Company and also in Bayer by using *cry1Ac* and *cry1Ab* to get rid of pests in rice field. Okra is also genetically modified by using *cry1Ac* that leads to develop insect resistant plant and this process was carried out in Maharashtra Hybrid Seeds Company.

12.4   IMPACT OF TRANSGENIC CROPS ON SOIL QUALITY

Soil plays an important role for the growth and development of plants as soil contain different species of microorganisms which uses extracts of plants and convert them into organic matter, humus, and maintain fertile soil. Plant extract and waste produced by the plants are utilized by the microorganisms for their survival. If there are any changes in the content of waste and extract of plants directly it will affect sometimes to the microorganisms and their functioning in maintaining soil fertility also and that finally affect the growth and development of plants. In recent times, since there is a steady increase in the usage of transgenic plants to get rid of pest damage, weeds, and stress, there are chances of transgenic genes entering in to the soil and sometimes it may affect the functioning of microorganisms. Therefore it is very much necessary to study the impact of transgenic crops on soil microflora by using different molecular techniques like DGGE and use of ELISA method, gives to differentiate the function of microorganism on transgenic and nontransgenic plants.

12.5   IMPORTANCE OF SOIL MICROBIAL DIVERSITY

Soil microbial community plays vital role in biogeochemical cycle (Wall and Virginia, 1999; Kirk *et al.*, 2004), which is responsible for cycling of nutrients available in soil and also it helps in maintaining the above ground ecosystem by contributing nutrients for plant growth (George *et al.*, 1995), soil structure and soil fertility (Cairnay, 2000; Dodd *et al.*, 2000; Ovreas, 2000; Yao *et al.*, 2000; O'Donnell *et al.*, 2001; Kirk *et al.*, 2004). According to Torsvik *et al.* (1998) approximately 1 % of the soil microorganisms can be cultured by standard laboratory practices. It is not confirmed that this 1 % is representative of the soil microbial diversity. Many fungal species cannot be cultured using basic laboratory protocols when compared to bacteria (Thorn, 1997; van Elsas *et al.*, 2000). All the activities in biotic environment is basically depend on microbial community in one way or other (Pace, 1997), while many anthropogenic activities like new agricultural practices, using various chemical fertilizers and pesticides, soil pollution by dumping of wastes, construction can potentially affects and alters soil microbial diversity which leads to changes in above and below ground ecosystem (Kirk *et al.*, 2004).

Soil microbes show, considerable influence upon the overall improvement of plant growth and plants having significant influence on soil microbes in the rhizosphere (Smith and Goodman, 1999; Wall and Virginia, 1999). Soil microbes play important role in various aspects of the terrestrial ecosystem such as improving soil fertility which in turn increase the growth of the plant and also maintaining the balance of nutrients in entire ecosystem by biogeochemical cycle (Collins *et al.*, 1990; Molin and Molin, 1997; Trevors, 1998; Wall and Virginia, 1999; O'Donnell *et al.*, 2001). Soil microorganisms in rhizosphere assist plants in the uptake of several vital nutrients for their growth from the soil (George *et al.*, 1995; Timonen *et al.*, 1996; Trolove *et al.*, 2003; Cocking, 2003).

Soil microorganisms depend on plant for its carbon and in turn provide phosphorous, nitrogen, and other minerals for the growth of plant by decomposing the soil organic matter (Singh *et al.*, 2004). Rhizosphere is having both beneficial microorganisms and pathogens which act as parasites on plants. Hence, rhizosphere is also acts as a battle field where the microflora and microfauna acts against soil borne pathogens like fungi, bacteria, oomycetes, and nematodes and have positive effect on plant growth by providing nutrients and health (George *et al.*, 1995; Timonen *et al.*, 1996; Raaijmakers *et al.*, 2009). Diversities of microorganisms based on its structure and functions in the rhizosphere vary due to the differences in plants root exudates (Brimecombe *et al.*, 2001; Fang *et al.*, 2005).

12.6 IMPACT OF TRANSGENIC CROPS ON SOIL MICROBIAL DIVERSITY

Knowledge on the relationship between plants and soil microbes and impact of transgenic crop on soil microbes are necessary to understand the long term agronomic effect and to develop appropriate management practices for minimizing the negative impacts (Fang *et al.*, 2007). It is important to study the impact of transgenic crops on soil microbial diversity not only for the basic research but also to understand the link between transgenic crop and its rhizosphere microbial community's structure and function. Potential impact of transgenic crops on soil microorganisms are based on the differences in quantity, chemical composition and botanical form of Bt toxin, other soil properties, abiotic factors and also changes in management practices (Fang *et al.*, 2007). However, it is not clearly known if there are any alterations in rhizosphere microbial community means to ecosystem structure and functioning. Hence, understanding of the link between microbes and plants is important for maintaining sustainable ecosystems (Kirk, 2004).

12.7 MOLECULAR TOOLS FOR THE ANALYSIS OF SOIL MICROBIAL DIVERSITY

Many microorganisms are uncultivable or difficult to culture by common laboratory protocols. Conventional methods are insufficient and time consuming for the identification of environmental microorganisms. Active and inactive microorganisms cannot be distinguished by microscopic examinations. Some nonculturable microorganisms cannot grow in culture media for identification (Colwell et al., 2004; Doi et al., 2006; Tani et al., 2006). For these reasons, rapid and accurate methods are required. To overcome this issue, now-a-days various methods have been developed to assess and study the microorganisms including physiological and DNA- and RNA- based molecular methods to meet the need of environmental, medical, and agricultural sciences (Giller et al., 1997; Rondn et al., 1999; van Elsas et al., 2000).

There are number of molecular based tools have been developed and widely used to study soil microbial community analysis which includes DGGE, SSCP, RFLP, ARDRA, RISA, ARISA, and TRFLP. In the past, analysis of microbial diversity has been carried out using standard plate count using selective media for specific groups of soil microorganisms. Even though, the conventional methods provided useful information on soil microbes, it has its own limitations as strong inherent biases that are caused by the selection of specific medium and the cultivation conditions (Prosser, 2002). In order to overcome these limitations culture-independent molecular methods have been introduced to analyze the structural and functional diversity of soil microorganisms (Head et al., 1998; O'Donnell and Gores, 1999). Most of these methods include extraction of nucleic acid from soil matrix for profiling microbial communities by PCR. These techniques having high stability of the genome when compared to other physiological techniques, so as to consider more reliable, sensitive, and less biased information on soil microorganisms (Lechevalial et al., 1977; Minnikin et al., 1984; Ikeda et al., 2006b). Among these techniques DGGE separates PCR products of same size but different sequences by chemical denaturation. RFLP separates PCR products by recognizing only the terminal fragment of restriction digestion (Singh et al., 2004). Conventional methods are insufficient and time consuming for the identification of environmental microorganisms. Active and inactive microorganisms cannot be distinguished by microscopic examinations. Some nonculturable microorganisms cannot grow in culture media for identification (Doi *et al*., 2006; Tani *et al*., 2006; Colwell *et al*., 2004). For these reasons, rapid and accurate methods are required.

12.7.1 DENATURING GRADIENT GEL ELECTROPORESIS ANALYSIS

Denaturing Gradient Gel Electroporesis (DGGE) is more sensitive tool to assess microbial population and allows simultaneous analysis of multiple samples (Heuex *et al*., 2002). It is the electrophoretic separation of equal length of PCR amplicon in a sequence-specific manner using a poly-acrylamide gel containing a denaturing gradient of urea and formamide (Vanhoutte *et al*., 2005). DGGE analyses are used for the separation of double – stranded DNA fragments that are identical in length, but differ in sequence based on its melting behavior (Lerman *et al*., 1984; Muyzer *et al*., 1993; Jackson and Churchill, 1999; Jackson *et al*., 2000). This technique is utilizing the stability of G-C pairing as opposed to A-T pairing. To improve the separation of the fragments, a GC rich DNA fragment can be added with one of the primers to modify the melting behavior of the target fragment. Basically, there are three steps to carry out DGGE analysis: (i) extraction of DNA from target sample; (ii) PCR-controlled amplification using specific oligonucleotide primers and (iii) separation of the amplicons using DGGE. Efficient DNA extraction is more vital and need to optimize based on the nature of the sample (Vanhoutte *et al*., 2005). Samples containing DNA are added to an electrophoresis gel with denaturing agent. At various stages the denaturing gel induces melting of the DNA; finally the DNA spreads through the gel which can be analyzed for single components. Generally DNA rich in G-C is more stable and remains double strand in low concentration of denaturing agent. Double stranded DNA molecules migrate better in acrylamide gel when compared to denatured DNA molecules. Based on this different fragments of DNA can be separated in acrylamide gel. After separation the DGGE gel stained with DNA binding fluorescent dyes (SYBR green, ethidium bromide, silver staining) and observed under UV light. Further the bands are digitally captured and analyzed using software packages. Standards can be used to compare the bands in gel. Generally, one band stands for one species; therefore the number of bands in the gel denotes the microbial diversity of the sample (Vanhoutte *et al*., 2005).

12.7.2 RESTRICTION FRAGMENT LENGTH POLYMORPHISM

Restriction Fragment Length Polymorphism (RFLP) could be a technique in which organisms can be differentiated by analysis of patterns derived from cleavage of their DNA. If two organisms dissent in the distance between sites of cleavage of a specific restriction endonuclease, the length of the fragments created can dissent once the DNA is digested with a restriction endonuclease. The similarity of the patterns generated can be used to distinguish species from one another. RFLP provides an easy and rapid alternative for phenotypic identification of the microorganism. It is also inexpensive and simple means of analysis of microbial community (Navarro *et al.*, 1992; Moyer *et al.*, 1994; Moyer *et al.*, 1996; Wintzingerode *et al.*, 1997; Baffoni *et al.*, 2013). Ekrem *et al.* (2001) did PCR-RFLP analysis of the intergenic spacer regions of 14 strains showing promise as a rapid tool for distinguishing the strains of *Streptomyces*. Michelini *et al.* (2014) identified the *Bifidobacterium* spp. using *hsp60* gene.

12.7.3 TERMINAL RESTRICTION FRAGMENT LENGTH POLYMORPHISM

Terminal restriction fragment length polymorphism (T-RFLP) analysis is a genetic fingerprinting method and also a standard, rapid genetic screening technique to characterize microbial communities. Because of its relative simplicity, this approach is widely applied to detect changes in the structure and composition of microbial communities (Liu *et al.*, 1997; Von Wintzingerode *et al.*, 1997; Clement *et al.*, 1998; Tiedje *et al.*, 1999; Hans-Peter Horz *et al.*, 2000; Osborn *et al.*, 2000; Johnson *et al.*, 2004; Kennedy *et al.*, 2005; Genney *et al.*, 2006; Pereira *et al.*, 2006; Thies, 2007; Schütte *et al.*, 2008). This technique is useful as it is quick and does not require any genome sequence information. Since, every bacterial species produces a terminal – restriction fragment (T-RF) of a definite molecular weight (Liu *et al.*, 1997; Fei Sjöberg *et al.*, 2013) and a multifaceted bacterial community that generates a series of DNA bands differing in size. A disadvantage is that the identity of the T-RF is not immediately clear and a database is required to "translate" fragment size to genus/species. This method goes with the steps of amplification of the DNA with fluorescently labeled PCR primers. The amplified products are then digested using one or more restriction enzymes, for example, *HhaI, RsaI*, and the digested products thus formed are visualized through gel electrophoresis or with an automated DNA sequencer. To analyze the restriction pattern one could either evaluate manually or with special softwares. Different species will have different terminal fragments length due to variation in the occurrence and site of the cutting sites. T-RFLP analysis has been applied to the analysis screening of a marine archaeal clone library to determine the different phylotypes (Moeseneder *et al.*, 2003). Lukow *et al.* (2000) studied the spatial and temporal changes in the bacterial community structure by using this technique. Similarly, Singh *et al.* (2006) used T-RFLP technique to study different components of the soil microbial community. However, most frequently, the technique is used to assess bacterial diversity of soils (Smalla *et al.*, 2007). The advancement made in T-RFLP analysis of 16S rRNA and genes permits researchers to make methodological and statistical alternatives suitable for the hypotheses of their research. Culman *et al.* (2008) assessed T-RFLP sample heterogeneity by measuring beta diversity, and recommended the use of binary data or relativized peak height for ordination analyses over relativized peak area. More recently, Thiyagarajan *et al.* (2010) showed that the T-RFLP technique can be successfully applied to monitor the changes in bacterial communities due to the environmental and anthropological disturbances.

12.7.4   rRNA INTERGENIC SPACER ANALYSIS

rRNA Intergenic Spacer Analysis (RISA) could be a technique of microorganism community analysis which give estimate of microorganism diversity and structure of microorganism community while not the bias obligatory by culture-based approaches or the labor attached small-subunit rRNA sequence clone library construction (Torsvik *et al.*, 1990; Richaume *et al.*, 1993; Tsuji, 1995; Borneman and Triplett, 1997; Selenska-Pobell *et al.*, 2001). RISA involves PCR amplification of a region of the rRNA sequence between the small (16S) and large (23S) subunits referred to as the intergenic spacer region with oligonucleotide primers targeted to conserved regions in the 16S and 23S genes. The 16S-23S intergenic region, which may encode tRNAs depending on the microbial species, displays vital heterogeneity in each length and nucleotide sequence. Both forms of variation are extensively used to distinguish microbial strains and closely connected species. RISA technique was used originally to characterize diversity in soils and a plenty of recently to look at microbial diversity among the rhizosphere and marine environments (Robleto *et al.*, 1998; Fisher and Triplett, 1999; Brown *et al.*, 2005; Arias, 2006). Researchers have recently improved the RISA technique by addition of an automated component with the aid of automated genetic analyzer. RISA was even applied in the study of structure of bacterial community in the oil contaminated sediments (Philippe *et al.*, 2006).

12.7.5   AUTOMATED rRNA INTERGENIC SPACER ANALYSIS

The limitations of the prevailing methodology led to develop an improved version of RISA which is now referred to as ARISA. In this automated approach, the preliminary steps of ARISA make use of length heterogeneity by utilizing

PCR amplification across the ITS region to produce DNA fragment lengths characteristic of individuals present within the sample. DNA extraction and PCR amplification are identical as in RISA, whereas in ARISA, PCR is conducted by means of fluorescence labeled oligonucleotide primer. Subsequently, electrophoretic step is executed with an automated system, and observation of these fluorescent DNA bands is made with the aid of laser. 1,400 bp long DNA fragments are generated by ARISA PCR. Reports suggest that ARISA is a rapid and effective method for assessing microbial community diversity and structure that can be especially useful at the fine spatial and temporal scales necessary in ecological studies (Scheinert *et al.*, 1996; Borneman *et al.*, 1997; Liu *et al.*, 1997; Massimiliano *et al.*, 2004; Yannarell *et al.*, 2004; Mark *et al.*, 2005). ARISA has been extensively used to analyze the genetic structures of several bacterial and fungal communities from samples of freshwater, marine water and other different environments. Indeed, the instrumental automatism of ARISA and therefore the simple analysis of its output data make it a really suitable technique for analyzing and comparing massive numbers of samples, the results generated being both reliable and reproducible.

12.7.6 SINGLE-STRAND CONFORMATION POLYMORPHISM

Single-Strand Conformation Polymorphism (SSCP) was invented by Orita *et al.* in 1989. Specific and scanning are two kinds of mutation detection methods: specific is used to recognize definite, well-characterized sequence variations. SSCP is the scanning mutation detection method. Mutation detection via SSCP analysis requires amplification of the DNA fragment of interest and denaturation of the double-stranded PCR product with heat and formamide, followed by gel electrophoresis under nondenaturing conditions. SSCP analysis comprises a rapid and easy to perform technique to recognize and distinguish closely related organisms, and has substantial potential for use in genetic screening of microbial community analysis and taxonomy. Earlier experiments used radioisotopes for the SSCP technique. Due to the difficulty in safety and handling, this technique was less professional but later radioisotopes were replaced by the alternative staining techniques such as the silver stain and the ethidium bromide (Ainsworth *et al.*, 1991; Yap and McGee, 1992). Moricca *et al.* (2000) developed a technique to differentiate among *S. cardinale* and *S. cupressi* on the basis of single-strand conformation polymorphism (SSCP). It has also been used to study many plant viral diseases (Amin, 1999; Sabek *et al.*, 1999; Rubio *et al.*, 1996; Amin *et al.*, 2009). Recent advancements in electrophoretic techniques like capillary electrophoresis have offered high through outputs with greater reliability and sensitivity to the current molecular techniques (Kourkine *et al.*, 2002).

12.7.7   AMPLIFIED RIBOSOMAL DNA RESTRICTION ANALYSIS

Amplified Ribosomal DNA Restriction Analysis (ARDRA) has proven to be a suitable and speedy method for classification studies of fungi. This technique is based on the PCR amplification of rDNA fragment, and then followed by the digestion of the amplicon with restriction endonucleases. The ensuing fragments are subsequently analyzed by agarose gel electrophoresis. The application of ARDRA on the 16S–23S rDNA region of the mycobacterial genome proved to be a rapid, simple, and reliable method for the differentiation of mycobacterial species including those that are regarded as opportunistic pathogens (Vaneechoutt *et al.*, 1992; Vaneechoutte *et al.*, 1994; Vaneechoutte *et al.*, 1995; Heyndrickx *et al.*, 1996; De Baere *et al.*, 2002; Frederic *et al.*, 2000; Blaszczyk *et al.*, 2011). ARDRA is able to identify variations among microbial communities from industrialized and domestic wastewater treatment plants, and these differences might be suitable to influent composition and temperature. Mekonnen *et al.* (2003) demonstrated the potential of the amplified ribosomal DNA-restriction analysis (ARDRA) for intra- and interspecies identification of the genus *Mycobacterium*.

12.7.8   LIMITATIONS OF MOLECULAR BASED METHODS

Molecular methods are having limitations like many other methods. Lysis efficiency of microorganisms varies between or within a microbial group. If the method used to lyse the cell is too harsh both Gram positive and Gram negative cells may be lysed and the DNA become sheared, if the method is too gentle Gram negative cells may get lysed (wintzingerode *et al.*, 1997; Trevors, 1998). In fungi, the lysis efficiency varies between spores and mycelia and within mycelial ages and it makes hurdles in molecular based microbial diversity studies. Along with RNA/DNA extraction, methods are also acting as a bias for molecular based studies. Some extraction methods like bead beating can shear the nucleic acid ultimately creating problems in further PCR analysis (Wintzingerode, 1997). It is more important to remove humic acid which will acts as a inhibitor substance from samples. Otherwise it can be coextracted and interfere in PCR analysis. During purification steps, there are concerns to get loss of DNA or RNA sequences again biasing molecular analysis. Differential amplifications of target genes can also bias PCR-based diversity studies. A major limitation of the established DNA fingerprinting techniques for soil microbes is that the complexity of rDNA fragments can exceed the resolving power of the current electrophoretic techniques due to the extreme complexity of soil ecosystem (Nakatsu *et al.*, 2000; Ikeda *et al.*, 2006b).

12.8   IMPACT OF TRANSGENIC COTTON: A CASE STUDY

In rhizosphere, several microbial groups are playing more important role for plant growth. Among these, diazotroph and chitinolytic are well characterized by the presence of nifH gene which is subunit of nitrogenase coding gene. Based on this analysis several bacteria of this group have been now recognized as plant growth promoters (Kennedy *et al.*, 2004; Ikeda *et al.*, 2006). According to Yuan and Crawford (1995) many chitinolytic bacteria are responsible for disease control in rhizosphere. These findings are implied based on the molecular assessment of functional genes of microbial diversity in the rhizosphere.

It is also well known that plant has significant influence on microbial diversity and spatial distribution (Wall and Virginia, 1999). Thus the cultivation of transgenic crops could have an influence on soil microbes through plant residues and altering structure of microbial community. However until recently, the analysis of soil microbes residing in the rhizhosphere of transgenic crops considered for the risk assessment of transgenic crops (Ikeda *et al.*, 2006b). Ikeda *et al.* (2006a) employed T-RFLP analysis for microbial community in the rhizosphere of transgenic tomato. They analyzed two of functional genes such as chitinase gene and *nifH* gene in rhizosphere. T-RFLP analysis revealed bands differing between the parental and transgenic tomato. From their results conclusion arrived as the T-RFLP analysis of functional genes may be feasible way to study the microbial community in rhizosphere.

By using DGGE and T-RFLP, it was found that there was a higher diversity of *amoA* genes and *nifH* genes in rhizosphere (Briones *et al.*, 2003; Cocking, 2003). Recent developments in molecular techniques create wide understanding of the relationship between plants and the microbes in rhizosphere (Gray and Head, 2001; Dahllof, 2002; De Long, 2002). Fang *et al.* (2005) examined the effects of transgenic corn on bacterial diversity by using DGGE and carbon substrate utilization. In this study they utilized both molecular assessment and physiological assessment to determine the impact of rhizosphere of transgenic corn on bacterial communities. They concluded that rhizosphere microbial diversity did not differ between transgenic and nontransgenic corn. The effects of transgenic corn residue on soil microbial communities and rates of carbon utilization were studied by Fang *et al.* (2007). They used DGGE for the analysis of microbial community and based on the patterns of DGGE they revealed that the Bt corn residue slightly alters the microbial diversity in rhizosphere than the nontransgenic rhizosphere.

There are several studies reported the effects of Bt toxin released through root exudates, biomass, and residues on soil microbial community. Generally, Bt toxin not showing any short term impact due to accumulation and persistence but long term effects on soil microbial community have not been evaluated extensively (Fang *et al.*, 2007). The effects of Bt corn residue on soil microbial communities and rates of carbon utilization were studied by Fang *et al.* (2007). They used DGGE and patterns of DGGE revealed that the Bt residue slightly alters the microbial communities in rhizosphere than the non-Bt isoline. This study also confirmed that the effect of Bt corn on total plate count. There are several studies reported the effect of Bt toxin released through root exudates, biomass, and residues on soil microbial community (Dunegan *et al.*, 1995; Betz, 2000; Head *et al.*, 2002; Zwahlen *et al.*, 2003 Fang *et al.*, 2007).

12.9 CONCLUSION AND FUTURE PERSPECTIVES

It is necessary to assess the technical performance of the transgenic crops and also the chances of these crops to meet societal and environmental issues. Governments should provide complete details of transgenic crops and their embedding into society and the ways implications such as risks and benefits are perceived. They should take efforts to involve experts, stakeholders, farmers, and public in discussion forum about transgenic crops (Butschi *et al.*, 2009). Tools of biotechnology provide wide knowledge and understanding on functions of various genes especially in agricultural crops. This offers solutions to solve multiple problems in agriculture. It is the right time to utilize biotechnological tools to achieve maximum agricultural productivity for increasing population.

KEYWORDS

- **Amplified ribosomal DNA restriction analysis**
- **Automated rRNA intergenic spacer analysis**
- **Bt toxin**
- **Denaturing gradient gel electrophoresis**
- **DNA fingerprinting**
- **Electrophoretic technique**
- **Microbial diversity**
- **Molecular analysis**
- **Molecular tools**
- **Polymerase chain reaction**

- **Restriction fragment length polymorphism**
- **Rhizosphere**
- **r-RNA intergenic spacer analysis**
- **Single-strand conformation polymorphism**
- **Soil microbial communities**
- **Soil quality**
- **Terminal restriction fragment length polymorphism**
- **Transgenic crops**

REFERENCES

Ainsworth, P.; Surh, L.; Coulter-Mackie, M. Diagnostic single strand conformation polymorphism, (SSCP): a simplified non-radioisotopic method as applied to Tay- Sachs B1 variant. *Nucleic Acids Res* 1991, 19, 405–406.

Amin, H. A. Molecular studies on certain viroid diseases affecting citrus in Egypt. *M.Sc. Thesis*, Faculty of Science, Ain Shams University, Egypt, 1999.

Arias, A. R.; Abernathy, J. W.; Liu, Z. Combined use of 16S ribosomal DNA and automated ribosomal intergenic spacer analysis to study the bacterial community in catfish ponds. *Letters in Applied Microbiology* 2006, 43, 287–292.

Atalan, E. Restriction Fragment Length Polymorphism Analysis (RFLP) of some Streptomyces strains from Soil. *Turk J Biol* 2001, 25, 397–404.

Baffoni, L.; Stenico, V.; Strahsburger, E.; Gaggìa, F.; Di Gioia, D.; Modesto, M. Identification of species belonging to the *Bifidobacterium* genus by PCR-RFLP analysis of a hsp60 gene fragment. *BMC Microbiol* 2013, 13, 149.

Betz, F. S.; Hammond, B. G.; Fuchs, R. L. Safety and advantages of *Bacillus thuringiensis* protected plants to control insect pests. *Regul Toxicol Pharmacol* 2000, 32, 156–173.

Blaszczyk, D.; Bednarek, H.; Machnik, G.; Sypniewski, D.; Soltysik, D.; Loch, T.; Galka, S. Amplified Ribosomal DNA Restriction Analysis (ARDRA) as a screening method for normal and bulking activated sludge sample differentiation. *Polish J Environ* 2011, 20(1), 29–36.

Borneman, J.; Triplett, E. W. Molecular microbial diversity in soils from Eastern Amazonia: evidence for unusual microorganisms and population shifts associated with deforestation. *Appl Environ Microbiol* 1997, 63, 2647–2653.

Brimecombe, M. J.; De Leji, F. A.; Lynch, J. M. The effect of root exudates on rhizosphere microbial populations. In *The rhizosphere: biochemistry and organic substrates at the soil-plant interface*, Pinton, R.; Varanini, Z.; Nannipieri, P., Ed.; Marcel-Decker. Inc: New York, 2001, pp. 95–140.

Briones, A. M.; Okabe, S.; Umemiya, Y.; Ramsing, N. B.; Reichardt, W.; Okuyama, H. Ammonia-oxidizing bacteria on root biofilms and their possible contribution to N use efficiency of different rice cultivars. *Plant Soil* 2003, 250, 335–348.

Brown, M. V.; Schwalbach, M. S.; Hewson, I.; Fuhrman, J. A. Coupling 16S-ITS rDNA clone libraries and automated ribosomal intergenic spacer analysis to show marine microbial diversity: development and application to time series. *Environmental Microbiology* 2005, 7, 1466–1479.

Brown, M. V.; Schwalbach, M. S.; Hewson, I.; Fuhrman, J. A. Coupling 16S-ITS rDNA clone libraries and automated ribosomal intergenic spacer analysis to show marine microbial diversity: development and application to a time series. *Environmental Microbiology* 2005, 7(9), 1466–1479.

Butschi, D.; Gram, S.; Haugen, J. M.; Meyer, R.; Sauter, A.; Steyaert, S.; Torgersen, H. Genetically modified plants and foods challenges and future issues in Europe. *European Parlimentary Technology Assessment Final Report* 2009, 92.

Cairney, J. W. G. Evaluation of mycorrhiza systems. Naturwissenschaften, 2000, 87, 467–475.

Cardinale, M.; Brusetti, L.; Quatrini, P.; Borin, S.; Puglia, A. M.; Rizzi, A.; Zanardini, E.; Sorlini, C.; Corselli, C.; Daffonchio, D. Comparison of different primer sets for use in automated ribosomal intergenic spacer analysis of complex bacterial communities. *Appl Environ Microbiol* 2004, 70(10), 6147–6156.

Clement, B. G.; Keh, L. E.; Debord, K. L.; Kitts, C. L. Terminal restriction fragment patterns (TRFPs), a rapid, PCR-based method for the comparison of complex bacterial communities. *Journal of Microbiological Methods* 1998, 31, 135–142.

Cocking, E. C. Endophytic colonization of plant roots by nitrogen fixing bacteria. *Plant Soil* 2003, 252, 169–175.

Collins, H. P.; Elliott, L. F.; Rickman, R. W.; Bezdicek, D. F.; Papendick, R. I. Decomposition and interaction among wheat residue components. *Soil Sci Soc Am J* 1990, 54, 780–785.

Colwell, R. R.; Grimes, D. J. Nonculturable microorganisms in the environment. ASM Press: Washington DC, USA, 2000.

Culman, S. W.; Gauch, H. G.; Blackwood, C. B.; Thies, J. E. Analysis of T-RFLP data using analysis of variance and ordination methods: A comparative study. *Journal of Microbiological Methods* 2008, 75, 55–63.

Dahllof, I. Molecualr community analysis of uncultured microorganisms. *ASM News* 2002, 70, 63–70.

De Baere, T.; de Mendonca, R.; Claeys, G.; Verschraegen, G.; Mijs, W.; Verhelst, R.; Rottiers, S.; Van Simaey, L.; De Ganck, C.; Vaneechoutte, M. Evaluation of amplified rDNA restriction analysis (ARDRA) for the identification of cultured mycobacteria in a diagnostic laboratory. *BMC Microbiol* 2002, 2, 4.

De Long, E. F. Microbial populations genomics and ecology. *Curr Opin Microbiol* 2002, 5, 520–524.

Dickie, I. A.; Xu, B.; Koide, R. T. Vertical niche differentiation of ectomycorrhizal hyphae in soil as shown by T-RFLP analysis. *New Phytol* 2002, 156, 527–535.

Dodd, J. C.; Boddington, C. L.; Rodriguez, A.; Gonzalez-Chavez, C.; Mansur, I. Mycelium of arbuscular mycorrhizal fungi (AMF) from different genera: form, function and detection. *Plant Soil* 2000, 226, 131–151.

Doi, T.; Abe, J.; Zhang, B.; Morita, S. Application of fluorescent *in situ* hybridization for observation of rhizosphere microorganisms. *Root Res* 2001, 1, 430–431.

Donegan, K. K.; Palm, C. J.; Fieland, V. J.; Porteous, L. A.; Ganio, L. M.; Schaller, D. L.; Buca, L. Q.; Seidler, R. J. Changes in levels, species and DNA finger prints of soil microorganisms associated with cotton expressing the *Bacillus thuringiensis* var. *kurstaki* endotoxin. *Appl Soil Ecol* 1995, 2, 111–124.

Dunbar, J.; Ticknor, L. O.; Kuske, C. R. Assessment of microbial diversity in four Southwestern United States soils by 16S rRNA gene terminal restriction fragment analysis. *Appl Environ Microbiol* 2000, 66(7), 2943–2950.

Fang, M.; Kremer, R. J.; Motavalli, P. P.; Davis, G. Bacterial diversity in rhizospheres of non trans-genic and transgenic corn. *Applied and Environmental Microbiology* 2005, 4132–4136.

Fang, M.; Motavalli, P. P.; Kremer, R. J.; Nelson, K. A. Assessing changes in soil microbial com-munities and carbon mineralization in Bt and non-Bt corn residue amended soil. *Applied Soil Ecology* 2007, 37, 150–160.

Fedi, S.; Tremaroli, V.; Scala, D.; Perez-Jimenez, J. R.; Fava, F.; Young, L.; Zannoni, D. T-RFLP analysis of bacterial communities in cyclodextrin-amended bioreactors developed for biodegra-dation of polychlorinated biphenyls. *Res Microbiol* 2005, 156(2), 201–210.

Fisher, M. M.; Triplett, E. W. Automated approach for ribosomal intergenic spacer analysis of mi-crobial diversity and its application to freshwater bacterial communities. *Appl Environ Microbiol* 1999, 65, 4630–4636.

Genney, D. R.; Anderson, I. C.; Alexander, I. J. Fine-scale distribution of pine ectomycorrhizas and their extramatrical mycelium. *New Phytol* 2006, 170, 381–390.

George, E.; Marschner, H.; Jakobsen, I. Role of arbuscular mycorrhizal fungi in uptake of phospho-rous and nitrogen from soil. *Crit Rev Biotechnol* 1995, 15, 257–270.

Gich, F. B.; Amer, E.; Figueras, J. B.; Abella, C. A.; Balaguer, M. D.; Poch, M. Assessment of microbial community structure changes by amplified ribosomal DNA restriction analysis (AR-DRA). *Int Microbiol* 2000, 3, 103–106.

Giller, K. E.; Beare, M. H.; Lavelle, P.; Izac, A. -M. N.; Swift, M. J. Agricultural intensification, soil biodiversity and agroecosystem function. *Appl Soil Ecol* 1997, 6, 3–16.

Girvan, M. S.; Bullimore, J.; Ball, A. S.; Pretty, J. N.; Osborn, A. M. Responses of active bacterial and fungal communities in soils under winter wheat to different fertilizer and pesticide regimens. *Appl Environ Microbiol* 2004, 70, 2692–2701.

Goszczynski, D. E.; Jooste, A. E. C. The application Single-strand conformation polymorphism (SSCP) technique for the analysis of molecular heterogeneity of grapevine virus A. *Vitis* 2002, 41(2), 77–82.

Gotmare, V. Research Highlights. Central Institute for Cotton Research, Nagpur, India. 2010, 26(1), 5.

Graeber, J. V.; Natziger, E. D.; Mies, D. W. Evaluation of transgenic Bt containing corn hybrids. *J Prod Agric* 1999, 12, 559–663.

Gray, N. D.; Head, I. M. Linking genetic identity and function in communities of uncultured bac-teria. *Environ Microbiol* 2001, 3, 481–492.

Hala, A.; Amin, A. M.; Abdel-Kader, A. E.; Aboul, A. Single Strand Conformational Polymor-phism (SSCP) strain analysis and partial nucleotide sequence of Maize Yellow Stripe Virus (MYSV) segment 3 genomic RNA. *Egypt J Phytopathol* 2009, 37(1), 71–82.

Hayashi, K. PCR-SSCP: a simple and sensitive method for detection of mutations in the genomic DNA. *PCR Methods Appl* 1991, 1, 34–38.

Head, G.; Surba, J. B.; Walson, J. A.; Martin, J. W.; Duan, J. J. No detection of *Cry* 1Ac protein in soil after multiple years of transgenic Bt cotton (Bollgard) use. *Environ Entomol* 2002, 31, 30–36.

Head, I. M.; Saunders, J. R.; Pickup, R. W. Microbial evolution, diversity, and ecology: a decade of ribosomal RNA analysis of uncultivated microorganisms. *Microb Ecol* 1998, 35, 1–21.

Heuer, H.; Kroppenstedt, R. M.; Lottmann, J.; Berg, G.; Smaller, K. Effects of T4 lysozyme re-lease from transgenic potato roots on bacterial rhizosphere communities are negligible relative to natural factors. *Appl Environ Microbiol* 2002, 68, 1325–1335.

Heyndrickx, M.; Vauterin, L.; Vandamme, P.; Kersters, K.; De Vos, P. Applicability of combined amplified ribosomal DNA restriction analysis (ARDRA) patterns in bacterial phylogeny and taxonomy. *J Microbiol Methods* 1996, 26, 247–259.

Horz, H. -P.; Rotthauwe, J. -H.; Lukow, T.; Liesack, W. Identification of major subgroups of ammonia-oxidizing bacteria in environmental samples by T-RFLP analysis of amoA PCR products. *Journal of Microbiological Methods* 2000, 39, 197–204.

Ikeda, S.; Omura, T.; Ytow, N.; Komaki, H.; Minamisawa, K.; Ezura, H.; Fujimura, T. Microbial community analysis in the rhizosphere of a transgenic tomato that overexpress 3-hydroxy-3-methyl glutaryl coenzyme A reductase. *Microbes Environ* 2006a, 21(4), 261–271.

Ikeda, S.; Ytow, N.; Ezura, H.; Minamisawa, K.; Fujimura, T. Soil microbial community analysis in the environmental risk assessment of transgenic plants. *Plant Biotechnol* 2006b, 23L, 137–151.

Jackson, C. R.; Churchill, P. F. Analysis of microbial communities by denaturing gradient gel elctrophoresis: Applications and limitations. *Recent Res Devel Microbiology* 1999, 3, 81–91.

Jackson, C. R.; Roden, E. E.; Curchill, P. F. Denaturing gel electrophoresis can fail to separate 16S rDNA fragments with multiple base differences. *Molecular Biology Today* 2000, 1(2), 49–51.

Johnson, D.; Vandenkoornhuyse, P. J.; Leake, J. R.; Gilbert, L.; Booth, R. E.; Grime, J. P.; Young, J. P. W.; Read, D. J. Plant communities affect arbuscular mycorrhizal fungal diversity and community composition in grassland microcosms. *New Phytol* 2004, 161, 503–515.

Kennedy, I. R.; Chaudhary, A. T. M. A.; Kecskes, M. L. Non symbiotic diazotrophs in crop farming systems: can their potential for plant growth promotion be better exploited? *Soil Biol Biochem* 2004, 36, 1224–1244.

Kennedy, N.; Edwards, S.; Clipson, N. Soil bacterial and fungal community structure across a range of unimproved and semiimproved upland grasslands. *Microb Ecol* 2005, 50, 463–473.

Kirk, J. L.; Beandette, L. A.; Hart, M.; Moutoglis, P.; Klironomos, J. N.; Lee, H.; Trevors, J. T. Methods of studying soil microbial diversity. *Journal of Microbiological Methods* 2004, 58, 169–188.

Kourkine, I. V.; Hestekin, C. N.; Barron, A. E. Technical challenges in applying capillary electrophoresis-single strand conformation polymorphism for routine genetic analysis. *Electrophoresis* 2002, 23, 1375–1385.

Kurabachew, M.; Enger, Ø.; Sandaa, R -A.; Lemma, E.; Bjorvatn, B. Amplified ribosomal DNA restriction analysis in the differentiation of related species of mycobacteria. *Journal of Microbiological Methods* 2003, 55, 83–90.

Lechevalier, M. P.; De Bievre, C.; Lechvalier, H. Chemotaxonomy of aerobic actinomycetes: phospholipid composition. *Biochem. Sys. Ecol* 1977, 5, 249–260.

Lerman, L. S.; Fischer, S. G.; Hurley, I.; Silverstein, K.; Lumelsky, N. Sequence-determined DNA separations. *Ann Rev Biophys Bioeng* 1984, 13, 399–423.

Liu, W. -T.; Marsh, T. L.; Cheng, H.; Forney, L. J. Characterization of microbial diversity by determining terminal restriction fragment length polymorphisms of genes encoding 16S rRNA. *Appl Environ Microbiol* 1997, 63, 4516–4522.

Liu, W.; Lu, H. H.; Wu, W.; Wei, Q. K.; Chen, Y. X.; Thies, J. E. Transgenic Bt rice does not affect enzyme activities and microbial composition in the rhizosphere during crop development. *Soil Biology and Biochemistry* 2008, 40(2), 475–486.

Lukow, T.; Dunfield, P. F.; Liesack, W. Use of the T-RFLP technique to assess spatial and temporal changes in the bacterial community structure within an agricultural soil planted with transgenic and non-transgenic potato plants. *FEMS Microbiol Ecol* 2000, 3, 241–247.

Mangal, M.; Malik, K.; Randhawa, G. J. Import of transgenic planting material: National scenario. *Current Science* 2003, 85(4), 454–458.

Marschner, P.; Baumann, K. Changes in bacterial community structure induced by mycorrhizal colonization in split-root maize. *Plant Soil* 2003, 251, 279–289.

Minnikin, D. E.; O'Donnell, A. G.; Goodfellow, M.; Alderson, G.; Athalye, M.; Schaal, A.; Parlett, J. H. An integrated procedure for the extraction of isoprenoid quinones and polar lipids. *J Microbiol Meth* 1984, 2, 233–241.

Moeseneder, M. M.; Winter, C.; Arrieta, J. M.; Hernd, G. J. Terminal-restriction fragment length polymorphism T-RFLP screening of a marine archaeal clone library to determine the different phylotypes. *Journal of Microbiological Methods* 2001, 44, 159–172.

Mohan, K. S.; Manjunath, T.M. Bt Cotton – India's first transgenic crop. *J Plant Biol* 2002, 29(3), 225-236.

Molin, J.; Molin, S. CASE: complex adaptive systems ecology. In *Advances in Microbial Ecology, Vol. 15*, Jones, J. G., Ed.; Plenum: New York, 1997, pp. 27–79.

Moyer, C. L.; Dobbs, F. C.; Karl, D. M. Estimation of diversity and community structure through restriction fragment length polymorphism distribution analysis of bacterial 16S rRNA genes from a microbial mat at an active, hydrothermal vent system, Loihi Seamount, Hawaii. *Appl Environ Microbiol* 1994, 60, 871–879.

Moyer, C. L.; Tiedje, J. M.; Dobbs, F. C.; Karl, D. M. A computer-simulated restriction fragment length polymorphism analysis of bacterial small subunit rRNA genes: efficacy of selected tetrameric restriction enzymes for studies of microbial diversity in nature. *Appl Environ Microbiol* 1996, 62, 2501–2507.

Muyzer, G.; De Waal, E. C.; Uitterlinden, A. G. Profiling of complex microbial populations by denaturing gradient gel electrophoresis analysis of polymerase chain reaction-amplified genes coding for 16S rRNA. *Appl Environ Microbiol* 1993, 59, 695–700.

Nakatsu, C. H.; Torsvik, V.; Øvreas, L. Soil community analysis using DGGE of 16s rDNA polymerase chain reaction products. *Soil Sci Soc Am J* 2000, 64, 1382–1388.

Navarro, E.; Simonet, P.; Normand, P.; Bardin, R. Characterization of natural populations of *Nitrobacter* spp. using PCR/RFLP analysis of the ribosomal intergenic spacer. *Arch Microbiol* 1992, 157, 107–115.

O'Donnell, A. G.; Gores, H. E. 16S rRNA methods in soil microbiology. *Cuerr Opin Biotechnol* 1999, 10, 225–229.

O'Donnell, A. G.; Seasman, M.; Macrae, A.; Waite, I.; Davies, J. T. Plants and fertilizers as drivers of change in microbial community structure and function in soils. *Plant Soil* 2001, 232, 135–145.

Obrycki, J. J.; Losey, J. E.; Tayler, O. R.; Jesse, L. C. H. Transgenic insecticidal corn: beyond insecticidal toxicity to ecological complexity. *Bioscience* 2001, 51, 353–361.

Orita, M.; Suzuki, Y.; Sekiya, T.; Hayashi, K. Rapid and sensitive detection of point mutations and DNA polymorphisms using the polymerase chain reaction. *Genomics* 1989, 5, 874–879.

Osborn, A. M.; Moore, E. R. B.; Timmis, K. N. An evaluation of terminal-restriction fragment length polymorphism (T-RFLP) analysis for the study of microbial community structure and dynamics. *Environmental Microbiology* 2000, 2, 39–50.

Ovreas, L.; Jensen, S.; Daae, F. L.; Torsvik, V. Microbial community changes in a perturbed agricultural soil investigated by molecular and physiological approaches. *Appl Environ Microbiol* 1998, 64, 2739–2742.

Pereira, M. G.; Latchford, J. W.; Mudge, S. M. The use of terminal restriction fragment length polymorphism (T-RFLP) for the characterization of microbial communities in marine sediments. *Geomicrobiol J* 2006, 23, 247–251.

Pray, C. E.; Huang, J.; Hu, R.; Rozelle, S. Five years of Bt-cotton in China-the benefits continue. *The Plant Journal* 2002, 31(4), 423–430.

Prosser, J. I. Molecular and functional diversity in soil microorganisms. *Plant Soil* 2002, 244, 9–17.

Qaim, M.; Cap, E. J.; Janvry, A. Agronomics and sustainability of transgenic cotton in Argentina. *AgBioForum* 2003, 6(2), 41–47.

Raaijmakers, J. M.; Paulitz, T. C.; Steinberg, C.; Alabouvette, C.; Moënne-Loccoz, Y. The Rhizosphere: a playground and battlefield for soilborne pathogens and beneficial microorganisms. *Plant Soil* 2009, 321, 341–361.

Ranjard, L.; Poly, F.; Lata, J. C.; Mougel, C.; Thioulouse, J.; Nazaret, S. Characterization of bacterial and fungal soil communities by automated ribosomal intergenic spacer analysis fingerprints: Biological and methodological variability. *Appl Environ Microbiol* 2001, 67, 4479–4487.

Richaume, A.; Steinberg, C.; Jocteur-Monrozier, L.; Faurie, G. Differences between direct and indirect enumeration of soil bacteria: The influence of soil structure and cell location. *Soil Biol Biochem* 1993, 25, 641–643.

Robleto, E. A.; Borneman, J.; Triplett, E. W. Effects of bacterial antibiotic production on rhizosphere microbial communities from a culture-independent perspective. *Applied and Environmental Microbiology* 1998, 64, 5020–5022.

Rondon, M. R.; Goodman, R. M.; Handelsman, J. The earth's bounty: assessing and accessing soil microbial diversity. *Trends Biotech* 1999, 17, 403–409.

Rubio, L.; Ayllon, M. A.; Guerri, J.; Pappu, H.; Niblett, C.; Moreno, B. Differentiation of citrus tristeza clasterovirus (CTV) isolates by single- strand conformation polymorphism analysis of the coat protein gene. *Ann Appl Biol* 1996, 129, 479–489.

Sabek, A. M.; Barakat, A. B.; Amin, H. A. Characterization of different citrus exocortis and cachexia isolates by single strand conformation polymorphism (SSCP). *AL-Azhar Egypt. J. Biotechnol* 1999, 7, 90–100.

Saeki, M.; Toyota, K. Effect of bensulfuron-methyl (a sulfonyluera herbicide) on the soil bacteria community of a paddy soil microcosm. *Biol Fertil Soils* 2004, 40, 110–118.

Scheinert, P.; Krausse, R.; Ullman, U.; Soller, R.; Krupp, G. Molecular differentiation of bacteria by PCR amplification of the 16S-23S rRNA spacer. *J Microbiol Methods* 1996, 26, 103–117.

Schütte, U. M. E.; Abdo, Z.; Bent, S. J.; Shyu, C.; Williams, C. J.; Pierson, J. D.; Forney, L.J. Advances in the use of terminal restriction fragment length polymorphism (T-RFLP) analysis of 16S rRNA genes to characterize microbial communities. *Appl Microbiol Biotechnol* 2008, DOI 10.1007/s00253-008-1565-4.

Selenska-Pobell, S.; Kampf, G.; Flemming, K.; Radeva, G.; Satchanska, G. Bacterial diversity in soil samples from two uranium waste piles as determined by rep-APD, RISA and 16S rDNA retrieval. *Antonie van Leeuwenhoek - International Journal of General and Molecular Microbiology* 2001, 79, 149–161.

Shyu, C.; Soule, T.; Bent, S. J.; Foster, J. A.; Forney, L. J. MiCA: a web based tool for the analysis of microbial communities based on terminal-restriction fragment length polymorphisms of 16S and 18S rRNA genes. *Microb Ecol* 2007, 53, 562–570.

Singh, B. K.; Millard, P.; Whiteley, A. S.; Murrell, J. C. Unraveling rhizosphere-microbial interactions: oppurtunities and limitations. *Trends in Microbiology* 2004, 12(8), 386–393.

Singh, B. K.; Munro, S.; Reid, E.; Ord, B.; Potts, J. M.; Paterson, E.; Millard, P. Investigating microbial community structure in soils by physiological, biochemical, and molecular fingerprinting methods. *Eur J Soil Sci* 2006, 57, 72–82.

Singh, B. K.; Nazaries, L.; Munro, S.; Anderson, I. C.; Campbell, D. C. Use of multiplex terminal restriction fragment length polymorphism for rapid and simultaneous analysis of different components of the soil microbial community. *Appl Environ Microbiol* 2006, 72, 7278–7285.

Singh, B. K.; Walker, A.; Morgan, J. A. W.; Wright, D. J. Effects of soil pH on the biodegradation of chlorpyrifos and isolation of a chlorpyrifos-degrading bacterium. *Appl Environ Microbiol* 2003, 70, 1821–1826.

Sjöberg, F.; Nowrouzian, F.; Rangel, I.; Hannoun, C.; Moore, E.; Adlerberth, I.; Wold, A. E. Comparison between terminal-restriction fragment length polymorphism (T-RFLP) and quantitative culture for analysis of infants' gut microbiota. *Journal of Microbiological Methods* 2013, 94, 37–46.

Smalla, K.; Oros-Sichler, M.; Milling, A.; Heuer, H.; Baumgarte, S.; Becker, R.; Neuber, G.; Kropf, S.; Ulrich, A.; Tebbe, C. C. Bacterial diversity of soils assessed by DGGE, T-RFLP and SSCP fingerprints of PCR-amplified 16S rRNA gene fragments: do the different methods provide similar results? *J Microbiol Methods* 2007, 69, 470–479.

Smith, K. P.; Goodman, R. M. Host variation for interactions with beneficial plant associated microbes. *Ann Rev Phytopathol* 1999, 37, 473–491.

Stenico, V.; Michelini, S.; Modesto, M.; Baffoni, L.; Mattarelli, P.; Biavati, B. Identification of *Bifidobacterium* spp. using *hsp60* PCR-RFLP analysis: An update. Anaerobe 2014, 26, 36–40.

Tang, J.; Tao, J.; Urakawa, H.; Corander, J. T-BAPS: a bayesian statistical tool for comparison of microbial communities using terminal restriction fragment length polymorphism (T-RFLP) data. *Stat Appl Genet Mol Biol* 2007, 6, 30

Tani, K.; Yamaguchi, N.; Nasu, M. Rapid detection of specific microbial cells by fluorescent staining techniques. *J Jpn Toxicol Environ Health* 1997, 43, 145–154.

Thies, J. E. Soil microbial community analysis using terminal restriction fragment length polymorphisms. *Soil Science Society of America Journal* 2007, 71, 579–591.

Thorn, G. The fungi in soil. In *Modern soil microbiology*, van Elsas, J. D.; Trevors, J. T.; Wellington, E. M. H., Ed.; Marcel Dekker: New York, 1997, pp. 63–127.

Tiedje, J. M.; Asuming-Brempong, S.; Nusslein, K.; Marsh, T. L.; Flynn, S. J. Opening the black box of soil microbial diversity. *Applied Soil Ecology* 1999, 13, 109–122.

Torsvik, V.; Daae, F. L.; Sandaa, R. -A.; Ovreas, L. Novel technique for analysing microbial diversity in natural and perturbed environments. *J Biotechnol* 1998, 64, 53–62.

Torsvik, V.; Goksoyr, J.; Daae, F. L. High diversity in DNA of soil bacteria. *Appl Environ Microbiol* 1990, 56, 782–787.

Traxler, G.; Godoy-Avila, S. Transgenic cotton in Mexico. *Ag Bio Forum* 2004, 7(1), 57–62.

Trevors, J. T. Molecular evolution in bacteria: cell division. *Rev Microbiol* 1998, 29, 237–245.

Tsuji, T.; Kawasaki, Y.; Takeshima, S.; Sekiya, T.; Tanaka, S. A. New fluorescence staining assay for visualizing living microorganisms in soil. *Appl Environ Microbiol* 1995, 61, 3415–3421.

Van Elsas, J. D.; Flois-Duarto, G.; Keizer-wolter, A.; Smit, E. Analysis of the dynamics of fungal communities in soil via fungal-specific PCR of soil DNA followed by denaturing gradient gel electrophoresis. *J Microbiol Methods* 2000, 43, 133–151.

Vaneechoutte, M.; De Beenhouwer, H.; Claeys, G.; Verschraegen, G.; De Rouck, A.; Paepe, N.; Elaichouni, A.; Portaels, F. Identification of *Mycobacterium* species by using amplified ribosomal DNA restriction analysis. *J Clin Microbiol* 1993, 31, 2061–2065.

Vaneechoutte, M.; Dijkshoorn, L.; Tjernberg, I.; Elaichouni, A.; de Vos, P.; Claeys, G.; Verschraegen, G. Identification of *Acinetobacter* genomic species by amplified ribosomal DNA restriction analysis. *J Clin Microbiol* 1995, 33, 11–15.

Vaneechoutte, M.; Rossau, R.; De Vos, P.; Gillis, M.; Janssens, D.; Paepe, N.; De Rouck, A.; Fiers, T.; Claeys, G.; Kersters, K. Rapid identification of bacteria of the Comamonadaceae with amplified ribosomal DNA-restriction analysis (ARDRA). *FEMS Microbiology Letters* 1992, 93(3), 227–233.

Wang, M.; Ahrne, S.; Antonsson, M.; Molin, G. T-RFLP combined with principal component analysis and 16S rRNA gene sequencing: an effective strategy for comparison of fecal microbiota in infants of different ages. *J Microbiol Methods* 2004, 59, 53–69.

Wintzingererode, F. V.; Gobel, U. B.; Stackebrandt, E. Determination of microbial diversity in environmental samples: pitfalls of PCR based rRNA analysis. *FEMS Microbiol Rev* 1997, 21, 213–229.

Yannarell, A. C.; Triplett, E. W. Within- and between-lake variability in the composition of bacterioplankton communities: investigations using multiple spatial scales. *Applied and Environmental Microbiology* 2004, 70, 214–223.

Yao, H.; He, Z.; Wilson, M. J.; Campbell, C. D. Microbial biomass and community structure in a sequence of soils with increasing fertillity and changing land use. *Microbiol Ecol* 2000, 40, 223–237.

Yap, E.; McGee, J. Nonisotopic SSCP detection in PCR products by ethidium bromide staining. *Trends Genet* 1992, 8, 49.

Yaun, W. M.; Crawford, D. I. Characterization of *Streptomyces lydicus* WYEC 108 as a potential biocontrol agent against fungal root and seed rots. *Appl Environ Microbiol* 1995, 61, 3119–3128.

Zwahlen, C.; Hilbeck, A.; Gugerli, P.; Nentwig, W. Degradation of the *Cry* 1Ab protein within transgenic *Bacillus thuringiensis* corn tissue in the field. *Mol Ecol* 2003, 12, 765–775.

CHAPTER 13

RECENT ADVANCES IN BIOTECHNOLOGY AND GENOMIC APPROACHES FOR ABIOTIC STRESS TOLERANCE IN CROP PLANTS

MIRZA HASANUZZAMAN, RAJIB ROYCHOWDHURY, JOYDIP KARMAKAR, NAROTTAM DEY, KAMRUN NAHAR, and MASAYUKI FUJITA

CONTENTS

13.1   Introduction 334

13.2   Biotechnological Approaches Used in Developing Abiotic
Stress Tolerance.. 335

13.3   Development and Application of Molecular Markers....................... 337

13.4   Molecular Breeding and Marker Assisted Selection 341

13.5   Stress Responsive Genes and Their Regulation 343

13.6   LEA-Protein 346

13.7   Regulatory Genes 346

13.8   Functional Genomics and Stress Response...................................... 347

13.9   Modern Transgenic Approaches for Abiotic Stress Tolerance 349

13.10 Metabolic Engineering for Stress Tolerance 351

13.11 Conclusion and Future Perspectives 356

Keywords.. 357

References... 357

13.1 INTRODUCTION

Abiotic stresses such as high temperature, low temperature, drought, flooding, salinity, heavy metals, radiation, ozone are the most threatening and limiting factors for agricultural productivity worldwide (Doupis *et al.*, 2011; Hasanuzzaman *et al.*, 2009; Hasanuzzaman *et al.*, 2010; Hasanuzzaman *et al.*, 2011; Hasanuzzaman *et al.*, 2012; Hasanuzzaman *et al.*, 2013a; Hasanuzzaman *et al.*, 2013b; Hasanuzzaman and Fujita, 2012). These stresses negatively influence the survival of crop plants, leading to losses of up to 70 % in biomass production and yields of staple food crops (Kaur *et al.*, 2008; Thakur *et al.*, 2010) and decreases of as much as 50 % in overall productivity (Rodríguez *et al.*, 2005; Acquaah, 2007). For these reasons, abiotic stresses threaten food security worldwide. The degree of severity of stress and the plasticity of the plants determine the types of morphological, anatomical, and physiological changes that adversely affect plant growth, metabolic profile, nutritional potential, developmental processes, and productivity (Altman, 2003). At the molecular level, abiotic stresses lead to an enhanced production of reactive oxygen species, which cause peroxidation of lipids, oxidation of proteins, damage to nucleic acids, enzyme inhibition, and-at extreme conditions-programmed cell death and even the death of the entire plant (Mittler, 2002; Sharma and Dubey, 2005; Sharma and Dubey, 2007).

Controlling the external environment in favor of better plant production ranges from very difficult to essentially impossible in most cases. Consequently, we use our limited aptitude in cultural practices or we try to improve conventional crop plants through breeding techniques or biotechnological approaches. The principal objective in plant breeding is to obtain plants that give higher yields and that show stability, better quality, and stress-resistant characteristics (Bray, 2004; Chaves and Oliveira, 2004). The current consensus is that stress tolerance is multigenic and quantitative in nature (Collins *et al.*, 2008) and therefore difficult to instill in plants through available methods. The molecular mechanisms underlying abiotic stress tolerances in plants are being unraveled, but a full understanding of the mechanisms by which plants perceive environmental signals and the ways that these signals are transmitted to cellular machinery to activate adaptive responses still remain elusive.

At present, sequencing and functional genomics are considered to be essential for understanding stress signal perception and transduction associated with molecular regulatory networks involved in stress responses (Heidarvand and Amiri, 2010; Ray *et al.*, 2010; Sanchez *et al.*, 2011). Polyomics approaches like transcriptomics, proteomics, metabolomics, ionomics, and micromics are involved in the molecular mechanisms are also important for discovering new genes, proteins, and secondary plant metabolites that are responsible for plants adaptation to stress (Mochida and Shinozaki, 2011). Biotechnological approaches are considered to represent one of the most reliable resources for investigating the factors underlying stress tolerance and for instilling desirable traits that will confer stress tolerance in important agricultural crops. Recent advances in biotechnology have dramatically changed the capabilities for gene discovery and functional genomics, rendering the possibility of full transcriptomic, proteomic, and metabolomic profiling of a cell (Cramer *et al.*, 2011). This type of biotechnological approach has pushed forward the identification of stress-responsive and tolerant genes.

Functional genomics approaches have helped to identify single to thousands of genes and have provided clues for functional characterization of stress-responsive genes and stress tolerance mechanisms (Sreenivasulu *et al.*, 2007). New approaches like microarray-based transcriptomic profiling of differential gene expression (Walia *et al.*, 2007) and combinations of genetic mapping and expression profiling (Pandit *et al.*, 2010) are now unraveling the intrinsic mystery of stress tolerance. Recent molecular marker technology is now identifying both inherited and quantitative traits that control individual genes (Roychowdhury *et al.*, 2012; Karmakar *et al.*, 2012). Metabolic engineering is helping to improve the cellular metabolic properties through modification of specific biochemical reactions and even through introduction of novel reactions through the use of recombinant DNA technology, conferring stress tolerance in several model plant species (Stephanopoulos, 1999; Gao *et al.*, 2001). Lastly, the use of the most astonishing genetic engineering and transgenetic techniques have allowed the introduction of specific desired traits for conferring the abiotic stress tolerance, as these approaches have no barrier in transferring useful genes, even across different species from the animal or plant kingdoms (Cramer *et al.*, 2011). This review considers the importance of biotechnological approaches as powerful tools for improving abiotic stress tolerance and explores the recent advances in biotechnological approaches with related paradigms.

13.2   BIOTECHNOLOGICAL APPROACHES USED IN DEVELOPING ABIOTIC STRESS TOLERANCE

As crop plants are complex organisms their productivity is highly influenced by different abiotic stresses due to continual unexpected global climatic changes in this agricultural modernization era (Cramer *et al.*, 2011). Plants must respond and adapt to a wide range of abiotic stresses to survive in various environmental conditions and they have acquired several stress tolerance mechanisms, which are different processes involving physiological and biochemical changes that result in adaptive or morphological changes (Urano *et al.*, 2010). Though cultivated crop is moderately tolerant to various abiotic stresses, the crop losses due to unfavorable environmental conditions can be unpredictably severe. So far, several efforts have been made to improve abiotic stress tolerance in the crop cultivars through cultural practices, breeding techniques, and biotechnological approaches. The objective of plant breeding for stress tolerance is to accumulate favorable alleles those contribute to tolerance in the plant genome. Genes that confer stress resistance can be sourced from germplasm collections, including wild relatives of crops that are held in genebanks or organisms that currently live in different harsh environmental habitats that have evolved to cope with those conditions (Nevo and Chen, 2010). Introgression of abiotic stress tolerance properties to cultivated plants from more tolerant wild relatives through classical breeding has been attempted with limited success due to: (a) the complex nature of abiotic stress tolerance (timing, duration, intensity, frequency), thereby its quantification and repeatability; (b) the undesirable genes are also transferred along with desirable traits; and (c) reproductive barriers limit the transfer of favorable alleles from diverse genetic resources. Biotechnology is a viable option for developing crop genotypes that can perform better under harsh environmental conditions. For instance, advances in genomics coupled with "omics" approaches and stress biology can provide useful genes or alleles for conferring stress tolerance. Superior genes or alleles where they have been identified in the same species can be transferred into elite genotypes through molecular marker-assisted selection (Varshney *et al.*, 2011).

The quantitative trait loci (QTLs) related with stress tolerance have significant role in understanding the stress response and generating stress-tolerant plants (Gorantla *et al.*, 2005). Methods of identifying genes underlying QTLs have progressed from map-based cloning approach to microarray-based transcriptomic profiling of differential gene expression (Walia *et al.*, 2007) or combination of genetic mapping and expression profiling (Pandit *et al.*, 2010). The recent developments in molecular marker analyses have made it feasible to analyze both simply inherited, as well as the quantitative traits, and identify the individual gene controlling the trait of interest. Molecular markers could be used to tag QTLs and to evaluate their contributions to the phenotype by selecting favorable alleles at these loci in a marker-aided selection scheme aiming to accelerate genetic advance (Lodha *et al.*, 2011; Karmakar *et al.*, 2012; Roychowdhury *et al.*, 2012). Advanced backcross QTL analysis can be used to evaluate mapped donor introgression in the genetic background of an elite recurrent parent (Turan *et al.*, 2012). Moreover, by using an approach such as genetic engineering and transgenesis, there is no barrier to transferring useful genes or alleles (that are known to be involved in stress response and putative stress tolerance) across different species from the animal or plant kingdoms. Such could provide powerful tools for improving abiotic stress tolerance coupled with the growing knowledge of stress physiology and biochemistry. Recent advances in biotechnology have dramatically changed the capabilities for gene discovery and functional genomics. For the first time, now it is available to obtain the information of a cell with its full transcriptomic, proteomic, and metabolomics profiling (Cramer *et al.*, 2011). Recently, transgenic approaches have been employed to produce plants with enhanced tolerance to various abiotic stresses by overexpressing genes involved in different tolerance-related physiological mechanisms (Pandey *et al.*, 2011; dos Reis *et al.*, 2012). These include genes encoding enzymes in the synthesis of solutes such as mannitol (Chan *et al.*, 2011), glycine betaine (Fan *et al.*, 2012), and polyamines (Cheng *et al.*, 2009), which contribute to osmoregulation and impart tolerance to abiotic stresses. Recent advances in genome-wide analyses have revealed complex regulatory networks that control global gene expression, protein modification, and metabolite composition. Genetic with epigenetic regulation, including changes in nucleosome distribution, histone modification, DNA methylation, and npcRNAs (nonprotein-coding RNA) play important roles in abiotic stress gene networks. Transcriptomics, metabolomics, bioinformatics, and high-through-put DNA sequencing have enabled active analyses of regulatory networks that control abiotic stress responses. Such analyses have markedly increased the understanding of global plant systems in adaptive responses and tolerance to abiotic stress conditions (Urano *et al.*, 2010). As a result, biotechnology approaches offer novel strategies for producing suitable crop genotypes that are able to resist drought, high temperature, flooding, salinity, and other abiotic stresses.

13.3 DEVELOPMENT AND APPLICATION OF MOLECULAR MARKERS

Molecular markers have demonstrated a potential role to detect tolerance or susceptibility for abiotic stresses in crop plants, genetic diversity and to aid the management of plant genetic resources (Reddy *et al.*, 2012). In contrast to morphological traits (morphological markers) and biochemical components (protein products from genes acts as biochemical markers), molecular markers can reveal differences among genotypes at the DNA level, providing a more direct, reliable, and efficient tool for germplasm characterization, conservation, and management. Till date, numerous types of molecular markers have been reported (Semagn *et al.*, 2006) that are as follows: allele-specific associated primers (ASAP), allele-specific oligo (ASO), allele-specific polymerase chain reaction (AS-PCR), amplified fragment length polymorphism (AFLP), anchored microsatellite primed PCR (AMP-PCR), anchored simple-sequence repeats (ASSR), arbitrarily primed polymerase chain reaction (AP-PCR), cleaved amplified polymorphic sequence (CAPS), degenerate oligonucleotide primed PCR (DOP-

PCR), diversity array technology (DArT), DNA amplification fingerprinting (DAF), expressed sequence tags (EST), intersimple sequence repeat (ISSR), inverse PCR (IPCR), inverse sequence-tagged repeats (ISTR), microsatellite primed PCR (MP-PCR), multiplexed allele-specific diagnostic assay (MAS-DA), random amplified microsatellite polymorphisms (RAMP), random amplified microsatellites (RAM), random amplified polymorphic DNA (RAPD), restriction fragment length polymorphism (RFLP), selective amplification of microsatellite polymorphic loci (SAMPL), sequence characterized amplified regions (SCAR), sequence specific amplification polymorphisms (S-SAP), sequence-tagged microsatellite site (STMS), sequence-tagged site (STS), short tandem repeats (STR), simple sequence length polymorphism (SSLP), simple sequence repeats (SSR), single nucleotide polymorphism (SNP), single primer amplification reactions (SPAR), single stranded conformational polymorphism (SSCP), site-selected insertion PCR (SSI), strand displacement amplification (SDA), and variable number tandem repeats (VNTR). Although some of these markers are very similar (for instance, ASAP, ASO, and AS-PCR), some synonymous (e.g., ISSR, RAMP, RAM, SPAR, AMP-PCR, MP-PCR, and ASSR) and some are identical (such as, SSLP, STMS, STR, and SSR), there are still a wide range of techniques for researchers to choose any of them as per their experimental design. Out of the vast lists of molecular markers, microsatellite or SSR (simple-sequence repeat) markers (Karmakar *et al.*, 2012; Roychowdhury *et al.*, 2012), RAPD (random amplified polymorphic DNA) (Williams *et al.*, 1990), AFLP (amplified fragment length polymorphism) (Vos *et al.*, 1995), RFLP (restriction fragment length polymorphism) (Botstein *et al.*, 1980) are efficiently used in today's research field related to stress tolerance in cereals, floricultural crops, legumes, biodiesel plants, horticultural crops (fruits and vegetables) and others. Amongst these, SSRs and RFLPs are codominant in nature and their genetic map location on crop genome is publicly disclosed; in case of AFLP and RAPD, they produce random amplification and are largely dominant markers in nature (Ram *et al.*, 2007).

PCR-based SSR markers are cost-effective, high diversity in crop plants due to their frequently available nature, codominant, and greater efficient to others (Chen *et al.*, 1997; Temnykh *et al.*, 2000). Compared to RFLPs, microsatellite markers detect a significantly higher degree of polymorphism in rice (Yang *et al.*, 1994) and are especially suitable for evaluating genetic diversity among closely related cultivars (Akagi *et al.*, 1997). Locus-specific microsatellite-based markers have been reported from many plant species such as lettuce (*Lactuca sativa* L.), barley (*Hordeum vulgare* L.) and rice (*Oryza sativa* L.) (Wu and Tanksley, 1993; Saghai Maroof *et al.*, 1994; van de Wiel *et al.*, 1999). Diversity based on phenological and morphological characters varies in different environments and its evaluation requires growing plants till its full maturity.

Markers those are based on expressed gene products, proteins or isozymes, are also influenced by the environment and reveal a low level of polymorphism and low abundance (Ravi *et al.*, 2003). In contrast, DNA-based molecular markers have proven to be powerful tools in the assessment of genetic variation and in the elucidation of genetic relationships within and among the species, characterized by abundance and untouched by environmental influence, that is, genotype-environment interaction and several types of abiotic stresses for the responsible environment (Powell *et al.*, 1996). Scientific experiments demonstrated the remarkable potential of microsatellite markers to discriminate between the crop genotypes, as compared to other molecular markers. Traditional crop cultivars (like rice, wheat, and maize) have a high level of genetic heterogeneity compared to modern high yielding varieties (HYV). This genetic variability is very important for the sustainability of small farmers, because despite the low yield capacity, these varieties represent high yield stability (Brondani *et al.*, 2006). High degree of similarity between different cereal and legume genomes in terms of gene content and gene order facilitates crop improvement and breeding for other crops as well (Garris *et al.*, 2005). The genetic relationship among the different diverse rice lines has been assessed by a number of workers (Dey *et al.*, 2005; Chattopadhyay *et al.*, 2008; Lodha *et al.*, 2011) who used various molecular markers for rice genomic DNA fingerprinting. Assessment of genetic diversity and tolerance governed genes, using molecular markers, is an essential component in germplasm characterization, identification, and conservation of crops. Results obtained in genetic diversity studies of crop landraces with SSR markers indicate that more genetic diversity exists in their gene pools. Selection increases the frequency of alleles or allelic combinations with favorable effects at the expense of others, eventually eliminating many of them. Many studies have also reported significantly greater allelic diversity of microsatellite markers than other available molecular markers (McCouch *et al.*, 2001). In the fingerprint data, lower polymorphism information contents (PIC values) may be the result of closely related genotypes (Rahman *et al.*, 2010). The number of alleles and its PIC values also depends upon the repeat motif and the repeat sequence of the markers. Temnykh *et al.* (2000) showed that CTT- and AT-rich repeat motifs amplified with higher efficiency leading to greater overall polymorphism. Lastly, the magnitude of similarity matrix value is directly proportional to the varietal genetic/evolutionary distance.

Using the basis on differential DNA amplification, polymorphism and/or molecular banding variability in RAPD is lower when compared to the other available molecular marker techniques; and this is due to the INDEL (insertion and/or deletion) nature within the genomic regions where short random oligo-primers will bind (Williams *et al.*, 1991). As the approach requires no prior knowledge of the genome that is being analyzed, it can be employed across species using universal primers. The RAPD analysis of NILs (nonisogenic lines) has been successful in identifying markers linked to disease resistance genes in tomato (*Lycopersicon* sp.) (Martin *et al.*, 1991), lettuce (*Lactuca* sp.) (Paran *et al.*, 1991) and common bean (*Phaseolus vulgaris*) (Adam-Blondon *et al.*, 1994). On the other hand, AFLP combines the power of RFLP with the flexibility of PCR-based technology by ligating primer recognition sequences (adaptors) to the restricted DNA and selective PCR amplification of restriction fragments using a limited set of primers. Usually 50–100 bands per assay are produced by the primer pairs used for AFLP. Number of amplicon per AFLP assay is a function of the number of selective nucleotides in the AFLP primer combination, the selective nucleotide motif, GC-content, physical genome size and complexity. The AFLP technique generates fingerprints of DNA from any source, and without any prior knowledge of DNA sequence. Most AFLP fragments correspond to unique positions on the genome and hence can be exploited greatly in genetic and physical mapping (Agarwal *et al.*, 2008).

Single nucleotide polymorphism (SNP) constitutes the most abundant and widely distributed molecular markers in the genome, which varies greatly among species. For example, maize has 1 SNP per 602–120 bp (Ching *et al.*, 2002). The SNPs are usually more prevalent in the noncoding regions of the genome. Within the coding regions, an SNP is either nonsynonymous resulting in an amino acid sequence change, or it is synonymous not being able to alter the amino acid sequence. The latter can result in phenotypic differences due to mRNA splicing (Richard and Beckman, 1995). Direct analysis of genetic variation at the DNA sequence has been made possible by improvements in sequencing technology and availability of an increasing number of EST sequence level. Majority of SNP genotyping assays are based on one or two of the following molecular mechanisms: allele-specific hybridization, primer extension, oligonucleotide ligation and invasive cleavage (Sobrino *et al.*, 2005). High throughput genotyping methods, including DNA chips, allele-specific PCR and primer extension approaches make SNPs attractive genetic markers. They are suitable for automation, rapid identification of crop cultivars and construction of ultrahigh-density genetic maps.

Sophisticated strategies have been developed and applied in various crop species for unlocking the genetic potential (allelic selection) from its nondomesticated genetic stock. One such very recent strategy is "allele mining"which dissects naturally occurring allelic variation at abiotic stress tolerance governed gene loci controlling key agronomic traits and exploits the DNA sequence of one genotype to isolate useful alleles from related genotypes (Latha *et al.*, 2004). As the map-based complete crop genome sequence is now freely available to researcher worldwide (Raghuvangshii *et al.*, 2009), the identification and isolation of novel and superior allele for agronomically important genes became a popular research area in modern crop improvement program (Ram Kumar *et al.*, 2010) for their stress tolerance property, particularly in drought, salinity, and oxidative stress conditions.

13.4 MOLECULAR BREEDING AND MARKER ASSISTED SELECTION

Most of the important traits, especially tolerance to abiotic stress and agronomically important characters, are much complex and polygenic in nature and are controlled by quantitative trait loci, in short QTL. Abiotic stresses, including drought, salinity, submergence, low temperature and the effects of several types of adverse soils, are a frequent constraint to crop production. Genetic sources of tolerance of these stresses are available in traditional cultivars and in some improved cultivars, but the complexity of the traits has hindered transfer of the tolerance genes into elite genotypes. However, QTL for these traits have been identified and marker-assisted selection (MAS) has been used for specific QTL introgression into sensitive cultivars (Collard and Mackill, 2007). As we know that abiotic stress tolerance is under polygenic control, crop breeding for achieving such properties in sustainable manner is so difficult with the presence of some barriers like the unknown inheritance pattern of multigenic-controlled trait(s), low genetic variability, low scope for selection parameters, etc. (Singh *et al.*, 2011). Several parameters, such as heritability of the target trait, population size and possibility of false QTL detection should be taken into consideration for the efficiency of QTL for MAS. The advantages of using MAS in crop improvement have been well documented (Mackill, 2007).

A simulation study conducted by Moreau *et al.* (1998) revealed the sevaral relationships between QTL and MAS in crops: (i) If the heritability is high, the genotypic values are well estimated by the phenotype, and the weight given to markers is equivalent to phenotypic selection. (ii) MAS is not effective at α (selection index) of 5 % and heritability < 0.15. (iii) The efficiency of MAS decreases as the number of QTL increases and the efficiency of MAS increases when individual QTL explain a large part of the genetic variance, (iv) The relative efficiency of MAS increases with population size (the population should be larger than 100 or 200 individuals) and if the distance between markers and QTL decreases.

Due to several constrains of conventional breeding approaches, a rapid progress has been made toward the development of molecular marker technologies and their application in linkage mapping, molecular dissection of the complex agronomical traits and marker-assisted breeding (Flowers *et al.*, 2000; Singh *et al.*, 2011). Application of molecular marker technology can greatly enhance the efficiency and accuracy of molecular breeding process (Singh, 2009). Molecular markers have proven to be useful in both basic and applied research such as DNA fingerprinting, varietal identification, diversity analysis, phylogenetic analysis, and marker-assisted breeding and map-based cloning of genes in rice (McCouch *et al.*, 2001). Different types of DNA markers including microsatellite DNA markers (simple sequence repeats, SSRs), AFLPs and ISSRs have been successfully used for genotype identification, diversity analysis and gene/QTL analysis. DNA markers are of great importance in plant breeding, especially for the selection of polygenic traits because they have several advantages like no G × E effect, no epistatic effect, desired homozygous plants can be easily picked up, and greater reliability to distinguish the homozygous lines from others at an early stage/generation. Focusing on specific crop and problem, development of transgenic production and/or marker-assisted selection will provide to make highly stress-tolerant crop cultivars. Molecular marker based *in situ* approaches were proved as potential tools for molecular breeding and hybridization performance in major crops such as wheat, maize, rice, mungbean, mustard. The marker based approaches are necessary for (i) finding and utilities of underexploited trait related gene(s) of crops (QTLs), (ii) engineering of required metabolic networks, and (iii) exploitation of hybrid vigor using specific markers. By using quantitative trait loci (QTL) methodologies in ancestral (wild and wild-relatives), traditional and hybrid crop cultivars, it is now possible to make a correlation and draw a linking line between the markers and the specific stress tolerance properties. By the deployment and development of map-based marker-aided breeding strategies (molecular breeding and molecular farming) for the classic improvement in major crops like wheat, rice, maize, *Brassica* for the abiotic stress linked trait(s), genes/QTLs have been mapped for several important agronomical traits such as tolerance against abiotic stresses including drought and salinity (Flowers *et al.*, 2000; Forster *et al.*, 2000).

Amongst the abiotic stresses, maximum progress has been made toward the drought related traits and there have been only a few studies to map QTLs for salinity tolerance (Flowers *et al.*, 2000; Koyama *et al.*, 2001; Lang *et al.*, 2001). Molecular markers such as RFLP (Ali *et al.*, 2000), RAPD (Xie *et al.*, 2000) and AFLP (Ali *et al.*, 2000), have been used for QTL mapping for salinity tolerance. Recently, a new class of PCR-based markers, inter simple sequence repeat (ISSR) markers, has been widely used to determine intra- and intergenomic diversity (Singh *et al.*, 2008; Singh *et al.*, 2009), since they reveal variation within unique regions of the genome at several loci simultaneously (Zietkiewicz *et al.*, 1994). ISSRs targets a subset of simple sequence repeats (SSRs) and amplifies the region between two closely spaced and oppositely oriented SSRs. Application of molecular marker technology can greatly enhance the efficiency and

accuracy of breeding process. Molecular marker technologies have revolutionized the genetic analysis of crop plants and its application has been suggested
for the molecular dissection of complex physiological traits such as salt tolerance (Flowers *et al.*, 2000). In several crops, comprehensive molecular marker/
linkage maps with a variety of DNA markers have been developed. Most of
the maps have been developed using mapping populations, which include the
recombinant inbred lines (RILs), double-haploid lines (DHLs), and backcross/
F2/F3 families (Lang *et al.*, 2001).

13.5 STRESS RESPONSIVE GENES AND THEIR REGULATION

Several abiotic and biotic stresses, as well as narrow genetic diversity in modern
cultivars of crop plants are the major constraints to further increases in productivity. With the development of a comprehensive molecular genetic map,
several genes have been identified corresponding to several traits of economic
importance. In addition to several genes of morphological and physiological
traits, numerous genes are for disease resistance like bacterial blight, blast, virus
diseases and for resistance to insects have been identified. Several genes and
quantitative trait loci (QTL) have been identified for abiotic stresses such as
drought, salinity, submergence, and cold (Jena and Mackill, 2008). The complex
plant response to abiotic stress involves many genes and biochemical-molecular
mechanisms. Under stressful conditions, such type of genes produces some major proteinaceous metabolites for cellular protection and also helps in genetic
regulation of other genes by the production of transcription factors and maintains the integrated circuits of stress-responsive signal transduction pathways.
In this condition, regulons (*cis*-acting elements in the promoter region of target
genes) play a vital regulatory role where transcription factors are the key players
(Todaka *et al.*, 2012). Various genes respond to drought stress in various species,
and functions of their gene products have been predicted from sequence homology with known proteins. Many drought-inducible genes are also induced by
salt stress and low temperature, which suggests the existence of similar mechanisms of stress responses. Genes induced during drought-stress conditions are
thought to function not only in protecting cells from water deficit by the production of important metabolic proteins but also in the regulation of genes for signal
transduction in the drought-stress response (Shinozaki *et al.*, 2003). Thus, these
gene products are classified into three major groups: (1) those that encode products that directly protect plant cells against stresses such as heat stress proteins
(HSPs) or chaperones, LEA proteins, osmoprotectants, antifreeze proteins, detoxification enzymes and free-radical scavengers (Wang *et al.*, 2003); (2) those
that are involved in signaling cascades and in transcriptional control, such as

MAPK, CDPK (Ludwig *et al.*, 2004) and SOS kinase (Zhu, 2001), phospholipases (Frank *et al.*, 2000), and transcriptional factors (Shinozaki and Yamaguchi-Shinozaki, 2000); (3) those that are involved in water and ion uptake and transport such as aquaporins and ion transporters (Blumwald, 2000). With the aid of gene transfer technologies in plant cell through genetic engineering, it is most important to know the molecular mechanism, plant's responses and underlying genetic manipulation of several stress-inducible genes in crops for abiotic stress tolerance (Seki *et al.*, 2003). There is several sensor molecules present in the cellular boundary to catch up the multiple stress signals and transmit it to the inside of the cell to induce the regulation mechanism via the regulators such as transcription factors and some accessory bioactive products (for instance, phytohormones, secondary messengers). In this way, several functional proteins are produced in this signal cascade and directly involve in the stress tolerance of plants with the integration of both negative and positive regulation of stress-responsive genes (Figure 13.1).

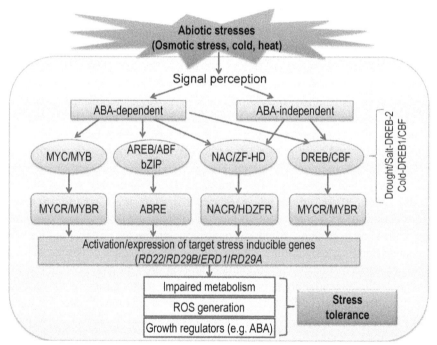

FIGURE 13.1 Overall hypothetical cellular model of an abiotic stress response with signaling and responsive genes for stress tolerance in plants.

Several bioactive compounds like LEA protein, carbohydrate transporter protein, metallothionein-like protein, enzymes (catalase, lipoxygenase), calmodulin and TFs (Myb, NAC, DREB, zinc finger) are encoded by many abiotic stress-responsive genes that are found in many crop plants. In the transcriptomics study on *Oryza sativa*, a number of such genes were identified for drought (62 genes), low temperature (36 genes) and salinity (57 genes) tolerance response with a differential response pattern as some genes are expressed in roots and some others are in leaves in two different plant genotypes (Rabbani *et al.*, 2003). Such genes and their protein profiles in form of enzymes, TFs, and others help the crops to maintain the native cellular physiology and protection for oxidative stress associated with osmotic stresses. Differential expression patterns of osmotic stress-responsive genes in plants were examined (Zhou *et al.*, 2007) and proved as organ specific expression. When plants are under stressful situations, transcriptional regulatory networks regulate numerous genes' expression pattern by either up-and/or down-regulation (Figure 13.2). In such platform, Todaka *et al.* (2012) explain the regulatory mechanism of different regulons (DREB, AREB, and NAC) that incorporates the modulation of a number of transcription factors (DREB1, DREB2, AREB, NAC).

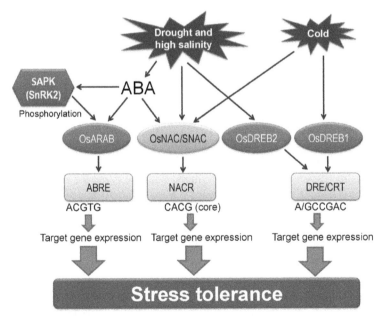

FIGURE 13.2 Overview of the representative transcriptional networks mediated by transcription factors and *cis*-elements under abiotic stresses in *Oryza sativa*. Ellipses represent transcription factors, boxes represent *cis*-elements (Todaka *et al.*, 2012).

13.6   LEA-PROTEIN

Late embryogenesis abundant (LEA) proteins represent another category of high molecular weight proteins that are abundant during late embryogenesis and accumulate during seed desiccation and in response to water stress (Galau *et al.*, 1987). Amongst the several groups of LEA proteins, those belonging to group 3 are predicted to play a role in sequestering ions that are concentrated during cellular dehydration. These proteins have 11-mer amino acid motifs with the consensus sequence TAQAAKEKAGE repeated as many as 13 times (Dure, 1993). Hyper hydrophilic property for binding with water molecule of LEA group-I proteins is comparable with LEA group-V one that helps in desiccation responsive ion sequestration. Xu *et al.* (1996) produced transgenic tolerant rice in salinity and desiccation stress environment with the incorporation of over-expressive barley *HVA1* gene that encodes a particular type of LEA group-III protein. In further, Rohilla *et al.* (2002) proposed that such overexpression of *HVA1* confers the cellular integrity and developmental growth traits for osmotic stress tolerance in *Oryza sativa* (rice) and *Triticum aestium* (wheat). Although, the reported water use efficiency (WUE) was extremely low when compared to other data reported in wheat cultigens, transgenic rice (TNG67) plants expressing a wheat LEA group 2 protein (PMA80) gene or the wheat LEA group 1 protein (PMA1959) gene resulted in increased tolerance to dehydration and salt stresses (Cheng *et al.*, 2002). Besides, protective chaperone like function of LEA proteins acting against cellular damage has been proposed, indicating the role of LEA proteins in anti-aggregation of enzymes under desiccation and freezing stresses (Goyal *et al.*, 2005).

13.7   REGULATORY GENES

In transcriptional point of view, poly-stress (for instance, high temperature, salinity, drought) responsive genes are induced with the presence of abscisic acid (ABA) (Mundy and Chua, 1988) toward the osmotic stress protection for the cell, tissue and/or whole plant body (Skriver and Mundy, 1990). For restoring the plant's stress-related functional physiology for tolerance, single gene transfer (encoding a particular protein product for stresses) is not too much sufficient for its main aim (Bohnert *et al.*, 1995). To reach this goal, scientists are more concern to target multiple stress linked genes and their encoded transcriptory products that can act as transcription factor for other stress-responsive genes and can regulate several other targeting genes for abiotic stress tolerance in crop plants (Chinnusamy *et al.*, 2005). Therefore, a category of preferable genes, are using in crop genetic engineering, that switch on transcription factors regulating the expression of several genes related to abiotic stresses. An attractive target category for manipulation and gene regulation is the small group of transcription factors that have been identified to bind to promoter regulatory elements in genes that are regulated by abiotic stresses (Shinozaki and Yamaguchi-Shinozaki, 1997). The transcription factors activate cascades of genes that act together in enhancing tolerance toward multiple stresses. Introduction of transcription factors in the ABA signaling pathway can also be a mechanism of genetic improvement of plant stress tolerance. Several stress-induced *cor* genes such as *rd29A*, *cor15A*, *kin1*, and *cor6.6* are triggered in response to cold treatment, ABA and water deficit stress (Thomashow, 1998). Similarly, a *cis*-acting element, dehydration responsive element (DRE) identified in *A. thaliana*, is also involved in ABA-independent gene expression under drought, low temperature and high salt stress conditions in many dehydration responsive genes like *rd29A* that are responsible for dehydration and cold-induced gene expression (Iwasaki *et al.*, 1997). Several cDNAs encoding the DRE binding proteins, DREB1A, and DREB2A are shown to specifically bind and activate the transcription of genes containing DRE sequences (Liu *et al.*, 1998). DREB1/CBFs are thought to function in cold-responsive gene expression, whereas DREB2s are involved in drought-responsive gene expression. The transcriptional activation of stress-induced genes has been possible in transgenic plants overexpressing one or more transcription factors that recognize regulatory elements of these genes.

13.8 FUNCTIONAL GENOMICS AND STRESS RESPONSE

Environmental or abiotic stresses such as high temperature, low water availability, salinity, heat, mineral toxicity, mineral deficiency frequently affect plants in agricultural systems, and represent major limitations to the yield and quality of crops. As a sequel to it, physiological and biochemical responses in plants vary and cellular aqueous and ionic equilibriums are disrupted. Also, hundreds of genes and their products respond to these stresses at transcriptional and translational level (Cushman and Bohnert, 2000; Sreenivasulu *et al.*, 2004; Yamaguchi-Shinozaki and Shinozaki, 2005; Umezawa *et al.*, 2006). Understanding the functions of these stress-inducible genes helps to unravel the possible mechanisms of stress tolerance. Functional genomics approaches triggered a major paradigm from single gene discovery to thousands of genes by using multi-parallel high-throughput techniques. Generation of expressed sequence tags (ESTs) from cDNA libraries prepared from abiotic stress-treated seedlings of various crops, complete genome sequence information of rice and Arabidopsis provided a valuable resource for gene discovery. Furthermore, expression profiling by microarrays, random, and targeted mutagenesis, complementation, and promoter-trapping strategies allow the identification of the key stress-responsive gene pools and in turn provide important clues for functional characterization of stress-responsive genes and stress tolerance mechanisms (Sreenivasulu *et al.*, 2007).

In response to abiotic stress, plants respond and adapt to those stresses by altering thousands of genes. As a result of these alterations several physiological, biochemical, and cellular processes will be modified. Transcriptome analysis using microarray technology is a useful tool to find out the expression of genes during abiotic stress at the global level. For doing functional study some related information are required such as gene discovery, high-throughput gene expression, functional characterization of genes of interest via high-throughput gene inactivation techniques (Sreenivasulu *et al.*, 2007).

An important genomic approach to identify abiotic stress-related genes is based on ESTs generated from different cDNA libraries from abiotic stress treated tissues collected at various stages of development. In addition, the clustering of EST sequences generated from abiotic stress-treated cDNA libraries provides information on gene number, gene content and possible number of gene families involved in stress responses. Putative functions of such stress-responsive genes can be assigned using Swissprot database (Sreenivasulu *et al.*, 2007). This type of analysis provides valuable information regarding stress-responsive genes (Wang *et al.*, 2004) as well as provides information related to underlying regulatory and metabolic networks. In contrast to digital *in silico* approaches, array-based (cDNA microarray) transcript profiling technologies and quantitative real time PCR (qRT-PCR) are novel approaches to identify expression of thousands of genes related to stress tolerance mechanisms (Chen *et al.*, 2002; Sreenivasulu *et al.*, 2006). Through this high-throughput expression study we can also identify which gene is up regulated or down regulated in specific stress conditions. This approach also enables the identification of new promoter elements/transcription factor binding sites in coexpressed gene sets, which in turn helps to explore regulatory networks controlling abiotic stress responses (Sreenivasulu *et al.*, 2002). Utilizing all the expression profile we can chose desirable genes for transformation. Even though a high number of abiotic stress-responsive candidate genes are identified but more than 40–50 % of identified stress-responsive gene functions remain to be characterized. In order to reveal their putative functions involved in abiotic stress tolerance, various high-throughput methods were developed for the confirmation and validation of gene function by gene inactivation. There are two main complementary approaches developed for identifying mutations in target genes, namely TILLING (Targeted Induced Local Lesions In Genomes) and T-DNA insertion mutant lines. Using these techniques various attempts were made in order to show the putative functions of abiotic stress-responsive genes in *Arabidopsis*, rice, maize, and barley (Xiong *et al.*, 1999; Liu *et al.*, 2000; Shi *et al.*, 2000; Zhu, 2001). In recent years, large amount of data have been accumulated on plant's responses to abiotic stresses based on functional genomics though more elaborate study is required to increase the database. All the valuable information gating from functional genomics database increase the knowledge for engineering stress-tolerant plants for their ultimate use in sustainable agriculture.

13.9 MODERN TRANSGENIC APPROACHES FOR ABIOTIC STRESS TOLERANCE

After introduction of the concept of the polygenic nature of abiotic stress and its tolerance instead of the monogenic with the help of genetic engineering, the regulatory network of transcription factors was emerged as a new tool for controlling the expression of many stress-responsive genes. Some of the examples of transgenic crop plants, tolerant to abiotic stress, are given in Table 13.1. There are several techniques (*Agrobacterium*-mediated, particle bombardment, plant virus mediated, chloroplast transformation, protoplast transformation) available for gene transfer in plants. Among them *Agrobacterium*-mediated and particle bombardment techniques are generally used for gene transfer in plants (Primrose and Twyman, 2006). In case of *Agrobacterium*-mediated gene transfer T-regions (Figure 13.3) of Ti or Ri plasmids are replaced by desirable stress-related genes, which are transferred in targeted plants. This insert size can be as big as 23 kbp or more. Hence, multiple genes can be transferred in one transformation, which is very essential for abiotic stress-related transgenic approaches (Stanton, 2003). On the other hand in case of particle bombardment different genes can be transferred to wide range of targeted plants (Primrose and Twyman, 2006).

TABLE 13.1 Responsible genes to develop transgenic crop plants for abiotic stress tolerance.

Gene	Species	Cellular Response	References
Mn-superoxide dismutase (*Mn-SOD*)	*O. sativa*	Salt tolerance	Tanaka *et al.*, 1999
Fe-superoxide dismutase (*Fe-SOD*)	*Z. mays*	Cold tolerance	Van Bruesegem *et al.*, 1999
Catalase (*cat*)	*O. sativa*	Chilling tolerance	Matsumura *et al.*, 2002
Proline (*p5cs*)	*O. sativa*	Drought, salt tolerance, and oxidative stress tolerance	Hong *et al.*, 2000
	T. aestivum	Salt tolerance	Sawahel and Hassan, 2002

TABLE 13.1 *(Continued)*

Gene	Species	Cellular Response	References
Choline dehydrogenase (*codA*)	*O. sativa*	Drought and salt tolerance	Takabe *et al.*, 1998
Mannitol dehydrogenase (*mtlD*)	*T. aestivum*	Drought and salt tolerance	Abebe *et al.*, 2003
waxy gene	*O. sativa*	Cold tolerance	Hirano and Sano, 1998
Glutamine synthetase (*gs*)	*O. sativa*	Salt and cold tolerance	Hoshida *et al.*, 2000
Arginine decarboxylase (*adc*)	*O. sativa*	Drought tolerance	Capell *et al.*, 1998
Calcium dependent protein kinase (*cdpk*)	*O. sativa*	Salt, drought, and cold tolerance	Saijo *et al.*, 2000
Dehydration response element binding factors (*dreb1A*)	*O. sativa*	Drought tolerance	Pellegrineshi *et al.*, 2002
Late embryogenesis protein (*hva1*)	*A. sativa*	Drought tolerance	Maqbool *et al.*, 2002
	T. aestivum	Drought tolerance	Sivamani *et al.*, 2000
	O. sativa	Drought and salt tolerance	Xu *et al.*, 1996
Heat shock protein (*Hsp90*, *Hsp104*)	*O. sativa*	Heat tolerance	Pareek *et al.*, 1995
WCS genes	*T. aestivum*	Cold tolerance	Oullet *et al.*, 1998
HKT1	*O. sativa*	Salt tolerance	Laurie *et al.*, 2002
Na^+/H^+ antiport	*O. sativa*	Salt tolerance	Ohta *et al.*, 2002
Pyruvate decarboxylase 1	*O. sativa*	Flooding tolerance	Quimio *et al.*, 2000
Alcohol dehydrogenase	*O. sativa*	Flooding tolerance	Minhas and Grover, 1999
Glycine betaine (*codA*)	*O. sativa*	Chilling, salt, heat, strong light, and freezing tolerance	Chen and Murata, 2002
CBF1	*A. thaliana*	Cold tolerance	Bhatnagar-Mathur *et al.*, 2008
AtMYC2	*A. thaliana*	Drought tolerance	Umezawa *et al.*, 2006

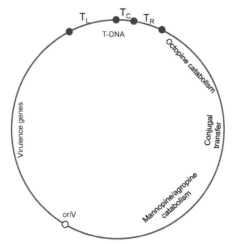

FIGURE 13.3   Schematic representation of a typical octopine-type Ti plasmid. The T-DNA is divided into three regions. TL (T-DNA left), TC (T-DNA centre), and TR (T-DNA right). The black circles indicate T-DNA border repeat sequences. *oriV*, the vegetative origin of replication of the Ti plasmid, is indicated by a white circle.

Nevertheless, with the proper incorporation of stress tolerance properties, transgenic plant production is not limited to the progress of transformant generation. Today's major target to evaluate the transgenic plants acclimatized under stressful environments, it is necessary to understand the effect (physiological, biochemical and molecular) of engineered gene for stress tolerance in the plant body and this will helps to overcome the present problems. Establishment of transgenic crops in natural conditions is the most challenging aspect of transgenic approaches. Development of transgenic plants using biotechnological and genomic tools has become important in plant abiotic stress. The use of novel approaches combining genetic, physiological, biochemical, and molecular techniques should provide excellent results in the near future. We can overcome this problem by increasing awareness and more elaborate study of transgenic crops.

13.10   METABOLIC ENGINEERING FOR STRESS TOLERANCE

Plants have evolved a number of adaptations to such abiotic stresses: some of these adaptations are metabolic and others structural. Accumulation of certain organic solutes (known as osmoprotectants) is a common metabolic adaptation found in diverse taxa. These solutes protect proteins and membranes against damage by high concentrations of inorganic ions. Some osmoprotectants also protect the metabolic machinery against oxidative damage. Many major crops lack the ability to synthesize the special osmoprotectants that are naturally accumulated by stress-tolerant organisms. Therefore, it was hypothesized that installing osmoprotectant synthesis pathways is a potential route to breed stress-tolerant crops. Recently metabolic engineering efforts in model species have been utilized to improve crop's abiotic stress tolerance properties. Metabolic engineering is used for the direct improvement of cellular properties through the modification of specific biochemical reactions or the introduction of new ones, with the use of recombinant DNA technology (Stephanopoulos, 1999). Some of the metabolic adaptations to stress have been manipulated in model plant species using metabolic engineering.

Osmoprotectants (proline, glycine betaine, β-alanine, δ-Mannitol, 3-Dimethylsulphoniopropionate and so on) and some compatible solutes (fructan, sorbitol, trehalose, and others) are mainly used for metabolic engineering to improve crop's abiotic stress tolerance property. Among them glycine betaine is the most used one. Recently, several successful attempts of metabolic engineering have been achieved (Table 13.2). For metabolic engineering experiments we have to understand the role of osmoprotectants, primary metabolic pathways from which osmoprotectant synthesis pathways branch. Future research avenues include the identification and exploitation of diverse osmoprotectants in naturally stress-tolerant organisms, and the use of multiple genes and reiterative engineering to increase osmoprotectant in response to stress. Here we tried to understand role of some osmoprotectant for better use in metabolic engineering.

TABLE 13.2 List of different important osmoprotectants, their host plant and responsible gene along with its ability to enhance abiotic stress tolerance.

Osmopro-tectants	Gene	Host Plant	Enhanced Tolerance	References
Glycine betaine	*Coda*	*Oryza sativa, Arabidopsis thaliana, Brassica juncea, Diospyros kaki*	Chilling, salt, heat, strong light, freezing	Hayashi *et al.*, 1997; Alia *et al.*, 1998
	Beta	*Nicotiana tabacum*	Salt	Lilius *et al.*, 1996
	betA/betB	*N. tabacum*	Chilling, salt	Holmström *et al.*, 2000
	betA (modified)	*O. sativa*	Drought, salt	Takabe *et al.*, 1998

TABLE 13.2 *(Continued)*

Osmopro-tectants	Gene	Host Plant	Enhanced Toler-ance	References
Fructan	*SacB*	*N. tabacum, Beta vulgaris*	Drought	Pilon-Smits *et al.*, 1995
Mannitol	*mt1D*	*A. thaliana, N. tabacum*	Salt, oxidative stress	Tarczynski *et al.*, 1992
δ-Ononitol	*imt1*	*N. tabacum*	Drought, salt	Sheveleva *et al.*, 1997
Proline	Anti-ProDH	*A. thaliana*	Freezing, salt	Nanjo *et al.*, 1999
	P5CS^{F127A}	*N. tabacum*	Salt	Hong *et al.*, 2000
	Mothbean P5CS	*N. tabacum, O. sativa, Glycine max*	Drought, salt, heat, oxidative stress	Slater *et al.*, 2011
Sorbitol	*S6PDH*	*Diospyros kaki*	Salt	Gao *et al.*, 2001
Trehalose	*TPS1*	*N. tabacum, Solanum tu-berosum*	Drought	Holmström *et al.*, 1996
	otsA	*N. tabacum*	Drought	Pilon-Smits *et al.*, 1998
	otsB	*N. tabacum*	Drought	Pilon-Smits *et al.*, 1998

The elevated level of proline significantly enhanced the ability of the transgenic plants to grow in salt condition that contained up to 200 mM NaCl. The increased level of proline also reduced the levels of free radicals, as determined by monitoring the production of malondialdehyde (MDA). Thus, it appears that, in addition to acting as an osmolyte, proline might play a role in reducing the oxidative stress that is brought on by osmotic stress (Nanjo *et al.*, 1999). Glycine betaine (*N,N,N*-trimethyl glycine) is a quaternary ammonium compound that occurs naturally in a wide variety of plants, animals, and microorganisms (Rhodes and Hanson, 1993). Glycine betaine is synthesized by a two-step oxidation of choline via betaine aldehyde. Choline has a vital role as the precursor for phosphatidylcholine, a dominant constituent of membrane phospholipids in eukaryotes. Despite this, a large proportion of free choline is diverted to glycine betaine in plants that naturally accumulate glycine betaine in response to stress. The synthetic routes to choline and glycine betaine as known in spinach are shown in Figure 13.4. Microbial choline-oxidizing enzymes have also been expressed in transgenic tobacco and *Arabidopsis thaliana* two species that do not naturally accumulate glycine betaine. The transgenic plants exhibited salinity and temperature stress tolerance (Alia *et al.*, 1998). Polyols such as mannitol is an osmoprotectant in algae, certain halophytic plants exposed to freezing (Yancey *et al.*, 1982). Figure 13.5 illustrates polyol synthesis. When expressed in transgenic tobacco and *Arabidopsis*, a gene encoding mannitol 1-phosphate dehydrogenase (mtlD) from *Escherichia coli* resulted in mannitol production and a salinity-tolerant phenotype (Thomas *et al.*, 1995).

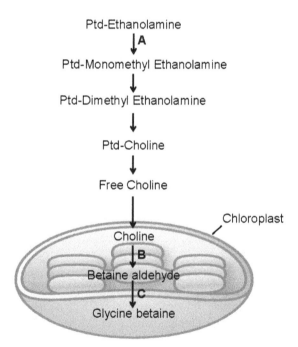

FIGURE 13.4 Interrelationships and compartmentation of choline and glycine betaine synthesis in *Spinacia oleracea*. Enzymes discussed in the text are numbered (A) to (C). (A) *S*-Adenosyl-L-methionine: Phosphotidylethanolamine N-methyl transferase, (B) Choline monooxygenase, and (C) Betaine aldehyde dehydrogenase.

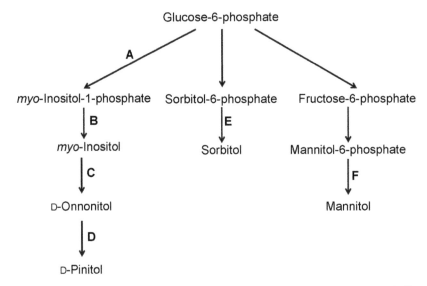

FIGURE 13.5 *Myo*-Inositol and polyol (mannitol) biosynthesis. Solid arrows indicate enzyme-catalyzed steps. Enzymes discussed in the text are numbered (A) to (F). (A) *myo*-Inositol-1-phosphate synthase, (B) *myo*-Inositol-1-phosphate phosphatase, (C) *myo*-inositol-*O*-methyltransferase, (D) D-ononitol epimerase, (E) sorbitol-6-phosphate dehydrogenase, and (F) mannitol-1-phosphate dehydrogenase. (C) and (D) are unique to the ice plant.

The fructan producing plants performed significantly better than controls under drought conditions, having a 55 % more rapid growth rate, 33 % greater fresh weight and 59 % greater dry weight than wild type plants. Under drought conditions, the transgenic plants accumulated fructan to concentrations as high as 0.35 mg g^{-1} fresh weight close to 7-fold higher than the level accumulated by the same plants under nonstress conditions (Pilon-Smits *et al.*, 1998). When δ-Ononitol transgenic plants were exposed to salt or drought stress, δ-ononitol (a sugar alcohol) accumulated to concentrations that exceeded 35 μmol g^{-1} fresh weight in the cytosol. Furthermore, the photosynthetic fixation of CO_2 was inhibited to a lesser extent during salt or drought stress in the transgenic plants that accumulated δ-ononitol than in wild type plants (Gao *et al.*, 2001). Under salt stress, photosynthetic activity declined less in the leaves of sorbitol-producing transgenic plants than in the leaves of wild type plants (Gao *et al.*, 2001). Trehalose transgenic plants were more drought-tolerant than controls, even though they exhibited substantial changes in morphology and accumulated higher levels of nonstructural carbohydrates (Holmström *et al.*, 1996). Structures of all the discussed compounds are given in Figure 13.6.

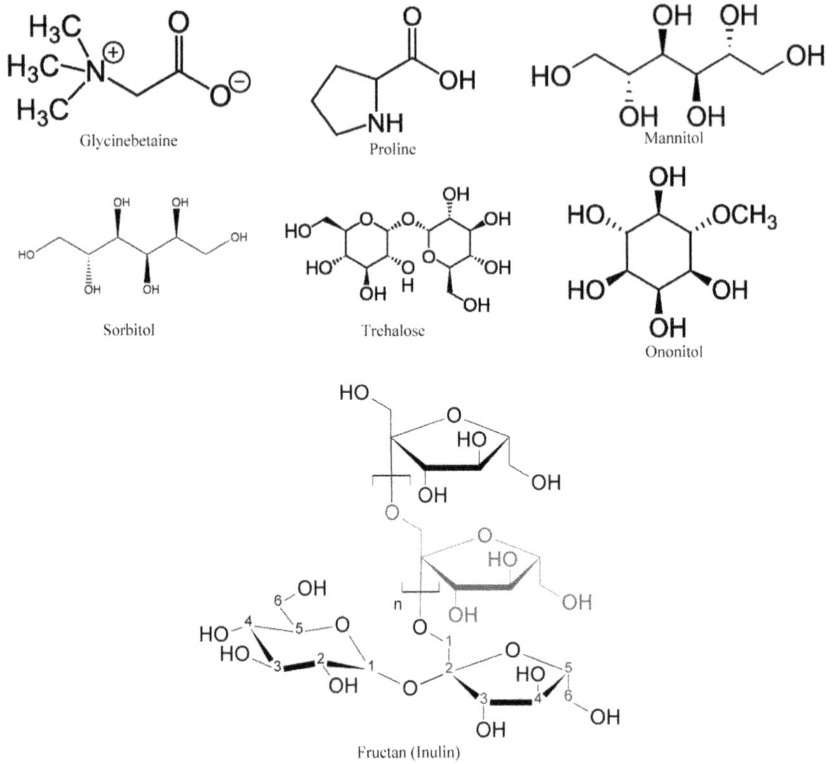

FIGURE 13.6   Chemical structures of some osmoprotectants that are primarily targeted for metabolic engineering.

13.11   CONCLUSION AND FUTURE PERSPECTIVES

Mendel's work on inheritance of phenotypic traits in plants laid the foundations of modern plant breeding. Scientific plant breeding from the early part of the Twentieth century onwards brought huge increases in crop yield, without which the current human population levels would already be unsustainable. In the present era of climate change, important crop plants now often suffer from various environmental stresses, making the development of stress tolerant cultivars the utmost task for plants breeders and plant physiologists. In the past several years, and because of the great interest in both basic and applied research, important progress has been made in the understanding of the mechanisms and processes underlying abiotic stress adaptation and defense in different plant species. The emergence of the novel "omics" technologies, such as genomics, proteomics, and metabolomics, is now allowing researchers to identify the genetics behind plant stress responses. However, a better understanding of the underlying physiological processes in response to different abiotic stresses could help to drive the selection of appropriate promoters or transcription factors for use in crop plant transformation. More importantly, more field trials are necessary for plants modified through these approaches, to confirm that they are in fact fit for use by farmers.

KEYWORDS

- **Abiotic stress tolerance**
- **Dehydration responsive element**
- **Genomics**
- **Metabolomics**
- **Molecular breeding**
- **Proteomics**
- **Quantitative trait loci**
- **Transcriptomics**
- **Water use efficiency**

REFERENCES

Abebe, T.; Guenzi, A. C.; Martin, B.; Cushman, J. C. Tolerance of mannitol-accumulating transgenic wheat to water stress and salinity. *Plant Physiol* 2003, 131, 1748-1755.

Acquaah, G. *Principles of plant genetics and breeding*. Blackwell: Oxford, 2007, p. 385.

Adam-Blondon, A. F.; Sevignac, M.; Bannerot, H.; Dron, M. SCAR, RAPD and RFLP markers linked to a dominant gene (Are) conferring resistance to anthracnose in common bean. *Theor Appl Genet* 1994, 88, 865-870.

Agarwal, M.; Shrivastava, N.; Padh, H. Advances in molecular marker techniques and their applications in plant sciences. *Plant Cell Rep* 2008, 27, 617-631.

Akagi, H.; Inagaki, Y. A.; Fujimura, T. Highly polymorphic microsatellites of rice consist of AT and a classification of closely related cultivars with these microsatellite loci. *Theor Appl Genet* 1997, 94, 61-67.

Ali, M. L.; Pathan, M. S.; Zhang, J.; Bai, G.; Sarkarung, S.; Nguyen, H. T. Mapping QTLs for root traits in a recombinant inbred population from two indica ecotypes in rice. *Theor Appl Genet* 2000, 101, 756-766.

Alia.; Chen, T. H. H.; Murata, N. Transformation with a gene for choline oxidase enhances the cold tolerance of *Arabidopsis* during germination and early growth. *Plant Cell Environ* 1998, 21, 232-239.

Altman, A. From plant tissue culture to biotechnology: scientific revolutions, abiotic stress tolerance, and forestry. *In Vitro Cell Dev Biol Plant* 2003, 39, 75-84.

Bhatnagar-Mathur, P.; Vadez, V.; Sharma, K. K. Transgenic approaches for abiotic stress tolerance in plants: retrospect and prospects. *Plant Cell Rep* 2008, 27, 411-424.

Blumwald, E. Sodium transport and salt tolerance in plants. *Curr Opin Cell Biol* 2000, 12, 431-434.

Bohnert, H. J.; Nelson, D. F.; Jenson, R. G. Adaptation to environmental stresses. *Plant Cell* 1995, 7, 1099-1111.

Botstein, D.; White, R. L.; Skolnick, M.; Davis, R. W. Construction of a genetic linkage map in man using restriction fragment length polymorphism. *American J Hum Genet* 1980, 32, 314-331.

Bray, E. A. Genes commonly regulated by water deficit stress in *Arabidopsis thaliana. J Exp Bot* 2004, 55, 2331-2341.

Brondani, C.; Borba, T. C. O.; Rangel, P. H. N.; Brondani, R. P. V. Determination of genetic variability of traditional varieties of Brazilian rice using microsatellite markers. *Genet Mol Biol* 2006, 29, 676-684.

Capell, T.; Escobar, C.; Liu, H.; Burtin, D.; Lepri, O.; Cristou, P. Overexpression of oat arginine decarboxylase cDNA in transgenic rice (*Oryza sativa* L.) affects normal development pattern *in vitro* and results in putrescine accumulation in transgenic plants. *Theor Appl Genet* 1998, 97, 246-254.

Chan, Z.; Grumet, R.; Loescher, W. Global gene expression analysis of transgenic, mannitol-producing, and salt-tolerant *Arabidopsis thaliana* indicates widespread changes in abiotic and biotic stress-related genes. *J Exp Bot* 2011, 62, 1-17.

Chattopadhyay, T.; Biswas, T.; Chatterjee, M.; Mandal, N.; Bhattacharyya, S. Biochemical and SSR marker based characterization of some Bengal landraces of rice suffixed with 'sail' in their name. *Ind J Genet* 2008, 68, 15-20.

Chaves, M. M.; Oliveira, M. M. Mechanisms underlying plant resilience to water deficits: prospects for water-saving agriculture. *J Exp Bot* 2004, 55, 2365-2384.

Chen, T. H. N.; Murata, N. Enhancement of tolerance of abiotic stress by metabolic engineering of betaines and other compatible solutes. *Curr Opin Plant Biol* 2002, 5, 250-257.

Chen, W.; Provart, N. J.; Glazebrook, J.; Katagiri, F.; Chang, H.; Eulgem, T.; Mauch, F.; Luan, S.; Zou, G.; Whitham, S. A.; Budworth, P. R.; Tao, Y.; Xie, Z.; Chen, X.; Lam, S.; Kreps, J. A.; Harper, J. F.; Si-Ammour, A.; Mauch-Mani, B.; Heinlein, M.; Kobayashi, K.; Hohn, T.; Dangl, J. L.; Wang, X.; Zhu, T. Expression profile matrix of *Arabidopsis* transcription factor genes suggests their putative functions in response to environmental stresses. *Plant Cell* 2002, 14, 559-574.

Chen, X.; Temnykh, S.; Xu, Y.; Cho, Y. G.; McCouch, S. R. Development of a microsatellite framework map providing genome-wide coverage in rice (*Oryza sativa* L.). *Theor Appl Genet* 1997, 95, 553-567.

Cheng, L.; Zou, Y.; Ding, S.; Zhang, J.; Yu, X.; Cao, J.; Lu, G. Polyamine accumulation in transgenic tomato enhances the tolerance to high temperature stress. *J Integr Plant Biol* 2009, 51, 489-499.

Cheng, W. H.; Endo, A.; Zhou, L.; Penney, J.; Chen, H. C.; Arroyo, A.; Leon, P.; Nambara, E.; Asami, T.; Seo, M. A unique short-chain dehydrogenase/reductase in *Arabidopsis* glucose signaling and abscisic acid biosynthesis and functions. *Plant Cell* 2002, 14, 2723-2743.

Ching, A. D. A.; Caldwell, K. S.; Jung, M.; Dolan, M.; Smith, O. S.; Tingey, S.; Morgante, M.; Rafalski, A. SNP frequency, haplotype structure and linkage disequilibrium in elite maize inbred lines. *BMC Genet* 2002, 3, 19.

Chinnusamy, V.; Jagendorf, A.; Zhu, J. K. Understanding and improving salt tolerance in plants. *Crop Sci* 2005, 45, 437-448.

Collard, B. C. Y.; Mackill, D. J. Marker-assisted selection: An approach for precision plant breeding in the twenty-first century. *Philos Trans R Soc Lond B Biol Sci* 2007, 17, 1-16.

Collins, N. C.; Tardieu, F.; Tuberosa, R. Quantitative trait loci and crop performance under abiotic stress: where do we stand? *Plant Physiol* 2008, 147, 469-486.

Cramer, G. R.; Urano, K.; Delrot, S.; Pezzotti, M.; Shinozaki, K. Effects of abiotic stress on plants: a systems biology perspective. *BMC Plant Biol* 2011, 11, 163.

Cushman, J. C.; Bohnert, H. J. Genomic approaches to plant stress tolerance. *Curr Opin Plant Biol* 2000, 3, 117-124.

Dey, N.; Biswas, S.; Ray Chaudhury, T.; Dey, S. R.; De, M.; Ghose, T. K. RAPD-based genetic diversity analysis of aromatic rice (*Oryza sativa* L.). *Plant Cell Biotechnol Mol Biol* 2005, 6, 133-142.

dos Reis, S. P.; Lima, A. M.; de Souza, C. R. B. Recent molecular advances on downstream plant responses to abiotic stress. *Int J Mol Sci* 2012, 13, 8628-8647.

Doupis, G.; Chartzoulakis, K.; Beis, A.; Patakas, A. Allometric and biochemical responses of grapevines subjected to drought and enhanced ultraviolet-B radiation. *Aust J Grape Wine Res* 2011, 17, 36-42.

Dure, L. 3rd. A repeating 11-mer amino acid motif and plant desiccation. *Plant J* 1993, 3, 363-369.

Fan, W.; Zhang, M.; Zhang, H.; Zhang, P. Improved tolerance to various abiotic stresses in transgenic sweet potato (*Ipomoea batatas*) expressing spinach betaine aldehyde dehydrogenase. *PLoS One* 2012, 7, e37344.

Flowers, T. J.; Koyama, M. L.; Flowers, S. A.; Sudhakar, C.; Singh, K. P.; Yeo, A. R. QTL: their place in engineering tolerance of rice to salinity. *J Exp Bot* 2000, 51, 99-106.

Forster, B. P.; Ellis, R. P.; Thomas, W. T. B.; Newton, A. C.; Tuberosa, R.; This, D., El-Enein, R. A.; Bahri, M. H.; Ben-Salem, M. The development and application of molecular markers for abiotic stress tolerance in barley. *J Exp Bot* 2000, 51, 19-27.

Frank, W.; Munnik, T.; Kerkmann, K.; Salamini, F.; Bartels, D. Water deficit triggers phospholipase D activity in the resurrection plant *Craterostigma plantagineum*. *Plant Cell* 2000, 12, 111-124.

Galau, G. A.; Bijaisoradat, N.; Hughes, D. W. Accumulation kinetics of cotton late embryogenesis-abundent (LEA) mRNAs and storage protein mRNAs: coordinate regulation during embryogenesis and role of abscisic acid. *Dev Biol* 1987, 123, 198-212.

Gao, M.; Tao, R.; Miura, K.; Dandekar, A. M.; Sugiura, A. Transformation of Japanese persimmon (*Diospyros kaki* Thunb.) with apple cDNA encoding NADP-dependent sorbitol-6-phosphate dehydrogenase. *Plant Sci* 2001, 160, 837-845.

Garris, A.; Tai, T.; Coburn, J.; Kresovich, S.; McCouch, S. R. Genetic structure and diversity in *Oryza sativa* L. *Genet* 2005, 169, 1631-1638.

Gorantla, M.; Babu, P. R.; Reddy, V. B. L.; Feltus, F. A.; Paterson, A. H.; Reddy, A. R. Functional genomics of drought stress response in rice: transcript mapping of annotated unigenes of an indica rice (*Oryza sativa* L. cv. Nagina 22). *Curr Sci* 2005, 89, 496-514.

Goyal, K.; Walton, L. J.; Tunnacliffe, A. LEA proteins prevent protein aggregation due to water stress. *Biochem J* 2005, 388, 151-157.

Hasanuzzaman, M.; Fujita, M. Heavy metals in the environment: Current status, toxic effects on plants and possible phytoremediation. In *Phytotechnologies: Remediation of environmental contaminants*, Anjum, N. A.; Pereira, M. A.; Ahmad, I.; Duarte, A. C.; Umar, S.; Khan, N. A., Ed.; CRC Press: Boca Raton, 2012, pp. 7-73.

Hasanuzzaman, M.; Fujita, M.; Islam, M. N.; Ahamed, K. U.; Nahar, K. Performance of four irrigated rice varieties under different levels of salinity stress. *Int J Integr Biol* 2009, 6, 85-90.

Hasanuzzaman, M.; Hossain, M. A.; da Silva, J. A. T.; Fujita, M. Plant responses and tolerance to abiotic oxidative stress: Antioxidant defenses is a key factors. In *Crop stress and its management: Perspectives and strategies*, Bandi, V.; Shanker, A. K.; Shanker, C.; Mandapaka, M., Ed.; Springer: Berlin, 2012, pp. 261-316.

Hasanuzzaman, M.; Hossain, M. A.; Fujita, M. Physiological and biochemical mechanisms of nitric oxide induced abiotic stress tolerance in plants. *Am J Plant Physiol* 2010, 5, 295-324.

Hasanuzzaman, M.; Hossain, M. A.; Fujita, M. Nitric oxide modulates antioxidant defense and the methylglyoxal detoxification system and reduces salinity-induced damage of wheat seedlings. *Plant Biotechnol Rep* 2011, 5, 353-365.

Hasanuzzaman, M.; Nahar, K.; Fujita, M. Plant response to salt stress and role of exogenous protectants to mitigate salt-induced damages. In *Ecophysiology and responses of plants under salt stress*, Ahmad, P.; Azooz, M. M.; Prasad, M. N. V., Ed.; Springer: New York, 2013a, pp. 25-87.

Hasanuzzaman, M.; Nahar, K.; Fujita, M.; Ahmad, P.; Chandna, R.; Prasad, M. N. V.; Öztürk, M. Enhancing plant productivity under salt stress - relevance of poly-omics. In *Salt stress in plants: Omics, signaling and responses*, Ahmad, P.; Azooz, M. M.; Prasad, M. N. V., Ed.; Springer: Berlin, 2013b, pp. 113-156.

Hasanuzzaman, M.; Nahar, K.; Fujita, M. Extreme temperatures, oxidative stress and antioxidant defense in plants, In *Abiotic stress: Plant Responses and Applications in Agriculture*, Vahdati, K.; Leslie, C., Ed.; InTech: Rijeka, 2013c, pp. 169-205.

Hayashi, H.; Alia.; Mustardy, L.; Deshnium, P.; Ida, M.; Murata, N. Transformation of *Arabidopsis thaliana* with the *codA* gene for choline oxidase; accumulation of glycine betaine and enhanced tolerance to salt and cold stress. *Plant J* 1997, 12, 133-142.

Heidarvand, L.; Amiri, R. M. What happens in plant molecular responses to cold stress? *Acta Physiol Plant* 2010, 32, 419-431.

Hirano, H. Y.; Sano, Y. Enhancement of *Wx* gene expression and the accumulation of amylase in response to cool temperatures during seed development in rice. *Plant Cell Physiol* 1998, 39, 807-812.

Holmström, K.; Somersalo, S.; Mandal, A.; Palva, T. E.; Welin, B. Improved tolerance to salinity and low temperature in transgenic tobacco producing glycine betaine. *J Exp Bot* 2000, 51, 177-185.

Holmström, K. O.; Mäntylä, E.; Welin, B.; Mandal, A.; Palva, E. T.; Tunnela, O. E.; Londesborough, J. Drought tolerance in tobacco. *Nature* 1996, 379, 683-684.

Hong, Z.; Lakkineni, K.; Zhang, Z.; Verma, D. P. S. Removal of feedback inhibition of 1 pyrroline-5-carboxylate synthetase (P5CS) results in increased proline accumulation and protection of plants from osmotic stress. *Plant Physiol* 2000, 122, 1129-1136.

Hoshida, H.; Tanaka, Y.; Hibino, T.; Hayashi, Y.; Tanaka, A.; Takabe, T. Enhanced tolerance to salt stress in transgenic rice that overexpresses chloroplast glutamine synthetase. *Plant Mol Biol* 2000, 43, 103–111.

Iwasaki, T.; Kiyosue, T.; Yamaguchi-Shinozaki, K. The dehydration- inducible rd17 (*cor47*) gene and its promoter region in *Arabidopsis thaliana*. *Plant Physiol* 1997, 115, 128.

Jena, K. K.; Mackill, D. J. Molecular markers and their use in marker-assisted selection in rice. *Crop Sci* 2008, 48, 1266-1276.

Karmakar, J.; Roychowdhury, R.; Kar, R. K.; Deb, D.; Dey, N. Profiling of selected indigenous rice (*Oryza sativa* L.) landraces of Rarh Bengal in relation to osmotic stress tolerance. *Physiol Mol Biol Plant* 2012, 18, 125-132 .

Kaur, G.; Kumar, S.; Nayyar, H.; Upadhyaya, H. D. Cold stress injury during the pod-filling phase in chickpea (*Cicer arietinum* L.): effects on quantitative and qualitative components of seeds. *J Agron Crop Sci* 2008, 194, 457-464.

Koyama, M. L.; Levesley, A.; Koebner, R. M. D.; Flowers, T. J.; Yeo, A. R. Quantitative trait loci for component physiological traits determining salt tolerance in rice. *Plant Physiol* 2001, 125, 406-422.

Lang, N. T.; Yanagihara, S.; Buu, B. C. QTL analysis of salt tolerance in rice (*Oryza sativa* L.). *Sabrao J Breed Genet* 2001, 33, 11-20.

Latha, R.; Rao, C. S.; Subramanium, S. R.; Eganathan, P.; Swaminathan, M. S. Approaches to breeding for salinity tolerance - a case study on *Porteresia coarctata*. *Ann Appl Biol* 2004, 144, 177-184.

Laurie, S.; Freeney, K. A.; Maathuis, F. J. M.; Heard, P. J.; Brown, S. J.; Leigh, R. A. A role for *HKT1* in sodium uptake by wheat roots. *Plant J* 2002, 32, 139-149.

Lilius, G.; Holmberg, N.; Bülow, L. Enhanced NaCl stress tolerance in transgenic tobacco expressing bacterial choline dehydrogenase. *Biotechnol* 1996, 14, 177-180.

Liu, J. P.; Ishitani, M.; Halfter, U.; Kim, C. S.; Zhu, J. K. The *Arabidopsis thaliana* SOS2 gene encodes a protein kinase that is required for salt tolerance. *Proc Natl Acad Sci USA* 2000, 97, 3730-3734.

Liu, Q.; Kasuga, M.; Sakuma, Y.; Abe, H.; Miura, S.; Yamaguchi-Shinozaki, K.; Shinozaki, K. Two transcription factors, DREB1 and DREB2, with an EREBP/AP2 DNA binding domainseparate two cellular signal transduction pathways in drought- and low temperature-responsive gene expression, respectively, in *Arabidopsis*. *Plant Cell* 1998, 10, 1391-1406.

Lodha, T.; Karmakar, J.; Roychowdhury, R.; Dey, N. Assessment of genetic diversity of some commonly grown rice genotypes of South Bengal using microsatellite markers associated with *saltol* QTL mapped on 1st Chromosome, *NBU J Plant Sci* 2011, 5, 35-39.

Ludwig, A.; Romeis, T.; Jones, J. D. CDPK mediated signalling pathways: specificity and crosstalk. *J Exp Bot* 2004. 55, 181-188.

Mackill, D. J. Molecular markers and marker-assisted selection in rice. In *Genomics assisted crop improvement. Vol. 2. Genomics applications in crops*, Varshney, R. K.; Tuberosa, R., Ed.; Springer: New York, 2007, pp. 147-168.

Maqbool, S. B.; Zhong, H.; El-Maghraby, Y.; Ahmad, A.; Chai, B.; Wang, W.; Sabzikar, R.; Sticklen, M. B. Competence of oat (*Avena sativa* L.) shoot apical meristems for integrative transformation, inherited expression, and osmotic tolerance of transgenic lines containing *hva1*. *Theor Appl Genet* 2002, 105, 201-208.

Martin, G. B.; Williams, J. G. K.; Tanksley, S. D. Rapid identification of markers linked to a Pseudomonas resistance gene in tomato by using random primers and near-isogenic lines. *Proc Natl Acad Sci* 1991, 88, 2336-2340.

Matsumura, T.; Tabayashi, N.; Kamagata, Y.; Souma, C.; Saruyama, H. Wheat catalase expressed in transgenic rice can improve tolerance against low temperature stress. *Physiol Plant* 2002, 116, 317-327.

McCouch, S. R.; Temnykh, S.; Lukashova, A.; Coburn, J.; DeClerck, G.; Cartinhour, S.; Harrington, S.; Thomson, M.; Septiningsih, E.; Semon, M.; Moncada, P.; Li, J. Microsetellite markers in rice: abundance, diversity and applications. In *Rice Genetics IV. IRRI*, Khush, G. S.; Brar, D. S.; Hardy, B., Ed.; Los Bãnos: Philippines, 2001, pp. 117-135 .

Minhas, D.; Grover, A. Towards developing transgenic rice plants tolerant to flooding stress. *Proc Ind Natl Sci Acad* 1999, 65, 33-50.

Mittler, R. Oxidative stress, antioxidants and stress tolerance. *Trends Plant Sci* 2002, 7, 405-410.

Mochida, K.; Shinozaki, K. Advances in omics and bioinformatics tools for systems analyses of plant functions. *Plant Cell Physiol* 2011, 52, 2017-2038.

Moreau, L.; Charcosset, A.; Hospital, F.; Gallis, A. Marker associated selection efficiency in populations of finite size. *Genetics* 1998, 148, 1353-1365.

Mundy, J.; Chua, N. H. Abscisic acid and water-stress induce the expression of a novel rice gene. *EMBO J* 1988, 7, 2279-2286.

Nanjo, T.; Kobayashi, T. M.; Yoshida, Y.; Kakubari, Y.; Yamaguchi-Shinozaki, K.; Shinozaki, K. Antisense suppression of proline degradation improves tolerance to freezing and salinity in *Arabidopsis thaliana*. *FEBS Lett* 1999, 461, 205-210.

Nevo, E.; Chen, G. X. Drought and salt tolerances in wild relatives for wheat and barley improvement. *Plant Cell Environ* 2010, 33, 670-685.

Ohta, M.; Hayashi, Y.; Nakashima, A.; Tsunetomi, N.; Hamada, A.; Tanaka, A.; Nakamura, T. Salt tolerance of rice is conferred by introduction of Na$^+$/H$^+$ antiporter gene from *Atriplex gmelini*. *Plant Cell Physiol* 2002, 43, S56-S56.

Oullet, F.; Vazquez-Tello, A.; Sarhan, F. The wheat *wcs120* promoter is cold inducible in broth monocotyledons and dicotyledons species. *FEBS Lett* 1998, 423, 324-328.

Pandey, S. K.; Nookaraju, A.; Upadhyaya, C. P.; Gururani, M. A.; Venkatesh J.; ,Kim, D. H.; Park, S. W. An update on biotechnological approaches for improving abiotic stress tolerance in tomato. *Crop Sci* 2011, 51, 1-22.

Pandit, A.; Rai, V.; Bal, S.; Sinha, S.; Kumar, V.; Chauhan, M.; Gautam, R. K.; Singh, R.; Sharma, P. C.; Singh, A. K.; Gaikwad, K.; Sharma, T. R.; Mohapatra, T.; Sing, N. K. Combining QTL mapping and transcriptome profiling of bulked RILs for identification of functional polymorphism for salt tolerance genes in rice (*Oryza sativa* L.). *Mol Genet Genom* 2010, 284, 121-136.

Paran, I.; Kesseli, R.; Michelmore, R. Identification of restriction fragment-length-polymorphism and random amplified polymorphic DNA markers linked to downy mildew resistance genes in lettuce, using near isogenic lines. *Genome* 1991, 34, 1021-1027.

Pareek, A.; Singla, S. L.; Grover, A. Immunological evidence for accumulation of two high molecular weight (104 and 90 kDa) HSPs in response to different stresses in rice and in response to high temperature stress in diverse plant genera. *Plant Mol Biol* 1995, 29, 293-301.

Pellegrineshi, A.; Ribaut, J. M.; Trethowan, R.; Yamaguchi-Shinozaki, K.; Hoisinggton, D. Progress in the genetic engineering of wheat for water limited conditions. *JIRCAS Report* 2002, 55-60.

Pilon-Smits, E. A.; Terry, N.; Sears, T.; Kim, H.; Zayed, A.; Hwang, S.; van Dun, K.; Voogd, E.; Verwoerd, T. C.; Krutwagen, R. W. H. H.; Gooddijn, O. J. M. Trehalose-producing transgenic tobacco plants show improved growth performance under drought stress. *J Plant Physiol* 1998, 152, 525-532.

Pilon-Smits, E. A. H.; Ebstamp, M. J. M.; Paul, M. J.; Jeuken, M. J. W. Improved performance of transgenic fructan-accumulating tobacco under drought stress. *Plant Physiol* 1995, 107, 125-130.

Powell, W.; Morgante, M.; Andre, C.; Hanafey, M.; Vogel, J.; Tingey, S.; Rafalski, A. The comparison of RFLP, RAPD, AFLP and SSR (microsatellite) markers for germplasm analysis. *Mol Breed* 1996, 2, 225-238.

Primrose, S. B.; Twyman, R. M. Principles of gene manipulation and genomics. Blackwell Publishing: Victoria, Australia, 2006, 7, 274-298.

Quimio, C. A.; Torrizo, L. B.; Stter, T. L.; Ellis, M.; Grover, A.; Abrigo, E. M.; Oliva, N. P.; Ella, E. S.; Carpena, A. l.; Ito, O.; Peacock, W. J.; Dennis, E.; Datta, S. K. Enhancement of submergence tolerance in transgenic rice overproducing pyruvate decarboxylase. *J Plant Physiol* 2000, 156, 516-521.

Rabbani, M. A.; Maruyama, K.; Abe, H.; Khan, M. A.; Katsura, K.; Ito, Y.; Yoshiwara, K.; Seki, M.; Shinozaki, K.; Yamaguchi-Shinozaki, K. Monitoring expression profiles of rice genes under cold, drought, and high salinity stresses and abscisic acid application using cDNA microarray and RNA gel-blot analyses. *Plant Physiol* 2003, 133, 1755-1767.

Raghuvanshi, S.; Kapoor, M.; Tyagi, S.; Kapoor, S.; Khurana, P.; Khurana, J.; Tyagi, A. Rice genomics moves ahead. *Mol Breed* 2009, 10, 9367-9377.

Rahman, M. S.; Sohag, M. K. H.; Rahman, L. Microsatellite based DNA fingerprinting of 28 local rice (*Oryza sativa* L.) varieties of Bangladesh. *J Bangladesh Agric Univ* 2010, 8, 7-17.

Ram Kumar, G.; Sakthivel, K.; Sundaram, R. M.; Neeraja, C. N.; Balachandran, S. M.; Shobha Rani, N.; Viraktamath, B. C.; Madhav, M. S. Allele mining in crops: Prospects and potentials. *Biotechnol Adv* 2010, 28, 451-461.

Ram, S. G.; Thiruvengadam, V.; Vinod, K. K. Genetic diversity among cultivars, landraces and wild-relatives of rice as revealed by microsatellite markers. *J Appl Genet* 2007, 48, 337-345.

Ravi, M.; Geethanjali, S.; Sameeyafarheen, F.; Maheswaran, M. Molecular marker based genetic diversity analysis in rice (*Oryza sativa* L.) using RAPD and SSR markers. *Euphytica* 2003, 133, 243-252.

Ray, S.; Dansana, P. K, Bhaskar, A.; Giri, J.; Kapoor, S.; Khurana, J. P.; Tyagi, A. K. Emerging trends in functional genomics for stress tolerance in crop plants. In *Plant stress biology: from genomics to systems biology*, Hirt, H., Ed.; Wiley: Weinheim, 2010.

Reddy, D. S.; Mathur, P. B.; Vadez, V.; Sharma, K. K. Grain Legumes (Soybean, Chickpea, and Peanut): Omics approaches to enhance abiotic stress tolerance. In *Improving crop resistance to abiotic stress*, Tuteja, N.; Gill, S. S.; Tiburcio, A. F.; Tuteja, R., Ed.; Wiley: New York, 2012, pp. 993-1030.

Rhodes, D.; Hanson, A. D. Quaternary ammonium and tertiary sulfonium compounds in higher plants. *Annu Rev Plant Physiol Plant Mol Biol* 1993, 44, 357-384.

Richard, I.; Beckman, J. S. How neutral are synonymous codon mutations? *Nat Genet* 1995, 10, 259.

Rodríguez, M.; Canales, E.; Borrás-Hidalgo, O. Molecular aspects of abiotic stress in plants. *Biotechnol Appl* 2005, 22, 1-10.

Rohila, J. S.; Jain, R. K.; Wu, R. Genetic improvement of Basmati rice for salt and drought tolerance by regulated expression of a barley Hva1 cDNA. *Plant Sci* 2002, 163, 525-532.

Roychowdhury, R.; Karmakar, J.; Dey, N. PCR-compatible genomic DNA isolation from different tissues of rice (*Oryza sativa*) for SSR fingerprinting. *EurAsian J BioSci* 2012, 6, 85-90.

Saghai Maroof, M. A.; Biyashev, R. M.; Yang, G. P.; Zhang, Q.; Allard, R. W. Extraordinarily polymorphic microsatellite DNA in barley: species diversity, chromosomal locations, and population dynamics. *Proc Natl Acad Sci USA* 1994, 91, 5466-5470.

Saijo, Y.; Hata, S.; Kyozuka, J.; Shimamoto, K.; Izui, K. Over-expression of a single Ca^{2+}-dependent protein kinase confers both cold and salt/drought tolerance. *J Exp Bot* 2000, 51, 81-88.

Sanchez, D. H.; Pieckenstain, F. L.; Szymanski, J.; Erban, A.; Bromke, M.; Hannah, M. A.; Kraemer, U.; Kopka, J.; Udvardi, M. K. Comparative functional genomics of salt stress in related model and cultivated plants identifies and overcomes limitations to translational genomics. *PLoS One* 2011, 6, e17094.

Sawahel, W. A.; Hassan, A. H. Generation of transgenic wheat plants producing high levels of osmoprotectant proline. *Biotech Lett* 2002, 24, 721-725.

Seki, M.; Kamei, A.; Yamaguchi-Shinozaki, K.; Shinozaki, K. Molecular responses to drought, salinity and frost: common and different paths for plant protection. *Curr Opin Biotechnol* 2003, 14, 194-199.

Semagn, K.; Bjornstad, A.; Ndjiondjop, M. N. An overview of molecular marker methods for plants. *Afr J Biotechnol* 2006, 5, 2540-2568.

Sharma, P.; Dubey, R. S. Drought induces oxidative stress and enhances the activities of antioxidant enzymes in growing rice seedlings. *Plant Growth Regul* 2005, 46, 209-221.

Sharma, P.; Dubey, R.S. Involvement of oxidative stress and role of antioxidative defense system in growing rice seedlings exposed to toxic levels of aluminium. *Plant Cell Rep* 2007, 26, 2027-2038.

Sheveleva, E.; Chmara, W.; Bohnert, H. J.; Jensen, R. G. Increased salt and drought tolerance by D-ononitol production in transgenic *Nicotiana tabacum* L. *Plant Physiol* 1997, 115, 1211-1219.

Shi, H. Z.; Ishitani, M.; Kim, C. S.; Zhu, J. K. The *Arabidopsis thaliana* salt tolerance gene SOS1 encodes a putative Na^+/H^+ antiporter. *Proc Natl Acad Sci USA* 2000, 97, 6896-6901.

Shinozaki, K.; Yamaguchi-Shinozaki, K. Gene expression and signal transduction in water-stress response. *Plant Physiol* 1997, 115, 327-334.

Shinozaki, K.; Yamaguchi-Shinozaki, K. Molecular responses to dehydration and low temperature: differences and cross-talk between two stress signalling pathways. *Curr Opin Plant Biol* 2000, 3, 217-223.

Shinozaki, K.; Yamaguchi-Shinozaki, K.; Seki, M. Regulatory network of gene expression in the drought and cold stress responses. *Curr Opin Plant Biol* 2003, 6, 410-417.

Singh, D.; Kumar, A.; Kumar, A.; Chauhan, P.; Kumar, V.; Kumar, N.; Singh, A.; Mahajan, N.; Sirohi, P.; Chand, S.; Ramesh, B.; Singh, J.; Kumar, P.; Kumar, R.; Yadav, R. B.; Naresh, R. K. Marker assisted selection and crop management for salt tolerance: A review. *Afr J Biotechnol* 2011, 10, 14694-14698.

Singh, D.; Kumar, A.; Sirohi, A.; Purohit, D. R.; Dhaliwal, H. S.; Kumar, V.; Chand, S. Inter-SSR and SSR based molecular Profiling of basmati and non basmati indica rice. *Prog Agric* 2009, 9, 192-202.

Singh, D.; Kumar, A.; Sirohi, A.; Rajpurohit, D. R.; Dhaliwal, H. S. Inter- SSRs and SSRs based molecular profiling of Basmati and Non-Basmati *Indica* Rice. In: Souvenir and paper for oral presentation under Quantitative Trait Category in International Major and Mini Symposia on American Society of Plant Biologists, Merida, Mexico, 2008, pp. 133-134.

Sivamani, E.; Bahieldin, I. A.; Wraith, J. M.; Al-Niemi, T.; Dyer, W. E.; Ho, T. D.; Wu, R. Improved biomass productivity and water use efficiency under water deficit conditions in transgenic wheat constitutively expressing the barley *HVA1* gene. *Plant Sci* 2000, 155, 1-9.

Skriver, K.; Mundy, J. Gene expression in response to abscisic acid and osmotic stress. *Plant Cell* 1990, 5, 503-512.

Slater, A.; Nigel, W. S.; Mark, R. F. *Plant Biotechnology the genetic manipulation of plants*. Oxford University Press: Oxford, United Kingdom, 2011, 2-221.

Sobrino, B.; Briona, M.; Carracedoa, A. SNPs in forensic genetics: a review on SNP typing methodologies. *Forensic Sci Int* 2005, 154, 181-194.

Sreenivasulu, N.; Kavi Kishor, P. B.; Varshney, R. K.; Altschmied, L. Mining functional information from cereal genomes- the utility of expressed sequence tags (ESTs). *Curr Sci* 2002, 83, 1-9.

Sreenivasulu, N.; Radchuk, V.; Strickert, M.; Miersch, O.; Weschke, W.; Wobus, U. Gene expression patterns reveal tissue-specific signaling networks controlling programmed cell death and ABA-regulated maturation in developing barley seeds. *Plant J* 2006, 47, 310-327.

Sreenivasulu, N.; Sopory, S. K.; Kavi Kishor, P. B. Deciphering the regulatory mechanisms of abiotic stress tolerance in plants by genomic approaches. *Gene* 2007, 388, 1-13.

Sreenivasulu, N.; Varshney, R. K.; Kavi Kishor, P. B.; Weschke, W. Tolerance to abiotic stress in cereals: a functional genomics approach. In *Cereal Genome*, Gupta, P. K.; Varshney, R. K., Ed.; Springer: New York, 2004, pp. 483-514.

Stanton, B. G. *Agrobacterium*-mediated plant transformation: the biology behind the "Gene-Jockeying" tool. *Micro Mol Biol Rev* 2003, 67, 16-37.

Stephanopoulos, G. Metabolic fluxes and metabolic engineering. *Met Eng* 1999, 1, 1-11.

Takabe, T.; Hayashi, Y.; Tanaka, A.; Takabe, T.; Kishitani, S. Evaluation of glycine betaine accumulation for stress tolerance in transgenic rice plants. In *Proceedings of the International Workshop on Breeding and Biotechnology for Environmental Stress in Rice*, Hokkaido National Agriculture Experiment Station and Japan International Science and Technology Exchange Center, Japan, 1998, 63-68.

Tanaka, Y.; Hibino, T.; Hayashi, Y.; Tanaka, A.; Kishitani, S.; Takabe, T.; Yokata, S.; Takabe, T. Salt tolerance of transgenic rice over expressing yeast mitochondrial Mn-SOD in chloroplasts. *Plant Sci* 1999, 148, 131-138.

Tarczynski, M. C.; Jensen, R. G.; Bohnert, H. J. Expression of a bacterial *mtlD* gene in transgenic tobacco leads to production and accumulation of mannitol. *Proc Natl Acad Sci USA* 1992, 89, 2600-2604.

Temnykh, S.; Park, W. D.; Ayres, N.; Cartinhour, S.; Hauck, N.; Lipovich, L.; Cho, Y. G.; Ishii, T.; McCouch, S. R. Mapping and genome organization of microsatellite sequences in rice (*Oryza sativa* L.). *Theor Appl Genet* 2000, 100, 697-712.

Thakur, P.; Kumar, S.; Malik, J. A.; Berger, J. D.; Nayyar, H. Cold stress effects on reproductive development in grain crops: an overview. *Environ Exp Bot* 2010, 67, 429-443.

Thomas, J. C.; Sepahi, M.; Arendall, B.; Bohnert, H. J. Enhancement of seed germination in high salinity by engineering mannitol expression in *Arabidopsis thaliana*. *Plant Cell Environ* 1995, 18, 801-806.

Thomashow, M. F. Role of cold responsive genes in plant freezing tolerance. *Plant Physiol* 1998, 118, 1-7.

Todaka, D.; Nakashima, K.; Shinozaki, K.; Yamaguchi-Shinozaki, K. Toward understanding transcriptional regulatory networks in abiotic stress responses and tolerance in rice. *Rice* 2012, 5, 1-9.

Turan, S.; Cornish, K.; Kumar, S. Salinity tolerance in plants: Breeding and genetic engineering. *Aust J Crop Sci* 2012, 6, 1337-1348.

Umezawa, T.; Fujita, M.; Fujita, Y.; Yamaguchi-Shinozaki, K.; Shinozaki, K. Engineering drought tolerance in plants: discovering and tailoring genes to unlock the future. *Curr Opin Plant Biotechnol* 2006, 17, 113-122.

Urano, K.; Kurihara, Y.; Seki, M.; Shinozaki, K. 'Omics' analyses of regulatory networks in plant abiotic stress responses. *Curr Opin Plant Biol* 2010, 13, 132-138.

Van Bruesegem, F.; Slooten, L.; Stassart, J. M.; Botterman, J.; Moens, T.; Van Montagu, M.; Inze, D. Effects of overproduction of tobacco MnSOD in maize chloroplasts on foliar tolerance to cold and oxidative stress. *J Exp Bot* 1999, 50, 71-78.

van de Wiel, C.; Arens, P.; Vosman, B. Microsatellite retrieval in lettuce (*Lactuca sativa* L.). *Genome* 1999, 42, 139-149.

Varshney, R. K.; Bansal, K. C.; Aggarwal, P. K.; Datta, S. K.; Crauford, P. Q. Agricultural biotechnology for crop improvement in a variable climate: hope or hype? *Trends Plant Sci* 2011, 16, 363-371.

Vos, P.; Hogers, R.; Bleeker, M.; Reijans, M.; van de Lee, T.; Hornes, M.; Frijters, A.; Pot, J.; Peleman, J.; Kuiper, M. AFLP: A new technique for DNA fingerprinting. *Nucleic Acids Res* 1995, 23, 4407-4414.

Walia, H.; Wilson, C.; Zeng, L.; Ismail, A. M.; Condamine, P.; Close, T. J. Genome-wide transcriptional analysis of salinity stressed japonica and indica rice genotypes during panicle initiation stage. *Plant Mol Biol* 2007, 63, 609-623.

Wang, W. X.; Barak, T.; Vinocur, B.; Shoseyov, O.; Altman, A. Abiotic resistance and chaperones: possible physiological role of SP1, a stable and stabilizing protein from Populus. In *Plant Biotechnology 2000*, Vasil, I. K., Ed.; Kluwer: Dordrecht, 2003, pp. 439-443.

Wang, Z.; Li, P.; Fredricksen, M.; Gong, Z.; Kim, C. S.; Zhang, C.; Bohnert, H. J.; Zhu, J.; Bressan, R. A.; Hasegawa, P. M.; Zhao, Y.; Zhang, H. Expressed sequence tags from *Thellungiella halophila*, a new model to study plant salt-tolerance. *Plant Sci* 2004, 166, 609-616.

Williams, J. G. K.; Kubelik, A. R.; Livak, K. J.; Rafalski, J. A.; Tingey, S. V. DNA polymorphism amplified by arbitrary primers are useful as genetic markers. *Nucleic Acids Res* 1990, 18, 6531-653.

Williams, J. G. K.; Kubelik, A. R.; Livak, K. J.; Rafalski, J. A.; Tingey, S. V. DNA polymorphisms amplified by arbitrary primers are useful as genetic markers. *Nucleic Acids Res* 1991, 18, 6531-6535.

Wu, K. S.; Tanksley, S. D. Abundance, polymorphism and genetic mapping of microsatellites in rice. *Mol Gen Genet* 1993, 241, 225- 235.

Xie, J. H.; Zapata-Arias, F. J.; Shen, M.; Afza, R. Salinity tolerant performance and genetic diversity of four rice varieties. *Euphytica* 2000, 116, 105-110.

Xiong, L. M.; Ishitani, M.; Zhu, J. K. Interaction of osmotic stress, temperature, and abscisic acid in the regulation of gene expression in *Arabidopsis*. *Plant Physiol* 1999, 119, 205-211.

Xu, D.; Duan, X.; Wang, B.; Hong, B.; Ho, T. H. D.; Wu, R. Expression of a late embryogenesis abundant protein gene, *HVA1*, from barley confers tolerance to water deficit and salt stress in transgenic rice. *Plant Physiol* 1996, 110, 249-257.

Yamaguchi-Shinozaki, K.; Shinozaki, K. Organization of cis-acting regulatory elements in osmotic and cold-stress-responsive promoters. *Trends Plant Sci* 2005, 10, 88-94.

Yancey, P. H.; Clark, M. E.; Hand, S. C.; Bowlus, R. D.; Somero, G. N. Living with water stress: Evolution of osmolyte systems. *Science* 1982, 217, 1214-1222.

Yang, G. P.; Maroof, M. A. S.; Xu, C. G.; Zhang, Q.; Biyashev, R. M. Comparative analysis of microsatellite DNA polymorphism in landraces and cultivars of rice. *Mol Gen Genet* 1994, 245, 187-194.

Zhou, L.; Wang, J. K.; Yi. Q.; Wang, Y. Z.; Zhu, Y. G.; Zhang, Z. H. Quantitative trait loci for seedling vigor in rice under field conditions. *Field Crops Res* 2007, 100, 294-301.

Zhu, J. K. Cell signaling under salt, water and cold stresses. *Curr Opin Plant Biol* 2001, 4, 401-406.

Zietkiewicz, E.; Rafalski, A.; Labuda, D. Genome fingerprinting by simple sequence repeat (SSR)-anchored polymerase chain reaction amplification. *Genomics* 1994, 20, 176-183.

CHAPTER 14

MOLECULAR AND GENOMIC APPROACHES FOR MICROBIAL REMEDIATION OF PETROLEUM CONTAMINATED SOILS

JEYABALAN SANGEETHA, DEVARAJAN THANGADURAI, MUNISWAMY DAVID, and JADHAV SHRINIVAS

CONTENTS

14.1 Introduction 368
14.2 Characterization of Oil and Oil Wastes 372
14.3 Toxicity of Petroleum Oil Contamination 374
14.4 Taxonomic Diversity of Oil Degrading Microorganisms 375
14.5 Environmental Factors Influencing Biodegradation of Petroleum
       Hydrocarbons 379
14.6 Microbial Seeding 386
14.7 Molecular and Genomic Perspectives 387
14.8 Conclusion 390
Keywords 391
References 391

14.1   INTRODUCTION

Microorganisms have been considered to be responsible for many biochemical transformations, including degradation of recalcitrant compounds in soil. Numerous studies have been carried out to describe microbe-microbe and microbe-hydrocarbon interactions by extrapolating from the detailed laboratory studies with isolates from hydrocarbon-contaminated environments. Microbial biodegradation of petroleum hydrocarbons has been investigated from as early as the 1950's and 1960's. The ability of microorganism to use hydrocarbons as a food source has evolved into their use for the biodegradation of petroleum hydrocarbons. *Pullularia pullulans* was tested for its ability to utilize a series of n-alkanes for its growth and the subsequent 10 day experiment indicated adaptation of the organism to utilize hydrocarbons as a sole source of carbon (Skinner *et al.*, 2009).

Crude oil is a homogeneous but complex mixture of hundreds of different hydrocarbons, which widely vary in their characteristics (Jonathan *et al.*, 2003). It is expected that certain microorganisms are able to degrade these hydrocarbons in various forms. The physical state of petroleum hydrocarbon has a marked effect on their biodegradation. Bioremediation of petroleum from anthropogenic sources have been discovered through the metabolic pathways of alkane, cycloalkane, and aromatic hydrocarbon utilization (Atlas and Bartha, 1992). Biodegradation of hydrocarbons by natural microbial populations has been the main means of eliminating hydrocarbon pollutants from the environment (NRC, 1985).

The relation between hydrocarbon degradation and biosurfactant (rhamnolipid) production by a new *Bacillus subtilis 22BN* strain was investigated by Christova *et al.* (2004). The strain was isolated for its capacity to utilize *n*-hexadecane and naphthalene and at the same time to produce surface active compound at high concentrations. The strain is a good degrader of both hydrocarbons used with degradability of 98.3 % and 75 % for *n*-hexadecane and naphthalene, respectively. Evaluation of inoculums addition to stimulate *in situ* bioremediation of oily sludge contaminated soil at an oily refinery where the indigenous population of hydrocarbon degrading bacteria in the soil was very low as demonstrated by Mishra *et al.* (2001). In this study, out of six treatments, the application of a bacterial consortium and nutrients resulted in maximum biodegradation of total petroleum hydrocarbon in 120 days. The alkane fraction of TPH was reduced by 94.2 %, the aromatic fraction of TPH was reduced by 91.9 % and NSO and asphaltenes fraction of TPH were reduced by 85.2 % in 1 year.

Petroleum refineries in their operation generates huge amount of oily sludge which contains reasonable amount of oil as unrecovered during refining. Storage of such a huge amount of oily sludge and its management has become an important issue due to its hazardous nature. Bioremediation of organic waste is becoming an increasingly important method of waste treatment. The advantages of this option include inexpensive equipment, environmentally friendly nature of the process and simplicity (Odokuma and Dickson, 2003). Compared to physicochemical methods, bioremediation offers an effective technology for the treatment of oil pollution because the majority of molecules in the crude oil and refined products are biodegradable and oil degrading microorganisms are ubiquitous. Saturated and low-molecular-weight aromatic fractions of oil are more easily biodegradable than high-molecular-weight aromatic fractions (Atlas and Bartha, 1992).

Biodegradations of n-alkanes with molecular weights upto n-C44 have been reported (Haines and Alexander, 1974). Double bonds and branching makes a hydrocarbon more resistant to degradation. For instance, N-alkanes from the *Amoco Cadiz* oil spill, degraded about twice as fast as branched alkanes (Atlas *et al.*, 1981). Peripheral degradation pathways convert hydrocarbons into intermediates of the central intermediary metabolism, e.g. tricarboxylic acid cycle. Biosynthesis of cell biomass results from the central precursor metabolites like acetyl-CoA (Atlas, 1984). A test tube model of an oil spill was built by Brown (http://www.accessexcellence.org/AE/AEC/AEF/1994/brown_oil.php) in order to experiment with conditions needed for bioremediation. He concluded that *Acinetobacter calcoaceticus* RAG 1 a marine bacterium can utilize hydrocarbon in oil as a source of carbon. Definition of bioremediation and background on how bioremediation works and many factors of the microbial degradation of oil were discussed by Gordon and Spring (www.geocities.com/capecanaveral/lab/2094/bioremed.html).

It was observed that petroleum refinery wastewater contains a wide range of organics and depending on the type of production, their concentration range from several hundred mg/l to several thousand mg/l (Tanjore and Viraraghavan, 1994). Effluent treatment practices at Indian Oil Corporation Limited (IOCL) refineries in relation to progressive improvement, performance standard, technological aspects and water resource conservation were discussed by Chakravarthy (1995). Primary concerns in environmental management in refinery are water, air, solid wastes and occupational health of employees. IOCL are meeting the prescribed treated effluents quality and quantity standards. Measures are being taken to improve performance as well as to reduce fresh water consumption. A list was made containing bacteria and fungi genera that are able to degrade a wide spectrum of pollutants, proceeding from marine water as well as the soil by Bouchez-Naitali *et al.* (1999). Amongst the filamentous fungi *Trichoderma* sp. and *Morfierella* sp. are the most commonly isolated from the soil. *Aspergillus* sp. and *Penicillium* sp. have frequently been isolated from marine and terrestrial environments.

Bioremediation was explained and the list of oil biodegradation microorganisms was prepared by Daniel (1996), which includes *Pseudomonas alcaligens, Micrococcus* sp., *Corynebacterium* sp., *Mycobacterium* sp., *Nocardia* sp., *Candida* sp., and *Penicillium* sp. The products of oil refining and sources of wastewater discharging from oil refineries have been explained by Manivasakam (1997). Large volumes of water are used in refineries for cooling, steam generation and for refining and subsequently discharged as waste. And also oil dispose is coming from leaks, spills, tank draw off and other sources. The quantity of oil disposed as waste is estimated to be about 3 % of the weight of crude oil processed. Less degradation in a heavy fuel oil than in a light fuel oil was found in the microbial degradation of hydrocarbons in weathered crude and fuel oils (Leahy and Colwell, 1990). Leahy and Colwell (1990) reported major differences in the susceptibility for the degradation of each component within the context of the different hydrocarbons mixtures of the oil tested. Investigation of the microbial community composition associated with benzene oxidation under *in situ* Fe (III)-reducing conditions in a petroleum-contaminated aquifer located in Bemidji, Minnesota, suggested that the microbial community composition in the Bemidji aquifer may have played a key role in anaerobic benzene degradation and that *Geobactereceae*, in particular, were associated with benzene degrading activity (Juliette *et al.*, 1999).

The National Environmental Engineering Research Institute (NEERI), Nagpur, India has developed a technology by which chemically treated saw dust and some microorganisms can be efficiently used to clear oil spills and treat petroleum refinery effluents. Two major genus *Pseudomonas* and *Bacillus* recently gained importance due to their use in oil industry. The activity of the indigenous microflora in the agricultural soil steadily receiving petroleum refinery effluent at Mathura, India was studied by Ashok and Mussarat (1999). In addition proteolytic, cellulolytic bacteria *Rhizobium* sp. and *Actinomycetes* sp. were detected. Hence microorganisms have been considered to be responsible for many biochemical transformations and degradation of pollutants in polluted soil.

The techniques that are used for cleaning up oils, such as skimmers, booms, screw conveyer, and adsorbents are listed by Bhargava and Sharma (2000). The Tata Energy Research Institute developed a biological method of using microorganisms to clean up oil contaminated sites. Four bacterial strains from petroleum polluted soil of oil field situated at Moran, Assam, India were isolated by Deka (2001). Strains were identified as *Pseudomonas aeruginosa, Pseudomonas stutzeri, Bacillus aneurirolylicus* and *Serratia marcescens*. Results revealed that the degradation of crude oil was increasing gradually with increasing the incubation time of the bacteria. Moreover, degradation of crude oil depends on the physicochemical properties of the soil. Generally microorganisms isolated from soils of pH ranging 6.5 to 7.5, were efficient, whereas, bacteria isolated from soils of pH below 6.5 were not efficient. Further, supplementation of inorganic nutrients and oxygen help the oil-degrading bacteria to grow and multiply at a much faster rate, thus speeding up the process of biodegradation.

Phenol and its derivatives are among the most frequently found pollutants in rivers, industrial effluents, and landfill run off waters (Abd-El-Haleem *et al.*, 2003). Phenol biodegradation by *Acinetobacter* sp. strain W-17 in the presence of high concentration of nitrogen components (NH_4 and NO_3) is enhanced. Biodegradability of fuels such as gasoline or diesel oil using a reference microflora taken from an urban waste water treatment plant was determined by Marchal *et al.* (2003). Gasoline exhibited a high intrinsic biodegradability (96 %) but that of a commercial diesel oil was significantly lower (60–73 %). The biodegradation rate was close to 100 % when linear alkanes were most abundant. The population density and activity of microbial community associated with the sediment and rhizosphere of an intertidal freshwater wetland were dominated by *Scirpus pungens* (Greer *et al.*, 2003) and was monitored before and following the application of weathered Mesa light crude oil and fertilizers.

Bacterial strains were isolated from hydrocarbon contaminated soils for the development of a product COBE 10, applicable in soil bioremediation (Ilyina, 2003). Initially 82 bacterial strains were isolated in selective agar. After monitoring absorbance change of mineral media containing petroleum as sole carbon source inoculated by isolated strains of about 30 strains were selected. All the strains were involved in the degradation process and evaluated for their hydrocarbons degradation potential in natural and artificial medium in laboratory conditions. Finally, based on degradation rate in a day, 6 strains were selected. Seven carrier materials were tested to select suitable vehicle for final formulation of COBE 10. The *Staphylococcus* sp. is a strong primary utilizer of the base oil and has potential for application in bioremediation processes involving oil based drilling fluids (Nweke and Okpokwasili, 2003). The cell has strong affinity for the base oil and was able to produce extracellular biosurfactant capable of emulsifying the base oil. These are the mechanisms used by microorganisms to take up substrates with low water solubility.

The microorganisms *Bacillus* sp., *Micrococcus* sp., *Proteus* sp., *Penicillium* sp., *Aspergillus* sp., and *Rhizopus* sp. are able to utilize and degrade the crude oil constituents (Okerentuba and Ezeronye, 2003). They concluded that single culture was observed to be better crude oil degraders than the mixed cultures. Further, soil microcosm experiments revealed that the four bacterial isolates were able to biodegrade Bonny light crude oil in the soil system, although the degree of activities varied. A bio-preparation is a compound of naturally occurring bacteria and enzymes used to bioremediate hydrocarbons, hence supporting the hypothesis that a biopreparation can enhance the oil degrading activity of *Pseudomonas* at low temperatures.

A laboratory screening of several natural bacterial consortia and laboratory tests to establish the performance in degradation of hydrocarbons contained in oily sludge from the Jordan Oil Refinery Plant was made by Mrayyan and Battikhi (2004). This is the first report concerning biological treatment of total petroleum hydrocarbon by bacteria isolated from the oil refinery plant, where it laid the ground for full integrated studies recommended for hydrocarbon degradation that assists in solving sludge problems. Pollution abatement of petrochemical waste water by bioprocess using bacteria showed that the use of *Pseudomonas* sp. is an effective treatment solution for petrochemical waste water (Hossain *et al.*, 2004). Pollution, solid wastes, oil sludge pollution from refineries were reported by Radhika (2004) and shown the process description of the available treatment and their feasibility status.

Two species of *Pseudomonas* such as *P. aeruginosa* and *P. fluorescence* for their bioremediation potential on refinery effluent with respect to phenol bioremediation in a batch reactor were studied at Nigerian Refinery and Petrochemical effluents with the hope of isolating and using the organisms for the bioremediation of waste waters (Ojumu *et al.*, 2005). Results revealed with respect to phenol biodegradation in a batch reactor, *P. aeruginosa* completely mineralize phenol at the 60th hour of cultivation; whereas only 75 % of phenol was degraded by *P. fluorescence*. Complete degradation was achieved at the 84th hour of cultivation.

14.2   CHARACTERIZATION OF OIL AND OIL WASTES

Natural gas and petroleum are formed by the decomposition of plant and animal material that has been buried for long periods of time in earth's crust, an environment with little oxygen (Bruice, 2004). "Petroleum hydrocarbons" is a generic term applied to the various organic chemicals that comprise geological petroleum deposits or products refined from this source. Chemically, petroleum hydrocarbons are, in general, more reduced than organic molecules characteristics of living tissue and contain fewer and less complex bonding arrangement with elements other than carbon and hydrogen (Fuhr *et al.*, 1988). The details of petroleum obtained from underground deposits that have been tapped by drilling oil wells were given by Pauling (1988). According to his statement, crude oil is a dark-colored, highly viscous liquid which is composed of large molecules of hydrocarbons. He also discussed about the process of petroleum refining. Large deposits of hydrocarbons (natural gas and petroleum) are buried in the earth, and these supplies are tapped by oil and gas wells. Petroleum consisting of more than a hundred different compounds varying in carbon content from 6 to 10 carbon atoms per molecule constitutes various qualities of gasoline. The best hydrocarbon for use in gasoline for internal combustion engines contains 8 carbon atoms per molecule (octane) (Sienko and Plane, 1976). The chemical constituents of crude oil are extremely variable. In general, the five components of crude oil is present but their relative abundances varies with factors such as the age and depth of the petroleum deposit, which affect primarily the extent of cracking and fate of hetero elements such as sulfur. In this scheme, end members are represented by young-shallow and old-deep crudes. The former will tend to have high aromatic and sulfur contents and mean molecular weight. Old-deep crudes, the most economically desirable type, are expected to have low aromatic and sulfur contents because cracking is maximal and most S will have been converted to H_2S. Sulfur is usually the third most abundant element in crude oil, normally accounting for 0.05 to 5 %, but up to 14 % in heavier oils (Speight, 1991). In crude oil, sulfur is mainly found in the form of condensed thiophenes and physicochemical methods including hydrodesulfurization are in use to remove sufur (Shennan, 1996).

Crude oil contains about 0.5 to 2.1 % nitrogen, with 70 to 75 % consisting of pyrroles, indoles, and carbazole nonbasic compounds (Speight, 1991; Benedik et al., 1998). The characteristics of oily sludge were shown by Ururahy et al. (1998). Fortifying media with necessary nutrients is more important as the oily sludge is having limited nutrients especially, nitrogen and phosphorous. Generally, nitrogen is present as complex structure and hence, microorganisms are not able to utilize the source. Although the concentration of asphaltenes is not very high, the predominance of aromatics and high concentration of polycyclic aromatic hydrocarbons, usually found in this kind of residue, confers on it high toxicity. It was concluded that water and dirt in crude oil cause corrosion and scaling on pipelines and reactors, and a maximum sediment and water content of 0.5 to 2.0 % is required for pipeline quality oil (Lee, 1999).

It was reported that crude oil petroleum is a complex mixture which contains hundreds of compounds belonging to the categories of paraffins (25 %), cycloparaffins (20 %), aromatics (5 %) and naphthalene aromatics (De, 2000). Priority pollutant metal is another category of pollutants of concern, as various crude oils have varying levels of metal contaminants. The metal of most concern in crude oil is mercury, the concentration of which can vary widely from one crude oil source to another. Other toxins that were reported in the final effluent were the metal toxins of cadmium, antimony, and lead. The refineries generally believe that the sources for these toxins are tied to crude oil or plant metallurgy. Many hydrocarbons are poorly accessible to bacteria due to their low solubility in aqueous systems compatible with microbial life (Gutierrez *et al.*, 2013). Heavily contaminated soils contain a separate nonaqueous-phase liquid, which may be present as droplets or films on soil particle surfaces. Many hydrocarbons are insoluble in water and remain partitioned in the nonaqueous-phase liquid (Stelmack *et al.*, 1999).

14.3 TOXICITY OF PETROLEUM OIL CONTAMINATION

The hydrocarbon contamination of habitats constitutes public health and socio-economic hazards (Smith and Dragun, 1984). The hydrocarbons so discharged may also pose serious aquatic toxicity problems (Ikenaka *et al.*, 2013). Hydro-carbons may even affect the physiological process of microorganisms (Delille and Delille, 1999). It is speculated that the discharged refinery effluent or waste in the long run will be a potential hazard due to accumulation of mutagenic and carcinogenic aromatic hydrocarbon in soil (Ashok and Musarrat, 1999). Seepage during storage is also another major problem because it is responsible for ground water contamination. Polycyclic aromatic hydrocarbons (PAH) are relatively stable constituents of petroleum, and from the environmental aspects, they are probably the most important analytes because many of these com-pounds are potential or proven carcinogens. Unfortunately, their low aqueous solubility, limited volatility and recalcitrance toward degradation allows PAHs to accumulate to levels at which they may exert toxic effects upon the environ-ment (Senn and Johnson, 1987; Pavlova and Ivanova, 2003).

An oil production plant in China treated 327.7 to 371.2 and 151.0 g/kg of total hydrocarbon content (THC) in oily sludge and oil polluted soil respectively using biopreparation. The THC decreases effectively after three times applica-tion of bioprocess by 46–53 % in sludge and soil. Application of treated sludge shows decreased toxicity on *Festuca arundinace* (Ouyang *et al.*, 2005). Mice introduced to one PAH during pregnancy faced difficulty in reproducing its off-springs and those were also had birth defects. Further, many researchers have also shown that PAHs may also cause harmful effects on body fluid, skin, and immunity (www.atsdr.cdc.gov/toxfaq.html).

PAH is also an effective carcinogen and people who are exposed long time to PAHs and other related compounds having developed cancer. Laboratory ani-mals developed lung cancer when exposed to PAHs compounds while breath-ing, stomach cancer (fed with contaminated feed) and skin cancer (http://www.atsdr.cdc.gov/toxfaqs/tf.asp?id=121&tid=25). Additionally, PAHs shows puta-tive estrogenic and anti-estrogenic properties in human body. Human exposed to PAHs by ingestion of contaminated meat or vegetables can cause intestine en-terocytes and liver hepatocytes. In these cells, PAHs act as ligand to human aryl hydrocarbon (Ah) receptor, which is having necessary role in the toxic response of aromatic hydrocarbons by the regulation of typical human biotransformation enzymes (van Maanen *et al.*, 1994; Hankinson, 1995). Bisphenol-A (BPA) stimulates prostate cancer cells19 and causes breast tissue changes resembling early stage breast cancer in both mice and humans (Munoz-de-Toro, 2005). Significant developmental, reproductive and immune effects from low-level exposure in numerous animals were studied by Vom Saal and Hughes (2005). In humans, higher BPA levels in urine have been associated with ovarian dysfunction (Takeuchi, 2004). Nonylphenols (NPEs) may be related to increased hormone-dependent cancers and also related to decreasing quality and quantity of sperm in humans (Guenther, 2002).

The effect of exposure of oily sludge derived chemicals on human has been studied *in vitro* test using mammalian cell cultures. Genotoxicity assays such as DNA damage, DNA strand break, chromosomal aberration, p^{53}protein induction and apoptosis are used to study the potential genotoxic risk of exposure to PAHs to human beings (Krishnamoorthy *et al.*, 2006). Toxicity from hydrocarbon exposure can lead to different syndromes, especially on which organ system is predominantly involved. Organ systems that can be affected by hydrocarbons include the pulmonary especially aspiration which leads severe pneumonitis, neurological, cardiac, gastrointestinal, hepatic, renal, dermatologic, and hematological systems. Hydrocarbon pneumonitis results from a direct toxic effect by the hydrocarbon on the lung parenchyma. Intestinal inflammation, intra-alveolar hemorrhage and edema, hyperemia, bronchial necrosis and vascular necrosis are the severe results of hydrocarbon aspiration (www.emedicine.com/emer/emergTOXICOLOGY.htm). Certain hydrocarbons are highly lipophilic, may affect the central nervous system directly which acts as blood-brain barrier. Hypercarbia which lead decreased level of arousal, also one of the health impacts of PAHs. The chlorinated hydrocarbons, in particular carbon tetrachloride, are quite hepatotoxic and some PAHs show cardiotoxicity. Anemia, hemolysis, multiple myeloma, and acute myelogenous leukemia is also the result of prolonged exposure to hydrocarbons especially benzene (Levine and Johnson, 2006).

14.4 TAXONOMIC DIVERSITY OF OIL DEGRADING MICROORGANISMS

It has been shown that among 200 microbial genera, encompassing 500 species were identified as capable of biodegradation hydrocarbon (Kostka *et al.*, 2001). A great deal of work has been carried out in trying to rationalize the persistence of PAH in the environment. It is evident that a vast number of microbial species having a capability to utilize or degrade diverse of both low and high molecular weight hydrocarbons (Zobell *et al.*, 1946; Juhasz and Naidu, 2000). Yeasts that can utilize hydrocarbon and strains of *Candida, Rhodosporidium, Rhodotorula, Saccharomyces, Sporobolomyces,* and *Trichosporon* were isolated by Ahearn *et al.* (1971). There is tremendous diversity in organisms that can biodegrade hydrocarbons (Nilanjana and Preethy, 2011). This ability to degrade petroleum hydrocarbon is not restricted to a few microbial genera. A list of 22 bacterial genera, 14 fungal genera and 1 algal genus was prepared by Batha and Atlas (1977). The most important genera of hydrocarbon utilizers are *Pseudomonas, Achromobacter, Micrococcus, Nocardia, Vibrio, Acinetobacter, Brevibacter, Corynebacterium,* and *Flavobacter*.

Oscillatoria sp., *Microcoleus* sp., *Anabaena* sp., *Agmenellum* sp., *Coccochloris* sp., *Nostoc* sp., *Aphanocapsa* sp., *Chlorella* sp., *Dunaliella* sp., *Chlamydomonas* sp., *Ulva* sp., *Cylindretheca* sp., *Amphora* sp., *Porphyridium* sp., and *Petolania* were found to be capable of oxidizing naphthalene (Cerniglia *et al.*, 1978). Their results indicated that the ability to oxidize aromatic hydrocarbons is widely distributed among the cyanobacteria and algae. Microbial changes during oil decomposition in soil were examined by Nilanjana and Preethy (2011). They found an increase from 60–82 % in oil utilizing fungi and an increase from 3 to 50 % in oil degrading bacteria after a fuel oil spill. The most commonly reported genera of hydrocarbon degraders include *Pseudomonas, Acetobacter, Nocardia, Vibrio,* and *Achromobacter* (Floodgate, 1984).

Microbial species including *Nocardia amarae* (Cairns *et al.*, 1982), *Bacillus subtilis* (Janiyani *et al.*, 1994), *Micrococcus* sp. (Das, 2001), *Torulopsis bombicola* (Duvnjak and Kosaric, 1987) and *Pseudomonas* sp. and *Acinetobacter* sp. containing mixed bacterial cultures (Nadarajah *et al.*, 2001; Nadarajah *et al.*, 2002) exhibited de-emulsification capabilities. A list of 16 bacterial genera that can degrade polyaromatic hydrocarbon and 13 genera containing species that metabolize hydrocarbons other than methane were prepared by Cerniglia (1992) and Watkinson and Morgan (1990) respectively. Bacteria, yeasts, and filamentous fungi have been considered as transforming agents because of their ability to degrade a wide variety of xenobiotics substances, commonly found in wastes from the oil industry (Prince and Sambasivam, 1993). Biological treatment is having several advantages such as permanent destruction of residues by mineralization, eliminating further contamination, ecofriendly, and acceptable by public. *Pseudomonas cepacia, Pseudomonas aureofaciens, Pseudomonas picketti, Flavobacterium indologenes, Xanthomonas maltophilia* and *Ochrobactrum anthropi* are some microbial species found suitable for diverse group of hydrocarbons.

Drilling wastes have been reported to be degraded by bacterial isolates identified species of *Staphylococcus, Acinetobacter, Alcaligenes, Serratia,*

Clostridium, Enterobacter, Nocardia, Bacillus, Actinomyces, Micrococcus, and *Pseudomonas* (Benka-Coker and Olumagin, 1995). The Ampol Kurnell refinery has 15 years' experience with landfarming for the treatment and final disposal of oily wastes. These wastes are mainly generated during the cleaning of storage tanks and the waste water treatment plant. Land farming has proved to be a cost effective and environmentally sound method bor dealing with these wastes. However, successful landfarming does require the correct integration of site selection, design, operation, and environmental controls. Various mixtures of alkane utilizing *Acinetobacter* sp. and a *Rhodococcus* sp. an alkylbenzene-degrading *Pseudomonas putida* and a phenanthrene-utilizing *Sphingomonas* sp. have been evaluated in an attempt to elucidate how alkane and aromatic degrading microorganisms interact (Komukai-Nakamura *et al.*, 1996; Aitken *et al.*, 1998; Hamme *et al.*, 2003). The degradation of Arabian light crude oil and a combination of the *Acinetobacter* sp. and *P. putida* was an effective mixture of the four microorganisms, degrading 40 % of the saturates and 21 % of the aromatics were monitored by Korda *et al.* (1997). On a contrary laboratory study were also made on Alaska North Slope crude oil with 12 commercially available microbial cultures which showed higher biodegradation rates after 38 days (Aldrett *et al.*, 1997). A similarly test of 10 different commercial microbial products on Alaskan crude in flask microcosms was carried out by Venosa *et al.* (1991). The Alaskan microorganism showed better degradation than the seeded microorganism suggesting that bioaugmentation maybe more effective in controlled environmental conditions than in the field. Hill has demonstrated a special process which utilizes a strain of *Pseudomonas* which cost-effectively remove contaminants from oil industry wastewater. This process has been further tested by Saskatchewan Research Council (SRC) using the *Pseudomonas* strain and another type of oil eating bacteria found in bioremediation lagoons.

The current state of knowledge of microorganisms from petroleum reservoirs, including mesophilic and thermophilic sulfate reducing bacteria, methanogens, mesophilic, and thermophilic fermentative bacteria and iron reducing bacteria were recently been reviewed (Magot *et al.*, 2000). Sixty four species of filamentous fungi were found from five flare pits in northern and western Canada and tested their ability to degrade crude oil using gas chromatographic analysis of residual hydrocarbons following incubation (April *et al.*, 2000). Their study showed that filamentous fungi may play an integral role in the *in situ* bioremediation of aliphatic pollutants in flare pit soils (Annweiler *et al.*, 2000).

A study was carried out to assess the effects of various hydrocarbon substrates, and chemical surfactant capable of enhancing crude-oil biodegradation, on the community structure of mixed bacterial inoculums in batch culture (Hamme *et al.*, 2000). Cultures were exposed to 20 g/l Bow River crude oil with and without surfactant. A group of six genera dominated the cultures: *Acineto-bacter, Alcaligens, Ochrobactrum, Pseudomonas, Stenotrophomonas, and* the pattern of degradation of BTEX mixtures by a fungus capable of growth on aromatic hydrocarbons were investigated (*Yersinia*; Boldu *et al.*, 2002). The deuteromycete *Cladophialophora* sp. strain T1, which grows on toluene, was selected as a model fungus. The soil fungi *Cladophialophora* sp. could contribute significantly to bioremediation of BTEX pollution. Further, the saprophytic nature of *Pseudomonas* species is considered as one of the important role to carry out the bioremediation of hydrocarbons and these bacteria showed significant results (Kumar, 2002). *Pseudomonas aeruginosa, Pseudomonas stutzeri, Bacillus aneurinolyticus* and *Serratia marcescens* from oil polluted soil are isolated by Deka (2001) and the prevalence of seven genotypes involved in the degradation of n-alkanes (*Pseudomonas putida* GPo1 *alkB, Acinetobacter* sp. *alkM, Rhodococcus* sp. *alkB2*), aromatic hydrocarbons (*P. putida* xyIE), and polycyclic aromatic hydrocarbons (*P. putida ndoB* and *Mycobacterium* sp. strain PYR-1 *nidA*) was determined in 12 oil contaminated and 8 pristine Alpine soils from Tyrol (Austria) (Margesin *et al.*, 2003).

Species of *Pseudomonas, Bacillus, Klebsiella, Actinomyces, Aeromonas, Nocardia, Xanthomonas*, and *Streptomyces* are mentioned as the well known microbial genera that have been associated with hydrocarbon degradation (Nweke and Okpokwasili, 2003). It is well known that a number of bacteria utilize a variety of hydrocarbons in nature and that bacterial oxidation rate may be as much as 10 times than auto-oxidation rate (Zobell, 1964). The results of the work showed that *Staphylococcus* sp. has potential application in the bioremediation of sites polluted by oil-based drilling fluid base oil. Many microbial species which were involved in the bioremediation process namely, *Chromobacterium, Flavobacterium, Bacillus, Vibrio, Citrobacter, Enterobacter, Micrococcus, Klebsiella, Planococcus, Pseudomonas,* and *Campylobacter* were isolated and fungal isolates were *Aspergillus, Penicillium,* and *Rhizopus* sp. (Okerentugba and Ezeronye, 2003).

The development of an active indigenous bacterial consortium that could be of relevance in bioremediation of petroleum contaminated systems in Nigeria was carried out by Okah (2003); three hydrocarbon degrading bacteria strains were isolated namely, *Pseudomonas aeruginosa, Burkholderia cepaciae,* and *Stenotrophomonas maltipholia*. The soil microcosm experiment revealed that the four bacterial isolates were able to biodegrade Bonny light crude oil in the soil system, although the degree of activities varied. The biodegradation of hydrocarbon is accompanied by a diverse range of taxonomy that can utilize hydrocarbons. Hydrocarbons are naturally occurring organic compounds and that it is not surprising that microorganisms have evolved the ability to utilize these compounds. When natural ecosystems are contaminated with hydrocarbons, the indigenous microbial populations of differing taxonomic relationships degraded the contaminating hydrocarbon (Raghavan, 1998).

Petroleum oil sludge sample was collected from the waste disposal site of an automobile service station at Mayiladudhurai and bacteria as well as fungi were isolated from this automobile waste contaminated soil. The sample was serially diluted, inoculated into the nutrient agar and PDA for isolation of bacteria and fungi respectively. After incubation bacteria such as *Bacillus cereus*, *Pseudomonas aeruginosa*, *Ps. Putida*, and *Serratia*, fungi such as *Aspergillus niger*, *Penicillium granulatum*, *Mucor* sp., *Fusarium* sp. and *Aspergillus terreus* were identified ((Bhuvaneswari and Ravindhran, 2008).

The white rot fungus *Polyporus* sp. S133 collected from petroleum contaminated soil was tested for its ability to grow and degrade crude oil, obtained from petroleum industry. The ability of *Polyporus* sp. S133 pregrown on wood meal to degrade crude oil was measured. Maximal degradation (93 %) was obtained when *Polyporus* sp. S133 was incubated in 1000 ppm of crude oil for 60 days, as compared to 19 % degradation rate in 15000 ppm. Increased concentration of crude oil decreased the degradation rate (Hadibarata and Tachibana, 2009).

14.5   ENVIRONMENTAL FACTORS INFLUENCING BIODEGRADATION OF PETROLEUM HYDROCARBONS

A vast amount of information exists on the biochemical activities of bacteria and fungi grown in pure culture at high substrate concentrations in laboratory media. This research has created a foundation for the understanding of the nutrition, genetics, and catabolic potential of microorganisms. Yet, in nature, bacteria and fungi are exposed to enormously different conditions. They may have an insufficient supply of inorganic nutrients, growth factors, and temperature and pH values at their extremes of tolerance and toxins that retard their growth or result in loss of viability. They may benefit from the activities of other microorganisms or be consumed by species residing in the same habitat. In the last decade, investigations have shown that environmental factors enhancing soil microbial activity, such as fertilizing, plowing, and drainage increase the breakdown of the organic pollutants (Harmsen, 1991). A common theme of early reviews focused on the examination of factors, including nutrients, physical state of the oil, oxygen, temperature, salinity, and pressure influencing petroleum biodegradation rates, with a view of developing environmental applications (Atlas, 1981).

Oily sludge and sea water contaminated by oil were subjected to biodegradation with *Nocardia*, *Pseudomonas*, and *Bacillus*; it was reported from the study that *Pseudomonas* was not efficient in degrading hydrocarbons in sea water, rather remaining organisms were degraded hydrocarbons presented in sea water at low concentration within 48 hours. With the availability of nutrition, hydrocarbon degradation occurred efficiently even up to 86 % in mixed cultures. The increase of degradation capability was noticed to be directly proportional to the increase in incubation period to the maximum of over 70 %. The percentage degradation was also correlated with increase in cell mass. Addition of indigenous bacteria followed by fertilizer application and tilling recorded the highest rate of hydrocarbon loss (Harmsen, 1991). This development revealed that nutrient enhanced bioremediation could achieve the desired level of hydrocarbon loss.

Addition of nutrients, mineral fertilizers, different agricultural byproducts and molasses along with bacterial inoculation has been reported to enhance the degradation process (Olivara et al., 1997). The degradation of oil was increasing gradually with increasing the process time of the microbe (Deka, 2005). Generally, microbial species identified from soil of pH ranging from 6.5 to 7.5 were efficient, whereas, microbes isolated from soils of pH below 6.5 were not efficient. Limitation of the speed of the biodegradation is not depending on the number microorganisms but the amount of nitrogen and phosphorous as utilizable source. Availability of nutrients is often the controlling factor for biodegradation (Ron and Rosenberg, 2002). Biostimulation is another approach that is widely used to overcome these limitations. Biostimulation enhances the growth of microorganisms through the addition of nutrients and other growth stimulating cosubstrates (Venosa and Zhu, 2003). The role of these limiting factors for microbial activities on petroleum hydrocarbon degradation was described (Atlas, 1984; Ron and Rosenberg, 2002). The limiting factors were reported to be nutrient requirements (oxygen, nitrogen and phosphate), air, temperature, and pH.

The technical viability of biological treatment of oily sludge by stimulating native microorganisms was studied by Ururahy et al. (1998). Those microorganisms able to utilize oily compounds as carbon and energy source for their growth. Oily sludge with concentrations of 5 % (v/v) and 10 % (v/v), as carbon ratio of 100:1 were studied. Increased microbial population observed at 5 % (v/v) and substrate inhibition and toxic effect took place in the 10 % (v/v) concentration. In this microbial population Pseudomonas predominated and aeration on the microbial activity also evaluated. Two yeast species were also identified and two filamentous fungi were isolated. Further, the rate of microbial degradation of crude oil or oil waste depends on a variety of factors, some of them are the physical conditions and the nature, concentration, and ratios of various structural classes of hydrocarbons present, the bioavailability of the substrate, and the properties of the biological system involved (Yuste et al., 2000).

Resistance and toxic compounds present in the oil, low water temperatures, deficiency of nutrients, decreased dissolved oxygen and scarcity of hydrogen degrading microorganisms are some of the limiting factors of degradation (Yuste et al., 2000; Okerentugba and Ezeronye, 2003). Biodegradation in fact is highly dependent on the composition and incubation temperature. This knowledge came from the work done on biodegradability of seven different crude oils (Atlas, 1975). At 20°C, lighter oils had greater abiotic losses and were more susceptible to biodegradation than heavier oils. Hence the temperature influences the biodegradation of hydrocarbons by affecting the physical and chemical composition and rate of hydrocarbon metabolism by microorganisms (Leahy and Colwell, 1990).

Temperature is a major environmental factor influencing bioremediation rates as well as directly affecting bacterial metabolism and growth rates; temperature can have profound effects on the soil matrix and on the physicochemical state of the contaminants (Baker, 1994). In simple bioremediation systems, which require little or no microbiological expertize, process-limiting factors often relate to nutrient or oxygen availability or the lack of relatively homogenous conditions throughout the contaminated medium. Microbial growth and degradation processes operating under such conditions are typically variable and suboptimal, leading at best to prolonged degradation cycle (Mueller et al., 1998). The rate of degradation decreases with decreasing temperature due to a reduction in enzymatic activities. Higher temperature tends to increase the rate up to a maximum temperature of 30 to 40°C (Arnosti et al., 1998). Therefore, the bioremediation in the Arctic shorelines where the temperature varies from 3 to 7°C, even though slow release and soluble organic fertilizers were used to enhance the rates (Prince et al., 2003). Extensive biodegradation of Metula crude oil by mixed cultures of marine bacteria at 3°C was reported (Colwell and Walker, 1977).

Seasonal changes in the functional changes of bacteria over 9 months in uncontaminated, contaminated, and treated (Inipol EAP22 fertilizer) plots in Antartica were examined (Delille et al., 1997). Total bacteria saprophytes and hydrocarbon degrading bacteria were assayed. In all the cases, changes in total bacterial abundance, reaching a minimum in winter were correlated with seasonal variations. Later it was found that surfactant levels close to the critical misallocation in soil inhibited mineralization and shifted the community from Rhodococcus and Nocardia populations to Pseudomonas and Alcaligens species able to degrade both surfactant and hydrocarbon (Colores et al., 2000).

The application of Bacillus subtilis 09 surfactant (Bs) in soil bioremediation was considered by Cubitto et al. (2004); it was performed at laboratory scale and so it only considers some of all the biotic and abiotic factors involved in the natural environment. Although at the end of the experiment the Bs treated and control microcosms had similar hydrocarbon concentrations; the application of an adequate concentration of Bs treated and control microcosms had similar hydrocarbon concentrations, the application of an adequate concentration of Bs diminished the concentration of some hydrocarbon fractions in less time. Some have found that applying surfactants inhibits biodegradation and reduction in substrate bio-availability when bound into surfactant micelles (Volkering *et al.*, 1995). Further, surfactant may alter the interactions between the cells and the substrate (Stelmack *et al.*, 1999). Hence, it was concluded that surfactant toxicity may alter the composition of the microbial populations responsible for hydrocarbon mineralization (Colores *et al.*, 2000).

Studies on Arctic terrestrial ecosystem showed that spilling a crude oil stimulated the microflora but did not distinguish between nonhydrocarbon and hydrocarbon utilizing microorganisms (Onwurah *et al.*, 2007), and it was concluded that addition of nitrogen and phosphate speeded up the utilization of the alkane fraction of the crude oil that had been spilled on soil. Chromatographic separations of recovered oils indicted that extensive degradation of the alkane fractions was taking place, but no data were given for quantities of residual oil remaining with time. They concluded from the chromatographic data that fertilizer speeded up degradation. They also suggested that application of oil-utilizing bacteria to the natural flora was beneficial.

Nutrient amendment to oil-contaminated beach sediments is a critical factor for the enhancement of indigenous microbial activity and biodegradation of petroleum hydrocarbons in the intertidal marine environment (Xu and Obbard, 2003). They had investigated the stimulatory effect of the slow-release fertilizers Osmocote (Os) and Inipol EAP 22 combined with inorganic nutrients on the bioremediation of oil-spiked beach sediments using an open irrigation system with artificial seawater over a 45 days period. Osmocote is surrounded by water soluble inorganic N and P which is covered with semipermeable membrane. It has been used for oil cleanup on beach substrate.

Addition of nutrients to increase the biodegradation rates has been reviewed in various works (Bragg *et al.*, 1994). Nitrogen and phosphorus are the essential nutrients that are incorporated in the microbial biomass and they are elements which comprise the physical structure of cells. Microbes are dependent on nutrients for survival which is building blocks of life to produce necessary enzymes to breakdown xenbobiotics (www.geocites.com/CapeCanaveral/Lab/2094/bioremed.html). Microbial cells are composed of carbon (C), nitrogen (N) and phosphorous (P) at an average ratio of 50:14:3. Sufficient amount of these nutrients must be available in a usable form and in proper proportions for unrestricted microbial growth to occur. Temperature directly influences the rate of biodegradation by controlling the rates of enzyme catalyzed reactions.

Biodegradation can occur under a wide-range of pH however, a pH of 6.5 to 8.5 is generally optimal for biodegradation in most aquatic and terrestrial systems and values ranging from 5 to 9 are considered acceptable. It was also mentioned that nutrients availability, especially nitrogen and phosphorous appears to be the most common limiting factor (Pritchard *et al.*, 1992). In simple bioremediation systems, which require little or no microbiological expertize, process-limiting factors often relate to nutrient or oxygen availability or the lack relatively homogenous conditions are typically variable and suboptimal, leading prolonged degradation cycles (Mueller *et al.*, 1998).

The biodegradation of petroleum hydrocarbons in oil-contaminated marine foreshore environments can be accelerated by the application of nutrients (Mearns, 1997). Approximately 1.5 mg NO_3-N L^{-1} in interstitial pore water of beach sediments is sufficient to maintain optimal degradation activity by the microbial biomass (Venosa *et al.*, 1997), and this level could be maintained by the daily application of inorganic nutrients. It is suggested that fishmeal and related products, composed mainly of protein, are other types of organic fertilizer can accelerate oil biodegradation (Santas and Santas, 2000). It is concluded that the inorganic nutrients such as nitrogen and phosphorous are most often limiting in the bioremediation of hazardous organic compounds (Lin and Mendelssohn, 1998). Although not all bioremediation systems respond to nutrient additions, the addition of N fertilizer has been observed to enhanced bioremediation of oil contamination in aquatic systems and soils.

A laboratory experiment was made on the application of various nutrients to soil to optimize biodegradation of xylene, anthracene, phenanthrene, and n-hexadecane (Graham *et al.*, 1995). Biodegradation was enhanced by nutrient addition, and each type of hydrocarbon reacted differently to various levels of fertilization. The degradation of crude oil was increasing gradually with increasing the incubation time of the bacteria (Deka, 2003). Moreover, degradation of crude oil depends on the physicochemical properties of the soil. Generally, bacteria isolated from soil pH ranging from 6.5 to 7.5, were efficient, whereas, bacteria isolated from soils of pH below 6.5 were not efficient (Deka, 2003). The various factors influencing bioprocess procedure like, pH, time, and temperature have been studied by Hossain and Anatharaman (2004). Their result showed the optimum time for maximum reduction of all the petroleum pollutant constituents is 144 h and pH is range from acidic to basic in nature. The optimum pH is 7.5, slightly basic medium of the bacterial environment. Other pH values can reduce the pollution parameters to some little bit less extent comparing pH of 7.5 (Hossain and Anatharaman, 2004). Special emphasis was made on the importance of temperature in the degradation of oil that it takes a long time for oil to disappear in Arctic regions (Nissen, 1970).

Studies reported that mixed and pure bacterial cultures were shown to be capable of growing on and degrading a wide variety of hydrocarbon sources, including the heavy fuel oil, at temperatures as low as 5°C (Kanaly and Harayama, 2000). In the case of the *Nocardia* sp., the rate of growth on various hydrocarbons decreased, on the average, 2.2 times when the temperature was lowered from 15 to 5°C rather than 28°C. And also in his studies the concentration of nitrogen controlled the amount of growth, but did not affect the growth of the *Nocardia* sp., and the degradation of oil. At low temperature the viscosity of oil increases, the volatility of straight chain alkane is reduced, thus delaying the onset of biodegradation. Rates of degradation are generally believed to decrease with decreasing temperature. It is due to decreased rates of enzyme activity.

About 18 genera were isolated and grouped into two main clusters; cluster 1 containing 12 isolates grown at 43°C, and cluster 2 containing six isolates grown at 37°C. Three natural bacterial consortia with ability to degrade total petroleum hydrocarbons (TPH) were prepared from these isolates. Experiments were conducted in Erlenmeyer flasks under aerobic conditions, with TPH removal percentage varying from 5.9 to 25.1 %, depending upon consortia type and concentration. TPH removal rate was enhanced after addition of nutrients to incubated flasks. The highest TPH reduction (37 %) was estimated after addition of a combination of nitrogen, phosphorus, and sulfur (Mrayyan and Battikhi, 2004).

Most heterotrophic bacteria and fungi favor a pH near neutrality with fungi being more tolerant of acidic condition (Atlas, 1988). Extreme in pH would therefore expect to have negative influence on the ability of microbial population to degrade hydrocarbons. A pH of 7.0 has been shown to be optimal for biodegradation (Walter *et al.*, 1991). Biodegradation of gasoline were doubled when the pH of soil was adjusted from 4.5 to 7.4 (Marchal *et al.*, 2003). Rates dropped when pH was raised further. Yet the pH is dependent upon a number of other factors including the mineral content, the composition and the presence of acidic organic matter.

Several investigators have reported that concentration of available nitrogen and phosphorous in sea water are severely limiting microbial degradation (Atlas and Bartha, 1972; Floodgate, 1984). In contrast other investigators have reached the opposite conclusion. In oil slick, there is a mass of carbon available for microbial growth within limited area. Since microorganisms require N and P for incorporation into biomass, the availability of these nutrients within the same area as the hydrocarbon is critical (Ammary, 2004). The concentration of nitrogen and phosphorous needed for the biodegradation of oil or other materials present at the high concentration either throughout the environment or within the oil itself, is usually assumed to reflect the amount of those elements that must be incorporated into the biomass that would be formed as the microorganisms use the organic materials as carbon source for growth. The concentration of nitrogen and phosphorous needed for the biodegradation of oil must be incorporated into the biomass. That would be used by the microorganisms as carbon source for their growth (Arosetin and Alexander, 1992). Nitrogen and phosphorous may be limiting in soils and acceleration of the biodegradation of crude oil can be enhanced by addition of urea phosphate, NPK fertilizers and ammonium phosphate salts (Amenaghawon, 2013).

The effects of mixed nitrogen sources on biodegradation of phenol immobilized *Acinetobacter* sp. strain W-17 was studied by Abd-El-Haleem *et al.* (2003). Phenol biodegradation by *Acinetobacter* sp. strain W-17 in the presence of high concentration of nitrogen components (NH_4 and NO_3) is enhanced. It is seen that a positive impact on biodegradation of some aliphatic chlorinated xenobiotics when the culture medium was supplemented with minerals (Henery and Gribic-Galic, 1995). The process of oil biodegradation by bacteria was fast and efficient if all the factors presented. The oil-contaminated water containing bacteria, inorganic nutrients and source of oxygen demonstrated the environment in which biodegradation proceeded most efficiently. The inorganic nutrients and oxygen helped the oil-degrading bacteria to grow and multiply at a much faster rate, thus speeding the process of biodegradation.

From a nutritional point of view, the oily sludge contains limited quantities of nitrogen and phosphate denoting the need to fortify the medium with these nutrients. Furthermore, little of the nitrogen is available, since it appears as part of complex structures barely accessible to microorganisms (Cerniglia, 1992). Several other biotic and abiotic factors also affect the efficacy of degradation of PAHs and other organic pollutants in soil. The soil porosity is also affecting the movement of microorganisms through soil to the site of contamination for mineralization of organic pollutants in soil (Ashok and Musarrat, 1999).

Over the past one decade, Gas Technology Institute (GTI), USA and others have conducted research on the bioremediation of organic contaminants in soil. Most of this work has been associated with remediation of former Manufactured Gas Plant (MGP) sites, soils from industrial areas and oil production areas have also been studied. The results have shown that: (1) organic contaminants are biodegraded by indigenous soil microorganisms to a concentration that no longer decreases, or that decreases very slowly, with continued treatment; (2) reductions below this concentration are limited by the availability of the contaminants to the microorganisms; and (3) the residual contaminants that remain after biological treatment, regardless of the extent of treatment, are significantly less leachable and significantly less available to other organisms, as measured by simple indicator toxicity tests with bacteria and invertebrates.

Periodic monitoring of refinery sludge and the recipient soil is strongly suggested for the quantitative assessment of the potentially hazardous contaminants entering into the soil system. This is important not only in tracing the overall impact of land spreading but also in determining when sludge reapplication should be made and how successful certain management practices will be, in achieving the goal of rapid, environmentally safe hydrocarbon degradation and contaminant demobilization (Ashok and Musarrat, 1999). Bioremediation is predominantly oxidation process. Bacterial enzymes will catalyze the insertion of oxygen into the hydrocarbon so that the molecule can subsequently be consumed by cellular metabolism. Because of this oxygen is one of the most important requirements for the biodegradation of oil.

14.6   MICROBIAL SEEDING

Oil degrading microorganisms seem to be ubiquitous, with their number typically limited by the hydrocarbon supply (Namazi et al., 2008). Another approach is to add exogenous microorganisms with known degradative activities, either natural isolates or engineered (Narasimhan et al., 1983). Seeding, introduction of allochthonous microorganisms is important to increase the rate of degradation of xenobiotics in the environment, when autochthonous microorganisms may not be effective in degrading the pollutants present in the environment Raghavan, 1998). Terrestrial ecosystems differ from aquatic ecosystems in that soils contain higher concentrations of organic and inorganic matter and generally, large number of microorganisms and are more variable in terms of physical and chemical conditions (Bossert et al., 1984). The presence of indigenous microorganisms which are highly adapted to a particular environment would negatively influence the ability of seed microorganisms to compete successfully and survive; for this reason soils are not widely considered to be amenable to improvements in rates of biodegradation through seeding alone (Bossert et al., 1984).

Bushnell-Hass broth generally is used to evaluate the ability of microorganisms to decompose hydrocarbons (Bushnell and Hass, 1941). The medium was used to enumerate total heterotrophs and hydrocarbon degrading microorganisms. Oil spillage into the environment is of a great concern (Lehmann, 1998). Seeding with oil-degrading microbes is an important approach for bioremediation of oil spills. The oil industry generates large quantities of oily and viscous residues, which are formed during production, transportation, and refining. These residues, called oily sludges, are composed of oil, water, solids, and their characteristics, such as varied composition, make them highly recalcitrant and very difficult to reutilize. One of the difficulties in the study of oil-degrading bacteria is the choice of hydrocarbon. Utilization of purified hydrocarbons has been identified in many studies (Leddy *et al.*, 1995; Ramos *et al.*, 1995). Degradation of hydrocarbons by environmental microflora involves microorganisms having specialized metabolic capacities. In polluted environments, specialized microorganisms are abundant because of the adaptation of the microflora to pollutant (Forsyth *et al.*, 1995).

14.7 MOLECULAR AND GENOMIC PERSPECTIVES

Petroleum and derivatives have been well thought-out one of the major environmental noxious wastes. Because of their worldwide and immense creation, transportation, and consumption as primary energy source and raw material for several products like plastic, solvents, pharmaceuticals, cosmetics, fuel, synthetic rubber among others (Mazzeo *et al.*, 2011). Ecological pollution by petroleum and derivatives spills has been commonly reported as a corollary of accidents in loading, discharging, transportation, production of byproducts and combustion. Putative genotoxic and mutagenic effects of different concentrations of BTEX and their biodegraded products were studied through analyses of chromosomal aberrations and micronuclei (MN) in meristematic and F1 cells of *Allium cepa* roots (Mazzeo *et al.*, 2011). They have shown that the BTEX biodegradation process by bacteria was efficient in reducing damages observed in the genetic material of *A. cepa* cells.

Molecular analysis of the bacterial genome for strain identification is a routine procedure. Denaturing gradient gel electrophoresis is generally carried out to find out the similarities and differences in the bacterial community. Metagenomics is the study of inherited material isolated in a straight line from ecological samples. The ground may also be referred to as environmental genomics, ecogenomics or community genomics. While conventional microbiology and microbial genome sequencing and genomics depend upon sophisticated clonal cultures, early environmental gene sequencing cloned specific genes (often the 16S rRNA gene) to produce a profile of diversity in a natural sample. Such work revealed that the vast majority of microbial biodiversity had been missed by cultivation-based methods (Hugenholz *et al.*, 1998). Modern coming out of "metagenomics" favors us for the examination of microbial communities devoid of tedious cultivation efforts. Metagenomics attitude is parallel to the genomics with the difference that it does not contract with the single genome from a clone or microbe cultured or characterized in laboratory, but moderately with that from the whole microbial community present in an environmental sample. It is the community genome. Global understanding by metagenomics depends essentially on the possibility of isolating the entire bulk DNA and identifying the genomes, genes, and proteins more relevant to each of the environmental sample under investigation (Vieites *et al.*, 2010).

14.7.1   GENOMIC SHUFFLING

During recent trends, technique for strain enhancement varied from arbitrary mutagenesis to highly balanced methods of genetic and metabolic engineering (Kumar *et al.*, 2012). Genomic shuffling tends to be unique advance for strain development. This has been broadly used for scaling up of metabolites by bacterial species, increasing nutrient uptake and strain resistance. This method bring about the benefit of multiparental crossing allowed by DNA shuffling together with the recombination of entire genomes normally associated with conventional breeding, or through protoplast fusion that increases the recombination process. Genome shuffling often referred to as an important breakthrough in potent strain development and metabolic engineering (Zhan *et al.*, 2002; Leja *et al.*, 2011), thus aids in recombination between diverse selected and improved populations by accelerating direct evolution (Dai *et al.*, 2005; Singleton *et al.*, 2009; Pinel *et al.*, 2011).

Techniques like genomic shuffling may prove most efficient in the degradation of PAH's because these degradation are mediated by multiple genes. Altering the expression of one or more genes may result in enhanced degradation rates (Kumar *et al.*, 2012). This method shall be useful in the development of mutitrait phenotypes. Otherwise conventional strain improvement techniques prove difficult over gene shuffling as it would improve a complex trait while still maintaining other characteristics of the strains. Gene shuffling renders improvement in gene pool among selective organisms through extensive recombination among selected strains and produce "multi-parental complex progeny". Thus, resulting in larger combinatorial library of the original genetic diversity (Kumar *et al.*, 2012).

14.7.2   GENETIC ENGINEERING

Recent advances in the study of bacterial genetics responsible for the degradation of aromatic hydrocarbons have attracted a great deal of attention. The ultimate aim of current studies is to gain a good knowledge of the pathways by which these organisms degrade the aromatic hydrocarbons. Much of the work has been carried out toward enhancing the capabilities of these microbes to degrade major class of environmental pollutants. Recently genes for several aromatic hydrocarbon degradation pathways have been cloned (Zylstra and Gibson, 1991). Studies with the cloned genes have also focused on the mechanisms for induction of the biochemical pathways. In addition, hybrid pathways have been constructed for novel routes of degradation with the use of cloned genes (Ramos *et al.*, 1990).

Two level approaches have been followed to increase the physiologic efficiency of a microorganism for a defined biodegradation application: (i) management of the precise catabolic pathway, and (ii) management of the host cell. For the improved rate of removal of pollutants from the environment, and to enhance the uptake of substrates, alteration of some key enzymes involved in the regulatory mechanisms controlling the expression of catabolic genes are required (Timmis, 1999; Lorenzo, 2001). Rationally directed molecular evolution technique like gene shuffling and proteomics can be used for the mRNA stability and altering the protein activity (Timmis, 1999; Furukawa, 2000). In addition, genetic manipulation allows the development of new hybrid pathways. Manipulating physiological modules and assembling the genes of different origins in a same host cell will provide greater pathway expansion, uptake of new substrate and favors in creating of bacteria with multiple pathways (Reineke, 1998; Timmis, 1999; Pieper, 2000; Rieger, 2002). The formation of toxic end products of degradation can be prevented by misrouting the pollutants. This can be achieved by balanced combination of catabolic pathways by allowing the complete metabolism of xenobiotics (Timmis, 1999; Pieper, 2000).

Some PAH's have very limited bioavailability. This largely contributes to the persistence of these hydrocarbons in the nature for very long period. This is true for hydrocarbons present in fossils which often attract microorganisms and yet they never undergo degradation. Genetic engineered organisms can be deployed to overcome this sort of problems. Microorganisms can be engineered to produce more biosurfactants and chemotaxis (Reineke, 1998; Dua, 2002). Surfactant production has been combined with the ability to selectively cleave carbon-sulfur bonds in the sulfur-containing compounds present in oil (biodesulfurization) and, hence, efficient recombinant biodesulfurizers have been obtained (Gallardo, 1997).

14.7.3 METAGENOMICS

Metagenomics is a molecular tool that overcomes the limitations imposed by the classical approach, enabling a broader perspective of the taxonomic and functional variety of environmental microorganisms and access to their metabolic potential (Handelsman *et al.*, 1998). Metagenomics can pick up policies for supervising the influence of xenobiotics on environment and for cleaning up polluted ecosystem. Enhanced understanding of how microbial communities manage with xenobiotics improves assessments of the potential of contaminated sites to recover from pollution and increases the chances of bioaugmentation or biostimulation trials to succeed. Shotgun sequencing has been employed to unravel numerous environmental samples and number of metagenomic projects has been taken up during recent decades (Evanova *et al.*, 2010). A function-driven metagenomic approach to identify diverse and potentially novel hydrocarbon biodegraders in petroleum reservoirs was carried out by Vasconcellos *et al.* (2010). Aerobically and anaerobically degraded petroleum sample enrichments were used to construct a fosmid library by metagenomic DNA. And, hexadecane was used to screen the library for hydrocarbon-degrading fosmid clones.

14.8   CONCLUSION

Evidently, hydrocarbon degradation research is advancing on many fronts, speeding up by new knowledge of cellular structure and function gained through molecular and protein engineering techniques, combined with more conventional microbial methods. Enhanced schemes for bioremediation of petroleum components are being commercialized with proper financial and ecological benefits. The immense adaptability of microorganisms presents simpler, economic, and more environmental friendly approaches to diminish environmental pollution than nonbiological options. Till date only about 5 % is known about microbial diversity on the planet, even though they represent more than half of the biomass. This implies a great deal of unexplored microbial world. Hence, huge potentials of microbes toward sustainability of the natural world, still awaits further exploration. Microorganisms have adopted different tactics to utilize PAH's present at the contamination sites. Many microorganisms produce biosurfactants to increase the bioavailability of the poorly available substrates. Some even produce chemotaxis toward PAH compounds, which could lead to improved degradation. Bacteria isolated from natural localities have the capacity to degrade narrow window of PAH's. During the course of degradation, the ubiquitous coexistence of bacteria and fungi in soil and there metabolic cooperation suggest that bacterial-fungal interactions may be of importance for PAH degradation. Combining genetic engineering tools such as gene conversion, gene duplication and transposition, one can produce novel strains with desirable properties for bioremediation applications based on the knowledge of PAH degradation by microorganisms.

KEYWORDS

- **Bioavailability**
- **Biodegradation**
- **Bioremediation**
- **Biostimulation**
- **Biosurfactants**
- **Chemotaxis**
- **Genomics**
- **Metagenomics**
- **Microbial strains**
- **Micronuclei**
- **Microorganisms**
- **Polyaromatic hydrocarbons**
- **Soil pollution**
- **Total petroleum hydrocarbons**

REFERENCES

Abd-El-Haleem, D.; Beshay, U.; Abdelhamid, A. O.; Moawad, H.; Saki, Z. Effects of mixed nitrogen sources on biodegradation of phenol by immobilized *Acinetobacter* sp. strain W-17. *African Journal of Biotechnology* 2003, 2(1), 8-12.

Ahearn, D. G.; Meyers, S. P.; Standard, P. G. The role of yeasts in the decomposition of oils in marine environments. *Dev Ind Microbiol* 1971, 12, 126-134.

Aitken, M. D.; Stringfellow, W. T.; Nagel, R. D.; Kazuna, C.; Chen, S. H. Characteristics of Phenanthrene degrading bacteria isolated from soils contaminated with polycyclic aromatic hydrocarbons. *Can J Microbiol* 1998, 44, 743-752.

Aldrett, S.; Bonner, J. S.; Mills, M. A.; Autenrieth, R. L.; Stephens, F. L. Microbial degradation of crude oil in marine environments tested in a flask experiment. *Wat Res* 1997, 31(11), 2840-2848.

Allard, A. S.; Neilson, A. H. Bioremediation of organic waste sites: a critical review of microbiological aspects. *Internat Biodeter Biodegrad* 1997, 39, 253-285.

Alloway, B. J.; Ayres, D. C. *Chemical Principles of Environmental Pollution*, Blackie Academics and Professional: London, 1997, pp. 252-257.

Amenaghawon, N. A.; Asegame, A. P.; Obahiagbon, K. O. Investigation into the use of urea and NPK fertilizers in the bioremediation of domestic waste water. *World Essays J* 2013, 1(2), 52-57.

Ammery, B. Y. Nutrients requirements in biological industrial wastewater treatment. *Afr J Biotechnol* 2004, 3(4), 236-238.

Amthor, J. S. *Sorghum* seedling growth as a function of sodium chloride salinity and seed size. *Ann Bot* 1983, 52, 915-917.

Annweiler, E.; Richnow, H. H.; Antranikian, S.; Hebenbrock, S.; Grams, C.; Franke, S.; Francke, W.; Michaelis, W. Naphthalene degradation and incorporation of naphthalene-derived carbon into biomass by the thermophile *Bacillus thermoleovorans*. *Appl Environ Microbiol* 2000, 66, 518-523.

April, T. M.; Foght, J. M.; Currah, R. S. Hydrocarbon-degrading filamentous fungi isolated from flare pit soils in northern and western Canada. *Can J Microbiol* 2000, 46(1), 38-49.

Arnosti, C.; Jorgensen, B. B.; Sagemann, J.; Thamdrup, B. Temperature dependence of microbial degradation of organic matter in marine sediments: polysaccharide hydrolysis, oxygen consumption and sulphate reduction. *Mar Ecol Prog Ser* 1998, 165, 59-70.

Arosetin, B. M.; Alexander, M. Surfactants at low concentration stimulated biodegradation of sorbed hydrocarbons in samples of aquifer sands and soil slurries. *Environ Toxicol Chem* 1992, 11, 1227-1233.

Arvin, E.; Jensen, B. K.; Gundersen, A. T. Substrate interactions during aerobic biodegradation of benzene. *Applied and Environmental Microbiology* 1989, 3221-3225.

Ashok, B. T.; Musarrat, J. Mechanical, physico-chemical and microbial analysis of oil refinery waste receiving agricultural soil. *Indian J Environ Hlth* 1999, 41(3), 207-216.

Atlas, R. M. Bioremediation of petroleum pollution. *International Biodeterioration and Biodegradation* 1995b, 35, 317-327.

Atlas, R. M. Effects of temperature and crude oil composition on petroleum biodegradation. *Appl Microbiol* 1975, 30, 396-403.

Atlas, R. M. Microbial degradation of petroleum hydrocarbons: an environmental perspectives. *Microbiol Rev* 1981, 45, 180-209.

Atlas, R. M. Microbiology: fundamentals and applications, Macmillian Publishing Co: London, UK 1988, pp. 352-353.

Atlas, R. M. Petroleum biodegradation and oil spill bioremediation. *Marine Pollution Bulletin* 1995a, 31(4-12), 178-182.

Atlas, R. M. Petroleum Microbiology, Macmillan Publishing Company: NewYork, 1984.

Atlas, R. M.; Bartha, R. Biodegradability testing and monitoring the bioremediation of xenobiotics pollutants. In *Microbial Ecology: Fundamentals and Applications*, 4th ed.; Pearson Education: Singapore, 2005.

Atlas, R. M.; Bartha, R. Biodegradation of petroleum in sea water at low temperature. *Can J Microbiol* 1972, 18, 1851-1855.

Atlas, R.; Bartha, R. Hydrocarbon biodegradation and oil spill bioremediation. In *Advances in Microbial Ecology*, Marshall, K. C., Ed.; Plenum Press: New York, USA 1992, pp. 287-338.

Bahl, A.; Bahl, B. S. Alkanes. A text book of organic chemistry, S. Chand and Company Ltd: New Delhi, 2004, pp. 191-199.

Baker, K. H.; Herson, D. S. *Bioremediation*, Mc-Graw-Hill: New York, USA 1994, pp. 9-60.

Balba, M. T.; Al-Awandhi, N.; Al-Daher, R. Bioremediation of oil contaminated soil: microbiological methods for feasibility assessment and field evaluation. *J Microbiol Methods* 1998, 32, 155-164.

Barbeau, U.; Kuhad, R. C.; Khanna, S. Mineralization of [^{14}C] octadecane by *Acenetobacter calcoaceticus* S19. *Can J Microbiol* 1998, 44, 681-686.

Bartha, R.; Atlas, R. M. The microbiology of aquatic oil spills. *Adv Appl Microbiol* 1977, 22, 225-266.

Benedik, M. J.; Gibbs, P. R.; Riddle, R. R.; Wilson, R. C. Microbial denitrogenation of fossil fuels. *Trends Biotechnol* 1998, 16, 392-395.

Benka-Coker, M. O.; Olumagin, A. Waste drilling fluid-utilising microorganisms in a tropical mangrove swamp oilfield location. *Bioresour Technol* 1995, 53, 211-215.

Bhargava, D. S.; Sharma, S. P. Environmental Protection Information. *Indian J Environmental Protection* 2000, 20(7), 555-558.

Boldu, F. X. P.; Vervoort, J.; Rotenhuis, J. T. C.; van Groenestijn, J. W. Substrate interaction during the biodegradation of benzene, toluene, ethylbenzene and xylene (BTEX) hydrocarbons by the fungus *Cladophialophora* sp. strain T1. *Applied and Environmental Microbiology* 2002, 68(6), 2660-2665.

Boonchan, S.; Britz, M. L.; Stanley, A. Degradation and mineralization of high-molecular weight polycyclic aromatic hydrocarbons by defined fungal bacterial coculture. *Applied and Environmental Microbiology* 2000, 66(3), 1007-1019.

Boopathy, R. Factors limiting bioremediation technologies. *Bioresource Technology* 2000, 74, 63-67.

Bosch, R.; Garcia-Valdes, E.; Moore, E. R. B. Genetic characterization and evolutionary implications of a chromosomally encoded naphthalene degradation upper pathway from *Pseudomonas stutzeri* AN10. *Gene* 1999, 236, 149-157.

Bossert, I.; Bartha, R. The fate of petroleum in soil ecosystems. In *Petroleum Microbiology*, Atlas, M., Ed.; Macmillan Publishing Co: New York, USA 1984, pp. 434-476.

Bouchez-Naitali, M.; Rakatozafy, H.; Marchal, R.; Levean, J. Y.; Vandecasteele, J. P. Diversity of bacterial strains degrading hexadecane in relation to the mode of substrate uptake. *Journal of Applied Microbiology* 1999, 26, 421-428.

Bradley, P. M. Microbial degradation of chloroethanes in groundwater systems. *Hydrogeology J* 2000, 8, 104-111.

Bragg, J. C.; Prince, R. C.; Harner, J. E.; Atlas, R. M. Effectiveness of bioremediation for the *Exxon Valdez* oil spill. *Nature* 1994, 368, 413-418.

Breed, R. S.; Murray, E. G. D.; Smith, N. R. Bergey's Manual of Determinative Bacteriology, The Williams and Wilkins Co.: Baltimore, 1957.

Brown, E. J.; Braddock, J. F. Sheen screen; a miniaturized most probable number method for the enumeration of oil-degrading microorganisms. *Applied and Environmental Microbiology* 1990, 52, 149 -156.

Brown, K. W.; Brawand, H.; Thomas, J. C.; Evans, G. B. Impact of simulated land treatment with oily sludges on ryegrass emergence and yield. *Agronomy J* 1982, 74, 257-261.

Bruice, P. Y. Organic Chemistry, Pearson Education: New Delhi, 2004, pp. 329-330.

Bushnell, L. D.; Hass, H. F. The utilization of hydrocarbons by microorganisms. *J Bacteriol* 1941, 41, 653-673.

Cacciatore, D. E.; Mc Neil, M. A. Principles of soil bioremediation. *Biocycle* 1995, 36, 61-64.

Cairns, W. L.; Cooper, D. G.; Zajic, J. E.; Wood, J. M.; Kosaric, N. Characterization of *Nocardia amarae* as a potent biological coalescing agent of water in oil emulsions. *Appl Environ Microbiol* 1982, 43, 362-366.

Calabrese, E. J.; Kenyon, E. M. Air toxics and risk assessment. Lewis Publishers: Chelsea, MI, 1991.

Caret, R. L.; Dennistin, K. J.; Topping, J. J. Foundations of inorganic, organic and biological chemistry. N.M.C. Brown Publishers: London, 1995, pp. 223-225.

Carlile, M. J.; Sarah, C.; Watkinson. The fungi, Academic Press: London, UK 1994, pp: 50-51, 348-350.

Cerniglia, C. E. Biodegradation of polycyclic aromatic hydrocarbons. *Biodeg* 1992, 3, 351-368.

Cerniglia, C. E.; Hebert, R. L.; Doge, R. H.; Szabuszlo, P. J.; Gibson, D. T. Some approaches to studies on the degradation of aromatic hydrocarbons by fungi. In *Microbial degradation of pollutants in marine environments*, Bourquin, A. L.; Pritchard, P. H., Ed.; Environmental Research Laboratory: Gulf Breeze, Florida, 1978, pp. 360-369.

Chakrabarthy, A. M. Microorganisms having multiple compatible degradative energy generating plasmids, and preparation thereof, U.S. Patent 4259, 1981, 444.

Chakravarthy, B. B. Effluent Management Practices at IOCL Refineries. *Journal IAEM* 1995, 22, 68-76.

Champagnat, A. Proteins from petroleum fermentations: A new source of food. *Impact* 1964, 14, 119-133.

Champagnat, A.; Llewelyn, D. A. B. Protein from petroleum. *New Sci* 1962, 16, 612-613.

Chauhan, S. V. S.; Kumar, S. Effect of tannery effluent on total phenolics and some isozyme profile in *Allium cepa* roots. *Journal of Environment and Pollution* 1997, 4(2), 85-99.

Christova, N.; Tuleva, B.; Damyanova, B. N. Enhanced hydrocarbon biodegradation by a newly isolated *Bacillus subtilis* strain. *Z Naturforsch* 2004, 59, 205-208.

Clayton, G. D.; Clayton, F. E. Patty's Industrial Hygiene and Toxicology, 3rd ed.; Vol. 2B, John Wiley and Sons: New York, 1981.

Colores, G. M.; Macur, R. E.; Ward, D. M.; Inskeep, W. P. Molecular analysis of surfactant-driven microbial population shifts in hydrocarbon contaminated soil. *Appl Environ Microbiol* 2000, 66, 2959-2964.

Colwell, R. R.; Walker, J. D.; Ecological aspects of microbial degradation of petroleum in marine environment. *Crit Rev Microbiol* 1977, 5, 423-445.

Colwell, R.; Walker, J. D. Ecological aspects of microbial degradation of petroleum in the marine environment. *Microbiol* 1997, 50, 846-850.

Cubitto, M. A.; Moran, A. C.; Commendatore, M.; Chiarello, M. N.; Baldini, M. D.; Sineriz, F. Effects of *Bacillus subtilis* 09 biosurfactant on the bioremediation of crude oil-polluted soils. *Biodegradation* 2004, 15, 281-287.

Dai, M. H.; Ziesman, S.; Ratcliffe, T.; Gill, R. T.; Copley, S. D. Visualization of protoplast fusion and quantitation of recombination in fused protoplasts of auxotrophic strains of *Escherichia coli*. *Metabolic Engineering* 2005, 7(1), 45-52.

Dalyan, U.; Harder, H.; Hoepner, T. Hydrocarbon biodegradation in sediments and soils. A systematic examination of physical and chemical conditions-Part 11 pH values. *Wissenschaft Science and Technik* 1990, 43, 337-342.

Daniel. J. C. Some recent aspects of water pollution in environmental aspects of microbiology, Bright Sun Publications: Chennai, India, 1996, pp. 294-297.

Das, M. Characterisation of deemulsification capabilities of a *Micrococcus* sp. *Biores Technol* 2001, 79, 15-22.

De Lorenzo, V. Cleaning up behind us. *EMBO Rep* 2001, 2, 357-359.

De, A.K. *Water Pollution in Environmental Chemistry*, New Age International Publications: New Delhi, India 2002, pp. 192-195.

Deka, S. Bacterial strains degrading crude oil from petroleum polluted soil of Assam. *Pollution Research* 2001, 20(4), 517-521.

Deka, S.; Sarma, P. K.; Bhattacharyya, K. G. Degradation of hydrocarbon of refinery sludge. *IJEP* 2005, 25(11), 1029-1032.

Delille, D.; Delille, B. Field observations on the variability of crude oil impact on indigenous hydrocarbon-degrading bacteria from sub-Antarctic intertidal sediments. *Marine Environmental Research* 1999, 49, 403-417.

Delille, P.; Bassers, E.; Dessommes, A. Seasonal variation of bacteria in sea ice contaminated by diesel fuel and dispersed crude oil. *Microb Ecol* 1997, 33, 97-105.

Denome, S. A.; Olson, E. S.; Young, K. D. Identification and cloning of genes involved in specific desulfurization of dibenzothiophene by *Rhodococcus* sp. strain IGTS8. *Appl Environ Microbiol* 1993, 59, 2837-2843.

Dibble, J. T.; Bartha, R. Effect of environmental parameters on the biodegradation of oil sludge. *Appl Environ Microbiol* 1979, 37, 729-739.

Donaldson, E. C.; Chilingarian, G. V.; Yen, T. F. The need for microbial enhanced oil recovery. In *Microbial Enhanced Oil Recovery*, Donaldson, E. C.; Chilingarian, G. V.; Yen, T. F., Ed.; Elsevier Science Publishers: New York, 1989, pp. 1-15.

Douglas, G. S.; McCarthy, K. J.; Dahlen, D. T.; Seavey, J. A.; Steinhauer, W.; Prince, R. C.; Elmendorf, D. L. The use of hydrocarbon analyses for environmental assessment and remediation. *J Soil Contam* 1992, 1, 197-216.

Dua, M.; Singh, A.; Sethunathan, N.; Johri, A. K. Biotechnology and bioremediation: successes and limitations. *Appl Microbiol Biot* 2002, 59, 143-152.

Duvnjak, Z.; Kosaric, N. De-emulsification of petroleum water in oil emulsion by selected bacterial and yeast cells. *Biotechnol Lett* 1987, 9, 39-42.

Eckford, R. E.; Fedorak, P. M. Planktonic nitrate reducing bacteria and sulfate reducing bacteria in some western Canadian oil field water. *J Ind Microbiol Biotechnol* 2002, 29, 83-92.

Ehrlich, H. L. Geomicrobiology, Marcel Dekker, Inc: New York, 1995.

Englert, C. J.; Kenzie, E. J.; Dragun, J. Bioremediation of petroleum products in soil. *Princ Prac Pet Contam Soils, API* 1993, 40, 111-129.

Eriksson, M.; Dalhammar, G.; Borg-Karlson, A. K. Aerobic degradation of a hydrocarbon mixture in natural uncontaminated potting soil by indigenous microorganisms at 20°C and 6°C. *Appl Microbiol Biotechnol* 1999, 51, 532-535.

Floodgate, G. C. Hydrocarbon degrading bacteria. *Appl Environ Microbiol* 1984, 64, 3633-3640.

Forsyth, J. V.; Tsao, Y. M.; Bleam, R. D. Bioremediation: when is bioaugmentation need? In *Bioaugmentation for site remediation*, Hinchee, R. E.; Fredrickson, J.; Alleman, B. C., Ed.; Battelle Press: Columbus, Ohio, 1995, pp. 1-14.

Fuenmayor, S. L.; Wild, M.; Boyes, A. L.; Williams, P. A. A gene cluster encoding steps in conversion of naphthalene to gentisate in *Pseudomonas* sp strain U2. *J Bacteriol* 1998, 180, 2522-2530.

Fuhr, B. J.; Holloway, L. R.; Reichert, C. Component-type analysis of shale oil by liquid and thin-layer chromatography. *J Chromatogr Sci* 1988, 26, 55-59.

Furukawa, K. Engineering dioxygenases for efficient degradation of environmental pollutants. *Curr Opin Biotech* 2000, 11, 244-249.

Gallardo, M. E.; Ferrández, A.; de Lorenzo, V.; García, J. L.; Díaz, E. Designing recombinant *Pseudomonas* strains to enhance biodesulfurization. *J Bacteriol* 1997, 179, 7156-7160.

Genouw, G.; de Naeyer, F.; van Meenen, P.; van de Werf, H.; de Nijs, W.; Verstraete, N. Degradation of oil sludge by land farming – a case study at the Ghent harbour. *Biodegradation* 1994, 5, 34-46.

Graham, D. W.; Smith, V. H.; Law, K. P. Application of variable nutrient supplies to optimize hydrocarbon biodegradation. In *Bioremediation of recalcitrant organics*, Battelle Press: Columbus, OH, 1995, pp. 331-340.

Greer, C. W.; Fortin, N.; Roy, R.; Whyte, L. G.; Lee, K. Indigenous sediment microbial activity in response to nutrient enrichment and plant growth following a controlled oil spill or a fresh water wetland. *Bioremediation* 2003, 15, 69-80.

Guenther, K. Endocrine disrupting nonylphenols are ubiquitous in food. *Environmental Science and Technology* 2002, 36, 1676-1680.

Guo, W.; Li, D.; Tao, Y.; Gao, P.; Hu, J. Isolation and description of a stable carbazole-degrading microbial consortium consisting of *Chryseobacterium* sp. NCY and *Achromobacter* sp. NCW. *Curr Microbiol* 2008, 57(3), 251-257.

Gutierrez, T.; Berry, D.; Yang, T.; Mishamandani, S.; McKay, L.; Teske, A.; Aitken, M. D. Role of bacterial exopolysaccharides (EPS) in the fate of the oil released during the deepwater horizon oil spill. *PLoS One*, 2013, 8(6), e67717, doi: 10.1371/journal.pone.0067717

Hadibarata, T; Tachibana, S. Microbial degradation of crude oil by fungi pre-grown on wood meal. In: *Interdisciplinary studies on environmental chemistry – Environmental Research in Asia.* Y. Obayashi, T. Isobe, A. Subramanian, S. Suzuki and S. Tanabe. Tokyo, Terra Scientific Publishing Company, pp. 317-322.

Haines, J. R.; Alexander, M. Microbial degradation of high molecular weight alkanes. *Appl Environ Microbiol* 1974, 28, 1084-1085.

Hamme, J. D. V.; Odumeru, J. A.; Ward, O. P. Community dynamics of a mixed bacterial culture growing on petroleum hydrocarbons in batch culture. *Can J Microbiol Rev Can Microbial* 2000, 46(5), 441-450.

Hamme, J. D. V.; Singh, A.; Ward, O. P. Recent advances in petroleum microbiology. *Microbiology and Molecular Biology Reviews* 2003, 67(4), 503-549.

Handelsman, J.; Rondon, M. R.; Brady, S. F.; Clardy, J.; Goodman, R. M. Molecular biological access to the chemistry of unknown soil microbes: A new frontier for natural products. *Chemistry and Biology* 1998, 5, R245-R249.

Hankinson, O. The aryl hydrocarbon receptor complex. *Annu Rev Pharmacol Toxicol* 1995, 35, 307-340.

Harmsen, J. Possibilities and limitations of landfarming for cleaning contaminated soils. In *On site Bioreclamation*, Hinchee, R. E.; Ollenbuttel, R. F., Ed.; Butterworth Heinemann Publishing: Stoneham, 1991, pp. 255-272.

Henery, S.; Grbicalic, D. Effect of mineral media on trichloroethylene oxidation by aquifer methanotrophs. *Microb Ecol* 1995, 20, 151-169.

Herbes, S. E.; Southworth, G. R.; Gehra, C. W. Organic contamination in aqueous coal conversion effluents: Environmental consequences and research priorities In *Trace substances in the Environment: A symposium*, Hemphill, D. D., Ed.; University of Missouri: Columbia, MO, 1976.

Hossain, S. K. M.; Anantharaman, N. Studies on biological pollution abatement of petrochemical industry wastewater. *IJEP* 2004, 24(11), 845-848.

Hugenholz, P.; Goebel, B. M.; Pace, N. R. Impact of culture-independent studies on the emerging phylogenetic view of bacterial diversity. *J Bacteriol* 1998, 180(18), 4765-4774.

Hung, P. E.; Fritz, V. A.; Walters, L. Infusion of Shrunken-2 sweet corn seed with organic solvents: Effects on germination and vigor. *HortScience* 1992, 27, 467-470.

Hutzinger, O.; Veerakamp, W. Xenobiotic chemicals with pollution potential. In *microbial degradation of xenobiotic and recalcitrant chemicals*, Lesinger, T.; Hutter, R.; Cooke, A. M.; Nuecsh, J., Ed.; Academic Press: London, 1981, pp. 3-45.

Ikenaka, Y.; Sakamoto, M.; Nagata, T.; Takahashi, H.; Miyabara, Y.; Hanazato, T.; Ishizuka, M.; Isobe, T.; Kim, J. W.; Chang, K. H. Effects of polycyclic aromatic hydrocarbons (PAHs) on an aquatic ecosystem: acute toxicity and community-level toxic impact tests of benzo[a]pyrene using lake zooplankton community. *J Toxicol Sci* 2013, 38(1), 131-136.

Ivanova, N.; Tringe, S. G.; Liolios, K.; Liu, W. T.; Morrison, N. A call for standardized classification of metagenome projects. *Environ Microbiol* 2010, 12, 1803-1805.

Iwamoto, T.; Nasu, M. Current bioremediation practice and perspective. *J Bioscience and Bioengineering* 2001, 92, 1-8.

Janiyani, K. L.; Purohit, H. J.; Shanker, R.; Khanna, P. Deemulsification of oil in water emulsions by *Bacillus subtilis*. *World J Microbiol Biotechnol* 1994, 10, 452-456.

Jobson, A.; Cook, F. D.; Westlake, D. W. S. Microbial utilization of crude oil. *Appl Microbiol* 1972, 23, 1082-1089.

Juhsaz, A. L.; Naidu, R. Bioremediation of high molecular weight polycyclic aromatic hydrocarbons: a review of the microbial degradation of benzo(a)pyrene. *Int Biodet Biodeg* 2000, 45, 57-88.

Kanaly, R. A.; Harayama, S. Biodegradation of high molecular weight polycyclic aromatic hydrocarbons by bacteria. *J Bacteriol* 2000, 182(8), 2059-2067.

Keymeulen, R.; Landgehove, H. W.; Schamp, N. Determination of monocyclic aromatic hydrocarbons in plant cuticles by Gas Chromatography-Mass Spectrometry. *J Chromatog* 1991, 541, 83-88.

Khan, A. A.; Wang, R. F.; Cao, W. W.; Doerge, D. R.; Wennerstrom, D.; Cerniglia, C. E. Molecular cloning, nucleotide sequence, and expression of genes encoding a polycyclic aromatic ring dioxygenase from *Mycobacterium* sp. strain PYR-1. Appl *Environ Microbiol* 2001, 67, 3577-3585.

Kiyohara, H.; Torioe, S.; Kaida, N.; Asaki, T.; Iida, T.; Hayashi, H.; Takizawa, N. Cloning and characterization of a chromosomal gene cluster, *pah*, that encodes the upper pathway for phenanthrene and naphthalene utilization by *Pseudomonas putida* OUS82. *J Bacteriol* 1994, 176, 2439-2443.

Komukai-Nakamura, S.; Sugiura, K.; Yamauchi-Inomata, Y.; Toki, H.; Venkateswaran, K.; Yamamoto, S.; Tanaka, H.; Harayama, S. Construction of bacterial consortia that degrade Arabian light crude oil. *Jr Ferment Bioeng* 1996, 82, 570-574.

Korda, A.; Sanatas, P.; Tenente, A.; Santas, R. Petroleum hydrocarbon bioremediation: sampling and analytical techniques, *in situ* treatment and commercial microorganisms currently used. *Appl Microbiol Biotechnol* 1997, 48, 677-686.

Kostka, J.E.; Prakash, O.; Overholt, W.A.; Green, S.J.; Freyer, G.; Canion, A.; Delgardio, J.; Norton, N.; Hazen, T.C.; Huettel, M. Hydrocarbon-degrading bacteria and the bacterial community response in Gulf of Mexico beach sands impacted by the deepwater horizon oil spill. *Appl Envl Microbiol* 2011, 77(22), 7962-7974.

Krishnamurthi, K.; Devi, S. S.; Chakrabarthi, T. The genotoxicity of priority polycyclic aromatic hydrocarbons (PAHs) containing sludge samples. *Toxicology Mechanisms and Methods* 2006, 17(1), 1-12.

Kumar, M.; Singh, M. P.; Tuli, D. K. Genome shuffling of *Pseudomonas* sp. for improving degradation of polycyclic aromatic hydrocarbons. *Advances in Microbiology* 2012, 2, 26-30.

Kurkela, S.; Lehvasllaiho, H.; Palva, E. T.; Teeri, T. H. Cloning, nucleotide sequence and characterization of genes encoding naphthalene dioxygenase of *Pseudomonas putida* strain NCIB9816. *Gene* 1988, 73, 355-362.

Kwon, K.; Lee, H.; Jung, H.; Kang, J.; Kim, S. *Yeosuana aromativorans* gen. nov., sp. nov., a mesophilic marine bacterium belonging to the family Flavobacteriaceae, isolated from estuarine sediment of the South Sea, Korea. *Int J Syst Evol Microbiol* 2006, 56(4), 727-732.

Lakshminarayan, B.; Kishore, N.; Reddy, S. R. Influence of some industrial effluents on AM colonization. *Poll Res* 2006, 25(2), 347-352.

Lalitha, K.; Balasubramanian, N.; Kalavathy, S. Studies of impact of chromium on *Vigna unuiculata* (L.) walp. var (long). *J Swamy Bot Cl* 1999, 16, 17-20.

Laurie, A. D.; Lloyd-Jones, G. The *phn* genes of *Burkholderia* sp. strain RP007 constitute a divergent gene cluster for polycyclic aromatic hydrocarbon catabolism. *J Bacteriol* 1999, 181, 531-540.

Leahy, J. G.; Colwell, R. R. Microbial degradation of hydrocarbons in the environment. *Microbiol Rev* 1990, 54(3), 305-315.

Leahy, J. G.; Colwell, R. R. Microbial degradation of hydrocarbons in the environment. *Microbiol Rev* 1990, 54(3), 305-315.

Leddy, M. B.; Phillips, D. W.; Ridgway, H. F. Catabolite mediated mutations in alternate toluene degradative pathways in *Pseudomonas putida. J Bacteriol* 1995, 177, 4713-4720.

Lee, R. F. Agents which promote and stabilize water-in-oil emulsions. *Spill Sci Technol Bull* 1999, 5, 117-126.

Lehmann, V. Bioremediation: A solution for polluted soils in the south. *Biotech Dev Monitor* 1998, 34, 13-17.

Leja, K.; Myszka, K.; Czaczyk, K. Genome shuffling: a method to improve biotechnological processes. *Journal of Biotechnology, Computational Biology and Bionanotechnology* 2011, 92(4), 345-351.

Li, J.; Bai, R. Effect of a commercial alcohol ethoxylate surfactant (C_{11-15}-E_7) on biodegradation of phenanthrene in a saline water medium by *Netunomonas naphthovorans*. *Biodegradation* 2005, 16, 57-65.

Lin, Q. X.; Mendelssehn, I. A. The combined effects of phytoremediation and biostimulation in enhancing habitat restoration and oil degradation of petroleum contaminated wetlands. *Ecol Eng* 1998, 10, 263-274.

Lundahl, P.; Cabridenc, R. Molecular structure-biological properties, relationship in anionic surface active agents. *Water Res* 1978, 12, 25-30.

Luo, Y. M.; Rimmer, D. L. Zinc copper interaction affecting plant growth on a metal contaminated soil. *Environmental Pollution* 1995, 88, 78-83.

Maanen, J. M. S.; Moonen, E. J. C.; Maas, L. M.; Kleinjans, J. C. S.; van Schooten, F. J. Formation of aromatic DNA adducts in white blood cells in relation to urinary excretion of 1-hydroxypyrene during consumption of grilled meat. *Carcinogenesis* 1994, 15, 2263-2268.

MacNaughton, S. J.; Stephen, J. R.; Venosa, A. D.; Davis, G. A.; Chang, Y. J.; White, D. C. Microbial population changes during bioremediation of an experimental oil spill. *Appl Environ Microbiol* 1999, 65, 3566-3574.

Magot, M.; Olivier, B.; Patel, B. K. C. Microbiology of petroleum reservoirs. *Antonie Van Leewenhoek* 2000, 77, 103-116.

Maguire, J. D. Speed of germination aid in selection and evaluation for the seedling emergence and vigour. *Crop Sci* 1962, 2, 176-177.

Manivasakam, N. Industrial effluents - origin, characteristics, effects analysis and treatment. Sakthi Publications: Coimbatore, India, 1997, pp. 151-161.

Marchal, R.; Penet, S.; Solano-Serena, F.; Vandecasteele, J. P. Gasoline and diesel oil biodegradation. *Oil Gas Sci Tech* 2003, 58(4), 441-448.

Margesin, R.; Labbe, D.; Schinner, F.; Greer, C. W.; Whyte, L. G. Characterization of hydrocarbon degrading microbial populations in contaminated and Pristine Alpine soils. *Applied and Environmental Microbiology* 2003, 69(6), 3085-3092.

Mazzeo, D. E. C.; Fernandes, T. C. C.; Marin-Morales, M. A. Cellular damages in the *Allium sepa* test system, caused by BTEX mixture prior and after bio-degradation process. *Chemosphere* 2011, 85(1), 13-18.

McDonald, M. B. Physical seed quality of soy bean. *Seed Sci Technol* 1985, 13, 601-628.

Mearns, A. J. Cleaning of oiled shores; putting bioremediation to the test. *Spill. Sci Technol Bull* 1997, 4, 209-217.

Mehlman, M. A. Dangerous and cancer-causing properties of products and chemicals in the oil refining and petrochemical industry. VIII. Health effects of motor fuels: Carcinogenicity of gasoline-scientific update. *Environ Res* 1992, 59, 238-249.

Merdinger, E.; Merdinger, R. P. Utilization of n-alkanes by *Pullularia pullulans*. *Applied Microbiology* 1970, 20(4), 651-652.

Mills, A.; Brenil, L. C.; Colwell, R. R. Enumeration of petroleum degrading marine and estuarine microorganisms by the probable number method. *Can J Microbiol* 1978, 24, 552-557.

Mishra, S.; Jyot, J.; Kuhad, R. C.; Lal, B. Evaluation of inoculum addition to stimulate *in situ* bioremediation of oily sludge contaminated soil. *Applied and Environmental Microbiology* 2001, 67(4), 1675-1681.

Mizesko, M. C.; Grrewe, C.; Grabner, A.; Miller, M. S. Alterations at the Ink4a locus in the transplacentally induced murine lung tumors. *Cancer Lett* 2001, 172, 59-66.

Mrayyan, B.; Battikhi, M. Biodegradation of total petroleum hydrocarbon (TPH) in Jordanian petroleum sludge. *World Review of Science, Technology and Sustainable Development* 2004, 1(2), 138-150.

Mueller, R.; Antranikian, G.; Maloney, S.; Sharp, R. Thermophilic degradation of environmental pollutants. *Adv Biochem Eng Biotechnol* 1998, 61, 155-169.

Munoz-de-Toro, M. Perinatal exposure to bisphenol-A alters peripubertal mammary gland development in mice. *Endocrinology* 2005, 146, 4138-4147.

Murugesh, S.; Kumar, S. G. Bioremediation technology to clean up crude oil spills using *Pseudomonas* species. *Journal of Ecotoxicology and Environmental Monitoring* 2002, 12, 133-137.

Nadarajah, N.; Singh, A.; Ward, O. P. De-emulsification of petroleum oil emulsion by a mixed bacterial culture. *Proc Biochem* 2001, 37, 1135-1141.

Nadarajah, N.; Singh, A.; Ward, O. P. Evaluation of a mixed bacterial culture for de-emusification of water in oil petroleum oil emulsions. *World J Microbiol Biotechnol* 2002, 18, 435-440.

Nadeau, R. R.; Singhvi, J.; Ryabik, Y.; Lin, I.; Syslo, J. Monitoring bioremediation for bioremediation efficacy: the Marrow Marsh experience. *Proceedings of the 1993 oil spill conference.* American Petroleum Institute: Washington D.C., 1993, pp. 477-485.

Nag, P.; Paul, A. K.; Mukherji, S. K. Heavy metal effects in plant tissues involving chlorophyll, Chlorophyllare Hill reaction activity and gel electrophoretic patterns of soluble proteins. *Ind J Exp Biol* 1981, 19, 702-716.

Namazi, B. A.; Shojaosadati, S. A.; Najafabadi, H. S. Biodegradation of used engine oil using mixed and isolated cultures. *Int J Environ Res* 2008, 2(4), 431-440.

Narasimhan, M. J. Jr.; Thirumalachar, M. J. Fungus to crude petroleum oil spillages: Pollution control, U.S. Patent 1983, pp. 4, 415, 661.

Nissen, T. V. Mikroorganismer og Kulbrinkter. *Dansk Kemi* 1970, 51, 118-123.

NRC (National Research Council). Oil in the Sea: Inputs, Fates, and Effects, National Academy Press: Washington, D.C, 1985.

Nweke, C. O.; Okpokwasili, G. C. Drilling fluid base oil biodegradation potential of a soil *Staphylococcus* species. *African Journal of Biotechnology* 2003, 2(9), 293-295.

Odokuma, L. O.; Dickson, A. A. Bioremediation of a crude oil polluted tropical rain forest soil. *Global Journal of Environmental Sciences* 2003, 2(1), 29-40.

Ojumu, T. V.; Bello, O. O.; Sonibare, J. A.; Solomon, B. O. Evaluation of microbial systems for bioremediation of petroleum refinery effluents in Nigeria. *African Journal of Biotechnology* 2005, 4(1), 31-35.

Okerentuba, P. A.; Ezeronye, O. U. Petroleum degrading potentials of single and mixed microbial cultures isolated from rivers and refinery effluent in Nigeria. *African Journal of Biotechnology* 2003, 2(9), 288-292.

Olivara, N. L.; Esteves, J. L.; Commendatore, M. G. Alkane biodegradation by a microbial community from contaminated sediments in Patagonia, Argentina. *Int Biodeterior Biodegrade* 1997, 40, 75-79.

Onwurah, I. N. E.; Ogugua, V. N.; Onyike, N. B.; Ochonogor, A. E.; Otitoju, O. F. Crude oil spills in the environment, effects and some innovative clean-up biotechnologies. *Int J Envir Res* 2007, 4, 307-320.

Ouyang, W.; Liu, H.; Murygina, V.; Yu, Y.; Xiu, Z.; Kalyuzhnyi, S. Comparison of bio-augmentation and composting for remediation of oily sludge: A field-scale study in China. *Process Biochemistry* 2005, 40, 3763-3768.

Pauling, L. Organic Chemistry in General Chemistry, Dover Publications: New Delhi, India, 1988, pp. 744-748.

Pavlova, A.; Ivanova, R. Determination of petroleum hydrocarbons and polycyclic aromatic hydrocarbons in sludge from waste water treatment basins. *J Environ Monit* 2003, 5, 319-323.

Peach, K. Allemeine MaBnohmen and Bestimmungan beider Aufarbeitung Vonp Flanzenmeterial. In *Modern methods of plant analysis*, Vol. 1. Peach, K.; Tracey, M. V., Ed.; Springer Verlag: Berlin, 1954, pp. 21-23.

Phillips, G. J. M.; Stewart, J. E. Effect of environmental parameters on bacterial degradation of Bunker C oil, crude oils, and hydrocarbons. *Applied Microbiology* 1974, 28(6), 915-922.

Pieper, D. H.; Reineke, W. Engineering bacteria for bioremediation. *Curr Opin Biotech* 2000, 11, 262-270.

Pinel, D.; D'Aoust, F.; Cardayre, S. B.; Bajwa, P. K.; Lee, H.; Martin, V. J. *Saccharomyces cerevisiae* genome shuffling through recursive population mating leads to improved tolerance to spent sulfite liquor. *Applied and Environmental Microbiology* 2011, 77(14), 4736-4743.

Pinholt, Y.; Struwe, S.; Kjoller, A. Microbial changes during oil decomposition in soil. *Holarctic Ecol* 1949, 2, 195-200.

Platt, A.; Shingler, V.; Taylor, S. C.; Williams, P. A. The 4-hydroxy 2-Oxovolerate aldolase and acetaldehyde dehtdrogenase (acylating) encoded by the nahM and nahO genes of the naphthalene catabolic plasmid pWW60-22 provide further evidence of conservation of meta- cleavage pathway gene sequences. *Microbiology* 1995, 141, 2223-2233.

Pollard, S. J.; Hrudey, S. E.; Fuhr, B. J.; Alex, R. F.; Holloway, L. R.; Tosto, F. Hydrocarbon wastes at petroleum- and creosote- contaminated site: Rapid characterization of component classes by thin layer chromatography with flame ionization detection. *Environ Sci Technol* 1992, 26, 2528-2534.

Prakash, N. K. U. Indoor molds isolation and identification. Color Wings (M) Pvt. Ltd: Chennai, India, 2004, pp. 59-69.

Prince, M.; Sambasivam, Y. Bioremediation of petroleum wastes from the refining of lubricant oils. *Environ Progress* 1993, 12, 5-11.

Prince, R. C. Petroleum spill bioremediation in marine environments. *Critical Reviews in Micro-biology* 1993, 19, 217-242.

Prince, R. C.; Garrett, R. M.; Rothenburger, S. J. Biodegradation of fuel oil under laboratory and Arctic marine conditions. *Spill Science and Technology Bulletin* 2003, 8(3), 297-302.

Pritchard, P. H.; Mueller, J. G.; Rogers, J. C.; Kremer, F. V.; laser, J. A. Oil spill bioremediation: experiences, lessons and results from the Exxon Valdez oil spill in Alaska. *Biodegradation* 1992, 25, 315-335.

Propst, T. L.; Lochmiller, R. L.; Qualls, C. W. Jr.; McBee, K. *In situ* (mesocosm) assessment of immunotoxicity risks to small mammals inhabiting petrochemical waste sites. *Chemosphere* 1999, 38, 1049-1067.

Radhika, S. R. Petroleum Refinery. *Environment Science and Engineering* 2004, 2(2), 77-81.

Raghavan, P. U. M. Characterization of naturally occurring *Pseudomonas putida* for bioremediation of oil-contaminated sites. Ph.D Thesis, Bharathidasan University: Trichy, India, 1998.

Ramana, K. V. R.; Das, V. S. R. Physiological studies on the influence of salinity and alkalinity changes in the growth, respiration, carbohydrates and fats during seedling growth of radish (*Raphanus sativus* L.). *Indian J Plant Physiol* 1978, 21, 93-105.

Ramos, J. L.; Michan, C.; Rojo, F.; Dwyer, D.; Timmis, K. Signal-regulator interactions, genetic analysis of the effector binding site of xyls, the benzoate-activated positive regulator of *Pseudomonas* TOL plasmid meta-cleavage pathway operon. *Journal of Molecular Biology* 1990, 211(2), 373-382.

Ramos, L. J.; Duque, E.; Hneertas, M. J.; Haridour, A. Isolation and expansion of the catabolic potential of a *Pseudomonas putida* strain to grow in the presence of high concentrations of aromatic hydrocarbons. *J Bacteriol* 1995, 177, 3911-3916.

Ramsay, B. A.; Cooper, D. G.; Margaritis, A.; Zajic, J. E. *Rhodochorous* bacteria: biosurfactant production and demulsifying ability. In *Microbial enhanced oil recovery*, Zajic, J. E.; Cooper, D. G.; Jack, T. R.; Kosaric, N., Ed.; Penn Well: Tulsa, Okla, 1983, pp. 61-65.

Reddy, P. G.; Singh, H. D.; Roy, K.; Bartha, J. N. Predominant role of hydrocarbon solubilization in the microbial uptake of hydrocarbons. *Biotechnol Bioeng* 1982, 24, 1241-1269.

Reineke, W. Development of hybrid strains for the mineralization of chloroaromatics by patchwork assembly. *Annu Rev Microbiol* 1998, 52, 287-331.

Rieger, P. G.; Meier, H. M.; Gerle, M.; Vogt, U.; Groth, T.; Knackmuss, H. J. Xenobiotics in the environment: present and future strategies to obviate the problem of biological persistence. *J Biotechnol* 2002, 94, 101-123.

Ron, E. Z.; Rosenberg, E. Biosurfactants and oil bioremediation. *Current Opinion in Biotefchnology* 2002, 13, 249-252.

Saito, A.; Iwabuchi, T.; Harayama, S. A novel phenanthrene dioxygenase from *Nocardioides* sp. strain KP7: expression in *Escherichia coli*. *J Bacteriol* 2000, 182, 2134-2141.

Salle, A. J. Fundamental principles of bacteriology, 7th ed.; Mc Graw-Hill Publishing Company: New Delhi, India, 2006.

Salunke, K. J.; Karande, S. M.; Apparao, B. J. Effect of pulp and paper mill effluent on carbohydrates metabolism during early seedling growth in Mungbean. *Pol Res* 2007, 26(1), 67-69.

Santas, R.; Santas, P. Effects of wave action on the bioremediation of crude oil saturated hydrocarbons. *Mar Pollut Bull* 2000, 40, 434-439.

Saraswathy, N. Induced salt stress in green gram (*Vigna radiate* (L.)Wilczek) and its alleviation by growth regulators. PhD Thesis, Bharathidasan University: Trichy, India, 2002.

Sax, N. I. Dangerous properties of industrial materials, 7th ed.; Van Nostrand Rheinhold: New York, USA, 1989.

Saxena, M. M. Environmental analysis water, soil and air. Agro Botanical Publishers: India, 1994, pp. 121-126.

Senn, R. B.; Johnson, M. S. Interpretation of Gas Chromatographic Data in subsurface hydrocarbon investigation. Groundwater Monitoring Report. *Winter* 1987, 1(3), 58-63.

Shanmugavel, P. Impact of sewage, paper and dye industry effluents on germination of greengram and maize seeds. *J Ecobiol* 1993, 5(1), 69-71.

Sharma, O. P. Text book of Fungi, Tata McGraw-Hill: New Delhi, 2005, pp. 116-163.

Shell, M. A.; Wender, P. E. Identification of the nahR gene product and nucleotide sequence required for its activation of the *sal* operon. *J Bacteriol* 1986, 166, 9-14.

Shennan, J. L. Microbial attack on sulfur-containing hydrocarbons: implications for the biodesulfurization of oils and coals. *J Chem Technol Biotechnol* 1996, 67, 109-123.

Shennan, J. L.; Levi, J. D. *In situ* microbial enhanced oil recovery. In *Biosurfactants and biotechnology*, Kosaric, N.; Cairns, W. L.; Gray, N. C. C., Ed.; Marcel Dekker: New York, USA, 1987, pp. 163-180.

Sienko, M. J.; Plane, R. A. Chemical principles and properties. McGraw Hill Kogakusha: Tokyo, 1976. pp. 462-463.

Simon, M. J.; Ossuland, T. D.; Saunders, R.; Ensley, B. D.; Suggs, S.; Harcourt, A.; Suen, W. C.; Ibson, D. T. C.; Zylstra, G. J. Sequence of genes encoding naphthalene dioxygenase in *Pseudomonas putida* strain G7 and NCIB 9816-4. *Gene* 1993, 127, 31-37.

Singleton, D. R.; Ramirez, L. G.; Aitken, M. D. Characterization of a polycyclic aromatic hydrocarbon degradation gene cluster in a phenanthrene-degrading acidovorax strain. *Applied and Environmental Microbiology* 2009, 75(9), 2613-2620.

Skinner, K.; Cuiffetti, L.; Hyman, M. Metabolism and cometabolism of cyclic ethers by a filamentous fungus, a *Graphium* sp. *Appl Environ Microbiol* 2009, 75(17), 5514-5522.

Smith, L. R.; Dragun, J. Degradation of volatile chlorinated aliphatic priority pollutants in ground water. *Environ Int* 1984, 10, 291-298.

Speight, J. G. The chemistry and technology of petroleum. Marcel Dekker: NewYork, USA, 1991.

Stelmack, P. L.; Murrray, R. G.; Pickard, M. A. Bacterial adhesion to soil contaminants in the presence of surfactants. *Appl Environ Microbiol* 1999, 65, 163-168.

Stewart, A. L.; Gray, N. C. C.; Cairns, W. L.; Kosaric, N. Bacteria induced deemulsification of water in oil petroleum emulsions. *Biotechnol. Lett* 1993, 5, 725-730.

Stewart, T. L.; Fogler, H. S. Biomass plug development and propagation in porous media. *Biotechnol Bioeng* 2001, 72, 353-363.

Story, S. P.; Parker, S. H.; Kline, J. D.; Tzeng, T. R. J.; Mueller, J. G.; Kline, E. L. Identification of four structural genes and two putative promoters necessary for utilization of naphthalene, phenanthrene, and fluoranthene by *Sphingomonas paucimobilis* var. EPA505. *Gene* 2000, 260, 155-169.

Sumathi, K.; Sundaramoorthy, P.; Baskaran, L.; Ganesh, K. S.; Rajasekaran, S. Effect of dye industry effluent on seed germination and seedling growth of green gram (*Vigna radiate* L.) varieties. *Eco Env Cons* 2008, 14(4), 543-547.

Takeuchi, T. Positive relationship between androgen and the endocrine disruptor, Bisphenol-A, in normal women and women with ovarian dysfunction. *Endocrine Journal* 2004, 51, 165-169.

Takizawa, N.; Kaida, N.; Torigoe, S.; Moritani, T.; Sawada, T.; Satoh, S.; Kiyohara, H. Identification and characterization of genes encoding polycyclic aromatic hydrocarbon dihydrodiol dehydrogenase in *Pseudomonas putida* OUS82. *J Bacteriol* 1994, 176, 2444-2449.

Tanjore, S.; Viraragavan, T. Treatment of petrochemical waste waters. *Journal IAEM* 1994, 21, 30-35.

Thion, C.; Cebron, A.; Beguiristain, T.; Leyval, C. Inoculation of PAH-degrading strains of *Fusarium solani* and *Arthrobacter oxydans* in rhizospheric sand and soil microcosms: microbial interactions and PAH dissipation. *Biodegradation* 2013, 24(4), 569-581.

Timmis, K. N.; Pieper, D. H. Bacteria designed for bioremediation. *Trends Biotechnol* 1999, 17, 200-204.

Treadway, S. L.; Yanagimachi, K. S.; Lankenau, E.; Lessard, P. A.; Stephanopoulos, G.; Sinskey, A. J. Isolation and characterization of Indene bioconversion genes from *Rhodococcus* strain 124. *Appl Microbiol Biotechnol* 1999, 51, 786-793.

Trivedy, R. K.; Goel, P. K.; Trisal, C. L. Practical methods in ecology and environmental science, Enviro Media Publications: Karad, India, 1998, pp. 115-137.

Tsai, P. J.; Sheih, H. Y.; Lee, W. J.; Lai, S. O. Health-risk assessment for workers exposed to polycyclic aromatic hydrocarbons (PAH) in a carbon black manufacturing industry. *Sci Total Environ* 2001, 278, 137-150.

Upadhyaya, S.; Shukla, S.; Bhadauria, S. Growth responses of *Lens esculenta* under petroleum contaminated soil in field and pot experiments. *Nature Environment and Pollution Technology* 2008, 7(3), 403-414.

Ururahy, A. F. P.; Marins, M. D. M.; Vital, R. L.; Gabardo, I. T.; Pereira, N. Jr. Effects of aeration on biodegradation of petroleum waste. *Rev Microbiol* 1998, 29(4), 254-258.

Vasanthy, M.; Thamaraiselvi, C.; Narmatha, N. Effect of untreated distillery effluent on growth, characteristics of *Brassica juncea*. *Pol Res* 2006, 25(2), 363-365.

Vasconcellos, S. P. D.; Angolini, C. F. F.; García, I. N. S.; Dellagnezze, M. B.; Silva, C. C. D. Screening for hydrocarbon biodegraders in a metagenomic clone library derived from Brazilian petroleum reservoirs. *Organic Geochemistry* 2010, 41, 675-681.

Venosa, A. D.; Haines, J. R.; Nisamaneepong, W.; Govind, R.; Pradhan, S.; Siddique, B. Screening of commercial innocula for efficacy in stimulating oil biodegradation in closed laboratory system. *J Hazard Mater* 1991, 28, 131-144.

Venosa, A. D.; Suidan, M. T.; King, D.; Wrenn, B. A. Use of hopane as a conservative biomarker for monitoring the bioremediation effectiveness of crude oil contaminating a sandy beach. *J Ind Microbiol Biot* 1997, 18, 131-139.

Venosa, A. D.; Zhu, X. Biodegradation of crude oil contaminating marine shorelines and freshwater wetlands. *Spill Science and Technology Bulletin* 2003, 8, 163-178.

Verschueren, K. Handbook of Environmental Data on Organic Chemicals, Van Nostrand Rheinhold Co: New York, USA, 1983.

Vieites, J. M.; Guazzaroni, M. E.; Beloqui, A.; Golyshin, P. N.; Ferrer, M. Molecular methods to study complex microbial communities. *Methods Mol Biol*. 2010, 668, 1-37.

Volkering, F.; Breure, A. M.; Van Andel, J. G.; Rulkens, W. H. Influence of nonionic surfactants on bioavailability and biodegradation of polycyclic aromatic hydrocarbons. *Appl Environ Microbiol* 1995, 61, 1699-1705.

Vom Saal, F. S.; Hughes, C. An extensive new literature concerning low-dose effects of bisphenol A shows the need for a new risk assessment. *Environmental Health Perspectives* 2005, 113, 926-933.

Walker, J. D.; Colwell, R. R. Enumeration of petroleum degrading microorganisms. *Appl Environ Microbiol* 1976, 31, 198-207.

Walter, U.; Beyer, M.; Klein, J.; Rehm, H. J. Degradation of pyrene by *Rhodococcus* sp. UW1. *Appl Microbiol Biotechnol* 1991, 34, 671-676.

Wang, L.; Wang, W.; Lai, Q.; Shao, Z. Gene diversity of CYP153A and AlkB alkane hydroxylases in oil-degrading bacteria isolated from the Atlantic Ocean. *Environ Microbiol* 2010, 12(5), 1230-1242.

Watkinson, R. J.; Morgan, P. Physiology of hydrocarbon-degrading microorganisms. *Biodegradation* 1990, 1, 79.

Xu, R.; Obbard, J. P. Effect on amendments on indigenous hydrocarbon biodegradation in oil-contaminated beach sediments. *J Environ Qual* 2003, 32, 1234-1243.

Yen, K. M.; Serdar, C. M. Genetics of naphthalene catabolism I Pseudomonads. *CRC. Crit. Rev. Microbiol.* 1988, 15, 247-268.

Yuste, L.; Corebella, M. E.; Turiegano, M. J.; Karlson, U.; Puyet, A.; Rojo, F. Characterization of bacterial strains able to grow on high molecular mass residues from crude oil processing. *FEMS Microbiol Ecol* 2000, 32, 69-75.

Zar, J. H. The analysis of variance. In *Biostatistical analysis*, Pearson Education Pvt. Ltd: London, 2006, pp. 177-270.

Zhan, Y. Z.; Perry, K.; Vinci, V. A.; Powell, K.; Stemmer, W. P. C. Genome shuffling leads to rapid phenotypic improvement in bacteria. *Nature* 2002, 415, 644-646.

Zobell, C. E. Action of microorganisms on hydrocarbons. *Bacterial Rev* 1946, 10, 1-49.

Zobell, C. E. Microbial modification of crude oil in the sea. In *Proceedings of joint conference on prevention and control of oil spills*, API: Washington, D.C., USA, 1969, pp. 317-326.

Zobell, C. E. The occurrence of effects and fates of oil polluting the sea. *Adv Water Pollut Res* 1964, 3, 85-118.

Zylstra, G. J.; Gibson, D. T. Aromatic hydrocarbon degradation: a molecular approach. *Genetic Engineering* 1991, 13, 183-203.

INDEX

A

A priori, 26, 34, 41, 44, 45, 208, 228
ABA signaling, 254, 255, 257, 258, 262, 269, 347
ABA synthesis, 257
Abdominal pain, 77
Abelson Murine Leukemia Virus, 60
Abiotic stresses, 334, 341
Abscisic acid, 255, 270–272, 274–276, 346
ABySS, 128
Acinetobacter, 330, 369, 371, 376–378, 385
Actin organization, 262, 269
Activator protein 1, 7
Acute neurodegeneration, 9
Affected sibling pairs, 23
African Americans, 144
Agarose gel electrophoresis, 321
Agrobacterium-mediated, 349
Akaike information criterion, 28, 39, 213
AKT signaling cascade, 156
Alcaligenes, 376
Allele-count test, 29
Allopatric model, 101, 103, 105
Alpha-methylacylcoenzyme A racemase, 161
Alzheimer's disease, 3, 177
American Academy of Family Physicians, 96
Ammonium phosphate salts, 385
Amplified fragment length polymorphisms, 282, 286
Amplified Ribosomal DNA Restriction Analysis, 321
Amyloid beta protein precursor, 4
Amyotrophic lateral sclerosis, 178
Anal prolapse, 72
Analysis of variance, 30– 32
Analysis plan, 19, 21, 23, 26, 49
Androgen (ligand), 156
Anemia, 48, 56, 77

ANOVA, 31, 32, 44
Antifreeze proteins, 343
Antimetabolite 1-beta-D-arabinofuranosyl-cytosine, 63
Antimetabolite, 63
Antimicrobials, 122
Antioxidant response element, 7
Antiviral therapy, 177
Apolipoprotein E, 4
Apoptosis, 9, 15, 17, 63–68, 73–76, 79–81, 156, 159, 178, 182, 375
Arabidopsis gene, 295
Arabidopsis genome, 256, 259, 263, 264, 268, 288
Arabidopsis nucleosome assembly protein 1, 267
Arabidopsis thaliana farnesylated protein 3, 264
Arabidopsis thaliana geranylgeranylated protein 1, 263
Arabidopsis thaliana, 263, 264, 272, 274–276, 278, 279, 288, 352, 354
Armitage test, 25
Armitage-Cochran test, 29
Arsenic trioxide, 79
Aspirin, 10
Astrocytes, 6, 8, 11
    cerebellum, 6
    cortex, 6
    hippocampus, 6
Ataxia Telangiectasia mutated, 63
Azoospermia factors, 97

B

Bacterial artificial chromosome, 285
Bacteriome, 123
Bacteriophage, 157
Bad, 66
Bateson, 21
Bax, 65, 66, 68

Bayes factors, 214
Bayesian analysis, 233, 235, 236, 241
Bayesian inference, 208, 230, 231, 235,
    236, 238
Bayesian information criterion, 28, 214
Bayesian phylogenetic inference, 234
B-cell lymphopenia, 74
BCR-ABL proteins, 60
Beacon Biosoft, 184
Betaine aldehyde dehydrogenase, 354
Bilirubin, 2, 6, 8, 10, 11
Biliverdin, 2, 6, 11
Biodiversity studies, 204
Biological Concept of Species, 111
BIONJ, 222
BLAST algorithm, 127
Bone morphogenetic proteins, 74
Bonferroni correction, 28, 29
Breakpoint cluster region, 60
Broccoli, 10
Bt toxin, 312, 315, 323

C

C-Abl, 60–82
Caenorhabditis elegans, 101, 176, 184
Cafestol, 10
Caffeic acid phenethyl ester, 10
Calmodulin, 8, 254, 260, 345
Calmodulin-like proteins, 265
Camin-Sokal parsimony, 226
Camptothecin, 63
Cancer biology, 177
Carbon monoxide, 6
Carboxyl methylation, 252, 254, 256, 263,
    265
Carnosol, 10
CART method, 39
Case-control design, 23, 30, 48
Case-control TDT, 30
Cationic liposomes, 180
Causative loci, 22
CD81, 192–197
Cell cycle arrest, 63, 68
Cell cycle progression, 69, 70, 76, 81
Cell damages, 4
Cell death, 2, 3, 12, 15, 16, 63–65, 67, 74,
    78–80, 177, 178, 334

Cell movement, 63, 70, 71, 267
Cell proliferation, 62, 64, 65, 68, 75, 76,
    80, 267
Central Institute of Cotton Research, 313
Central nerve system, 3
Cereals, 10, 338
Cerebellum, 6
Chaperones, 261, 343
Character matrix, 206, 207, 218, 219
Chi-square, 27–29, 34, 46
Chi-square distribution, 29
Chi-square subset, 47
Chi-square testing, 28
Chloroplast transformation, 349
Choline monooxygenase, 354
Chronic, 9, 76, 77, 134, 154, 190
Chronic myelogenous leukemia, 60, 86, 88,
    89, 177
Chronic myeloid leukemia, 154
Circoviridae, 131
Circulating recombinant forms, 133
Cisplatin, 63
Classification and regression trees, 33, 36
Clustering algorithms, 33
    HapMiner, 33
    CLADHC, 33
    EM clustering, 33
Clathrin-mediated endocytosis, 196
Claudin-1, 192, 193
Clinical treatment, 10
    see, neurodegeneration
Cluster analysis, 42
Clustering algorithms, 42
Clustering methods, 43, 219
clusters, 43, 44, 45, 241, 260, 264, 384
curse of dimensionality, 20, 26, 32
Cochran's Q test, 32
coenzyme Q10, 10
coffee, 10
cognitive impairment, 3
coiled-coil oligomerization domain, 76
collision-induced dissociation, 154
combinatorial approaches, 35, 37
Combinatorial Partitioning Method, 33, 37
Combinatorial types, 33
    Combinatorial Partitioning Method, 33

Detection of Informative Combined Effect, 33
   Multifactor Dimensionality Reduction, 33
   Patterning and Recursive Partitioning, 33
   Restricted Partition Method, 33
Comparative genomics, 204, 285
Contingency table, 29
Cordell, 20, 21, 22
Cortex, 6
Covarion models, 211, 213
CP motif. *See* cysteine and praline
Creatine lazaroids, 10
Criteria-based methods, 219
Crohn's disease, 131
   idiopathic bowel syndrome, 131
CrossMatch database, 284
Crude oil, 368, 373
C-terminal lobe, 61
C-terminal region, 73
C-type lectins, 192
Cubic splines, 30
Curcumin, 10
Cysteine and praline, 5
Cytarabine, 77, 79
Cytokines, 7, 74
Cytoplasm, 61, 66, 70, 71, 73, 76, 149, 194, 196, 256
Cytoskeleton, 61, 70, 71, 76

D

Data cleaning, 23
Data imputation, 24
Data mining, 32–35, 47
Data mining methods, 33, 34
   epistasis detection, 33
Data-mining approach, 21
DC-SIGN, 192, 193
DCT-1, 6
Dehydration responsive element, 347
Dementia, 3, 4
Demethylation, 257, 258
Denaturing Gradient Gel Electroporesis, 317
Dendritic cells, 192, 194
Department of Biotechnology, 312

Detection of Informative Combined Effect, 39
Detoxification enzymes, 343
Diacylglycerol, 62
Diarrhea, 77
Dietary antioxidants, 10
   cafestol, 10
      coffee, 10
   carnosol, 10
      rosemary, 10
   curcumin, 10
      turmeric, 10
   kahweol, 10
      coffee, 10
   α-lipoic acid, 10
      broccoli, 10
      see, spinach
   resveratrol, 10
      grape, 10
   selenium, 10
      see, cereals
      see, fish
   sulphoraphane, 10
      broccoli, 10
      sprouts, 10
   tomatoes, 10
Direct RNA sequencing, 127
Dirichlet multinomial mixture, 130
Disease-risk loci, 34
Disk Covering Methods, 241
Distribution of Members
Diversity Array Technologies, 286
Divide and conquer strategy, 36
DNA binding domain, 61, 65, 68, 72, 73
DNA damage, 61–63, 65, 66, 68, 79, 80–82, 135, 375
   causes, 63
      antimetabolite 1-beta-D-arabinofu-ranosylcytosine, 63
      camptothecin, 63
      cisplatin, 63
      doxorubicin, 63
      etoposide, 63
      ionizing irradiation, 63
      mitomycin C, 63

DNA fingerprinting, 321, 339, 342
DNA inversions, 204
DNA technology, 157, 311, 335, 352
DNA-binding domain, 157, 162
Dobzhansky-Muller (DM) model, 99
Dobzhansky-Muller heteromers, 96
Dollo parsimony, 226
Domains, 76
    coiled-coil oligomerization domain, 76
    GTPase exchange factor domain, 76
    RacGAP domain, 76
Down-stream of kinase, 71
Drosophila species, 96, 99, 100
Drosophila, 96, 99, 100
Dutch individuals, 49
Dyspnea, 77, 78

E

Ebselen, 10
Edema, 77, 78, 375
Edena, 128
Eigen analysis, 47
Electrophoresis, 125, 292, 317, 318, 320, 387
Electro-Spray Ionization, 151
Elevated liver enzymes, 77
EM algorithm, 45
Embolic clot, 3
Embryogenesis, 96, 98, 100, 101, 103–109, 111, 112, 346, 350
Embryogenesis plus fertility, 108
Endogenous antioxidative factor, 9
Endogenous gene silencing, 182
Endothelial cells, 9, 10, 15, 192, 193
Endotoxin, 7
Enzymes, 13, 63, 67, 122, 176, 257, 286, 287, 294, 295, 318, 337, 345, 346, 354, 372, 375, 382, 386, 389
Epidemiology, 21, 26, 30, 31, 34, 35, 37, 40, 41, 43, 128, 129, 132, 133, 135, 204, 215
Epidermal growth factor, 62, 156, 158
Epistasis, 19–23, 32, 33, 35, 49
Epistatic, 21, 22, 26, 39, 44, 45, 47–50, 342
Epstein-Barr virus, 131, 133
Escherichia coli, 5, 135, 160, 259, 263, 354
Ethyl ferulate, 10
Etoposide, 63

Evolutionary change, 111, 208, 225, 229
Evolutionary models, 208, 225, 226, 230, 231, 240
Expectation maximization, 43
Experimental therapeutics, 176, 182
Expressed sequence tag, 282, 294, 338, 347
Expression proteomics, 145
Extracellular loops, 193
    large extracelullar loop, 193
    small extracellular loop, 193
Extracellular signal-regulated kinase, 7

F

False Discovery Rate, 46
False discovery rate method, 28, 29
Family based association test, 30
Farnesyl diphosphate, 252
Farnesylcysteine lyase, 257, 258
Fatigue, 77
Felsenstein model, 210
Felsenstein zone, 225
Fenton reaction, 3
Ferritin, 6
Filter method, 49
First-generation sequencing, 125
Fish, 10
Fisher, 21, 24, 27, 319
Fisher's exact test, 29
Fitch algorithm, 226
Fitness landscape, 38
Flag-Tag, 143
Flaviviridae family, 190
Flavobacterium indologenes, 376
Floral meristems, 255, 265
Fluorescence (Förster) resonance energy transfer, 126, 160
Focused Interaction Testing Framework, 45
Food and Drug Administration, 176
Free-radical scavengers, 343
Friedman's two-way ANOVA, 27
F-tests, 28
    see, nested models
Fulminant hepatitis, 178
Functional proteomics, 145, 146, 156, 161
Functional redundancy, 255, 268

G

Galaxy, 128
Gammaherpes viruses (γHV), 133
   Epstein-Barr virus, 133
   Kaposi's sarcoma-associated herpesvirus, 133GAM-NGS, 128
Gas chromatography, 298
GC box binding, 7
Gene mutations, 4
Gene network redundancy, 108
Gene regulator, 6
Gene trees, 238, 240, 241
Gene-environment, 20–22, 28, 37, 41, 45, 46, 49
Gene-gene, 20–22, 28, 37, 41, 45, 49
General linear model, 31
General Time Reversible models, 210, 213
Generalized linear models, 30
Generalized TDT, 30
Genetic engineering, 335, 336, 344, 346, 349, 362, 365, 388, 390
Genome-wide association, 20
Genomic shuffling, 388
Genotype, 4, 22, 24, 38, 39, 47, 51–53, 56, 103, 107, 108, 119, 123, 133, 300, 340, 342
Geranylgeranyl diphosphate, 252
Geranylgeranylcysteine, 257, 258
GLM, 27, 30
Glutathione, 10, 67, 148, 260
Glutathione peroxidase 1, 63
Glutathione S-transferase, 160
Glutathione-agarose resin, 160
Glycosaminoglycans, 192
Golgi apparatus, 264
Grammatical Evolution Neural Networks, 41
Grape, 10
GTPase exchange factor domain, 76
GTPase-activating proteins, 262
Guanine nucleotide exchange factors, 262

H

Hardy – Weinberg equilibrium, 24
Hazards Regression, 27
HCV entry, 195, 196, 197
HCV infection, 190–195, 197
HCV life cycle, 190, 191
   assembly, 190
   release, 190
   replication, 190
   viral entry, 190
HCV tissue tropism, 193
HCVcc, 191–196
HCVpp, 191–196
Headache, 77
Heart, 5, 78, 79, 82, 123
Heat Map approach, 297
Heat stress proteins, 343
Heavy metals, 7, 264, 334
Heme ligand, 6
Heme oxygenase, 2
Heme-binding site, 5
Hemoglobin, 205
Hemoproteins, 2
Hepacivirus, 190
Heparan sulfate, 192, 195, 196
Hepatitis C virus, 140, 186, 188, 190, 196
Hepatocytes, 14, 72, 188, 193–195, 374
Hepatoma cells, 193, 194, 195
Hermann Joseph Muller, 99
Heterodimer, 7, 99, 103, 105–107
Heterodimeric DM genes, 102, 107
Heterodimeric enzyme, 252
Hidden Markov model, 211
High yielding varieties, 339
High-density lipoprotein, 194
High Resolution Plant Phenomics Centre, 296
High-throughput sequencing, 126, 128, 129
Hippocampal neurons, 9
Hippocampi, 8
Hippocampus, 6, 9
Hirshsprung's disease, 48
HIV, 132, 133, 178
HIV-1, 132
HKY model, 210
HMP Consortium, 123
HO Antioxidative System, 1
HO-3 transcript, 5, 6
   brain, 5
   heart, 5
   kidney, 5
   liver, 5
   prostate, 5

spleen, 5
testis, 5
thymus, 5
Homo sapiens, 98
Homology, 2, 94, 204, 238, 239, 343
Homoplasy, 223, 239
Hormones, 7
Host cell factors, 191, 197
Hot-deck approaches, 24
Housekeeping proteins, 149
Human body, 122, 123, 130, 374
Human cDNA library, 60
Human hepatoma cells, 192, 195
Human heptocytes, 193
Human immunodeficiency virus, 178
Human Microbiome Project, 122, 123, 130
Human population, 49, 54, 108, 110, 113, 356
Huntingtin genes, 177
Huntington's disease, 177
Hybrid male rescue, 100
Hydrodynamic transfection, 180
Hydrogen peroxide, 2
4-hydroxynonenal, 4
8-hydroxyguanosine, 4
Hypothetical ancestor, 216, 218
Hypoxia-insulted human endothelial cells, 8

I

Idiopathic bowel syndrome, 131
IGF-binding proteins, 147
Imatinib, 66, 73, 77, 78, 79
Immobilized metal affinity chromatography, 163
Impaired fecundity, 96, 97
In vivo ischemia, 9
Indian Council of Agricultural Research, 312, 313
Indian Oil Corporation Limited, 369
Indirect mutation targeting, 178
Infertility, 96, 97, 101, 110, 112
Ingroup, 218
Inositol-1,4,5-trisphophate, 62
Insulin, 10, 124, 156
Interaction Testing Framework, 46
Interferon (IFN) response, 181

Interior-branch test, 236
Interleukin-6, 156, 163
Internal nodes, 223
Interpopulation model, 105
Intracellular antioxidants, 2
Intracellular antioxidative systems, 2
Intrapopulation model hypothesis, 106
Intrinsic neuroprotective factors, 9
Investigational New Drug, 176
Ion Torrent sequencer, 126
Ionization techniques, 150
　Electro-Spray Ionization, 151ionizing irradiation, 63
　Matrix-Assisted Laser Desorption/Ionization, 150
Iron, 6, 377
Iron regulatory elements, 6
Iron regulatory protein, 6
Ischemia, 3, 9, 67
Isoproterenol, 10

J

Joint pain, 77
JR-Assembler, 128
Jukes and Cantor model, 210
Jun N-terminal kinase, 7, 64

K

Kahweol, 10
Kaposi's sarcoma-associated herpesvirus, 133
Kensington Pride, 282, 284, 287, 291, 299, 302
Kernel regression, 27, 30
Kidney, 5, 177
Kimura (K80) model, 210
King Abdulaziz University, 184
Knockout mice, 10, 72–75, 79, 80, 82, 96, 177
Kruskal-Wallis analysis, 27, 32

L

Large extracelullar loop, 193
Laser capture microdissection, 150
LDL receptor, 195
LEA proteins, 343, 346

Learning and memory, 9
Least-squares methods, 220
Lethal hybrid rescue, 100
Ligand-binding domain, 162
Likelihood ratio test, 237
Linear Regression, 27
Linux, 129
A-lipoic acid, 3, 10
Listeria monocytogenes, 131
Liver, 5, 6, 177, 178, 180, 190–194, 197, 374
Logistic Regression, 27
Logit function, 30
    see, case-control design
Log-likelihood value, 231
Longest common sequence, 128
Lou Gehrig's disease, 178
Lovastatin, 10
Low-density lipoprotein receptor, 192
L-SIGN, 192, 193
Lung cancer, 86, 144, 154, 374
Lysis buffers, 149

M

Machine-learning approaches, 40
MacOS X, 129
Macrophage phagocytosis, 66
Maharashtra Hybrid Seed Company, 312
Malondialdehyde, 353
Mammalian cell, 163, 375
Mammalian two-hybrid system, 157
Mangifera indica L., 282
Mango, 101, 120, 282
Mango Genomics database, 293–295, 297, 298, 300
Mango Genomics Information System, 283, 300
Mango Genomics Initiative, 283, 284, 286, 298
Mann-Whitney U test, 27, 32
MANOVA/MANCOVA, 27
MAP kinase kinase, 7
MAPK pathway, 65, 76
Markov process, 209
Markov stochastic process, 209
Marriage, 113

Mass spectrometry, 145, 146, 149–155, 158–160, 295, 298
Matrix-Assisted Laser Desorption/Ionization, 150
Maximum likelihood, 24
Maximum parsimony, 223
Maximum-likelihood method, 222, 226
Median test, 27, 32
MEDLINE abstracts, 47
Megaesophagus, 72
Melatonin, 10
Membrane-anchored Ubfold proteins, 267
Mendelian, 20, 23, 48
Mendelian disorders, 20
Metabolome, 123, 295
Metagenome, 123
Metagenomics, 123, 137, 387, 389
Metal ion chelators, 10
Metal responsive element, 7
Methicillin-resistant *Staphylococcus aureus*, 126, 135
Microbiome, 122–124, 130, 131
Microbiota, 122–124, 131, 135, 310
Mild Cognitive Impairment, 4
Mitochondria, 3, 61, 65, 66, 67, 256
Mitogen-activated protein kinase, 7, 64, 156
Mitomycin C, 63
Mobile genetic elements, 204, 206
    deletions formation, 204
    DNA inversions, 204
Molecular breeding, 342
Molecular clock, 205, 206, 217, 218, 221, 222
Molecular markers, 286, 289, 290, 292, 295, 297, 302, 303, 337, 338, 339, 340
Molecular systematics, 204
Moloney Murine Leukemia Virus, 60
Monte Carlo simulation, 214
Mosaic evolution, 96, 99, 106, 110, 111
MRNAs, 6, 145, 176, 179
    DCT-1, 6
    ferritin, 6
    transferrin receptor, 6
Multidimensional Protein Identification Technology, 151
Multilocus association, 23

Multiple imputations, 24
Mobile genetic elements, 204
Multivariate Adaptive Regression Splines, 36
Multivariate analysis of covariance, 32
Multivariate ANOVA, 32
Multivariate tests, 31
Murine cytomegalovirus, 131
Murine double minute, 69
Murine gammaherpesvirus 68, 131
Muscle cramps, 77
Musculoskeletal pain, 77
Mycobacterium, 134, 135, 138, 140, 250, 321, 370, 378
Mycobacterium leprae, 135
Mycobacterium tuberculosis, 134
Myriad, 7, 48

N

N-acetylcysteine, 10
Nam Doc Mai samples, 299
National Bureau of Plant Genetic Resources, 312
National Centre for Biotechnology Information, 282
National Environmental Engineering Research Institute, 370
Natural selection, 103, 106, 109, 111, 205
Natural species, 99, 106, 110, 112, 114, 115
Nausea, 77
Nearest-neighbor interchanges, 231
Necrotrophic pathogens, 268
Neighboring loci, 24
Neighbor-joining method, 221, 222, 229, 241
Nested models, 28, 213
Neural networks, 34, 40, 41
    Genetic Programming NN, 33
    Grammatical Evolution NN, 33
    Parameter Decreasing Method, 33
Neurobiology, 177
Neurodegeneration, 4, 9, 10
Neurodegenerative Diseases, 1, 3, 4
Neurological diseases, 177
Neuronal cell death, 2
Neutropenia, 77
Newton-Raphson method, 229

Next-generation sequencing, 123, 124, 130, 132, 137, 139, 141, 142, 182, 286, 300
Nextgeneration sequencing, 125, 286
Nicaraven, 10
Nickel-nitrilotriacetic acid, 159
Nicotiana tabacum geranylgeranylated protein 4, 264
N-methyl-D-aspartate receptor, 3
NO donors, 10
Nocardia, 370, 376–379, 381, 384
Nodes, 36, 40, 41, 215–217, 220, 228, 229, 236
    synaptic connections, 40
Nonfluorescent sequencing systems using nanopores or nanoedges, 126
N-terminal Cap, 61, 76
N-terminal domain, 160
N-terminal region, 191
Nuclear export signal, 61
Nuclear localization signals, 61
Nucleoporin96, 99
Nucleotides, 20, 56, 71, 98, 101, 128, 134, 176, 186, 188, 204, 206, 208–210, 215, 217, 219, 220, 223–229, 240, 253, 264, 282, 289, 291–293, 319, 338, 340

O

Ochrobactrum anthropi, 376
Odysseus-Homeobox gene, 99
Oligospermia, 97
Oncogene c-myc/max heterodimer binding site, 7
D-ononitol, 355
    sugar alcohol, 355
Onto-Tools suite, 47
Operation Taxonomic Units, 216
Oral biology, 177
Ordered Subset Analysis, 23
Organic nitrates, 10
Orione, 128
Oryza sativa, 338, 345, 346, 352
Osmoprotectants, 343, 351, 352, 356
Osteoblast, 73, 74
Osteoporosis, 72, 73
Osteoporotic phenotype, 73
Outgroup, 218, 240

Oxidative attack, 2
Oxidative neurotoxicity, 10
Oxidative stress, 1, 2, 4, 6–8, 10, 11, 66, 67, 69, 74, 79, 82, 341, 345, 349, 353

P

PAH compounds, 390
Papillomaviridae, 131
Parameter Decreasing Method, 33, 41
Particle bombardment, 349
Pathogen diversity mapping, 204
Patterning and Recursive Partitioning, 39
Paxillin, 70
PCR primers, 125, 290, 291, 318
Pearson chi-square, 29
Pearson goodness-of-fit test, 24
Pedigree Disequilibrium Test, 30
Perfect binding sequence, 68
Permutation testing, 29, 38, 46
Peroxisomal fatty acid metabolism, 66
Peroxynitrite, 3
Petroleum reservoirs, 377, 390
Phage display, 157, 158
Phage-coat protein, 157
Phagocytosis, 70
Pharmacologic inducers, 10
Phenotype, 21, 22, 24, 38, 43, 44, 51, 53, 106, 108, 123, 177, 255, 256, 257, 266, 267, 269, 300, 336, 341, 354
Phenyl-N-tert-butyl nitrone, 3
Phosphatidylinositol-4,5-bisphosphate, 62
Phospholipase C-γ1, 62
Phospholipid scramblase 1, 63
Phosphorylation, 8, 15, 16, 62, 63–77, 145, 155, 156, 162, 163, 182
Phylogenetic inference, 215, 226, 227, 236
Phylogenetic networks, 236
Phylogenetic tree, 44, 135, 205, 215–221, 227, 228, 230, 231, 236, 239
Phylogeny, 51, 208, 212, 216, 221, 224, 229–233, 235, 237, 238, 240, 292
Phytosterols, 270
Phytozome, 288
Piceatannol paclitaxel, 10
plant accelerator, 296
Plant development, 263
Plant Metabolic Network, 288, 295

Plant proteins, 259, 260, 288
Plant virus mediated, 349
Platelet-derived growth factor, 62
Poisson distribution, 205
Poisson Regression, 27
Polycyclic aromatic hydrocarbons, 374
Polymerase chain reaction, 24, 337
Polymorphic gene alleles, 112
Polyomaviridae, 131
Popperian sense, 113
Population genetics model, 105–108
Population stratification, 25
Prenyl screening, 259, 264
Prenylcysteine methylation, 254
Primary rat hepatocytes, 7
Principle components analysis, 26, 47
Probucol, 10
Prostate cancer, 144, 150
Prostate-specific antigen, 144, 147
Protein farnesyltransferase, 252
Protein geranylgeranyltransferase type I, 252
Protein geranylgeranyl transferase type II, 252
Protein kinase, 8, 63, 86, 88, 89, 92, 94, 159, 163, 165, 169, 171, 182, 350
Protein oxidations, 4
    See cell damages
Protein prenylation, 254, 255, 259, 260, 265, 269
Protein tyrosine kinases, 63
Protein tyrosine phosphatases, 72
Protein-membrane interactions, 253
Protein-protein interactions, 155, 157, 160, 162, 165, 166, 168, 169, 171–173, 253
Proteolysis, 70, 252, 253, 254, 256
Proteome, 123, 150, 153, 282, 284
Proto-oncogene, 80, 92
Protoplast transformation, 349
Prototype, 123
Protozoa, 127
Prune back, 36
Pseudo-controls, 22, 39
Pseudocontrols, 22
Pseudomonas aureofaciens, 376
Pseudomonas cepacia, 376
Pseudomonas picketti, 376

Pseudoreplicate data sets, 237
PubMed search, 48

Q

Quality Control, 23
Quantitative TDT, 30
Quantitative trait loci, 55, 336, 341, 342, 343
Quasi species, 122, 133
Queensland Facility for Advanced Bioinformatics, 285
Queensland Mango Genomics Initiative, 282, 283

R

Rab geranylgeranyl transferase, 252
RAB proteins, 259, 260, 263, 264
RABs, 251, 263
RacGAP domain, 76
Radiation therapy, 144
Radical prostatectomy, 144, 149
Random amplification of polymorphic DNA, 282, 289
Random Forests, 36
Rapamycin, 10
Rashes, 77
Reactive oxygen species, 14, 66, 334
Real time PCR, 8
Reduced representation libraries, 286, 294
Regression model, 25, 31
Regression trees, 36
Reproductive isolation, 96, 97, 113
Restricted Cubic Splines Regression, 27
Restricted Partition Method, 33, 37
Restriction fragment length polymorphisms, 282, 318, 338
Resveratrol, 10, 126
Retinoblastoma protein, 69
Rhizosphere, 310, 315, 316, 319, 322, 323, 371
Ridge regression, 28
RNA activation (RNAa) phenomenon, 183
RNA interference, 2, 176
RNAi therapeutics, 176–184
RNA-induced silencing complex, 179
RNAome, 184
ROP genes, 262

Rosemary, 10
Rosolic acid, 10
RRNA Intergenic Spacer Analysis, 319

S

S. aureus, 134
Saccharomyces cerevisiae, 171, 173, 253
Scirpus pungens, 371
Second human genome, 122
Secondary metabolism, 269
Secondary mitochondrial dysfunction, 3
Second-generation systems, 126
Selenium, 10
Seqanswer, 128
Set Association, 33, 45
Severe Acute Respiratory Syndrome virus, 178
Shoot apical meristem, 265
Short Quartet Method, 222
Short tandem repeats, 289, 338
Shotgun, 390
Shrinkage estimation, 28
Sibling pairs, 23, 27
Sib-TDT, 30
Sign test, 27, 32
Signal transducer and activator of transcription, 181
Simple sequence repeat, 285, 289, 290, 342
Simple sequence repeats, 338, 342
Simvastatin, 10
Single imputation, 24
Single nucleotide polymorphism, 20, 101, 285, 291, 340
Single-copy gene, 5
   see, HO-3 transcript
Single-molecule real-time sequencing, 126
Single-nucleotide polymorphisms, 134
Single-strand conformation polymorphism, 320
SiRNA delivery, 177, 179, 180
SiRNA technique, 9
Skin problems, 77
Small extracellular loop, 193
Small interfering RNAs, 176
Small nuclear RING finger protein, 157
SOAPdenovo, 128
Soil bioremediation, 371, 381

Soil microbial communities, 322, 323
Soil quality, 310
Speciation genes, 111
Species tree, 239, 240
Spin-trap scavenging agents, 10
Spinach, 10, 354
Spleen, 5, 72, 74
Split decomposition method, 222
Sprouts, 10
Stable isotope labeling by amino acids in cell culture, 154
Staphylococcus sp., 126, 135, 138, 139, 140, 142, 371, 376, 378
State space, 38, 234
    fitness landscape, 38
Strains, 370
    Bacillus aneurirolylicus, 370
    Pseudomonas aeruginosa, 370
    Pseudomonas stutzeri, 370
    Serratia marcescens, 370
Streptococcus pyogenes, 136
Streptococcus sanguinis, 123
Stressful, 6, 343, 345, 351
Stroke, 3, 9
Substantia nigra, 9
Substitution models, 208, 210, 212, 233, 235, 241
Substitution process, 212, 227, 229
Subtree pruning regrafting, 231
Sulforaphane, 10
Sulphoraphane, 10
Superoxide dismutase, 2, 178, 349
Superoxide, 2, 3, 178, 349
SureSelect eArray design tool, 300
Surface-enhanced laser desorption/ionization, 150
Surface-enhanced Raman spectroscopy, 126
Surrounding loci, 24
Sympatric model, 103, 106
Synaptic connections, 40

T

T lymphocytes, 192
Taipan, 128
Tandem affinity purification, 158
Tandem mass spectrometry, 153, 154, 158, 159

Targeted induced local lesions in genomes, 348
Targeting aberrant splicing isoforms, 178
T-cell lymphopenia, 74
Testis, 5
Tetraspanins, 193
TGS technologies, 126
    fluorescence resonance energy transfer, 126
    nonfluorescent sequencing systems using nanopores or nanoedges, 126
    single-molecule real-time sequencing, 126
    surface-enhanced Raman spectroscopy, 126
    true single-molecule sequencing, 126
Theodosius Dobzhansky, 99
Thermal-stressed human erythroblastic cell line YN-1-0-A, 8
Thioether linkage, 252, 253, 256
Thiol-reactive substances, 7
Third-generation sequencing, 126, 130
Thrombocytopenia, 77
Embolic clot, 3
Thymomas, 60
    see, mice
Thymus, 5, 60, 72, 74
Thymus atrophy, 72
Time-of flight, 151
Titanium dioxide, 163
Tomatoes, 10
Tommy Atkins, 302
Total hydrocarbon content, 374
Traditional statistical approaches, 26
Transcription activation domain, 157
Transcriptome, 123, 124, 126, 133, 285, 348
Transcriptomics, 334, 345
Transferrin receptor, 6
Transforming growth factor β, 62
Transgenic crops, 310, 311, 313, 314, 315, 322, 323, 351
Transmission disequilibrium test, 29
Tree analysis using new technology, 241
Tree topology, 212, 213, 216, 220–224, 226, 228, 229, 231, 232, 233–235, 238, 239
Tree-based method, 33, 34
    classification and regression trees, 33

mutivariate adaptive regression splines, 33
 random forests, 33
Tree-building method, 38, 218, 219, 220
Trios, 23
True single-molecule sequencing, 126
T-test, 27
Turmeric, 10
Two-dimensional difference gel electrophoresis, 149
Two-dimensional gel electrophoresis, 146
Two-step type, 33
 set association, 33
 focused interaction testing framework, 33

U

Ultra-deep pyrosequencing, 133
Unclamping, 62
Untranslated region, 7
Unweighted pair-group method using arithmetic averages, 221
Uric acid, 10
Urinary tract infection, 135
UV radiation, 7, 63

V

Varicella-Zoster virus, 79
Velvet, 128
Very low-density lipoproteins, 195
Vesicle docking, 261
Vibrio cholerae, 127
Virome, 123, 131, 132
Vitamins, 10
 coenzyme Q10, 10
 creatine lazaroids, 10

 ebselen, 10
 glutathione, 10
 melatonin, 10
 metal ion chelators, 10
 N-acetylcysteine, 10
 nicaraven, 10
 spin-trap scavenging agents, 10
 uric acid, 10
Volatile anesthetics, 7
Vomiting, 77
V-SNARE proteins, 264

W

Wald-Wolfowitz runs test, 27, 32
Water use efficiency, 346
Weighbor, 222
Weighted parsimony, 225
White box, 35
Wilcoxon's matched pairs test, 27, 32
Wild-type culture, 10
Windows, 129
Wiskott-Aldrich syndrome, 71
Wrappers, 28

X

Xanthomonas maltophilia, 376
Xenology, 239

Y

Yeast two-hybrid, 8, 101, 156
Yersinia pestis, 131
Yin and Yang, 184

Z

Z- score, 44